VOLUME SEVENTY SIX

Current Topics in MEMBRANES

Basement Membranes

CURRENT TOPICS IN MEMBRANES, VOLUME 76

Series Editors

ROBERT BALABAN
National Heart, Lung and Blood Institute
National Institutes of Health
Bethesda, Maryland, USA

SIDNEY A. SIMON
Department of Neurobiology
Duke University Medical Centre
Durham, North Carolina, USA

VOLUME SEVENTY SIX

CURRENT TOPICS IN MEMBRANES

Basement Membranes

Edited by

JEFFREY H. MINER
Washington University,
St Louis, MO, USA

ELSEVIER

Amsterdam • Boston • Heidelberg • London
New York • Oxford • Paris • San Diego
San Francisco • Singapore • Sydney • Tokyo

Academic Press is an imprint of Elsevier

Academic Press is an imprint of Elsevier
225 Wyman Street, Waltham, MA 02451, USA
525 B Street, Suite 1800, San Diego, CA 92101-4495, USA
125 London Wall, London EC2Y 5AS, UK
The Boulevard, Langford Lane, Kidlington, Oxford OX5 1GB, UK

First edition 2015

ISBN: 978-0-12-804015-7
ISSN: 1063-5823

For information on all Academic Press publications
visit our website at http://store.elsevier.com/

CONTENTS

CONTRIBUTORS

Marcel V. Alavi
Departments of Ophthalmology and Anatomy, Institute for Human Genetics, UCSF School of Medicine, San Francisco, CA, USA

Corina M. Borza
Division of Nephrology, Department of Medicine; Vanderbilt Center for Kidney Disease, Vanderbilt University Medical Center, Nashville, TN, USA

Xiwu Chen
Division of Nephrology, Department of Medicine, Vanderbilt University Medical Center, Nashville, TN, USA

Matthew R. Clay
Department of Biology, Duke University, Durham, NC, USA

Madeleine Durbeej
Department of Experimental Medical Science, Lund University, Lund, Sweden

Douglas B. Gould
Departments of Ophthalmology and Anatomy, Institute for Human Genetics, UCSF School of Medicine, San Francisco, CA, USA

Cristina Has
Department of Dermatology, University Medical Center Freiburg, Freiburg, Germany

Sally Horne-Badovinac
Committee on Development, Regeneration, and Stem Cell Biology; Department of Molecular Genetics and Cell Biology, The University of Chicago, Chicago, IL, USA

Adam J. Isabella
Committee on Development, Regeneration, and Stem Cell Biology, The University of Chicago, Chicago, IL, USA

Cassandre Labelle-Dumais
Departments of Ophthalmology and Anatomy, Institute for Human Genetics, UCSF School of Medicine, San Francisco, CA, USA

Rachel Lennon
Wellcome Trust Centre for Cell-Matrix Research, Faculty of Life Sciences, University of Manchester; Institute of Human Development, Faculty of Human Sciences, University of Manchester; Department of Paediatric Nephrology, Central Manchester University Hospitals NHS Foundation Trust (CMFT), Manchester Academic Health Science Centre (MAHSC), Manchester, UK

Mao Mao
Departments of Ophthalmology and Anatomy, Institute for Human Genetics, UCSF School of Medicine, San Francisco, CA, USA

Kevin J. McCarthy
Department of Pathology, Louisiana State University Health Sciences Center-Shreveport, Shreveport, LA, USA

Alexander Nyström
Department of Dermatology, University Medical Center Freiburg, Freiburg, Germany

Ambra Pozzi
Division of Nephrology, Department of Medicine, Vanderbilt University Medical Center; Department of Cancer Biology, Vanderbilt University Medical Center; Department of Molecular Physiology and Biophysics, Vanderbilt University Medical Center; Vanderbilt Center for Kidney Disease, Vanderbilt University Medical Center; Department of Medicine, Veterans Administration Hospital, Nashville, TN, USA

Michael Randles
Wellcome Trust Centre for Cell-Matrix Research, Faculty of Life Sciences; Institute of Human Development, Faculty of Human Sciences, University of Manchester, Manchester, UK

Kiyotoshi Sekiguchi
Laboratory of Extracellular Matrix Biochemistry, Institute for Protein Research, Osaka University, Suita, Osaka, Japan

David R. Sherwood
Department of Biology, Duke University, Durham, NC, USA

Masashi Yamada
Laboratory of Extracellular Matrix Biochemistry, Institute for Protein Research, Osaka University, Suita, Osaka, Japan

Peter D. Yurchenco
Department of Pathology & Laboratory Medicine, Robert Wood Johnson Medical School, Rutgers University, Piscataway, NJ, USA

Roy Zent
Division of Nephrology, Department of Medicine, Vanderbilt University Medical Center; Department of Cancer Biology, Vanderbilt University Medical Center; Department of Cell and Developmental Biology, Vanderbilt University Medical Center; Vanderbilt Center for Kidney Disease, Vanderbilt University Medical Center; Department of Medicine, Veterans Administration Hospital, Nashville, TN, USA

PREFACE

It is just as little ironic that I find myself compiling and editing a book focused on basement membranes, a topic about which I was never officially taught anything in graduate school or in college. In my experience, I find that students are easily underwhelmed by contemplating the function of molecules found outside of cells—in the domain of the extracellular matrix—when so much more seems to be going inside of cells. The realm of biological inquiry can appear boundless to a young scientist searching for a thesis or a postdoctoral training laboratory. What could be appealing about basement membranes as a topic for 4–6 years, if not a lifetime, of study?

I once told a nonscientist acquaintance at a party that I worked on basement membranes. He replied "Oh, so you work in construction?" As silly as the question seemed to me at the time and for years thereafter, in hindsight I now realize that for 23 years I have indeed been studying a type of biological "construction." Basement membranes are required for the construction of all multicellular organisms, exclusive of plants and biofilms (as I was once taught by a sharp student); they are required for the architecture of tissues, organs, and for the construction of "highways" upon which cells migrate; and they sometimes serve as well-constructed barriers, whether to block the passage of macromolecules in the kidney filter or to prevent the movement of cells from one tissue compartment into another.

Sometimes basement membranes can be defective, either due to genetic, mechanical, or environmental insults that impact the structure and/or function of the basement membrane or its components, in the same way that defective construction materials or a tornado can impact the integrity of a building. For basement membranes, the consequences of such defectiveness can be either lethal for the organism at an early developmental stage or cause an array of ongoing functional problems that we commonly refer to as diseases. And it is the diverse group of basement membrane-based diseases that has gotten (and kept) me, and I presume many others, so excited about the field. I hope this book will help to excite others and expand the cadre of investigators focused on basement membranes.

The opening chapter by Peter Yurchenco discusses the importance of laminins for the initiation of basement membrane assembly. A milestone in the basement membrane field was the seminal discovery that in knockout mice lacking laminin γ1, a component of most laminin heterotrimers, there were no basement membranes. No other protein has been shown to be as important for constructing a basement membrane.

Chapter 2 by Madeleine Durbeej focuses on the importance of one particular laminin trimer, laminin-211 (α2β1γ1), in the function and health of skeletal muscle. Although the Duchenne's form of muscular dystrophy with defects in dystrophin may be the best known and most common (due to X-linkage), studies of congenital muscular dystrophy due to mutation of the gene encoding laminin α2 (termed MDC1A) have provided important insights into how the linkage between the myofiber basement membrane and the actin cytoskeleton (via dystrophin and dystroglycan) is so critical. I use this story when teaching the importance of cell/matrix interactions to first year PhD students.

In Chapter 3, Mao Mao, Marcel Alavi, Cassandre Labelle-Dumais, and Doug Gould focus on another major component of all basement membranes, type IV collagen. There are six type IV collagen chains, two of which (α1 and α2) are considered ubiquitous and are required for life. However, heterozygous mutations in these two chains have been shown to cause defects in multiple organs in mice and humans, especially in the vasculature, brain, and skeletal muscle. In contrast, mutations that impact the α3, α4, and α5 chains cause kidney disease and eye and hearing defects.

The focus of Chapter 4, coauthored by Cristina Has and Alex Nyström, is the epidermal basement membrane. As with the skeletal muscle fiber basement membrane, laminin plays a critical role in cell/matrix interactions, but in the skin it is laminin-332 (α3β3γ2) that is so important for establishing and maintaining tight adhesion between the epidermis and its basement membrane through the establishment of hemidesmosomes. In addition, a special form of collagen, type VII, is critical for linking the basement membrane to the underlying dermis. Defects in either of these connections can cause severe skin blistering and in some cases, death.

In Chapter 5, coauthors Michael Randles and Rachel Lennon present the details of a state-of-the-art methodology for investigating the composition of extracellular matrices. They focus on methods for extracting and enriching extracellular matrix (ECM) components from tissue and the use of proteomics (mass spectrometry) to identify novel components of the ECM in multiple tissues. They also discuss how these methods have been used to investigate ECM composition in both health and disease.

The focus of Chapter 6 by Masashi Yamada and Kiyotoshi Sekiguchi turns to receptors for basement membrane laminins, with a particular emphasis on integrins. Defining exactly how cells bind to and respond to basement membranes is an important goal for a complete understanding of cell/matrix interactions. This chapter summarizes the three known laminin-binding integrins and discusses the sites on the laminin α, β, and γ chains that are critical for mediating the adhesion of integrins to the basement membrane. Also discussed are the tetraspanins, transmembrane proteins that can modulate the affinity of integrins for their laminin ligands.

In Chapter 7, Corina Borza, Xiwu Chen, Roy Zent, and Ambra Pozzi address the basement membrane and integrin proteins that are known to be important in the kidney. They discuss mouse models of two human kidney glomerular basement membrane diseases, Alport syndrome and Pierson syndrome. They also summarize data about how the collagen binding integrins α1β1 and α2β1 influence the production of collagen in normal and diseased kidney glomeruli and how important integrin β1 is for development and function of the kidney.

Kevin McCarthy focuses on the heparan sulfate proteoglycans perlecan and agrin in Chapter 8. These widely expressed macromolecules have been shown to be important both in basement membranes and in nonbasement membrane contexts. The different types of glycosaminoglycan side chains that can be appended to the core proteins and extent of sulfation that occurs make these molecules and their functions especially complex and interesting.

Chapters 9 and 10 cover basement membranes in two invertebrates. Adam Isabella and Sally Horne-Badovinac tackle the role of basement membranes in *Drosophila* development in Chapter 9. They discuss the powerful genetic tools that allow the study of the important roles of basement membranes in live, maturing *Drosophila* oocytes. They also discuss how basement membranes influence axonal pathfinding, BMP signaling, and stem cells.

In Chapter 10, Matthew Clay and David Sherwood summarize what is known about the importance of basement membranes in the nematode *C. elegans*. Worms carrying a variety of mutations in genes encoding basement membrane proteins have shown their importance in multiple processes, including those that involve the nervous system, muscle cells, and the germ stem cell niche. In addition, the worm has been especially useful for investigating how cells penetrate a basement membrane.

As might be obvious from the preceding chapter summaries, when planning this volume I chose to invite authors who share my own appreciation for the importance of specific basement membrane proteins in human disease, while at the same time recognize that it is basic studies of basement membranes, in both higher and lower organisms, that will bring us a complete understanding of how they assemble and the mechanisms by which cells interact and respond to them.

I wish to express my sincere thanks to all the authors who generously gave of their precious time to write their outstanding contributions. I also appreciate the invitation from Sid Simon and Bob Balaban to compile and edit this volume and the dedication of the staff at Elsevier—Mary Ann Zimmerman, Helene Kabes, and Roshmi Joy—who ensured the timeliness and the quality of the product you hold in your hands (or read on your screen). Finally, I wish to thank the many funding organizations around the world who have supported and will continue to support studies of basement membranes.

Jeffrey H. Miner
Washington University School of Medicine
St. Louis, Missouri, USA

PREVIOUS VOLUMES IN SERIES

Current Topics in Membranes and Transport

Volume 23 Genes and Membranes: Transport Proteins and Receptors* (1985)
Edited by Edward A. Adelberg and Carolyn W. Slayman

Volume 24 Membrane Protein Biosynthesis and Turnover (1985)
Edited by Philip A. Knauf and John S. Cook

Volume 25 Regulation of Calcium Transport across Muscle Membranes (1985)
Edited by Adil E. Shamoo

Volume 26 Na^+-H^+Exchange, Intracellular pH, and Cell Function* (1986)
Edited by Peter S. Aronson and Walter F. Boron

Volume 27 The Role of Membranes in Cell Growth and Differentiation (1986)
Edited by Lazaro J. Mandel and Dale J. Benos

Volume 28 Potassium Transport: Physiology and Pathophysiology* (1987)
Edited by Gerhard Giebisch

Volume 29 Membrane Structure and Function (1987)
Edited by Richard D. Klausner, Christoph Kempf, and Josvan Renswoude

Volume 30 Cell **Volume** Control: Fundamental and Comparative Aspects in Animal Cells (1987)
Edited by R. Gilles, Arnost Kleinzeller, and L. Bolis

Volume 31 Molecular Neurobiology: Endocrine Approaches (1987)
Edited by Jerome F. Strauss, III, and Donald W. Pfaff

Volume 32 Membrane Fusion in Fertilization, Cellular Transport, and Viral Infection (1988)
Edited by Nejat Düzgünes and Felix Bronner

Volume 33 Molecular Biology of Ionic Channels* (1988)
Edited by William S. Agnew, Toni Claudio, and Frederick J. Sigworth

Volume 34 Cellular and Molecular Biology of Sodium Transport* (1989)
Edited by Stanley G. Schultz

Volume 35 Mechanisms of Leukocyte Activation (1990)
Edited by Sergio Grinstein and Ori D. Rotstein

*Part of the series from the Yale Department of Cellular and Molecular Physiology

Volume 36 Protein–Membrane Interactions* (1990)
Edited by Toni Claudio

Volume 37 Channels and Noise in Epithelial Tissues (1990)
Edited by Sandy I. Helman and Willy Van Driessche

Current Topics in Membranes

Volume 38 Ordering the Membrane Cytoskeleton Trilayer* (1991)
Edited by Mark S. Mooseker and Jon S. Morrow

Volume 39 Developmental Biology of Membrane Transport Systems (1991)
Edited by Dale J. Benos

Volume 40 Cell Lipids (1994)
Edited by Dick Hoekstra

Volume 41 Cell Biology and Membrane Transport Processes* (1994)
Edited by Michael Caplan

Volume 42 Chloride Channels (1994)
Edited by William B. Guggino

Volume 43 Membrane Protein–Cytoskeleton Interactions (1996)
Edited by W. James Nelson

Volume 44 Lipid Polymorphism and Membrane Properties (1997)
Edited by Richard Epand

Volume 45 The Eye's Aqueous Humor: From Secretion to Glaucoma (1998)
Edited by Mortimer M. Civan

Volume 46 Potassium Ion Channels: Molecular Structure, Function, and Diseases (1999)
Edited by Yoshihisa Kurachi, Lily Yeh Jan, and Michel Lazdunski

Volume 47 Amiloride-Sensitive Sodium Channels: Physiology and Functional Diversity (1999)
Edited by Dale J. Benos

Volume 48 Membrane Permeability: 100 Years since Ernest Overton (1999)
Edited by David W. Deamer, Arnost Kleinzeller, and Douglas M. Fambrough

Volume 49 Gap Junctions: Molecular Basis of Cell Communication in Health and Disease
Edited by Camillo Peracchia

Volume 50 Gastrointestinal Transport: Molecular Physiology
Edited by Kim E. Barrett and Mark Donowitz

Volume 51 Aquaporins
Edited by Stefan Hohmann, Søren Nielsen and Peter Agre

Volume 52 Peptide–Lipid Interactions
Edited by Sidney A. Simon and Thomas J. McIntosh

Volume 53 Calcium-Activated Chloride Channels
Edited by Catherine Mary Fuller

Volume 54 Extracellular Nucleotides and Nucleosides: Release, Receptors, and Physiological and Pathophysiological Effects
Edited by Erik M. Schwiebert

Volume 55 Chemokines, Chemokine Receptors, and Disease
Edited by Lisa M. Schwiebert

Volume 56 Basement Membranes: Cell and Molecular Biology
Edited by Nicholas A. Kefalides and Jacques P. Borel

Volume 57 The Nociceptive Membrane
Edited by Uhtaek Oh

Volume 58 Mechanosensitive Ion Channels, Part A
Edited by Owen P. Hamill

Volume 59 Mechanosensitive Ion Channels, Part B
Edited by Owen P. Hamill

Volume 60 Computational Modelling of Membrane Bilayers
Edited by Scott E. Feller

Volume 61 Free Radical Effects on Membranes
Edited by Sadis Matalon

Volume 62 The Eye's Aqueous Humor
Edited by Mortimer M. Civan

Volume 63 Membrane Protein Crystallization
Edited by Larry DeLucas

Volume 64 Leukocyte Adhesion
Edited by Klaus Ley

Volume 65 Claudins
Edited by Alan S. L. Yu

Volume 66 Structure and Function of Calcium Release Channels
Edited by Irina I. Serysheva

Volume 67 Advances in Adrenergic Receptor Biology
Edited by Qin Wang

Volume 68 Membrane Fusion
Edited by Leonid V. Chernomordik and Michael M. Kozlov

Volume 69 Metal Transporters
Edited by Svetlana Lutsenko and José M. Argüello

Volume 70 Co-Transport Systems
Edited by Mark O. Bevensee

Volume 71 Store-Operated Calcium Channels
Edited by Murali Prakriya

Volume 72 Functional Organization of Vertebrate Plasma Membrane
Edited by Vann Bennett

Volume 73 Exchangers
Edited by Mark O. Bevensee

Volume 74 Thermal Sensors
Edited by León D. Islas and Feng Qin

Volume 75 Lipid Domains
Edited by Anne K. Kenworthy

CHAPTER ONE

Integrating Activities of Laminins that Drive Basement Membrane Assembly and Function

Peter D. Yurchenco
Department of Pathology & Laboratory Medicine, Robert Wood Johnson Medical School, Rutgers University, Piscataway, NJ, USA
E-mail: yurchenc@rwjms.rutgers.edu

Contents

Abstract

Studies on extracellular matrix proteins, cells, and genetically modified animals have converged to reveal mechanisms of basement membrane self-assembly as mediated by γ1 subunit-containing laminins, the focus of this chapter. The basic model is as follows: A member of the laminin family adheres to a competent cell surface and typically polymerizes followed by laminin binding to the extracellular adaptor proteins nidogen, perlecan, and agrin. Assembly is completed by the linking of nidogen and heparan sulfates to type IV collagen, allowing it to form a second stabilizing network polymer. The assembled matrix provides structural support, anchoring the extracellular matrix to the cytoskeleton, and acts as a signaling platform. Heterogeneity of function is created in part by the isoforms of laminin that vary in their ability to polymerize and to interact with integrins, dystroglycan, and other receptors. Mutations in laminin subunits, affecting expression or LN domain-specific functions, are a cause of human diseases that include those of muscle, nerve, brain, and kidney.

Current Topics in Membranes, Volume 76
ISSN 1063-5823
http://dx.doi.org/10.1016/bs.ctm.2015.05.001

1. INTRODUCTION

Basement membranes are cell-adherent extracellular scaffolds that are essential for embryonic morphogenesis and adult functions. These scaffolds are anchored to the cytoskeleton through receptors and present a dense array of tethered ligands that act as a signaling platform. Members of the laminin family, each a heterotrimeric glycoprotein, constitute key components of basement membranes. Laminins, particularly those possessing the γ1 subunit, form the initial polymer cell scaffold and are responsible for many biological activities. The significance of these activities is revealed by mutations preventing the expression of different laminin subunits. These mutations have consequences ranging from death in the peri-implantation period to diseases of kidney, peripheral nerve, brain, and skin. An understanding of the critical roles of laminins requires an understanding of their assembly, structural organization, and structure–function relationships. This understanding is expected to provide a rational basis to develop therapies to correct basement membrane disorders and insights for the design of biomaterials that mimic the biological properties of basement membranes for tissue repair and replacement. In the course of this chapter, the author will describe the role of γ1-laminins in basement membrane assembly as we currently understand it with particular attention to the role of the LN domains, and discuss the molecular basis of two diseases resulting from laminin mutations.

2. THE LAMININ FAMILY

The first laminin to be discovered and characterized was isolated from extracts of the Engelbreth-Holm-Swarm (EHS) tumor (Engel et al., 1981; Timpl et al., 1979). It was later renamed laminin-111 (Lm111) based on its α1β1γ1 subunit composition (Aumailley et al., 2005). This laminin (Figure 1), widely expressed during embryogenesis, is restricted to a few tissues such as brain and kidney after birth. Lm111 and similar laminins sharing the common and almost universally expressed γ1 subunit serves as a model for understanding many laminin functions. The N-terminal moieties of the laminin α1, β1, and γ1 subunits form three short arms. Each short arm is capped with a unique globular structure, the LN domain, connected to a semiflexible rod consisting of LE domains interrupted by two to three internal globular domains. The three subunits of the laminin are joined through a long coiled-coil domain forming the long arm. An α1 subunit C-terminal

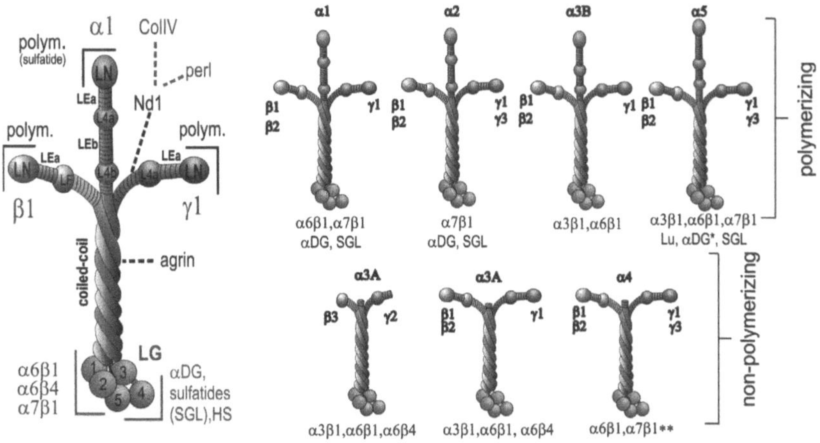

Figure 1 ***Laminin Family.*** Laminin subunits shown are the α1, α2, α3, α4, and α5, β1, β2, and β3 and the γ1, γ2, and γ3 chains (green, red, and blue respectively). Functional loci are shown adjacent to domain designations for laminin-111 on the left side (polym, polymerization domain; Nd1, nidogen-1-binding site; αDG, α-dystroglycan; αxβy integrins; SGL, sulfated glycolipids); Lu, Lutheran glycoprotein receptor; HS, heparan sulfates). Laminin isoforms (right side) are grouped by their domain similarities. Polymerizing laminins each possess three different LN domains, while non-polymerizing laminins lack one or more LN domains. Laminin-411 (**) is unusual among laminins in that it binds weakly to integrins and poorly to dystroglycan. (See color plate)

segment extends beyond the base of the coiled-coil domain to form five globular LG domains, each a beta-sandwich, with a short hingelike segment separating the third and fourth globule (Harrison et al., 2007).

The various functional binding activities map to different structural loci within the laminin heterotrimer (reviewed in Yurchenco, 2011 and Yurchenco and Patton, 2009). Self-assembly is mediated by the α-,β-, and γ-LN domains (McKee, Harrison, Capizzi, & Yurchenco, 2007; Yurchenco & Cheng, 1993; Yurchenco, Tsilibary, Charonis, & Furthmayr, 1985). Nidogens-1 and -2 bind to the third γ1LEb3 domain (Fox et al., 1991; Poschl et al., 1996; Stetefeld, Mayer, Timpl, & Huber, 1996). The N-terminal (NtA) domain of agrin interacts with a γ1 sequence (EQQ-TADQLLARASAAKALAE) that must be integrated within the coiled-coil for high-affinity agrin binding (Denzer et al., 1998; Kammerer et al., 1999; Mascarenhas et al., 2003). Integrins α6β1, α6β4, and α7β1 bind to LG domains 1–3 and the adjacent C-terminal γ1 segment extending beyond the coiled-coil, while α-dystroglycan (αDG) binds primarily to LG4 and, to a lesser degree, to LG1–3 (Gee et al., 1993). The αLG affinity and domain specificity depends on the particular laminin-α subunit (discussed in Yurchenco and Patton, 2009). For example, laminin-211 possesses two distinct dystroglycan-binding regions in LG1–3 and LG4–5 (Smirnov et al., 2002).

Since the discovery of laminin-111, five alpha (α1, α2, α3, α4, α5), four beta (β1, β2, β3, β4), and three gamma (γ1, γ2, γ3) laminin subunits have been identified in mammals. Furthermore, the laminin-α3 subunit exists as two splice variants differing by the absence (α3A) or presence (α3B) of a short arm. There are 45 possible α–β–γ combinations for laminin subunits α1, 2, 3, 4, 5, β1, 2, 3 and γ1, 2, 3 (little is known about Lmβ4); however, only 16 combinations of subunits have been identified to date (Macdonald, Lustig, Steinmetz, & Kammerer, 2010). This number results from restrictions in the allowed charge-dependent associations within the heptad repeats of the coiled-coil domain (Beck, Dixon, Engel, & Parry, 1993). The laminins can be classified according to their domain composition and fall broadly into polymerizing and non-polymerizing groups. Polymerizing laminins all possess a set of three different LN domains located at the N-termini of the three subunits. Non-polymerizing laminins lack an α-LN domain (α4 laminins and several α3A laminins) or, in the case of laminin-3A32, lack both α- and γ-LN domains (Cheng, Champliaud, Burgeson, Marinkovich, & Yurchenco, 1997). The laminins can be further subdivided based on the repertoire of receptor interactions through their LG domains. Of these, the α5-laminins (especially Lm-511) exhibit the broadest degree of integrin-binding interactions, bind to dystroglycan and the Lutheran (Lu) receptor, and represent the most common laminins after birth (Miner, Lewis, & Sanes, 1995). In contrast, the α2-laminins (principal laminins along the neuromuscular axis) interact only with the α7β1 integrin and dystroglycan. Laminin-411, a major laminin of blood vessel endothelium, skeletal muscle, and peripheral nerve endoneurium is unusual in that it lacks an α-LN domain, binds weakly to integrins (Nishiuchi et al., 2006), and binds poorly to dystroglycan (Talts et al., 2000). Laminin-3A32, a broad topic in its own right, is a specialized laminin found in epithelia that forms strong links with hemidesmosomes (reviewed in Walko, Castanon, and Wiche, 2015).

The invertebrates *Caenorhabditis elegans* and *Drosophila melanogaster* each possess only two laminins roughly equivalent to Lm111 and Lm511, both polymerizing laminins. Sponges may represent the most primitive metazoans to possess a heterotrimeric laminin with three different LN domains (Fahey & Degnan, 2012). *Monosiga brevicollis*, a choanoflagellate that is one of the closest living premetazoans, lacks laminins, other extracellular matrix proteins, and their receptors. Instead the species possesses laminin-like domains as part of larger complex transmembrane cadherin-like proteins. These proteins are members of the lefftyrin family and consist of an N-terminal LN domain adjacent to four epidermal growth factor (EGF)-like domains with or without an included laminin LG domain (Nichols, Roberts, Richter, Fairclough, & King, 2012). The lefftyrins might serve the function of mediating cell–cell adhesion, enabling the organisms to form rosettes from single cells.

3. LAMININ POLYMERIZATION

When laminin-111 is warmed in a neutral salt buffer containing calcium it becomes turbid and sedimentable, reflecting formation of a polymer (Yurchenco et al., 1985). This process of polymerization was found to be concentration dependent in a two-stage process. The first step is temperature-dependent but independent of divalent cation and represents the conversion of monomers to dimers and small oligomers that were visualized by electron microscopy of rotary-shadowed Pt/C replicas. The dimers appeared to be joined by short-arm to short-arm associations as well as by long-arm interactions. The latter were later found to be irrelevant for polymerization and likely resulted from bindings of an unpaired cysteine in the LG4–5 domains (Yurchenco & Cheng, 1993; Yurchenco, Cheng, & Schittny, 1990). The second stage of assembly requires divalent cation, in particular calcium, and is the step in which polymers form. The two steps of observed interactions and presence of a critical concentration of polymerization (70–140 nM for laminin-111) indicated that laminin polymerization occurs by a mechanism of nucleation–propagation (Yurchenco et al., 1985), i.e., a thermodynamically unfavorable nucleation binding followed by a calcium-dependent polymerization. The nucleation phase is now thought to occur by (1) formation of laminin β–γ LN bonds (KD = 22 μM by surface plasmon resonance, SPR; Purvis & Hohenester, 2012) followed by (2) calcium-dependent laminin αLN binding to β–γ LN domain dimers (1.4 μM by equilibrium gel filtration (EqGF) and 0.8 μM by SPR; Hussain, Carafoli, & Hohenester, 2011; Yurchenco & Cheng, 1993). Accretion of laminin monomers to the dimers, oligomers, and small polymers is thought to occur by binding of βLN and γLN domains of a free laminin as paired interactions to exposed available αLN domains (Figure 2(A)).

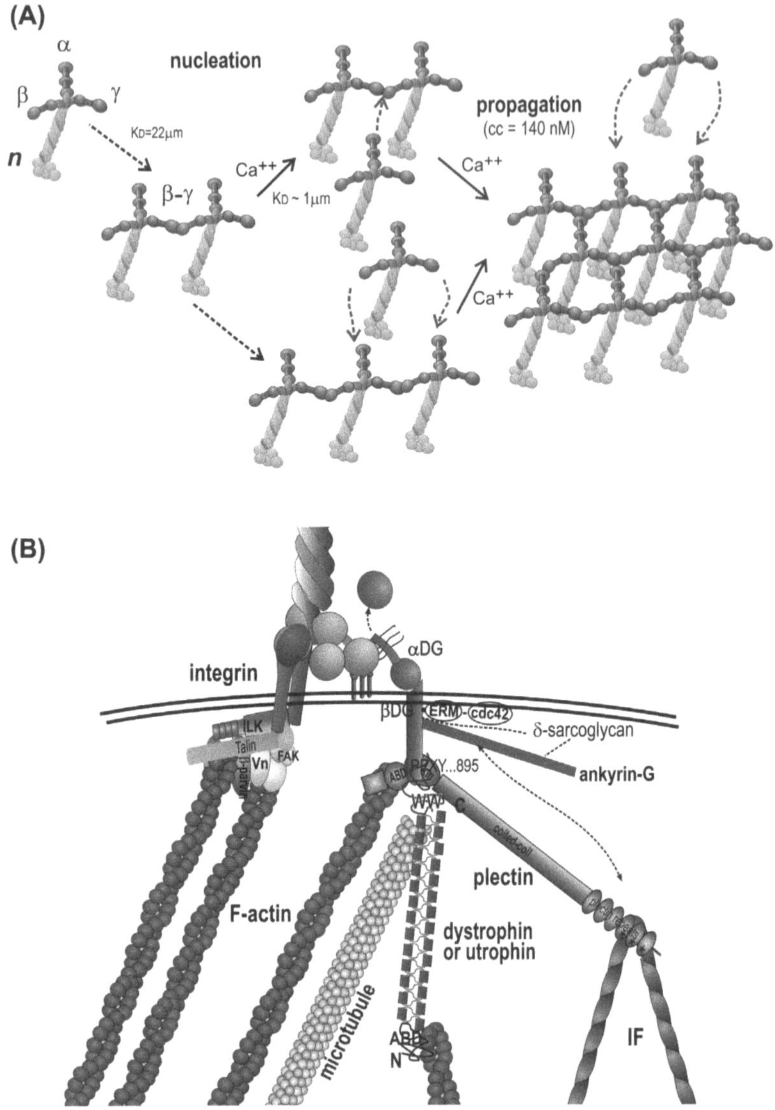

Figure 2 ***Laminin Polymerization and Anchorage.*** Panel A: Laminins that possess three different LN domains will polymerize through cooperative self-assembly. The first step (nucleation) involves thermodynamically unfavorable assembly of dimers and small oligomers through β-γ LN dimerization. These interactions do not require divalent cation. This is followed by a more favored step (propagation), dependent on the presence of calcium coordinated in the γ-LN domain, in which the α-LN domains of free laminins bind to the β-γ dimers and the free β and γ LN domains bind to the exposed α-LN domains of the growing sheet-like polymer. Panel B: Cartoon of interactions of laminin LG domains with integrin, the O-mannosyl carbohydrate of α-dystroglycan, and sulfated glycolipids (sulfatides) of cell surfaces. Integrins and dystroglycan, but not sulfated glycolipids, form linkages to the cytoskeleton.

At first, the organization of the polymer could not be determined from the rotary-shadowed images because of their complexity and tendency to pile up into dense aggregates. However, subsequent ultrastructural studies using high-angle unidirectional shadowing of freeze-dried laminin preparations with very thin metal coats revealed that the laminin polymer was a sheetlike array of interconnecting struts (Yurchenco, Cheng, & Colognato, 1992). Inter-laminin interactions were detected with native laminin fragments after incubation of fragment mixtures by rotary shadow Pt/C replication, EqGF, and polymerization inhibition assays (Yurchenco & Cheng, 1993; Yurchenco et al., 1992). The Pt/C replicas revealed heterotrimeric end-to-end complexes formed by the three different LN domains and the gel filtration experiments demonstrated that only fragments containing the three LN domains could bind to each other and inhibit polymerization. A "three-arm interaction model" was proposed based on these observations. The calcium-binding locus required for polymerization was thought to reside in the γ1 LN domain by analysis of $^{45}Ca^{++}$ binding overlays on laminin fragments. In agreement, the recent solution of the crystal structure of the laminin-γ1 LN—LEa2 domains revealed that a calcium ion is selectively bound to a short α-helix and coordinated by Asp106 and Thr114 (Carafoli, Hussain, & Hohenester, 2012).

Other members of the laminin family were found to self-assemble after the original discovery with laminin-111. Laminins-211 and -221, isolated from human placenta, were observed to polymerize by direct assay and by copolymerization with EHS-laminin (Cheng et al., 1997). It was deduced that α3B- and α5-laminins polymerize as evidenced by the ability of the corresponding recombinant short-arm N-terminal fragments to inhibit laminin-111 polymerization (Garbe, Gohring, Mann, Timpl, & Sasaki, 2002). Laminins lacking one or more LN domains were determined to be incapable of polymerization as judged by the inability to detect copolymerization with laminin-111 (Cheng et al., 1997).

The LN domain specificity of the three-arm interaction model was challenged in a study based on laminin fragment binding experiments by SPR and gel electrophoresis, and it was argued that there was considerable degeneracy in the allowed LN to LN domain binding (Odenthal et al., 2004). However, this alternative interpretation is not supported by more recent studies. First, loss-of-function analysis of recombinant mammalian laminin-111 polymerization bearing LN deletions and domain swaps revealed that only laminins with a full complement of α1, β1, and γ1 LN domains polymerized and assembled nascent basement membranes in culture

(McKee et al., 2007). Second, a gain-of-function evaluation of polymerization using a laminin-α1 short-arm-nidogen fusion protein that creates an artificial polymerization arm revealed that only polymerization of laminins lacking the α1 LN domain could be rescued (McKee, Capizzi, & Yurchenco, 2009). Third, extensive study of the binding of recombinant laminin short-arm segments by high-resolution gel filtration and SPR supported only the three-arm specific interaction model (Hussain et al., 2011; Purvis & Hohenester, 2012).

The determination of the crystal structures of the laminin LN and adjacent LEa domains for the α5, β1, and γ1 subunits represents a major new breakthrough for an understanding of laminin high-resolution structure and amino acid residues likely involved in polymerization (Carafoli et al., 2012; Hussain et al., 2011). The laminin-α5 LN structure has been described as a "β-jelly roll with elaborate insertions" attached "like a flower head" to the adjacent and more distal modified EGF-like (LE) domains (Hussain et al., 2011). The corresponding N-terminal domains of the β1 and γ1 subunits were found to be structurally quite similar despite only ~30% sequence identify (Carafoli et al., 2012). Each LN domain forms a pronounced angle with the LE domain stem, and each LN domain presents one face with one to two N-linked oligosaccharide chains and an opposite (bent out from the LE stalk) face that contains surface-exposed residues involved in laminin polymerization. A patch containing the sequence P-L-E-N-G-E (corresponding to residues 229–234) on the surface-exposed face (β6–β3–β8 sheet) of Lmα5 was strongly implicated in laminin polymerization since point mutagenesis of L230A, E231K, and E234K abolished the ability of the modified fragments to inhibit laminin-111 polymerization (Hussain et al., 2011). E202K, located on the opposite N-glycosylation β1-β2-β7-β4-β5 face, on the other hand, retained its ability to inhibit laminin polymerization. The structures, and implicated residues, suggest a model in which three LN surfaces contact each other to form the polymer node with the N-linked carbohydrate chains extending out from the side opposite the LN-contacting faces.

A forward genetic screen of *Drosophila* led to identification of hypomorphic point-specific mutations in the β1-LN domain that cause prominent defects in cardiac basement membrane generation and cardiac morphogenesis (Hollfelder, Frasch, & Reim, 2014). The distribution of defective laminin in the basement membranes around the myocardial tube lacked continuity and instead appeared as a series of disconnected dots. Other basement membranes were not altered in this manner. These conserved hypomorphic mutations (E215K, V226E, and G286R, the first two likely solvent accessible), distinct from stronger LAMB1 allelic truncating mutations, map to the β6-β3-β8 face. While the extent to which these mutations selectively diminish laminin polymerization remains to be determined, and while maternal laminin mRNA contributions might be a relevant factor in the pathology, the results support a key polymerization role of the LN domains as causative of the basement membrane defect. Furthermore, the findings suggest that different basement membranes respond to different degrees to β1LN mutations.

4. LAMININ ADHESION TO CELL SURFACES AND ANCHORAGE TO CYTOSKELETON

4.1 Laminin LG domain binding to cell surface receptors and signaling

Laminin LG domains bind to integrins and α-dystroglycan as well as to sulfated glycolipids. The integrins (e.g., α6β1 and α7β1 for laminins-111 and -511) that bind to laminin LG domains also require that the coiled-coil be present (Deutzmann et al., 1990; Sung, O'Rear, & Yurchenco, 1993). Recent mutational studies revealed that integrin binding to LG1–3 requires a short C-terminal segment of the γ1 subunit containing a critical glutamic acid that extends out beyond the coiled-coil (Ido et al., 2007). The last few residues containing this glutamic acid are missing in the γ3-laminin subunit, one of the most distinguishing features of this otherwise γ1-homologous subunit (Ido et al., 2008).

Laminin binding to its integrin is presumed to initiate outside-in signaling similar to other extracellular matrix (ECM) ligands by recruiting cytoplasmic adaptor proteins, phosphorylation of adaptor-binding proteins, and binding of adaptor proteins to F-actin (reviewed in Hu and Luo, 2013). From a study of fibronectin-activated FAK signaling that is conceptually applicable to laminins, signaling was found to increase linearly with ligand and receptor densities and to require the binding of integrin to a solid as opposed to soluble substrate (Shi & Boettiger, 2003). This is thought to be a fundamental aspect of ECM–integrin interactions, one raising the possibility that changes in the laminin polymer stiffness can alter signal strength.

The LG domains of most laminins, agrin, and perlecan also bind to the O-linked carbohydrate chains of α-dystroglycan (O-mannose-linked carbohydrate containing repeating xylose–glucuronic acid disaccharides; Inamori

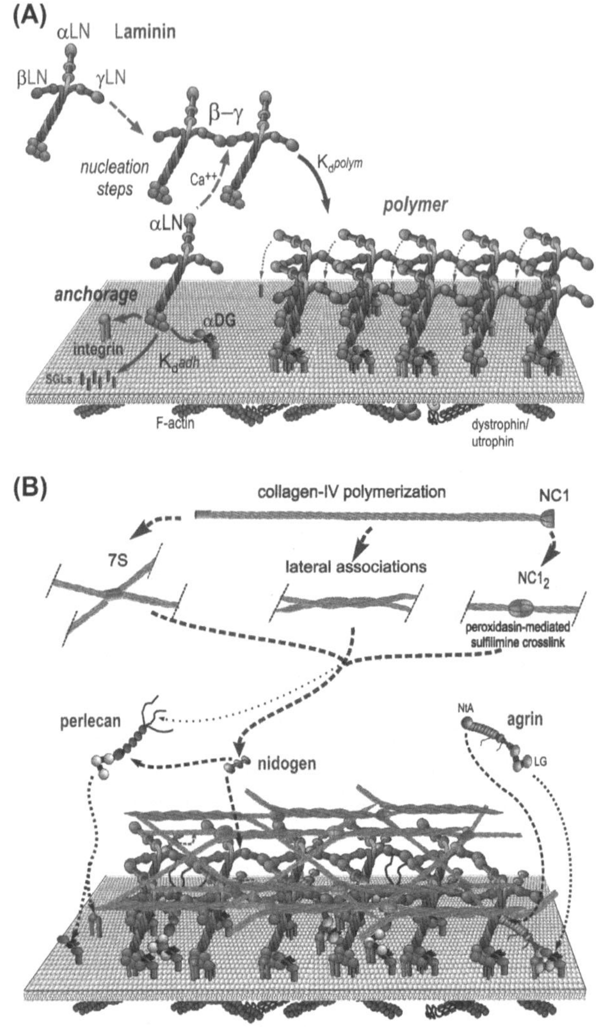

Figure 3 ***Model of Basement Membrane Assembly.*** Panel A: *Adhesion-facilitated laminin assembly*: Soluble laminins adhere to the surface of the plasma membrane by binding to cognate integrins, α-dystroglycan, and sulfated glycolipids. α-LN and β-LN domains of two laminins weakly bind together to form unstable dimers (and larger linear oligomers). The α-LN domain of a third laminin binds to each LN-dimer to form a ternary complex—this represents the polymerization node. Free laminin monomers accrete on the growing polymer by binding the incomplete ternary nodes to form a sheetlike polymer. Assembly is facilitated through the cooperativity of surface binding (increasing the local concentration) and polymerization (converting individual affinities into avidities). Laminin binding to integrins and/or dystroglycan anchors the nascent basement membrane to the underlying cytoskeleton, and initiates signaling through the integrins. Some of the laminins may only bind to sulfated glycolipids in a situation when these lipids are more abundant

et al., 2012). β-dystroglycan, a transmembrane protein linked noncovalently to the α subunit, binds to a series of cytoskeletal proteins that include dystrophin/utrophin, ankyrin-G, and plectin (Moore & Winder, 2010). Dystrophin and utrophin bind to F-actin, forming a linkage between basement membrane and the actin cytoskeleton (Matsumura et al., 1993). It has been found that the adaptor proteins of dystroglycan exhibit a host of binding interactions. β-dystroglycan binds to dystrophin and plectin, dystrophin binds to microtubules, and plectin binds to intermediate filaments (Belanto et al., 2014; Rezniczek et al., 2007). Thus dystroglycan can constitutively link up with all three cytoskeletal systems. Despite this repertoire of interactions, dystroglycan is important as a basement membrane link for some (e.g., Reichert's membrane, skeletal muscle, brain cortex, Schwann cell nodes of Ranvier) but not all (e.g., kidney) tissues (Jarad, Pippin, Shankland, Kreidberg, & Miner, 2011; Moore et al., 2002; Occhi et al., 2005; Saito et al., 2003; Williamson et al., 1997).

Laminin LG domains also bind to sulfated glycolipids, in particular the sulfatides (Taraboletti, Rao, Krutzsch, Liotta, & Roberts, 1990). These glycolipids are enriched in peripheral nerve Schwann cells (SCs) and kidney and can be found as part of the outer leaflet of the plasma membrane (Li et al., 2005; Trick, Decker, Groene, Schulze, & Wiegandt, 1999). Sulfatide expression in the developing peripheral nerve correlates with the appearance of the endoneurial basement membrane (Li et al., 2005). The extent of laminin binding to sulfatides is dependent upon the distribution density of the sulfate groups in the outer leaflet of the plasma membrane (McKee et al., 2009). Finally, α5-laminins are unique in that the LG1−3 domains bind to the Lutheran glycoprotein receptor present in epithelial and other cells (Kikkawa, Moulson, Virtanen, & Miner, 2002).

4.2 Laminin adhesion to cell surfaces initiate basement membrane assembly

A model of basement membrane assembly has developed from the findings of the author and colleagues in the field (Figure 3). Laminins accumulate on

than the transmembrane receptors. Panel B: *Adaptor proteins and type IV collagen assembly*: Nidogen binds to the laminin-γ1 short arm. The heparan sulfate proteoglycan (HSPG) perlecan binds to nidogen, dystroglycan, and integrin, while the HSPG agrin (usually the non-neural splice form) binds to the coiled-coil of laminin and to dystroglycan and integrin. Assembly is completed by the binding of type IV collagen to nidogen and heparan sulfate chains and its polymerization into a covalently stabilized network.

cell surfaces even in the absence of other basement membrane components (Colognato, Winkelmann, & Yurchenco, 1999; Li et al., 2005; McKee et al., 2007; Smirnov et al., 2002; Tsiper & Yurchenco, 2002). Nidogen-1 and type IV collagen coaccumulate with laminin on the cell surface, but fail to accumulate in the absence of laminin. This laminin requirement discerned from cell culture is in agreement with studies of knockout mice revealing that basement membranes can assemble in the absence of nidogens, type IV collagen, perlecan, or agrin, but not in the absence of γ1- or β1-laminins (reviewed in Miner and Yurchenco, 2004).

Basement membrane assembly is cell-surface selective. Electron micrographs of basement membranes at the epithelial–stromal junction have revealed the absence of a basement membrane on the stromal fibroblasts immediately adjacent to the epithelial basement membrane at distances that would allow diffusion of the secreted and soluble laminins to both locations. This suggests that cells must become "competent" for basement membrane assembly to occur, i.e., undergo cell surface changes that enable the binding of one or more basement membrane components. Since laminins are required for assembly, the obvious expectation is that competency is determined by the presence of suitable laminin-binding molecules, in particular cognate integrins, dystroglycan, and laminin-binding glycolipids. Sulfated glycolipids such as galactosyl sulfatide are among the cell surface molecules that can enable laminin cell-surface accumulation. In a study comparing cultured SCs with embryonic lung fibroblasts, laminin selectively accumulated on the SC surfaces (Li et al., 2005). This interaction was followed by a lateral surface coalescence of laminin, sulfatide, dystroglycan and c-Src, recruitment of utrophin, and activation of Src and focal adhesion kinase. Removal of the sulfate groups by aryl sulfatase prevented laminin surface assembly, whereas sulfatide loading of the cell surface (incorporation of sulfatide into the outer leaflet of the plasma membrane) restored assembly. Furthermore, sulfatide loading of the fibroblasts enabled laminin surface assembly. Although sulfated glycolipids can provide competency for LG domain-mediated adhesion, they are unable to anchor laminin to the underlying cytoskeleton, a critical function. This is accomplished by integrins and dystroglycan.

An incompletely resolved issue is the minimal requirement for laminin adhesion to a cell surface. Evidence to rule out the requirement for *both* bonds is that embryoid bodies lacking β1-integrins or dystroglycan assemble basement membranes in the presence of laminin and differentiate to form polarized epiblast (Li et al., 2002). This is supported by genetic data that separate knockout of relevant integrin subunits and dystroglycan generally do not prevent basement membrane assembly (reviewed in Yurchenco and Patton, 2009). Steric considerations, given the large size of integrin and dystroglycan, would suggest that double occupancy for a given laminin is problematic. Another uncertain issue is whether every laminin in an assembling basement membrane must bind either to an integrin or a dystroglycan. The low level of laminin surface assembly on SCs in the absence of sulfatides, but requirement for the presence of LG domains, suggests a model in which there is insufficient integrin and/or dystroglycan to accommodate all of the laminins on SCs (and perhaps other cells) and that surface adhesion for the majority of laminin molecules is provided by sulfated glycolipids (Li et al., 2005; McKee et al., 2009). These laminins interacting only with the sulfated glycolipids would not be anchored to the underlying cytoskeleton. A possible and proposed arrangement, therefore, is that only a subset of laminins is anchored through formal receptors to the underlying cytoskeleton with the remainder simply adhered to sulfated glycolipids. The ratios would vary in different tissues depending on the availability of laminin-binding surface components.

5. LAMININ POLYMERIZATION ON CELL SURFACES

Laminins polymerize on adhesive cell surfaces more efficiently and at lower concentration than they do in solution. The principle was first demonstrated on synthetic lipid bilayers analyzed by total internal reflection fluorescence microscopy (Kalb & Engel, 1991). Later studies using recombinant laminins with LN and LG deletions revealed that both LN and LG domains were required for efficient accumulation of laminin on cultured SCs. Laminin accumulation was detected at the cell surface at concentrations as low as 10 μM, well below the solution critical concentration of laminin-111 polymerization. Deletion of different LG domains resulted in a substantial decrease of laminin with relatively little reduction of type IV collagen accumulation (McKee et al., 2009, 2007). These and other considerations have led to a model of adhesion-facilitated laminin self-assembly ("receptor-facilitated laminin self-assembly"). The model holds that cell surfaces that bind to the laminin LG domains increase the local concentration at the cell–solvent interface, facilitating polymerization at that interface. A corollary is that laminin–laminin linkage converts individual laminin affinities to surface-binding molecules to ones of higher avidity. Laminin would be expected to accumulate as a monolayer polymer so long as no other components acted to bridge surface laminin to solution laminin.

Binding studies have revealed that adhesion of laminin to sulfatides, largely through the LG-domains, depends on the ratio of sulfatides to non-sulfated lipids of the binding surface (McKee et al., 2009). This creates a potential basis for differential binding to cell plasma membranes. Sulfatides also bind to the LN domain of different laminin-α subunits, albeit with modest affinity (reviewed in Yurchenco, 2011). Thus there is the possibility of a second laminin attachment site to cell surfaces (Figure 3(A), dotted arrows). If this site is so engaged, one would predict that the laminin molecules would reside largely on their side with the polymerization node attached to the sulfatide rather than as drawn with the short arms extending away from the cell surface. Currently there is no direct evidence for such a variation in the model. Indeed, at least in the glomerular basement membrane (GBM), this second attachment seems unlikely given the distribution of N-terminal moiety laminin domain epitopes detected across the double basement membrane by STORM immunofluorescence microscopy (Suleiman et al., 2013).

6. ADAPTOR PROTEINS THAT BIND TO LAMININS: NIDOGEN, PERLECAN, AND AGRIN

As laminins assemble a nascent basement membrane, they can then bind to other secreted basement membrane components, recruiting them to the cell surface (Figure 3(B)). One of these components is nidogen-1 (entactin). Nidogen-1 is a ~150 kDa sulfated glycoprotein consisting of three globular domains spaced by two rodlike domains (Carlin, Jaffe, Bender, & Chung, 1981; Timpl, Dziadek, Fujiwara, Nowack, & Wick, 1983). Nidogen binds strongly (K_D ~ 1 nM) to the short arm of the Lmγ1 subunit (Lmγ1 domain LEb3 by the new nomenclature) through its G3 domain. Nidogen also binds to type IV collagen at several sites along the collagen chain through its G2 and G3 domains with similarly high affinity. Thus nidogen can form a bridge between the laminin and collagen polymers (Fox, Mayer, Nischt, Aumailley et al., 1991). Mutagenesis analysis of the laminin LE rod segment and crystallization of both the laminin rod segment and nidogen-1 revealed the interacting residues and their spatial orientations (Poschl, Fox, Block, Mayer, & Timpl, 1994; Poschl et al., 1996; Stetefeld et al., 1996; Takagi, Yang, Liu, Wang, & Springer, 2003).

The nidogen G2 domain also forms a strong bond with immunoglobulin repeats (domain IV) of perlecan, a heparan sulfate proteoglycan of basement membranes (Battaglia, Mayer, Aumailley, & Timpl, 1992; Hopf, Gohring, Kohfeldt, Yamada, & Timpl, 1999; Hopf, Gohring, Mann, & Timpl, 2001; Hopf, Gohring, Ries, Timpl, & Hohenester, 2001). Nidogen can be considered a structural adaptor protein of basement membranes, playing a major role in the linking of laminin to type IV collagen and perlecan to laminin. This bridging role is seen prominently in tissue culture assembly of type IV collagen (McKee et al., 2007). However, genetic studies have revealed that nidogen cannot serve as the *only* linker in this regard (Bader et al., 2005; Kim & Wadsworth, 2000; Willem et al., 2002). A search for alternative/compensating linkers has led to the discovery that the heparan sulfate chains of perlecan (and presumably also agrin and type XVIII collagen) may provide this function (Behrens et al., 2012).

Perlecan and non-neural agrin are high molecular weight adaptor proteins whose heparan sulfate chains can serve as binding reservoirs for a large number of growth factors and morphogens that include fibroblast growth factors, vascular endothelial growth factors, and members of the TGFβ family. The C-terminal LG domains of perlecan bind to α-dystroglycan and to integrin α2β1 (Bix et al., 2004; Talts, Andac, Gorhing, Brancaccio, & Timpl, 1999). In this manner perlecan has the potential to bridge laminin through nidogen to integrin and dystroglycan and establish "collateral linkage" to the actin cytoskeleton. The N-terminal NtA domain of nonneural and neural agrin strongly binds the laminin-γ1 subunit in the coiled-coil domain (Denzer et al., 1998; Kammerer et al., 1999). The C-terminal LG domains bind to α-dystroglycan with high affinity and to the α3β1 integrin (Gesemann, Brancaccio, Schumacher, & Ruegg, 1998; McKee et al., 2012). This allows agrin to act as a collateral linker in a manner similar to perlecan. The ability of a shortened version of agrin ("miniagrin") and perlecan to link compensating laminins (especially laminin-411) to dystroglycan may explain its ability to ameliorate the muscle phenotype in the dyW model of laminin-α2-deficient congenital dystrophy (Meinen, Barzaghi, Lin, Lochmuller, & Ruegg, 2007; Moll et al., 2001).

7. TYPE IV COLLAGEN NETWORK

Type IV collagens self-assemble into a polymer through the formation of covalently stabilized N-terminal 7S complexes and C-terminal NC1 globular domain dimers with extensive entanglement through noncovalent lateral associations that create a finer network mesh (Timpl, Wiedemann, van Delden, Furthmayr, & Kuhn, 1981; Yurchenco & Furthmayr, 1984; Yurchenco & Furthmayr, 1986). Crystallization of the NC1 domain dimers (composed of six NC1 domain protomer fragments from each of the six polypeptide chains that make up the collagen triple helical dimer) revealed the existence of a unique covalent cross-link, the sulfilimine bond (−S=N−), located between Met93 and Hyl211 (Vanacore et al., 2009). This bond, whose formation is mediated by the enzyme peroxidasin, is conserved throughout metazoans and is essential for basement membrane stabilization (Bhave et al., 2012; Fidler et al., 2014). Type IV collagen with an $[\alpha 1_2 \alpha 2]$IV chain composition is present in nearly all basement membranes. Collagens with an [α3α4α5]IV composition, as found in the GBM, provide further stability through additional disulfide bonds (Gunwar et al., 1998).

8. LAMININ-DEFICIENT MUSCULAR DYSTROPHY AND NEUROPATHY

Laminins bearing the α2 subunit (Lm211, Lm221) are the principal laminins of the basement membranes of skeletal muscle (sarcolemma, myotendinous junction, and neuromuscular junction), peripheral nerve (endoneurial Schwann sheath), and an important laminin of brain capillaries forming the blood−brain barrier. Mutations in the *LAMA2* gene (coding for the Lmα2 subunit) are a major cause of congenital muscular dystrophy (designated type MDC1A; reviewed in Jimenez-Mallebrera, Brown, Sewry, and Muntoni, 2005) and a cause of a milder limb−girdle type dystrophy, the latter seen in a much smaller group of individuals. Patients with these dystrophies exhibit brain abnormalities seen as reduced white matter density in T2-weighted magnetic resonance images that are used in the diagnosis of the disease. Patients also may exhibit decreased motor and sensory conduction attributed to a peripheral nerve myelination abnormality. The severely affected patients generally lack laminin subunit expression due to truncations and premature stop codons. They present with weakness before one year of age, fail to meet developmental milestones, may suffer mental retardation, and often suffer from respiratory insufficiency that can result in death.

A mouse model for the severe form of the dystrophy is the dy3K/dy3K (LAMA2 knockout mouse). The mice are very small, extremely weak, and die after a few weeks of life (Miyagoe et al., 1997). The muscle pathology is one of myofiber degeneration and regeneration with fibrosis. It is seen in multiple muscles including those of the diaphragm and intercostal muscles.

The muscle abnormality is accompanied by defects of myelination, most apparent in the sciatic nerve. The myelination defect, in which there are many naked large and medium caliber axons, results from a failure of developing SCs to correctly extend out lamellipodia that sort axons into groups and finally a 1:1 ratio shortly after birth (radial axonal sorting; Martin & Webster, 1973; Webster, Martin, & O'Connell, 1973). The peripheral nerve abnormality becomes manifest in the postnatal period localized to the lower limbs, is recognized by distinctive lower limb flexion, and contributes to the weakness and subsequent development of lower limb extension contractures. Finally, there are defects of the brain microvasculature due to absence of the laminin-α2 subunit in the mouse model that may explain the brain abnormality, sometimes symptomatic, as detected by MRI in patients (Menezes et al., 2014).

Without Lm211, the SC endoneurial and muscle sarcolemmal basement membranes contain substantially increased compensatory Lm411 (with a modest increase of Lm511) and appear attenuated and/or discontinuous at the light immuno- and electron microscopic levels (Patton et al., 1999). Myofibers degenerate and myelination is reduced. The nerve defect may be explained by a loss of basement membrane adhesion to the cell surface and loss of Lm-polymerization-dependent basement membrane density that cannot be replaced by the non-polymerizing and poor receptor-binding Lm411. The myelination defect has been recapitulated in cultured dorsal root ganglia induced with non-polymerizing laminins (McKee et al., 2012). In skeletal muscle, the sarcolemma is destabilized by reduced laminin adhesion to the sarcolemma through loss of αDG-binding and polymerization (Colognato & Yurchenco, 1999; Han et al., 2009; Yurchenco, Cheng, Campbell, & Li, 2004).

Patients with the milder limb–girdle type of dystrophy that presents with symptoms after a year of age generally exhibit only reduced or near-normal expression of the laminin-α2 as determined from immunostaining of muscle biopsies (Figure 4). These patients can still develop the associated neurological problems seen in the more common cases of MDC1A. Differences in the staining intensities for the N-terminal (mAb 48H-2) and C-terminal 80 kDa fragment (mAb 1922) antibodies, in which mild to moderately decrease C-terminal and absent N-terminal staining was observed, have been reported for biopsies from a number of these patients (Gavassini et al., 2011). The selective N-terminal epitope loss suggests selective degradation of N-terminal domain(s) from the LN mutation, while the reduction in C-terminal epitope suggests reduced assembly of a secreted

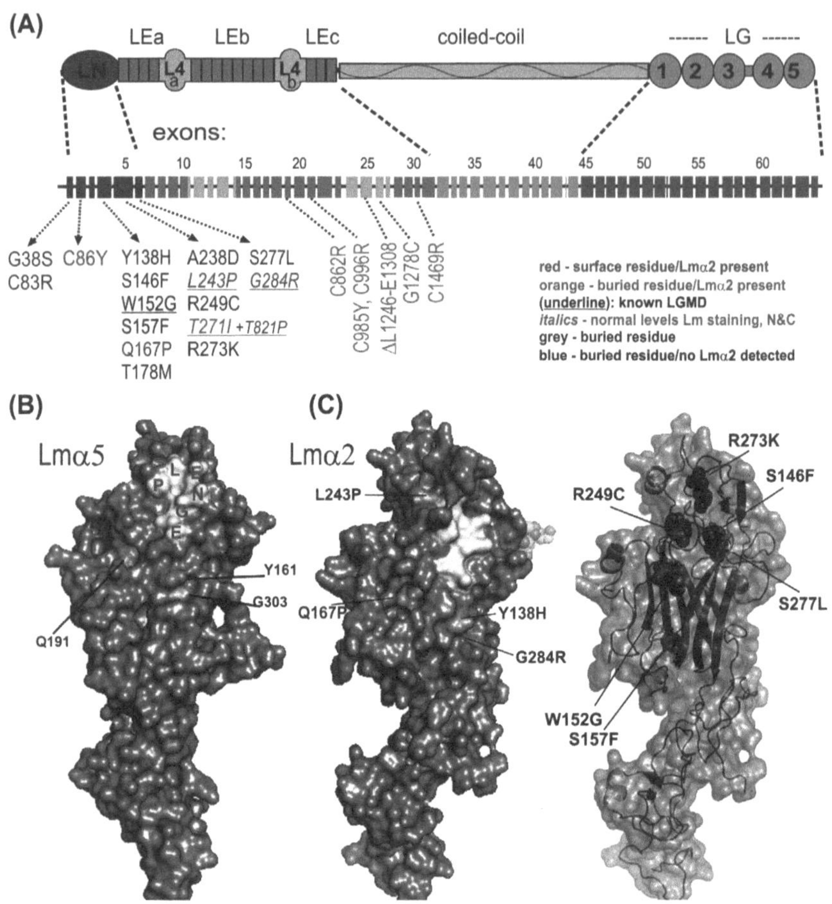

Figure 4 ***Laminin-α2-deficient muscular dystrophy mutations and Lmα-subunit short-arm structures.*** Panel A: *Map of missense single amino acid substitutions and in-frame deletions causing a muscular dystrophy with absent, reduced or normal expression of laminin-α2.* Mutations causing absence (blue text) or either reduced or normal (red, orange italics) laminin-α2 expression based on biopsy immunostaining are distinguished for the indicated mutations (Leiden Open Variation Database, www.dmd.nl/LAMA2_seqvar.html; Chan et al., 2014; Di Blasi et al., 2005; Gavassini et al., 2011; Geranmayeh et al., 2010; Marques et al., 2014; Oliveira et al., 2008; Rajakulendran, Parton, Holton, & Hanna, 2011). Panel B: *Lmα5–LN–LEa domains.* The mouse laminin-α5 LN and adjacent LEa domain structure has been determined by X-ray crystallography by Hohenester and colleagues (Hussain et al., 2011). The surface structure was rendered in the PyMOL Molecular Graphics System (v.1.5.0.4 Schrödinger, LLC) from the published coordinates (Protein Data Bank, 2Y38). The P-L-E-N-G-E polymerization patch is indicated in yellow. Residues that correspond to ones associated with laminin-α2

laminin unable to polymerize. Missense and short in-frame deletions are associated with the milder dystrophy. Many of these mutations cluster in the LN domain and are candidates for defects of laminin polymerization (Figure 3). The great majority of mutations, some on the surface and others in the interior of the domain, seem likely to disrupt the protein fold with some preventing protein secretion (e.g., W152G, S157F). Several of the mutations are located on the presumed polymerization face near the conserved P-L-E-N-G-E polymerization patch (Y138H, Q167P, L243P, and G284R). Analysis of these mutations may confirm a suspected polymerization basis of the disease and identify sequences that participate in forming the polymer node.

The dy2J/dy2J mouse is an animal model for the milder form of laminin-deficient muscular dystrophy. The protein defect arises from a splice-donor mutation in the LAMA2 gene causing an in-frame deletion within the LN domain (Sunada, Bernier, Kozak, Yamada, & Campbell, 1994). This deletion was found to destabilize the domain leading to increased susceptibility to proteolytic degradation and loss of polymerization activity (Colognato & Yurchenco, 1999). There is no evidence for a defect of secretion. These mice are weak, exhibit both dystrophic muscle changes and dysmyelination, but are long-lived. They ambulate by using their forelimbs with at best-limited contribution for the hindlimbs (that over time develop contractures). Immunostaining of the sarcolemmal and endoneurial basement membranes of muscle and nerve reveals only slight reductions of the laminin-α2 subunit in contrast to the basement membranes of the dy3K mouse where the subunit is absent.

The radial sorting defect was examined in peripheral nerves of dy2J/dy2J in the absence and presence of laminin-α4 (Yang et al., 2005). First, radial sorting and myelination in dy2J brachial nerves was found to be nearly normal (despite absence of basement membranes), while the sciatic nerve

deficiency are indicated in orange. Panel C: *Lmα2 LN–LEa domains, homology fit, with dystrophic mutations.* A homology fit using the mouse laminin-α2 sequence was generated in Modeller 9.11 *(kindly provided by David Briggs and Erhard Hohenester, Imperial College, London)* and rendered in PyMOL. The left side shows a surface distribution with solvent-accessible LAMA2 mutations (P-L-E-N-G-E polymerization patch in yellow). The right side shows a cartoon of the core structure with buried residues (blue and gray spheres) within a surface shell. C86Y, S146F, W152G, C199S, L243P, and R249C are mostly buried and seem likely to disrupt the domain fold, resulting either in secretion of a globally modified domain or in reduced/absent secretion. Other indicated mutations might selectively affect polymerization. (See color plate) exhibited a pronounced sorting defect. This difference may explain the pronounced but selective gait abnormality of the hindlimbs. Second, in dy2J, the basement membrane was noted to be substantially attenuated (even "absent") with moderately severe amyelination. Laminin-α4 deficiency had no observable effect on the endoneurial basement membrane structure or composition but caused considerable amyelination on its own. In combination with dy2J/dy2J, the sciatic nerve was very severely amyelinated. It is a challenge to explain the role of laminin-411 in light of the findings. One possibility is that laminin-411 acts as a "plasticizer," disrupting the laminin-211 polymer nodes by competing for the α2–LN interactions.

9. PIERSON SYNDROME

Pierson syndrome is a rare genetic disorder first described in 1963 and characterized by a congenital nephrotic syndrome and complex ocular maldevelopment, especially microcoria (reviewed in Matejas et al., 2010). Some patients develop muscular weakness/myasthenia and developmental retardation along with the renal and ocular defects. The disease results from mutations in the *LAMB2* gene coding for the laminin-β2 subunit found in the GBM, neuromuscular junction, and structures of the eye. The renal pathology in time manifests as diffuse mesangial sclerosis. An animal model of the disease is the LAMB2 knockout mouse that develops proteinuria with a normal appearing GBM and a defective neuromuscular junction (Noakes, Gautam, Mudd, Sanes, & Merlie, 1995; Noakes, Miner, et al., 1995). Analysis of the mice revealed that proteinuria precedes early foot-process fusion (Jarad, Cunningham, Shaw, & Miner, 2006). This observation greatly strengthened the concept that the GBM itself is a key filtration barrier separate from that of the slit diaphragm. The GBM in these mice also contains the homologous laminin-β1 subunit and expression of this subunit increases in the absence of the β2 subunit. Forced overexpression of the laminin-β1 subunit in the kidney was found to substantially correct the renal disease, evidence that β1 can compensate for β2 but that this compensation is insufficient in the Pierson syndrome model (Suh, Jarad, VanDeVoorde, & Miner, 2011).

A survey of mutations causing Pierson syndrome revealed that nonsense and truncating mutations are evenly distributed along the laminin gene (Matejas et al., 2011). There is also a cluster of missense and small deletion mutations confined to the genetic region corresponding to the LN domain (Figure 5). These mutations are associated with a higher average age at onset

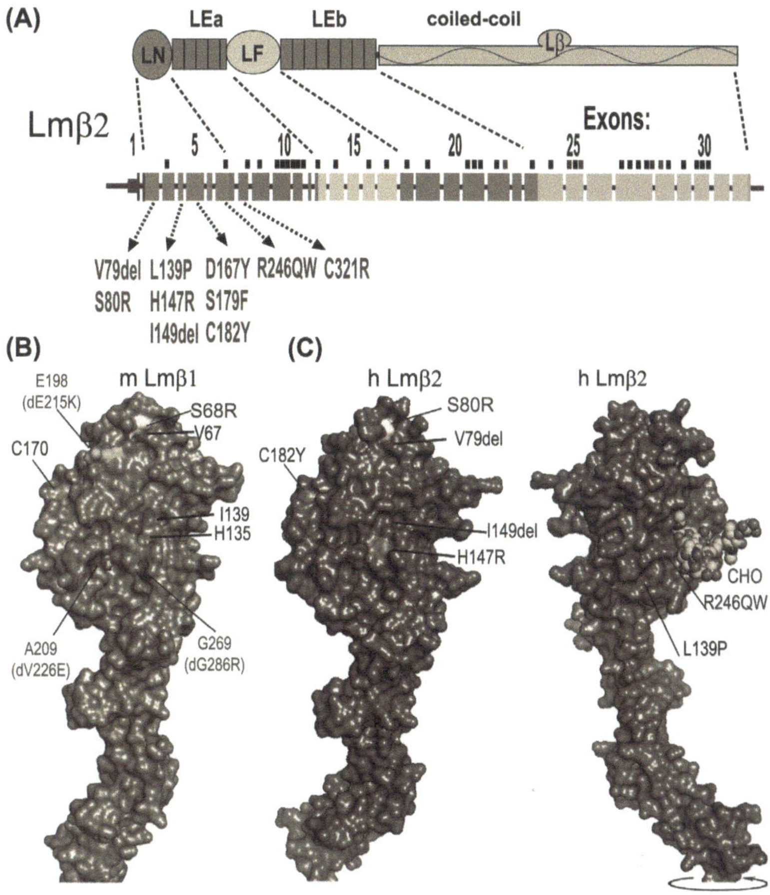

Figure 5 ***Pierson syndrome mutations and laminin-β-short-arm domain structures.*** Panel A: *Map of Pierson syndrome mutations.* Lmβ2 domains and exons are shown. Missense and short deletions, largely confined to LN domain, indicated with dotted arrows. Nonsense and other truncating mutations indicated with square dots over the exons. *(Map based on Matejas et al. (2010).)* Panel B: *Laminin-β1 LN–LEa domains.* The domains were rendered in PyMOL using the coordinates (PDB-4AQS) for the crystal structure of the mouse laminin-β1 tip (Purvis & Hohenester, 2012). Mutations corresponding to Pierson mutations labeled in red (likely ablating polymerization) and black (unknown effect). Surface and deeper (spheres) mutations causing failure of basement membrane in *Drosophila* labeled in green. Panel C: *Lmβ2 LN–LEa domains, homology fit, surface, with Pierson mutations indicated.* The human laminin-β1 sequence was homology fitted to Protein Data Bank structures using the Bioinformatics Toolkit (Max Planck Institute for Developmental Biology, HHpred). A rotated structure containing N-linked carbohydrate (CHO) is shown on the right. The R246Q and R246W mutations that cause a failure of secretion map adjacent to the asparagine near the carbohydrate. S80R, on the other hand, is located on the converse side on the same face as several Pierson and *Drosophila* mutations. (See color plate)

of proteinuria and the absence of neurological findings, suggesting that at least some of the mutations represent hypomorphic alleles in which the laminin is defective in its ability to polymerize. However, at least some of the mutations prevent or reduce secretion. Mutation R246W leads to significant reduction of laminin expression (Zenker et al., 2004). The mouse mutation corresponding to R246Q causes the nephrotic syndrome because of a failure of secretion (Chen, Kikkawa, & Miner, 2011). The R246 residue is located on the carbohydrate face of the LN domain opposite the identified polymerization face. On the other hand, the S80R mutation maps to what is considered the polymerization face. It was generated in an N-terminal segment of Lmβ1 (S68R) and was secreted normally by transfected cells. The modified protein was unable to inhibit laminin polymerization, suggesting that nephrotic syndrome arising from S80R is due to defective polymerization (Purvis & Hohenester, 2012). V79del, H147R, I149del, and C182Y also are located on the same surface as S80R. Cysteine mutations likely alter the fold, but need not necessarily cause a failure of secretion. H147 might represent a residue involved in polymerization. It will be important, going forward, to determine which residues are directly involved in polymerization, which residues are indirectly involved by altering the fold without depressing secretion, and which residues result in failures of secretion.

10. CONCLUSIONS

Our understanding of the mechanisms of basement membrane assembly in normal and disease states has been improved by a convergence of biochemical, structural, and genetic findings. In developing a model with predictive value to explain the binding interactions that enable assembly of basement membranes, the resulting molecular architectures, and the functional outcomes, the model should also help explain the etiology of diseases of basement membranes and provide insights into potential therapies.

If one considers the two disease deficiencies focused on in this chapter, those of laminin-α2 and laminin-β2, we see the existence of two different degrees of severity, one arising from total or near-total absence of laminin and the other from failure of a domain mediating polymerization. Approaches to treat MDC1A congenital muscular dystrophy have been under study for some time. The most successful intervention has been seen with transgenic expression of the homologous laminin-α1 subunit in muscle and nerve of the dy3K mouse (Gawlik, Miyagoe-Suzuki, Ekblom, Takeda, & Durbeej, 2004; Gawlik, Li, Petersen, & Durbeej, 2006). However, there is no obvious system to deliver the gene given its very large size. EHS laminin-111 protein injections seem to cause some improvement, but there are concerns over whether the protein itself repairs the fundamental structural defect (Rooney, Knapp, Hodges, Wuebbles, & Burkin, 2012). Significant amelioration has also been seen in the dyW hypomorphic mouse following transgenic expression of miniagrin (Moll et al., 2001). Miniagrin appears to work by providing collateral linkage to laminin-411, the main compensating laminin in the disease. Laminin-411 binds poorly if at all to α-dystroglycan, and weakly to the α6β1 and α7β1 integrins (reviewed in Yurchenco and Patton, 2009). However, laminin-411 is also a non-polymerizing laminin, suggesting that a repair of polymerization is required for optimization of laminin functions. The cDNA for miniagrin is small enough to fit in the capsid of adeno-associated virus (AAV), and it was shown that AAV can be successfully used to deliver the miniagrin gene to the laminin-α2-deficient dyW dystrophic mouse shortly after birth and affect an improvement in muscle histology and function (Qiao et al., 2005).

The engineering of a synthetic laminin short arm (αLNNd) that binds to the γ1-short arm and repairs polymerization defects due to αLN domain deletions/mutations offers another possible therapeutic approach. Experiments conducted on SCs indicate that it enables a dramatic improvement on basement membrane assembly mediated by non-polymerizing laminins and of myelination and SC proliferation in dorsal root ganglia organ cultures (McKee et al., 2009, 2012). The DNA coding for the protein, like that for miniagrin, is small enough to be delivered by an AAV vector that is under study for the treatment of Duchenne muscular dystrophy (McKee et al., 2009; Wang, Tapscott, Chamberlain, & Storb, 2011). αLNNd has potential to be an outstanding protein to repair mutations that inactivate laminin polymerization without abolishing secretion and might, in combination with miniagrin, prove beneficial for dystrophy arising from severe hypomorphic and null mutations. Similarly, it may be possible to generate a Lmβ1 synthetic arm analogous to αLNNd that can be used to repair the defective LN domain so long as the protein is expressed. If this can be done, the AAV approach described for laminin-α2 deficiency would be applicable for Pierson syndrome.

ACKNOWLEDGMENTS

The author's work is and has been supported by grant R01DK36425 from the National Institutes of Health. The author thanks Dr Erhard Hohenester (Imperial College, London) for valuable and elucidating discussions on the LN–LEa domain structures and the residues that may be involved in laminin polymerization.

REFERENCES

Aumailley, M., Bruckner-Tuderman, L., Carter, W. G., Deutzmann, R., Edgar, D., Ekblom, P., et al. (2005). A simplified laminin nomenclature. *Matrix Biology, 24*, 326–332.

Bader, B. L., Smyth, N., Nedbal, S., Miosge, N., Baranowsky, A., Mokkapati, S., et al. (2005). Compound genetic ablation of nidogen 1 and 2 causes basement membrane defects and perinatal lethality in mice. *Molecular and Cellular Biology, 25*, 6846–6856.

Battaglia, C., Mayer, U., Aumailley, M., & Timpl, R. (1992). Basement-membrane heparan sulfate proteoglycan binds to laminin by its heparan sulfate chains and to nidogen by sites in the protein core. *European Journal of Biochemistry, 208*, 359–366.

Beck, K., Dixon, T. W., Engel, J., & Parry, D. A. (1993). Ionic interactions in the coiled-coil domain of laminin determine the specificity of chain assembly. *Journal of Molecular Biology, 231*, 311–323.

Behrens, D. T., Villone, D., Koch, M., Brunner, G., Sorokin, L., Robenek, H., et al. (2012). The epidermal basement membrane is a composite of separate laminin- or collagen iv-containing networks connected by aggregated perlecan, but not by nidogens. *Journal of Biological Chemistry, 287*, 18700–18709.

Belanto, J. J., Mader, T. L., Eckhoff, M. D., Strandjord, D. M., Banks, G. B., Gardner, M. K., et al. (2014). Microtubule binding distinguishes dystrophin from utrophin. *Proceedings of the National Academy of Sciences of the United States of America, 111*, 5723–5728.

Bhave, G., Cummings, C. F., Vanacore, R. M., Kumagai-Cresse, C., Ero-Tolliver, I. A., Rafi, M., et al. (2012). Peroxidasin forms sulfilimine chemical bonds using hypohalous acids in tissue genesis. *Nature Chemical Biology, 8*, 784–790.

Bix, G., Fu, J., Gonzalez, E. M., Macro, L., Barker, A., Campbell, S., et al. (2004). Endorepellin causes endothelial cell disassembly of actin cytoskeleton and focal adhesions through alpha2beta1 integrin. *Journal of Cell Biology, 166*, 97–109.

Carafoli, F., Hussain, S. A., & Hohenester, E. (2012). Crystal structures of the network-forming short-arm tips of the laminin β1 and γ1 chains. *PLoS One, 7*, e42473.

Carlin, B., Jaffe, R., Bender, B., & Chung, A. E. (1981). Entactin, a novel basal lamina-associated sulfated glycoprotein. *Journal of Biological Chemistry, 256*, 5209–5214.

Chan, S. H., Foley, A. R., Phadke, R., Mathew, A. A., Pitt, M., Sewry, C., et al. (2014). Limb girdle muscular dystrophy due to *LAMA2* mutations: diagnostic difficulties due to associated peripheral neuropathy. *Neuromuscular Disorders, 24*, 677–683.

Chen, Y. M., Kikkawa, Y., & Miner, J. H. (2011). A missense *LAMB2* mutation causes congenital nephrotic syndrome by impairing laminin secretion. *Journal of the American Society of Nephrology, 22*, 849–858.

Cheng, Y. S., Champliaud, M. F., Burgeson, R. E., Marinkovich, M. P., & Yurchenco, P. D. (1997). Self-assembly of laminin isoforms. *Journal of Biological Chemistry, 272*, 31525–31532.

Colognato, H., Winkelmann, D. A., & Yurchenco, P. D. (1999). Laminin polymerization induces a receptor-cytoskeleton network. *Journal of Cell Biology, 145*, 619–631.

Colognato, H., & Yurchenco, P. D. (1999). The laminin alpha2 expressed by dystrophic dy(2j) mice is defective in its ability to form polymers. *Current Biology, 9*, 1327–1330.

Denzer, A. J., Schulthess, T., Fauser, C., Schumacher, B., Kammerer, R. A., Engel, J., et al. (1998). Electron microscopic structure of agrin and mapping of its binding site in laminin-1. *EMBO Journal, 17*, 335–343.

Deutzmann, R., Aumailley, M., Wiedemann, H., Pysny, W., Timpl, R., & Edgar, D. (1990). Cell adhesion, spreading and neurite stimulation by laminin fragment E8 depends on maintenance of secondary and tertiary structure in its rod and globular domain. *European Journal of Biochemistry, 191*, 513–522.

Di Blasi, C., Piga, D., Brioschi, P., Moroni, I., Pini, A., Ruggieri, A., et al. (2005). Lama2 gene analysis in congenital muscular dystrophy: new mutations, prenatal diagnosis, and founder effect. *Archives of Neurology, 62*, 1582–1586.

Engel, J., Odermatt, E., Engel, A., Madri, J. A., Furthmayr, H., Rohde, H., et al. (1981). Shapes, domain organizations and flexibility of laminin and fibronectin, two multifunctional proteins of the extracellular matrix. *Journal of Molecular Biology, 150*, 97–120.

Fahey, B., & Degnan, B. M. (2012). Origin and evolution of laminin gene family diversity. *Molecular Biology and Evolution, 29*, 1823–1836.

Fidler, A. L., Vanacore, R. M., Chetyrkin, S. V., Pedchenko, V. K., Bhave, G., Yin, V. P., et al. (2014). A unique covalent bond in basement membrane is a primordial innovation for tissue evolution. *Proceedings of the National Academy of Sciences of the United States of America, 111*, 331–336.

Fox, J. W., Mayer, U., Nischt, R., Aumailley, M., Reinhardt, D., Wiedemann, H., et al. (1991). Recombinant nidogen consists of three globular domains and mediates binding of laminin to collagen type IV. *EMBO Journal, 10*, 3137–3146.

Garbe, J. H., Gohring, W., Mann, K., Timpl, R., & Sasaki, T. (2002). Complete sequence, recombinant analysis and binding to laminins and sulphated ligands of the n-terminal domains of laminin [alpha]3b and [alpha]5 chains. *Biochemical Journal, 362*, 213–221.

Gavassini, B. F., Carboni, N., Nielsen, J. E., Danielsen, E. R., Thomsen, C., Svenstrup, K., et al. (2011). Clinical and molecular characterization of limb-girdle muscular dystrophy due to *LAMA2* mutations. *Muscle & Nerve, 44*, 703–709.

Gawlik, K., Miyagoe-Suzuki, Y., Ekblom, P., Takeda, S., & Durbeej, M. (2004). Laminin {alpha}1 chain reduces muscular dystrophy in laminin {alpha}2 chain deficient mice. *Human Molecular Genetics, 13*, 1775–1784.

Gawlik, K. I., Li, J. Y., Petersen, A., & Durbeej, M. (2006). Laminin alpha1 chain improves laminin alpha2 chain deficient peripheral neuropathy. *Human Molecular Genetics, 15*, 2690–2700.

Gee, S. H., Blacher, R. W., Douville, P. J., Provost, P. R., Yurchenco, P. D., & Carbonetto, S. (1993). Laminin-binding protein 120 from brain is closely related to the dystrophin-associated glycoprotein, dystroglycan, and binds with high affinity to the major heparin binding domain of laminin. *Journal of Biological Chemistry, 268*, 14972–14980.

Geranmayeh, F., Clement, E., Feng, L. H., Sewry, C., Pagan, J., Mein, R., et al. (2010). Genotype-phenotype correlation in a large population of muscular dystrophy patients with *LAMA2* mutations. *Neuromuscular Disorders, 20*, 241–250.

Gesemann, M., Brancaccio, A., Schumacher, B., & Ruegg, M. A. (1998). Agrin is a high-affinity binding protein of dystroglycan in non-muscle tissue. *Journal of Biological Chemistry, 273*, 600–605.

Gunwar, S., Ballester, F., Noelken, M. E., Sado, Y., Ninomiya, Y., & Hudson, B. G. (1998). Glomerular basement membrane. Identification of a novel disulfide-cross- linked network of alpha3, alpha4, and alpha5 chains of type IV collagen and its implications for the pathogenesis of alport syndrome. *Journal of Biological Chemistry, 273*, 8767–8775.

Han, R., Kanagawa, M., Yoshida-Moriguchi, T., Rader, E. P., Ng, R. A., Michele, D. E., et al. (2009). Basal lamina strengthens cell membrane integrity via the laminin g

domain-binding motif of alpha-dystroglycan. *Proceedings of the National Academy of Sciences of the United States of America, 106*, 12573–12579.
Harrison, D., Hussain, S. A., Combs, A. C., Ervasti, J. M., Yurchenco, P. D., & Hohenester, E. (2007). Crystal structure and cell surface anchorage sites of laminin alpha 1LG4-5. *Journal of Biological Chemistry, 282*, 11573–11581.
Hollfelder, D., Frasch, M., & Reim, I. (2014). Distinct functions of the laminin β LN domain and collagen IV during cardiac extracellular matrix formation and stabilization of alary muscle attachments revealed by ems mutagenesis in *Drosophila*. *BMC Developemental Biology, 14*, 26.
Hopf, M., Gohring, W., Kohfeldt, E., Yamada, Y., & Timpl, R. (1999). Recombinant domain IV of perlecan binds to nidogens, laminin-nidogen complex, fibronectin, fibulin-2 and heparin. *European Journal of Biochemistry, 259*, 917–925.
Hopf, M., Gohring, W., Mann, K., & Timpl, R. (2001). Mapping of binding sites for nidogens, fibulin-2, fibronectin and heparin to different ig modules of perlecan. *Journal of Molecular Biology, 311*, 529–541.
Hopf, M., Gohring, W., Ries, A., Timpl, R., & Hohenester, E. (2001). Crystal structure and mutational analysis of a perlecan-binding fragment of nidogen-1. *Nature Structural Biology, 8*, 634–640.
Hu, P., & Luo, B. H. (2013). Integrin bi-directional signaling across the plasma membrane. *Journal of Cellular Physiology, 228*, 306–312.
Hussain, S. A., Carafoli, F., & Hohenester, E. (2011). Determinants of laminin polymerization revealed by the structure of the alpha5 chain amino-terminal region. *EMBO Reports, 12*, 276–282.
Ido, H., Ito, S., Taniguchi, Y., Hayashi, M., Sato-Nishiuchi, R., Sanzen, N., et al. (2008). Laminin isoforms containing the gamma3 chain are unable to bind to integrins due to the absence of the glutamic acid residue conserved in the c-terminal regions of the gamma1 and gamma2 chains. *Journal of Biological Chemistry, 283*, 28149–28157.
Ido, H., Nakamura, A., Kobayashi, R., Ito, S., Li, S., Futaki, S., et al. (2007). The requirement of the glutamic acid residue at the third position from the carboxyl termini of the laminin gamma chains in integrin binding by laminins. *Journal of Biological Chemistry, 282*, 11144–11154.
Inamori, K., Yoshida-Moriguchi, T., Hara, Y., Anderson, M. E., Yu, L., & Campbell, K. P. (2012). Dystroglycan function requires xylosyl- and glucuronyltransferase activities of large. *Science, 335*, 93–96.
Jarad, G., Cunningham, J., Shaw, A. S., & Miner, J. H. (2006). Proteinuria precedes podocyte abnormalities in *Lamb2−/−* mice, implicating the glomerular basement membrane as an albumin barrier. *Journal of Clinical Investigation, 116*, 2272–2279.
Jarad, G., Pippin, J. W., Shankland, S. J., Kreidberg, J. A., & Miner, J. H. (2011). Dystroglycan does not contribute significantly to kidney development or function, in health or after injury. *American Journal of Physiology Renal Physiology, 300*, F811–F820.
Jimenez-Mallebrera, C., Brown, S. C., Sewry, C. A., & Muntoni, F. (2005). Congenital muscular dystrophy: molecular and cellular aspects. *Cellular and Molecular Life Sciences, 62*, 809–823.
Kalb, E., & Engel, J. (1991). Binding and calcium-induced aggregation of laminin onto lipid bilayers. *Journal of Biological Chemistry, 266*, 19047–19052.
Kammerer, R. A., Schulthess, T., Landwehr, R., Schumacher, B., Lustig, A., Yurchenco, P. D., et al. (1999). Interaction of agrin with laminin requires a coiled-coil conformation of the agrin-binding site within the laminin gamma1 chain. *EMBO Journal, 18*, 6762–6770.
Kikkawa, Y., Moulson, C. L., Virtanen, I., & Miner, J. H. (2002). Identification of the binding site for the lutheran blood group glycoprotein on laminin alpha 5 through expression of chimeric laminin chains in vivo. *Journal of Biological Chemistry, 277*, 44864–44869.

Kim, S., & Wadsworth, W. G. (2000). Positioning of longitudinal nerves in *C. elegans* by nidogen. *Science, 288*, 150–154.
Li, S., Harrison, D., Carbonetto, S., Fassler, R., Smyth, N., Edgar, D., et al. (2002). Matrix assembly, regulation, and survival functions of laminin and its receptors in embryonic stem cell differentiation. *Journal of Cell Biology, 157*, 1279–1290.
Li, S., Liquari, P., McKee, K. K., Harrison, D., Patel, R., Lee, S., et al. (2005). Laminin-sulfatide binding initiates basement membrane assembly and enables receptor signaling in schwann cells and fibroblasts. *Journal of Cell Biology, 169*, 179–189.
Macdonald, P. R., Lustig, A., Steinmetz, M. O., & Kammerer, R. A. (2010). Laminin chain assembly is regulated by specific coiled-coil interactions. *Journal of Structural Biology, 170*, 398–405.
Marques, J., Duarte, S. T., Costa, S., Jacinto, S., Oliveira, J., Oliveira, M. E., et al. (2014). Atypical phenotype in two patients with *LAMA2* mutations. *Neuromuscular Disorders, 24*, 419–424.
Martin, J. R., & Webster, H. D. (1973). Mitotic schwann cells in developing nerve: their changes in shape, fine structure, and axon relationships. *Developmental Biology, 32*, 417–431.
Mascarenhas, J. B., Ruegg, M. A., Winzen, U., Halfter, W., Engel, J., & Stetefeld, J. (2003). Mapping of the laminin-binding site of the n-terminal agrin domain (nta). *EMBO Journal, 22*, 529–536.
Matejas, V., Hinkes, B., Alkandari, F., Al-Gazali, L., Annexstad, E., Aytac, M. B., et al. (2010). Mutations in the human laminin β2 (*LAMB2*) gene and the associated phenotypic spectrum. *Human Mutation, 31*, 992–1002.
Matejas, V., Muscheites, J., Wigger, M., Kreutzer, H. J., Nizze, H., & Zenker, M. (2011). Paternal isodisomy of chromosome 3 unmasked by autosomal recessive microcoria-congenital nephrosis syndrome (Pierson syndrome) in a child with no other phenotypic abnormalities. *American Journal of Medical Genetics Part A, 155A*, 2601–2604.
Matsumura, K., Ohlendieck, K., Ionasescu, V. V., Tome, F. M., Nonaka, I., Burghes, A. H., et al. (1993). The role of the dystrophin-glycoprotein complex in the molecular pathogenesis of muscular dystrophies. *Neuromuscular Disorders, 3*, 533–535.
McKee, K. K., Capizzi, S., & Yurchenco, P. D. (2009). Scaffold-forming and adhesive contributions of synthetic laminin-binding proteins to basement membrane assembly. *Journal of Biological Chemistry, 284*, 8984–8994.
McKee, K. K., Harrison, D., Capizzi, S., & Yurchenco, P. D. (2007). Role of laminin terminal globular domains in basement membrane assembly. *Journal of Biological Chemistry, 282*, 21437–21447.
McKee, K. K., Yang, D. H., Patel, R., Chen, Z. L., Strickland, S., Takagi, J., et al. (2012). Schwann cell myelination requires integration of laminin activities. *Journal of Cell Science, 125*, 4609–4619.
Meinen, S., Barzaghi, P., Lin, S., Lochmuller, H., & Ruegg, M. A. (2007). Linker molecules between laminins and dystroglycan ameliorate laminin-alpha2-deficient muscular dystrophy at all disease stages. *Journal of Cell Biology, 176*, 979–993.
Menezes, M. J., McClenahan, F. K., Leiton, C. V., Aranmolate, A., Shan, X., & Colognato, H. (2014). The extracellular matrix protein laminin alpha2 regulates the maturation and function of the blood–brain barrier. *Journal of Neuroscience, 34*, 15260–15280.
Miner, J. H., Lewis, R. M., & Sanes, J. R. (1995). Molecular cloning of a novel laminin chain, alpha 5, and widespread expression in adult mouse tissues. *Journal of Biological Chemistry, 270*, 28523–28526.
Miner, J. H., & Yurchenco, P. D. (2004). Laminin functions in tissue morphogenesis. *Annual Review of Cell and Developmental Biology, 20*, 255–284.

Miyagoe, Y., Hanaoka, K., Nonaka, I., Hayasaka, M., Nabeshima, Y., Arahata, K., et al. (1997). Laminin alpha2 chain-null mutant mice by targeted disruption of the *Lama2* gene: a new model of merosin (laminin 2)-deficient congenital muscular dystrophy. *FEBS Letters, 415*, 33–39.
Moll, J., Barzaghi, P., Lin, S., Bezakova, G., Lochmuller, H., Engvall, E., et al. (2001). An agrin minigene rescues dystrophic symptoms in a mouse model for congenital muscular dystrophy. *Nature, 413*, 302–307.
Moore, C. J., & Winder, S. J. (2010). Dystroglycan versatility in cell adhesion: a tale of multiple motifs. *Cell Communication & Signaling, 8*, 3.
Moore, S. A., Saito, F., Chen, J., Michele, D. E., Henry, M. D., Messing, A., et al. (2002). Deletion of brain dystroglycan recapitulates aspects of congenital muscular dystrophy. *Nature, 418*, 422–425.
Nichols, S. A., Roberts, B. W., Richter, D. J., Fairclough, S. R., & King, N. (2012). Origin of metazoan cadherin diversity and the antiquity of the classical cadherin/beta-catenin complex. *Proceedings of the National Academy of Sciences of the United States of America, 109*, 13046–13051.
Nishiuchi, R., Takagi, J., Hayashi, M., Ido, H., Yagi, Y., Sanzen, N., et al. (2006). Ligand-binding specificities of laminin-binding integrins: a comprehensive survey of laminin-integrin interactions using recombinant alpha3beta1, alpha6beta1, alpha7beta1 and alpha6beta4 integrins. *Matrix Biology, 25*, 189–197.
Noakes, P. G., Gautam, M., Mudd, J., Sanes, J. R., & Merlie, J. P. (1995). Aberrant differentiation of neuromuscular junctions in mice lacking s-laminin/laminin beta 2. *Nature, 374*, 258–262.
Noakes, P. G., Miner, J. H., Gautam, M., Cunningham, J. M., Sanes, J. R., & Merlie, J. P. (1995). The renal glomerulus of mice lacking s-laminin/laminin beta 2: nephrosis despite molecular compensation by laminin beta 1. *Nature Genetics, 10*, 400–406.
Occhi, S., Zambroni, D., Del Carro, U., Amadio, S., Sirkowski, E. E., Scherer, S. S., et al. (2005). Both laminin and schwann cell dystroglycan are necessary for proper clustering of sodium channels at nodes of Ranvier. *Journal of Neuroscience, 25*, 9418–9427.
Odenthal, U., Haehn, S., Tunggal, P., Merkl, B., Schomburg, D., Frie, C., et al. (2004). Molecular analysis of laminin n-terminal domains mediating self-interactions. *Journal of Biological Chemistry, 279*, 44505–44512.
Oliveira, J., Santos, R., Soares-Silva, I., Jorge, P., Vieira, E., Oliveira, M. E., et al. (2008). *LAMA2* gene analysis in a cohort of 26 congenital muscular dystrophy patients. *Clinical Genetics, 74*, 502–512.
Patton, B. L., Connoll, A. M., Martin, P. T., Cunningham, J. M., Mehta, S., Pestronk, A., et al. (1999). Distribution of ten laminin chains in dystrophic and regenerating muscles. *Neuromuscular Disorders, 9*, 423–433.
Poschl, E., Fox, J. W., Block, D., Mayer, U., & Timpl, R. (1994). Two non-contiguous regions contribute to nidogen binding to a single EGF-like motif of the laminin gamma 1 chain. *EMBO Journal, 13*, 3741–3747.
Poschl, E., Mayer, U., Stetefeld, J., Baumgartner, R., Holak, T. A., Huber, R., et al. (1996). Site-directed mutagenesis and structural interpretation of the nidogen binding site of the laminin gamma1 chain. *EMBO Journal, 15*, 5154–5159.
Purvis, A., & Hohenester, E. (2012). Laminin network formation studied by reconstitution of ternary nodes in solution. *Journal of Biological Chemistry, 287*, 44270–44277.
Qiao, C., Li, J., Zhu, T., Draviam, R., Watkins, S., Ye, X., et al. (2005). Amelioration of laminin-{alpha}2-deficient congenital muscular dystrophy by somatic gene transfer of miniagrin. *Proceedings of the National Academy of Sciences of the United States of America, 102*, 11999–12004.
Rajakulendran, S., Parton, M., Holton, J. L., & Hanna, M. G. (2011). Clinical and pathological heterogeneity in late-onset partial merosin deficiency. *Muscle & Nerve, 44*, 590–593.

Rezniczek, G. A., Konieczny, P., Nikolic, B., Reipert, S., Schneller, D., Abrahamsberg, C., et al. (2007). Plectin 1f scaffolding at the sarcolemma of dystrophic (mdx) muscle fibers through multiple interactions with beta-dystroglycan. *Journal of Cell Biology, 176*, 965–977.
Rooney, J. E., Knapp, J. R., Hodges, B. L., Wuebbles, R. D., & Burkin, D. J. (2012). Laminin-111 protein therapy reduces muscle pathology and improves viability of a mouse model of merosin-deficient congenital muscular dystrophy. *American Journal of Pathology, 180*, 1593–1602.
Saito, F., Moore, S. A., Barresi, R., Henry, M. D., Messing, A., Ross-Barta, S. E., et al. (2003). Unique role of dystroglycan in peripheral nerve myelination, nodal structure, and sodium channel stabilization. *Neuron, 38*, 747–758.
Shi, Q., & Boettiger, D. (2003). A novel mode for integrin-mediated signaling: tethering is required for phosphorylation of FAK Y397. *Molecular Biology of the Cell, 14*, 4306–4315.
Smirnov, S. P., McDearmon, E. L., Li, S., Ervasti, J. M., Tryggvason, K., & Yurchenco, P. D. (2002). Contributions of the LG modules and furin processing to laminin-2 functions. *Journal of Biological Chemistry, 277*, 18928–18937.
Stetefeld, J., Mayer, U., Timpl, R., & Huber, R. (1996). Crystal structure of three consecutive laminin-type epidermal growth factor-like (LE) modules of laminin gamma1 chain harboring the nidogen binding site. *Journal of Molecular Biology, 257*, 644–657.
Suh, J. H., Jarad, G., VanDeVoorde, R. G., & Miner, J. H. (2011). Forced expression of laminin beta1 in podocytes prevents nephrotic syndrome in mice lacking laminin beta2, a model for Pierson syndrome. *Proceedings of the National Academy of Sciences of the United States of America, 108*, 15348–15353.
Suleiman, H., Zhang, L., Roth, R., Heuser, J. E., Miner, J. H., Shaw, A. S., et al. (2013). Nanoscale protein architecture of the kidney glomerular basement membrane. *eLife, 2*, e01149.
Sunada, Y., Bernier, S. M., Kozak, C. A., Yamada, Y., & Campbell, K. P. (1994). Deficiency of merosin in dystrophic dy mice and genetic linkage of laminin M chain gene to dy locus. *Journal of Biological Chemistry, 269*, 13729–13732.
Sung, U., O'Rear, J. J., & Yurchenco, P. D. (1993). Cell and heparin binding in the distal long arm of laminin: identification of active and cryptic sites with recombinant and hybrid glycoprotein. *Journal of Cell Biology, 123*, 1255–1268.
Takagi, J., Yang, Y., Liu, J. H., Wang, J. H., & Springer, T. A. (2003). Complex between nidogen and laminin fragments reveals a paradigmatic beta-propeller interface. *Nature, 424*, 969–974.
Talts, J. F., Andac, Z., Gorhing, W., Brancaccio, A., & Timpl, R. (1999). Binding of the G domains of laminin alpha1 and alpha2 chains and perlecan to heparin, sulfatides, alpha-dystroglycan and several extracellular matrix proteins. *EMBO Journal, 18*, 863–870.
Talts, J. F., Sasaki, T., Miosge, N., Gohring, W., Mann, K., Mayne, R., et al. (2000). Structural and functional analysis of the recombinant g domain of the laminin {alpha}4 chain and its proteolytic processing in tissues. *Journal of Biological Chemistry, 275*, 35192–35199.
Taraboletti, G., Rao, C. N., Krutzsch, H. C., Liotta, L. A., & Roberts, D. D. (1990). Sulfatide-binding domain of the laminin A chain. *Journal of Biological Chemistry, 265*, 12253–12258.
Timpl, R., Dziadek, M., Fujiwara, S., Nowack, H., & Wick, G. (1983). Nidogen: a new, self-aggregating basement membrane protein. *European Journal of Biochemistry, 137*, 455–465.
Timpl, R., Rohde, H., Robey, P. G., Rennard, S. I., Foidart, J. M., & Martin, G. R. (1979). Laminin—a glycoprotein from basement membranes. *Journal of Biological Chemistry, 254*, 9933–9937.
Timpl, R., Wiedemann, H., van Delden, V., Furthmayr, H., & Kuhn, K. (1981). A network model for the organization of type iv collagen molecules in basement membranes. *European Journal of Biochemistry, 120*, 203–211.

Trick, D., Decker, J., Groene, H. J., Schulze, M., & Wiegandt, H. (1999). Regional expression of sulfatides in rat kidney: immunohistochemical staining by use of monospecific polyclonal antibodies. *Histochemistry and Cell Biology, 111*, 143–151.
Tsiper, M. V., & Yurchenco, P. D. (2002). Laminin assembles into separate basement membrane and fibrillar matrices in schwann cells. *Journal of Cell Science, 115*, 1005–1015.
Vanacore, R., Ham, A. J., Voehler, M., Sanders, C. R., Conrads, T. P., Veenstra, T. D., et al. (2009). A sulfilimine bond identified in collagen IV. *Science, 325*, 1230–1234.
Walko, G., Castanon, M. J., & Wiche, G. (2015). Molecular architecture and function of the hemidesmosome. *Cell & Tissue Research, 360*, 363–378.
Wang, Z., Tapscott, S. J., Chamberlain, J. S., & Storb, R. (2011). Immunity and AAV-mediated gene therapy for muscular dystrophies in large animal models and human trials. *Front Microbiol, 2*, 201.
Webster, H. D., Martin, R., & O'Connell, M. F. (1973). The relationships between interphase schwann cells and axons before myelination: a quantitative electron microscopic study. *Developmental Biology, 32*, 401–416.
Willem, M., Miosge, N., Halfter, W., Smyth, N., Jannetti, I., Burghart, E., et al. (2002). Specific ablation of the nidogen-binding site in the laminin gamma1 chain interferes with kidney and lung development. *Development, 129*, 2711–2722.
Williamson, R. A., Henry, M. D., Daniels, K. J., Hrstka, R. F., Lee, J. C., Sunada, Y., et al. (1997). Dystroglycan is essential for early embryonic development: disruption of reichert's membrane in dag1-null mice. *Human Molecular Genetics, 6*, 831–841.
Yang, D., Bierman, J., Tarumi, Y. S., Zhong, Y. P., Rangwala, R., Proctor, T. M., et al. (2005). Coordinate control of axon defasciculation and myelination by laminin-2 and -8. *Journal of Cell Biology, 168*, 655–666.
Yurchenco, P. D. (2011). Basement membranes: cell scaffoldings and signaling platforms. *Cold Spring Harbor Perspectives in Biology, 3*, a004911.
Yurchenco, P. D., & Cheng, Y. S. (1993). Self-assembly and calcium-binding sites in laminin. A three-arm interaction model. *Journal of Biological Chemistry, 268*, 17286–17299.
Yurchenco, P. D., Cheng, Y. S., Campbell, K., & Li, S. (2004). Loss of basement membrane, receptor and cytoskeletal lattices in a laminin-deficient muscular dystrophy. *Journal of Cell Science, 117*, 735–742.
Yurchenco, P. D., Cheng, Y. S., & Colognato, H. (1992). Laminin forms an independent network in basement membranes. *Journal of Cell Biology, 117*, 1119–1133.
Yurchenco, P. D., Cheng, Y. S., & Schittny, J. C. (1990). Heparin modulation of laminin polymerization. *Journal of Biological Chemistry, 265*, 3981–3991.
Yurchenco, P. D., & Furthmayr, H. (1984). Self-assembly of basement membrane collagen. *Biochemistry, 23*, 1839–1850.
Yurchenco, P. D., & Furthmayr, H. (1986). Type IV collagen "7s" tetramer formation: aspects of kinetics and thermodynamics. *Annals of the New York Academy of Sciences, 460*, 530–533.
Yurchenco, P. D., & Patton, B. L. (2009). Developmental and pathogenic mechanisms of basement membrane assembly. *Current Pharmaceutical Design, 15*, 1277–1294.
Yurchenco, P. D., Tsilibary, E. C., Charonis, A. S., & Furthmayr, H. (1985). Laminin polymerization in vitro. Evidence for a two-step assembly with domain specificity. *Journal of Biological Chemistry, 260*, 7636–7644.
Zenker, M., Aigner, T., Wendler, O., Tralau, T., Muntefering, H., Fenski, R., et al. (2004). Human laminin beta2 deficiency causes congenital nephrosis with mesangial sclerosis and distinct eye abnormalities. *Human Molecular Genetics, 13*, 2625–2632.

CHAPTER TWO

Laminin-α2 Chain-Deficient Congenital Muscular Dystrophy: Pathophysiology and Development of Treatment

Madeleine Durbeej
Department of Experimental Medical Science, Lund University, Lund, Sweden
E-mail: madeleine.durbeej-hjalt@med.lu.se

Contents

Abstract

Laminin-211 is a major constituent of the skeletal muscle basement membrane. It stabilizes skeletal muscle and influences signal transduction events from the myomatrix to the muscle cell. Mutations in the gene encoding the α2 chain of laminin-211 lead to

Current Topics in Membranes, Volume 76
ISSN 1063-5823
http://dx.doi.org/10.1016/bs.ctm.2015.05.002

congenital muscular dystrophy type 1A (MDC1A), a life-threatening disease characterized by severe hypotonia, progressive muscle weakness, and joint contractures. Common complications include severely impaired motor ability, respiratory failure, and feeding difficulties. Several adequate animal models for laminin-α2 chain deficiency exist and analyses of different MDC1A mouse models have led to a significant improvement in our understanding of MDC1A pathogenesis. Importantly, the animal models have been indispensable tools for the preclinical development of new therapeutic approaches for laminin-α2 chain deficiency, highlighting a number of important disease driving mechanisms that can be targeted by pharmacological approaches. In this chapter, I will describe laminin-211 and discuss the cellular and molecular pathophysiology of MDC1A as well as progression toward development of treatment.

1. INTRODUCTION

Each skeletal muscle cell is surrounded by a basement membrane that is directly linked to the sarcolemma (muscle cell membrane). In addition, the basement membrane is also attached to an underlying interstitial extracellular matrix (rich in collagens) and collectively the basement membrane and the collagenous stroma are referred to as the myomatrix (Rutkowski et al., 2013; Yurchenco, 2011). The skeletal muscle basement membrane plays crucial roles during development and in adults. Its key functions involve maintaining muscle integrity, promoting myogenesis, and synaptogenesis as well as influencing skeletal muscle regeneration (Sanes, 2003). A major component of the extrasynaptic skeletal muscle basement membrane is laminin-211 (Patton, Miner, Chiu, & Sanes, 1997). The importance of laminin-211 in skeletal muscle is accentuated by the fact that absence or reduction of this laminin isoform leads to a very severe form of muscular dystrophy, MDC1A (laminin-α2 chain-deficient congenital muscular dystrophy) (Gawlik & Durbeej, 2011; Helbling-Leclerc et al., 1995). Here, I will mainly discuss laminin-211 and its functions in normal and diseased skeletal muscle.

2. LAMININ-211 AND CELL SURFACE RECEPTORS

Laminin-211 is a heterotrimeric cross-shaped molecule composed of laminin-α2, β1, and γ1 chains. It was originally identified by Eva Engvall and colleagues in 1988 as merosin and shown to be expressed in basement membranes of striated muscle, Schwann cells, and trophoblasts (Ehrig, Leivo, Argraves, Ruoslahti, & Engvall, 1991; Leivo & Engvall, 1988). Today, merosin is denoted as laminin-211 (Aumailley et al., 2005).

Laminin-211 is the major laminin isoform expressed in the neuromuscular system, including the skeletal muscle and Schwann cell basement membranes (Patton et al., 1997). The complete primary structure of laminin-α2 chain was determined in 1994 and the *LAMA2* gene was found to be located on chromosome 6q22-23. The gene is composed of 65 exons that encode a protein with a predicted molecular weight of 390 kDa (Ehrig et al., 1991; Vuolteenaho et al., 1994). After translation, laminin-α2 chain supposedly joins a heterodimer of laminin-β1 and -γ1 chain that is formed in the endoplasmic reticulum and the complete laminin-211 heterotrimer is then transported through the secretory pathway (Yurchenco et al., 1997). In general, little is known about how laminins are delivered to the skeletal muscle cell surface and in particular how this process is regulated. During or after assembly and secretion, the laminin-α2 chain is cleaved at amino acid 2580 into a 300 kDa N-terminal fragment and an 80 kDa C-terminal fragment that are noncovalently associated (Smirnov et al., 2002; Talts, Mann, Yamada, & Timpl, 1998). At the skeletal muscle cell surface, laminin-211 self-assembles into networks and it has been demonstrated that the laminin N-terminal domain (LN) of laminin-α2 chain is crucial for laminin-211 polymerization into supramolecular networks (Colognato & Yurchenco, 1999; Yurchenco, 2011). Assembly of a laminin network at the cell surface is additionally facilitated by anchorage to cell surface receptors (Li, Liquari, et al., 2005; Li, Shelton, & Engvall, 2005; Yurchenco, 2011). In order to form a highly cross-linked skeletal muscle basement membrane, the laminin-211 network is indirectly appended to the collagen IV network and to the heparan sulfate proteoglycan perlecan by nidogens (Hopf, Göhring, Mann, & Timpl, 2001; Yurchenco, 2011). Laminin-211 also binds other matrix proteins including muscle agrin and fibulins (Kammerer et al., 1999; Talts, Andac, Gohring, Brancaccio, & Timpl, 1999).

The C-terminus of the laminin-α2 chain is the other important business end of laminin-211. It is composed of a tandem of five laminin G-like (LG) domains, LG1–5, and they mediate cell adhesion to laminin 211 (Aumailley et al., 2005). The two major cell surface receptors expressed on skeletal muscle cells are dystroglycan and integrin α7β1 (Figure 1). Dystroglycan is a widely expressed protein that is composed of two subunits; the highly glycosylated extracellularly located α-dystroglycan that binds laminin-211 with high affinity and the transmembrane subunit β-dystroglycan that is attached to the intracellular protein dystrophin (Campbell & Kahl., 1989; Ibraghimov-Beskrovnaya et al., 1992). In skeletal muscle, dystroglycan forms the backbone of the multisubunit dystrophin-glycoprotein complex, which links

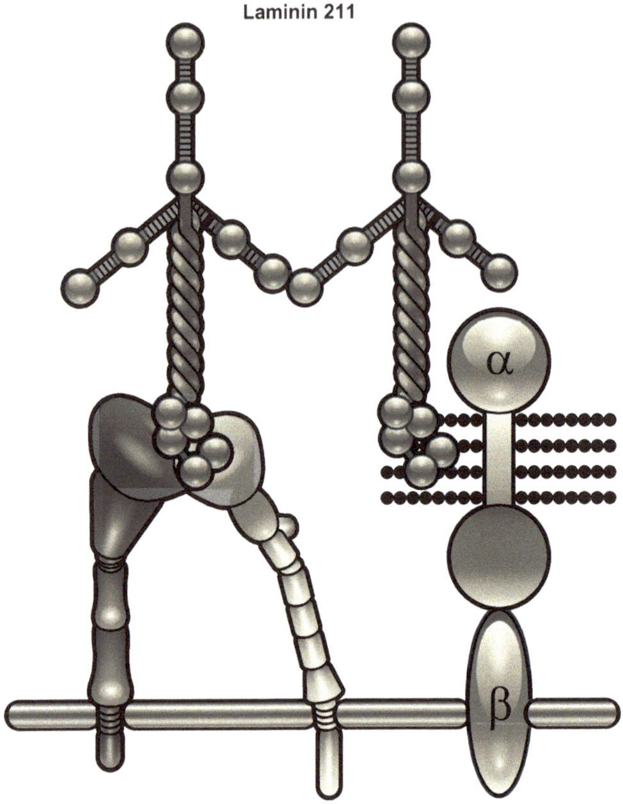

Figure 1 Laminin-211 binding to cell surface receptors dystroglycan and integrin α7β1 in skeletal muscle. Laminin-211 is a heterotrimeric protein composed of α2 (pink), β1 (green), and γ1 (blue) chains. Laminin-211 binds to dystroglycan and integrin α7β1 via the C-terminal laminin G-like domains of laminin-α2 chain. Dystroglycan is composed of two subunits; α- and β-dystroglycan. α-Dystroglycan is highly glycosylated and binds to laminin-211 via its O-linked glycan chain moieties (chains with black circles) (Yoshida-Moriguchi et al., 2010). α-Dystroglycan is also attached to transmembrane β-dystroglycan. Integrins are heterodimeric and consist of one α (purple) and one β (yellow) subunit. The predominant integrin in skeletal muscle is α7β1. Note that the figure is not drawn to scale. (See color plate)

laminin-211 to the intracellular components dystrophin and actin (Ervasti & Campbell, 1993). The dystrophin–glycoprotein complex further contains the sarcoglycan–sarcospan complex, which stabilizes dystroglycan at the muscle plasma membrane (Crosbie et al., 1999). Detailed mapping and structural analyses of the laminin-α2 chain LG1–5 domains have shown that α-dystroglycan binds to sites in LG4–5, and binding requires the calcium ion in LG4 and basic residues in the LG4–5 domains (Tisi, Talts, Timpl, & Hohenester, 2000; Wizemann et al., 2003). In addition, there is a second α-dystroglycan binding site in LG1–3 (Talts et al., 1999). Similarly, heparan sulfate proteoglycans (e.g., syndecan-1 and -2) and sulfated glycolipids bind to basic residues in LG4–5 domains, but the binding sites for α-dystroglycan and heparin/sulfatides are principally nonoverlapping (Katagiri et al., 2014; Wizemann et al., 2003).

The other important adhesion complex that links laminin-211 to the cytoskeleton in skeletal muscle is integrin α7β1 (von der Mark et al., 1991; Mayer, 2003; Song, Wang, Foster, Bielser, & Kaufman, 1992; Vachon et al., 1997). Laminin-211 binding to integrin α7β1 occurs through the LG1–3 domains with involvement of the coiled-coil domain sequences in the laminin-γ1 (in particular, the glutamic acid residue E1607) and β1 chains (Deutzmann et al., 1990; Ido et al., 2007; von der Mark et al., 1991).

Dystroglycan and integrin α7β1 are both important laminin-211 adhesion receptors, and their synergistic interplay is absolutely indispensable for normal muscle function (Allikian, Hack, Mewborn, Mayer, & McNally, 2004; Burkin, Wallace, Nicol, Kaufman, & Kaufman, 2001; Cohn et al., 2002; Guo et al., 2006; Mayer et al., 1997; Rooney et al., 2006; Sztal, Sonntag, Hall, & Currie, 2012). However, they also appear to have different functions in skeletal muscle. Both receptors contribute to force-production of muscles but only dystroglycan maintains sarcolemmal integrity and protects skeletal muscle from damage (Han et al., 2009). There is also an obvious compensatory, but nonredundant overlap in the function between laminin-α2 chain and single molecules in the dystrophin–glycoprotein complex (Gawlik, Holmberg, & Durbeej, 2014) but not between laminin-α2 chain and integrin α7β1 (unpublished data). Nevertheless, when the connection between the extracellular matrix and cytoskeleton is broken due to mutations in the genes encoding members of the dystrophin–glycoprotein complex, integrin α7, or laminin-α2 chain, different forms of muscular diseases, so-called muscular dystrophies, arise. They include Duchenne muscular dystrophy (DMD) caused by dystrophin mutations; various forms of limb-girdle muscular dystrophy caused by sarcoglycan mutations; a very rare form of congenital myopathy caused by mutations in the integrin α7 encoding gene and finally laminin-α2 chain-deficient congenital muscular dystrophy (Bönnemann et al., 1995; Hara et al., 2011; Hayashi et al., 1998; Helbling-Leclerc et al., 1995; Hoffman, Brown & Kunkel, 1987; Lim et al., 1995; Nigro et al., 1996; Noguchi et al., 1995; Roberds et al., 1994).

3. CLINICAL PHENOTYPE OF MDC1A

Muscular dystrophy is a general term that describes a group of inherited muscle disorders that cause progressive muscle weakness and muscle wasting. In muscular dystrophy it is believed that muscle fibers cannot sustain mechanical load, which leads to muscle cell membrane rupture, permitting unregulated calcium entry coupled to extensive muscle cell damage. Initially muscle regenerates to compensate for the muscle injury. Regeneration occurs with the help of muscle precursor cells, so-called satellite cells, but regeneration is often inefficient and eventually regeneration can no longer be maintained and muscle cells die, either through necrosis or apoptosis. Subsequently, muscle tissue is replaced by connective tissue and fat cells (Wallace & McNally, 2009).

Congenital muscular dystrophy with laminin-α2 chain deficiency, also known as MDC1A, is a very severe form of muscular dystrophy (Allamand & Guicheney, 2002; Voit & Tomé, 2004). The prevalence of MDC1A, which shows autosomal recessive inheritance, is not well known but it is estimated to be about 1–9/1,000,000 persons. Thus, MDC1A is an ultrarare disease and accounts for about 30% of the congenital muscular dystrophy cases in European countries but it is rarer in Asian countries (Darin & Tulinius, 2000; Gawlik & Durbeej, 2011; Mostacciuolo et al., 1996; Voit & Tomé, 2004). MDC1A has been known as a particular form of congenital muscular dystrophy for more than 20 years. Tomé et al. demonstrated specific absence of laminin-α2 chain in patients and the first causative mutations in the *LAMA2* gene were identified in 1995 (Helbling-Leclerc et al., 1995; Tomé et al., 1994). Today, we know of more than 350 different missense, nonsense, splice site, and deletion mutations that are scattered along the coding sequence containing 65 exons (www.dmd.nl). Genotype/phenotype analyses have shown that complete laminin-α2 chain deficiency in general leads to a severe phenotype whereas partial deficiency leads to a milder disease course, but it should be noted that this concept does not hold true for all individual patients (Geranmayeh et al., 2010; Xiong et al., 2015). The clinical manifestations of children with complete laminin-α2 chain deficiency include profound muscle hypotonia at birth, weak cry, generalized muscle weakness and proximal joint contractures, scoliosis, elevated creatine kinase (CK), and delayed motor milestones. Patients may achieve unsupported sitting, but very few children with complete laminin-α2 chain deficiency acquire independent ambulation (Quijano-Roy,

Sparks, & Rutkowski, 2012; Voit & Tomé, 2004). Feeding difficulties with failure to thrive are common, too, as patients have swallowing abnormalities and difficulty chewing (Philpot, Bagnall, King, Dubowitz, & Muntoni, 1999). Also, weakness of the respiratory muscles gives rise to recurrent chest infections and hypoventilation at night. Altogether, these are very serious complications that should be monitored with respiratory function testing, treatment with noninvasive ventilation support and assistance with coughing and gastrostomy. Still, respiratory tract infection is the most common cause of death, and 30% of the patients die within the first decade of life (Quijano-Roy et al., 2012; Voit & Tomé, 2004; Wang et al., 2010). Histologically, MDC1A skeletal muscle is characterized by a complete or partial deficiency of laminin-α2 chain and displays typical dystrophic characteristics with degenerating/regenerating fibers, atrophic fibers, early inflammation, and extensive connective tissue infiltration (Figure 2) (Quijano-Roy et al.,

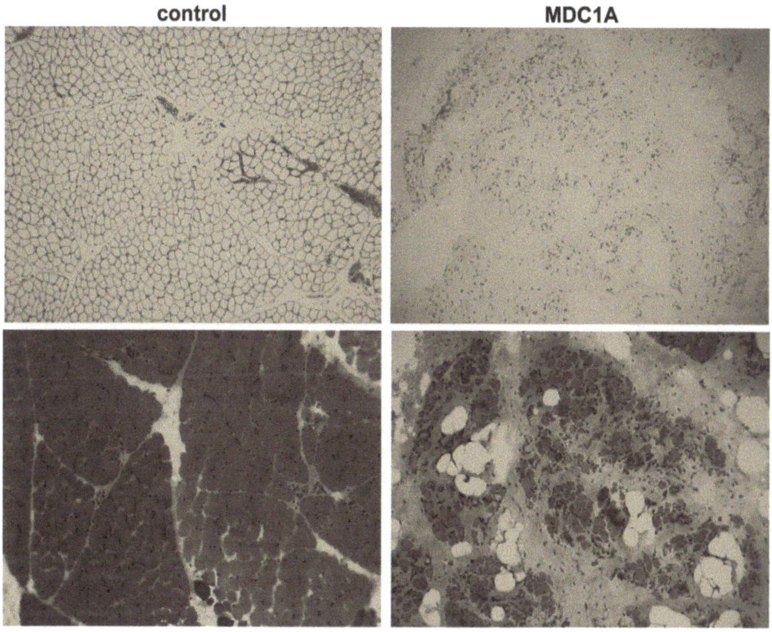

Figure 2 Immunohistochemical and histological assessment of MDC1A biopsy. Upper panel: sections were stained with monoclonal antibody 22B2 detecting the 300 kDa N-terminal fragment of laminin-α2 chain. Note complete absence of laminin-α2 chain in MDC1A muscle. Lower panel: hematoxylin and eosin staining reveals massive muscle degeneration, regenerating fibers with centrally located nuclei, fibrosis, and fat cell infiltration. (See color plate)

2012; Voit & Tomé, 2004). Skeletal muscle basement membranes are also disrupted or completely missing (Xu, Christmas, et al., 1994).

It is not only the skeletal muscle that is affected in MDC1A. Laminin-211 is widely distributed in the body including brain, Schwann cells, heart, testis, kidney, lung, intestine, stomach, placenta, and thymus (Sasaki, Giltay, Talts, Timpl, & Talts, 2002). Consequently, other tissues are, to various degrees, also affected by laminin-α2 chain absence. Characteristic white matter abnormalities, which can be detected by MRI, are usually present in MDC1A patients at 6 months of age. These white matter changes are not associated with any particular functional impairment but can aid in diagnosis. Structural brain changes have been reported in a small portion of children and seizures are present in about 30% of affected individuals (Quijano-Roy et al., 2012; Sunada, Edgar, Lotz, Rust, & Campbell, 1995; Voit & Tomé, 2004; Wang et al., 2010). Moreover, myelination defects in the peripheral nervous system may cause reduced peripheral nerve conduction velocity in patients. However, this malfunction is usually mild and clinically not significant (Deodato et al., 2002; Di Muzio et al., 2003; Quijano-Roy et al., 2012). Laminin-α2 chain expression is particularly high in heart (Sasaki et al., 2002), and echocardiography results are abnormal in around 30% of the patients. Still, cardiac failure is rarely reported in MDC1A (Quijano-Roy et al., 2012). Whereas the impact of laminin-α2 chain absence on the neuromuscular system has been well characterized, there is little known about nonneuromuscular effects in MDC1A patients. Yet, studies in laminin-α2 chain-deficient animals have revealed that ablation of laminin-α2 chain, besides neuromuscular disease, also leads to hearing loss, aberrant thymocyte development, impaired spermatogenesis and defective odontoblast differentiation (Häger, Gawlik, Nyström, Sasaki, & Durbeej, 2005; Pillers et al., 2002; Wagner et al., 2000; Yuasa et al., 2004). Thus, it may be important to investigate nonmuscle manifestations of MDC1A in more detail.

4. ANIMAL MODELS OF MDC1A

There are several mouse models for laminin-α2 chain deficiency (dy/dy, dy^{2J}/dy^{2J}, dy^{6J}/dy^{6J}, dy^{7J}/dy^{7J}, dy^{W}/dy^{W}, and dy^{3K}/dy^{3K} mice) and all of them except dy^{3K}/dy^{3K} mice are available from The Jackson Laboratories (www.jax.org) (Kuang, Xu, Vachon, & Engvall, 1998; Miyagoe et al., 1997; Patton, Wang, Tarumi, Seburn, & Burgess, 2008; Sunada, Bernier, Kozak,

Yamada, & Campbell, 1994; Sunada, Bernier, Utani, Yamada, & Campbell, 1995; Xu, Christmas, Wu, Wewer, & Engvall, 1994; Xu, Wu, Wewer, & Engvall, 1994). Overall, they satisfactorily model human disease and confirm the relationship between laminin-α2 chain expression and severity of disease (Guo et al., 2003). Each of the mouse models is more carefully described in Table 1. In addition, there are two zebrafish models of MDC1A (Table 1) (Gupta et al., 2012; Hall et al., 2007). On the whole, analyses of the various laminin-α2 chain-deficient animal models have led to identification of a number of important disease driving mechanisms that are discussed below.

5. PATHOGENESIS OF MDC1A

5.1 Altered extracellular matrix expression in MDC1A

In MDC1A, laminin-α2 chain expression is absent or significantly reduced (Voit & Tomé, 2004). This reduction leads to a compensatory upregulation of other laminin-α chains, namely laminin-α4 and -α5 chains in the muscle basement membrane (Patton et al., 1997; Ringelmann et al., 1999). Still, these laminin chains cannot fully compensate for laminin-α2 chain deficiency and the basement membrane remains disrupted. One plausible explanation for the lack of compensatory effects is the inability of laminin-α4 chain to self-polymerize and bind to α-dystroglycan (Talts et al., 2000). Indeed, if laminin-α4 chain binding to α-dystroglycan is facilitated by a bridging miniagrin (discussed further below) in laminin-α2 chain-deficient muscle, muscular dystrophy can be prevented in mice (Moll et al., 2001). While some laminin chains are upregulated, others are downregulated in laminin-α2 chain-deficient muscle. Laminin-β2 chain that is part of laminin-221, which is found both at the sarcolemma and in neuromuscular junction basement membranes, is severely reduced in the sarcolemma basement membrane (Cohn, Herrmann, Wewer, & Voit, 1997; Gawlik, Miyagoe-Suzuki, Ekblom, Takeda, & Durbeej, 2004).

Also, dystroglycan expression appears to be altered in MDC1A. The core protein of α-dystroglycan may be reduced in MDC1A muscle, but not glycosylated α-dystroglycan (or at least not the epitope recognized by monoclonal antibody IIH6) (Cohn et al., 1999; Jimenez-Mallebrera et al., 2009; Moll et al., 2001; Vachon et al., 1997). Instead, expression of glycosylated α-dystroglycan has in fact been demonstrated to be slightly increased, along with β-dystroglycan, in laminin-α2 chain-deficient skeletal muscle (Moll et al., 2001; Gawlik, Mayer, et al., 2006). Moreover, the expression

Table 1 Summary of laminin-α2 chain-deficient animal models

Animal model	Mutation	LM	Muscular dystrophy	Survival	References
dy/dy mouse	Unknown	Reduced expression of normal laminin-α2	Moderate	Less than 6 months	Xu, Christmas, et al. (1994), Sunada et al. (1994)
dy^{2J}/dy^{2J} mouse	Splice site mutation in LN domain	Slightly reduced expression of truncated laminin-α2 missing LN domain	Mild	More than 6 months	Xu, Wu, et al. (1994), Sunada, Bernier, et al. (1995)
dy^{6J}/dy^{6J} mouse[a]	?	?	Moderate ?	Less than 6 months	—
dy^{7J}/dy^{7J} mouse	Missense mutation in LN domain	Normal levels	Mild	Normal	Patton et al. (2008)
dy^{W}/dy^{W} mouse	Knock-out	Severely reduced expression of truncated laminin-α2 missing LN domain	Severe	10–15 weeks	Kuang et al. (1998)
dy^{3K}/dy^{3K} mouse	Knock-out	Complete deficiency	Very severe	3 weeks	Miyagoe et al. (1997)
candyfloss zebrafish	Nonsense mutation in LG4	Severe reduction of *lama2* mRNA	Severe	14–16 days postfertilization	Hall et al. (2007)
$lama2^{cl501/cl501}$ zebrafish	Splice site mutation in LN domain	Severely reduced	Severe	8–15 days postfertilization	Gupta et al. (2012)

LN, laminin N-terminal domain; ?, indicates that no information is available.
[a]The phenotype of dy^{6J}/dy^{6J} mice is described at jaxmice.jax.org/strain/003589.html.

of integrin $\alpha7\beta1$ is dysregulated in MDC1A (Cohn et al., 1999; Hodges et al., 1997; Vachon et al., 1997). While the synthesis of integrin $\alpha7$ subunit appears to be increased, there is an obvious impairment of its accumulation in the sarcolemma (Gawlik, Mayer, et al., 2006). Consequently, downstream intracellular signaling pathways are also interrupted in MDC1A. The dysregulation of a wide range of signaling pathways, which largely are unknown, lead to failed regeneration, apoptosis and muscle wasting. Also, inflammation is evident at the early stage of the disease and the excessive buildup of connective tissue (fibrosis) is the final step that destroys muscle. Each of these pathogenic events will be reviewed in more detail.

5.2 Regeneration in MDC1A

Satellite cells, which are located just outside the sarcolemma underneath the muscle basement membrane, control the repair of skeletal muscle. They become activated by cues induced by muscle injury or disease. In muscular dystrophy it is generally believed that the pool of satellite cells becomes depleted as a result of continuous cycles of degeneration and regeneration (Wallace & McNally, 2009). However, muscle regeneration per se could also be defective. In fact, it has been demonstrated that muscle regeneration is delayed and severely impaired after injury in laminin-$\alpha2$ chain-deficient muscle (Bentzinger, Barzaghi, Lin, & Rüegg, 2005; Kuang, Xu, Vilquin, & Engvall, 1999). Additionally, integrin $\alpha7\beta1$, which is secondarily reduced in laminin-$\alpha2$ chain-deficient muscle, is important for satellite cell activation and myoblast differentiation in response to muscle injury (Rooney, Gurpur, Yablonka-Reuveni, & Burkin, 2009). Thus, the laminin-$\alpha2$ chain-rich microenvironment surrounding satellite cells, which governs satellite cell behavior, is defective in MDC1A.

5.3 Apoptosis in MDC1A

Skeletal muscle cell apoptosis has been reported as a pathological hallmark in both laminin-$\alpha2$ chain-deficient patients and animals (Hayashi et al., 2001; Miyagoe et al., 1997; Yoon et al., 2013). Approximately 1–3% of MDC1A muscle fibers cultured in vitro are caspase-3-positive and around 1% of total nuclei are TUNEL-positive in dy^{3K}/dy^{3K} muscle (Miyagoe et al., 1997; Yoon et al., 2013). The signaling pathways leading to apoptosis are poorly understood but it was suggested already in 1997 that integrin $\alpha7\beta1$ is a major mediator of myofiber survival (Vachon et al., 1997). It has also been demonstrated that the GAPDH-Siah1-CBP/p300 signaling pathway, which activates several proapoptotic genes, is dysregulated in laminin-$\alpha2$

chain-deficient muscle (Erb et al., 2009). Furthermore, the proapoptotic protein Bax is a key promoter of abnormal cell death in MDC1A (Girgenrath, Dominov, Kostek, & Miller, 2004).

5.4 Proteasome activity and autophagy in MDC1A

MDC1A is also characterized by loss of muscle mass (Gawlik & Durbeej, 2011). Muscle atrophy occurs by a change in the balance between protein synthesis and protein degradation. The two major proteolytic systems in skeletal muscle are the ubiquitin-proteasome system and the autophagy-lysosome pathway. Briefly, both of these systems are mediated by atrogenes and autophagy-related genes, respectively, which are under the control of forkhead box O transcription factors that in turn are negatively regulated by Akt (Sandri, 2010). We have demonstrated that phosphorylation of Akt is significantly reduced in laminin-α2 chain-deficient mouse muscle and that both the ubiquitin-proteasome system and the autophagy-lysosome pathway are overactive in mouse and human MDC1A muscle (Carmignac, Svensson, et al., 2011; Carmignac, Quéré, & Durbeej, 2011; Körner, Fontes-Oliveira, Holmberg, Carmignac, & Durbeej, 2014). One of the atrogenes, MuRF1, can also be activated by nuclear factor κB (Cai et al., 2004) and it has been shown that the activation of its p65 subunit is also increased in laminin-α2 chain-deficient muscle (Mehuron et al., 2014). The relationship between laminin-α2 chain deficiency and enhanced proteasome/autophagy activity is yet to be determined. Disruption of the laminin—dystroglycan interaction in skeletal muscle cells in vitro resulted in decreased phosphorylation of Akt (Langenbach & Rando, 2002) and also the secondary reduction of integrin α7 subunit may lead to diminished Akt activation (Zhang, Li, Lian, & Li, 2010). Still, gene expression analyses suggested that neither dystroglycan nor integrin α7β1 is involved in the downstream proteasome/autophagy machinery (Carmignac, Svensson, et al., 2011; Carmignac, Quéré, et al., 2011; Körner et al., 2014).

5.5 Inflammation and fibrosis in MDC1A

A consequence of muscle damage is the invasion of inflammatory cells at the injury site, and the inflammatory response can significantly influence the outcome of fibrosis. Others as well as we have shown that inflammation in muscles from laminin-α2 chain-deficient mice and patients takes place at the early stage of disease and is one of the first features of muscle pathology (Gawlik et al., 2014; Mehuron et al., 2014; Pegoraro et al., 1996; Wardrop and Dominov, 2011). Macrophages are prominent in laminin-α2 chain-deficient muscles, and the expression of osteopontin, a potent proinflammatory cytokine produced by macrophages and by dystrophic muscle, is increased in MDC1A muscle as well as tumor necrosis factor-α and monocyte chemoattractant protein-1 (Gawlik et al., 2014; Mehuron et al., 2014). Also mRNAs for innate immune response regulators are elevated in diseased muscle (Jeudy, Wardrop, Alessi, & Dominov, 2011). However, the inflammatory response is transient, and the muscle phenotype evolves into a degenerative and fibrotic appearance (Gawlik et al., 2014). Probably for that reason inflammation in MDC1A has not yielded considerable attention. However, it could well be that the acute inflammation and its later arrest has a major role in muscle pathogenesis in MDC1A.

As already pointed out, laminin-α2 chain-deficient skeletal muscle is characterized by extensive connective tissue deposition, and the composition of the fibrotic lesions in laminin-α2 chain-deficient chain has been characterized to some extent. By gene and protein expression profiling and by immunofluorescence and Western blot analyses, it has been demonstrated that the expression of extracellular matrix proteins biglycan, collagen I, collagen III, collagen VI, connective tissue growth factor, fibrin, fibronectin, galectin-1, galectin-3, lumican, matrix metalloproteinases-2 and -9, periostin, tenascin-C, thrombospondin-4, and tissue inhibitor of metalloproteinases-1, are all significantly increased in laminin-α2 chain-deficient muscles (Doe et al., 2011; Gawlik et al., 2014; Häger et al., 2008; Jeudy et al., 2011; Mehuron et al., 2014; de Oliveira et al., 2014; Taniguchi et al., 2006). Moreover, angiotensin II receptor type 1 signaling and TGF-β levels are augmented in MDC1A muscle (Mehuron et al., 2014; Meinen, Lin, & Rüegg, 2012).

5.6 Transcriptomics and proteomics in MDC1A

A few transcriptomic and proteomic studies of laminin-α2 chain-deficient muscles have been performed in order to obtain novel insights into the molecular mechanisms underlying MDC1A. Microarray gene expression profiles of skeletal muscle from MDC1A patients showed that the most upregulated genes encode primarily extracellular matrix proteins (Taniguchi et al., 2006). We recently performed a comparative proteomic analysis of dy^{3K}/dy^{3K} muscle and obtained similar results (de Oliveira et al., 2014). We and others also presented gene expression profiling of *dy/dy* and dy^{3K}/dy^{3K} dystrophic muscle and showed that the most upregulated genes encode extracellular matrix proteins as well as specific isoforms of proteins that are transiently expressed during normal muscle development and regeneration (Häger et al., 2008; van Lunteren, Moyer, & Leahy, 2006). Those being downregulated, on the other hand, mainly participate in diverse metabolic processes and kinase activities (Häger et al., 2008). Also, the quantitative proteomic analysis revealed metabolic alterations and calcium dysregulation in laminin-α2 chain-deficient mouse muscle (de Oliveira et al., 2014). Altogether, results from these omic technologies imply that there is massive extracellular matrix remodeling in laminin-α2 chain-deficient muscle and that absence of laminin-α2 chain leads to metabolic crises in the muscle cell. Metabolic processes have to our knowledge never been analyzed in MDC1A and little is known about metabolic pathways in muscular dystrophy, in general. Although metabolic impairment may not the primary cause of muscle weakness in laminin-α2 chain deficiency, it may play a significant additional role potentially important for the time course of the disease. Similarly, abnormal calcium handling may contribute to MDC1A pathology.

Finally, the expression of selected microRNAs (miRNAs) was recently evaluated in laminin-α2 chain-deficient muscle and plasma. miRNAs are small nonprotein-coding RNAs that regulate gene expression by inhibiting translation/promoting mRNA degradation (Holmberg, Alajbegovic, Gawlik, Elowsson, & Durbeej, 2014). A role for miRNAs in muscle development and function and aberrant miRNA expression in muscular dystrophy has been demonstrated (Eisenberg, Alexander, & Kunkel, 2009). We showed that levels of muscle-specific miR-1, miR-133a, and miR-206 are altered in laminin-α2 chain-deficient skeletal muscle and plasma. Moreover, miRNA expression was partially normalized in response to therapy (Holmberg et al., 2014). These data indicate that miRNAs may play important roles in MDC1A pathology and that they could function as biomarkers for noninvasive monitoring of disease progression and response to experimental therapies.

6. AMELIORATION OF DISEASE IN MICE

The different mouse models have been very valuable tools for the preclinical development of novel therapeutic approaches. These approaches include correction of both the primary defects in cell—extracellular matrix adhesion as well as the secondary manifestations such as increased apoptosis, enhanced protein degradation, and pathological fibrosis. Although it is exceedingly difficult to compare studies that were performed in different mouse models with different severity and with somewhat diverse outcome measures, I have tried to score the skeletal muscle improvement, where 1 is medium; 2 is good; and 3 is excellent muscle restoration (see Table 2).

6.1 Extracellular matrix modulation

Since the extracellular matrix is affected in MDC1A, many approaches were aimed at targeting the expression of extracellular matrix proteins. Transgenic expression of the laminin-α1 and -α2 chains, miniagrin, cytotoxic T cell GalNac transferase, and integrin α7 subunit have all been found to compensate for laminin-α2 chain deficiency in mice (Doe et al., 2011; Gawlik et al., 2004; Kuang et al., 1998; Moll et al., 2001; Xu, Chandrasekharan, Yoon, Camboni, & Martin, 2007). Eva Engvall and colleagues were the first to describe a genetic correction in dy^{W}/dy^{W} and dy^{2J}/dy^{2J} mice with expression of the human α2 chain transgene under the control of a skeletal muscle-specific promoter (Kuang et al., 1998). Rüegg and coworkers created dy^{W}/dy^{W} mice overexpressing linker molecules (miniagrin, full-length agrin, agrin-perlecan fusion protein) that introduce a connection between laminin-α4 chain and dystroglycan in skeletal muscle. Resultant mice displayed a prolonged life span and significantly reduced muscle pathology (Meinen, Barzaghi, Lin, Lochmuller, & Rüegg, 2007; Moll et al., 2001). Also, adeno-associated virus-mediated overexpression miniagrin improved the overall phenotype and muscle function in dy^{W}/dy^{W} mice (Qiao et al., 2005).

The effects of laminin-α1 chain (structurally and functionally most similar to the α2 chain) overexpression in the neuromuscular system have also been extensively studied. We have shown that dy^{3K}/dy^{3K} mice overexpressing the laminin-α1 chain, mainly in skeletal muscle, heart, and Schwann cell basement membrane, have a near-normal life span and display significantly reduced muscular dystrophy, cardiomyopathy, and peripheral neuropathy (Gawlik et al., 2004; Gawlik, Li, et al., 2006; Gawlik, Åkerlund, Carmignac, Elamaa, & Durbeej, 2010; Gawlik & Durbeej, 2010). However, the use of laminin-α1 chain in gene therapy may be challenging (due to the large size of its cDNA) and therefore it was very interesting to learn that systemic delivery of laminin-111 attenuated muscle pathology and increased life span of dy^{W}/dy^{W} mice (Rooney, Knapp, Hodges, Wuebbles, & Burkin, 2012). Importantly, laminin-111 has received orphan drug status from the FDA for MDC1A. Moreover, Burkin and his team investigated if restoring integrin α7 expression could diminish muscular dystrophy in dy^{W}/dy^{W} mice. Indeed, enhanced integrin α7 expression decreased muscle pathology

Table 2 Summary of treatment approaches in laminin-α2 chain-deficient mice

Therapeutic approach	Protein/Compound	Mouse model	Outcome 1–3 (medium, good, and excellent muscle restoration)	References
Modulating ECM	Laminin-α2 transgene	dy^{W}/dy^{W}	3	Kuang et al. (1998)
		dy^{2J}/dy^{2J}	3	
	Laminin-α1 transgene	dy^{3K}/dy^{3K}	3	Gawlik et al. (2004), Gawlik, Li, et al. (2006), Gawlik and Durbeej (2010)
	Truncated laminin-α1 transgene	dy^{3K}/dy^{3K}	2 (diaphragm)	Gawlik et al. (2010)
	Miniagrin transgene	dy^{W}/dy^{W}	2–3	Moll et al. (2001), Bentzinger et al. (2005)
		dy^{3K}/dy^{3K}	2	
	Miniagrin AAV	dy^{W}/dy^{W}	2	Qiao et al. (2005)
	Agrin transgene	dy^{W}/dy^{W}	2	Meinen et al. (2007)
	Agrin-perlecan transgene	dy^{W}/dy^{W}	2	Meinen et al. (2007)
	Integrin α7 transgene	dy^{W}/dy^{W}	1–2	Doe et al. (2011)
	Galgt2 transgene/AAV	dy^{W}/dy^{W}	1–2	Xu et al. (2007)
	Laminin-111 protein	dy^{W}/dy^{W}	1–2	Rooney et al. (2012)
Triggering regeneration	IGF-1 transgene	dy^{W}/dy^{W}	1–2	Kumar et al. (2011)
Apoptosis suppression	Bcl-2 overexpression	dy^{W}/dy^{W}	1–2	Girgenrath et al. (2004)
	Bax inhibition	dy^{W}/dy^{W}	2	Girgenrath et al. (2004)
	Doxycycline	dy^{W}/dy^{W}	1	Girgenrath et al. (2009)
	Omigapil	dy^{W}/dy^{W}	1	Erb et al. (2009), Yu et al. (2013)
		dy^{2J}/dy^{2J}	1	

Proteasome inhibition	MG-132	dy^{3K}/dy^{3K}	1	Carmignac, Quéré, et al. (2011)
	Bortezomib	dy^{3K}/dy^{3K}	1	Körner et al. (2014)
Autophagy inhibition	3-Methyladenine	dy^{3K}/dy^{3K}	1	Carmignac, Svensson, et al. (2011)
Fibrosis and inflammation inhibition	Losartan	dy^{2J}/dy^{2J}	1	Elbaz et al. (2012)
	Losartan derivative	dy^{W}/dy^{W}	1	Meinen et al. (2012)
	Halofuginone	dy^{2J}/dy^{2J}	1	Nevo et al. (2010)
	Complement C3 deletion	dy/dy	1	Connolly et al. (2002)
Calcium modulation	Cyclophilin D deletion	dy^{W}/dy^{W}	1–2	Millay et al. (2008)
	Caldecrin	dy/dy	1	Tomomura et al. (2011)
Combinatorial treatment	Miniagrin + Bcl-2 overexpression	dy^{W}/dy^{W}	3	Meinen et al. (2011)
	Miniagrin + omigapil	dy^{W}/dy^{W}	2–3	Meinen et al. (2011)
	IGF-1 transgene + Bax inhibition	dy^{W}/dy^{W}	2–3	Yamaguchi et al. (2013)
	IGF-1 + Bax inhibition	dy^{W}/dy^{W}	2–3	Yamaguchi et al. (2013)
Exon skipping	Phosphorodiamidate morpholino oligomer	dy^{3K}/dy^{3K}	nd	Aoki et al. (2013)
Cell-based therapy	Myoblast transplantation	dy/dy	nd	Vilquin et al. (1999)
	Bone-marrow transplantation	dy/dy	1–2	Hagiwara et al. (2006)
	CD90+ cell transplantation	dy^{3K}/dy^{3K}	nd	Fukada et al. (2007)

AAV, adeno-associated virus; nd, not determined.

in mice, but to a lesser extent compared with transgenically expressed laminins or agrin (Doe et al., 2011).

Another way to modulate the myomatrix in dy^W/dy^W mice was tested by Xu et al. (2007). Cytotoxic T cell GalNAc transferase is an acetylgalactosaminyltransferase that creates the cytotoxic T cell carbohydrate antigen on selected glycoproteins and glycolipids. It is normally expressed at the neuromuscular junction but when overexpressed at the extrasynaptic membrane in dy^W/dy^W muscle, muscular dystrophy was reduced (Xu et al., 2007).

6.2 Triggering regeneration

In order to enhance the regenerative capacity in laminin-α2 chain-deficient mice, insulin-growth factor-1 (IGF-1) was selectively and transgenically overexpressed in skeletal muscle. IGF-1 is a central regulator of muscle regeneration and indeed, regeneration was improved in the $dy^W/dy^W + IGF\text{-}1tg$ mice and muscular dystrophy reduced (Kumar, Yamauchi, Girgenrath, & Girgenrath, 2011). In addition, overexpression of miniagrin and laminin-α1 chain also enhanced regeneration in dy^{3K}/dy^{3K} mice (Bentzinger et al., 2005; Van Ry, Minogue, Hodges, & Burkin, 2014).

However, boosting regeneration in laminin-α2 chain-deficient mice does not always improve the outcome. Myostatin is a secreted growth factor that inhibits muscle differentiation. Thus, inhibiting myostatin is a promising therapeutic strategy for muscle wasting diseases and has shown potential in animal models for DMD (Qiao et al., 2008). Absence of myostatin, on the other hand, had no effect on muscle pathology in dy^W/dy^W animals, even though increasing muscle regeneration (Li, Shelton, et al., 2005). In the same way, transgenic overexpression of disintegrin and metalloproteinase domain-containing protein 12 in dy^W/dy^W mice did not reduce muscular dystrophy, despite stimulating muscle regeneration (Guo, Shelton, Wewer, & Engvall, 2005).

6.3 Apoptosis inhibition

Several different approaches to reduce apoptosis in laminin-α2 chain-deficient mice have been assessed. Girgenrath and collaborators generated dy^W/dy^W mice with either inactivated expression of the proapoptotic protein Bax or overexpression of the antiapoptotic protein Bcl-2 (Girgenrath et al., 2004). Both of these interferences partly ameliorated muscular dystrophy in mice. Overexpression of Bcl-2 also hampered the early inflammatory response (Jeudy et al., 2011). Correspondingly, pharmacological treatment with the antibiotic doxycycline enhanced the life span of dy^W/dy^W animals and diminished muscle pathology (Girgenrath, Beermann, Vishnudas, Homma, & Miller, 2009). Likewise, administration of omigapil, which inhibits the GAPDH-Siah1-CBP/p300 signaling pathway, amended some of the pathological features in dy^W/dy^W and dy^{2J}/dy^{2J} animals (Erb et al., 2009; Yu et al., 2013). Based on these results, Santhera recently announced a Phase 1 clinical trial of omigapil in children with congenital muscular dystrophy.

6.4 Proteasome and autophagy inhibition

Because enhanced proteasomal degradation and autophagy, respectively, are features of laminin-α2 chain-deficient muscle (Carmignac, Svensson, et al., 2011; Carmignac, Quéré, et al., 2011), we imagined that proteasome and autophagy inhibition would lessen the myopathology but not entirely cure the disease. Indeed, treatment with the proteasome inhibitor MG-132 expanded the life span about twofold and improved muscle morphology of dy^{3K}/dy^{3K} mice (Carmignac, Quéré, et al., 2011) and so did the autophagy inhibitor 3-methyladenine (Carmignac, Svensson, et al., 2011). More recently, we explored the use of the selective proteasome inhibitor bortezomib in dy^{3K}/dy^{3K} mice. Bortezomib is currently used for treatment of relapsed multiple myeloma and mantle cell lymphoma. Indeed, it also lessened muscle pathology and partially normalized miR-1 and miR-133a expression in skeletal muscle and plasma (Körner et al., 2014).

6.5 Fibrosis and inflammation inhibition

Abundant muscle fibrosis is associated with disease progression in MDC1A patients and laminin-α2 chain-deficient mice. Accordingly, Nevo and colleagues recently evaluated antifibrotic compounds in two animal models of MDC1A (Elbaz et al., 2012). Halofuginone, a plant alkaloid analog that blocks TGFβ-mediated collagen synthesis, attenuated fibrosis in dy^{2J}/dy^{2J} mice (Nevo et al., 2010). Furthermore, losartan, which is an angiotensin II receptor blocker used for the treatment of hypertension, also mitigated fibrosis in dy^{2J}/dy^{2J} mice (Elbaz et al., 2012). Lastly, treatment with the angiotensin II receptor antagonist L-158809 (a potent derivative of losartan) also decreased fibrosis in dy^W/dy^W mice (Meinen et al., 2012). Inflammation has also been modulated in laminin-α2 chain-deficient mice. *Dy/dy* mice deficient in the C3 component of the complement system showed improved muscle strength up to 20 weeks of age (Connolly, Keeling, Streif, Pestronk, & Mehta, 2002).

6.6 Modulating abnormal calcium handling

The proteomic analyses of dy^{3K}/dy^{3K} muscle showed that the expression of numerous calcium-binding proteins is altered in dystrophic muscle. Furthermore, resting calcium levels in muscle fibers from *dy/dy* mice are elevated compared to normal fibers (Williams, Head, Bakker, & Stephenson, 1990). These data indicate a role for calcium in MDC1A pathology. Caldecrin is a serum calcium-decreasing factor isolated from pancreas. When *dy/dy* mice were treated with either recombinant caldecrin, or with a caldecrin-expression vector, some histopathological features were improved, but mice were not analyzed in great detail (Tomomura, Fujii, Sakagami, & Tomomura, 2011). Also, ablation of the mitochondrial calcium regulator cyclophilin D, which makes mitochondria resistant to cell death from overload of calcium and free radicals, reduced muscular dystrophy in dy^{W}/dy^{W} mice (Millay et al., 2008).

6.7 Combinatorial treatment

As expected, the transgenic approaches modulating the basement membrane have shown the highest efficacy in restoring skeletal muscle, while targeting the secondary downstream effects of laminin-α2 chain deficiency have been less effective (Table 2). Still, translating the transgenic approaches into effective gene therapy remains challenging. Furthermore, as laminin-α2 chain deficiency seems to affect many different biological pathways, it is quite likely that combinatorial treatment strategies may be relevant for MDC1A. Recently, two separate research groups evaluated different combinatorial approaches with great success. First, it was demonstrated that apoptosis inhibitors (either Bcl-2 overexpression or omigapil) and miniagrin have cumulative benefits in dy^{W}/dy^{W} mice (Meinen et al., 2011). Subsequently, it was demonstrated that apoptosis inhibition (Bax deletion) and overexpression of IGF-1 (either transgenically or systemically) also have additive effects in dy^{W}/dy^{W} mice (Yamauchi, Kumar, Duarte, Mehuron, & Girgenrath, 2013).

6.8 Other treatment approaches

A handful of other treatment approaches have also been tested in laminin-α2 chain-deficient mice and they include exon skipping and cell therapy. The exon-skipping technology involves mutation-specific treatment with the aim to restore the coding reading frame of an mRNA so that it can be read to produce a shorter but still functional protein. Preclinical and also clinical studies have demonstrated the potential of antisense oligonucleotide-mediated exon skipping therapy for DMD (Arechavala-Gomeza, Anthony, Morgan, & Muntoni, 2012). While this approach has been extensively tested, targeting different exons in numerous DMD models, it has only been tried once for MDC1A (Aoki et al., 2013). Phosphorodiamidate-mediated exon 4 skipping in dy^{3K}/dy^{3K} mice corrected the open reading frame and led to the production of a truncated laminin-α2 chain, devoid of a part of the LN domain. Thus, the very severe phenotype of dy^{3K}/dy^{3K} mice was presumably converted into a relatively mild muscular dystrophy phenotype, similar to that of dy^{2J}/dy^{2J} animals. However, the functional benefit was not further analyzed in the mice (Aoki et al., 2013).

Cell therapy has also been assessed in mouse models of MDC1A. Myoblast transplantations as well as transplantation of CD90-positive cells, which is an additional cell population resident in skeletal muscle with fibroblast-like characteristics, led to laminin-α2 chain production in MDC1A mice. However, it remains unclear if the dystrophic phenotype was improved (Fukada et al., 2007; Vilquin et al., 1999). Bone-marrow transplantation, on the other hand, improved the outcome in the *dy/dy* mouse model (Hagiwara et al., 2006).

7. CONCLUSION

MDC1A is an ultrarare disorder that affects a small number of patients. Nonetheless, its impact on patients, their families, and society is profound. MDC1A patients face premature death since there is no therapy available. Importantly, great advances have been made during the last decade in identifying targets to treat MDC1A. Laminin-α2 chain-deficient mice have positively responded to omigapil with inhibition of apoptosis (Erb et al., 2009; Yu et al., 2013). Consequently, Santhera recently announced a Phase 1 clinical trial of omigapil in children with congenital muscular dystrophy. Also, drugs that are already in clinical use such as losartan, and bortezomib have shown promise in laminin-α2 chain-deficient mouse models and could advance to MDC1A clinical trials quickly (Elbaz et al., 2012; Körner et al., 2014). Still, these drug-based approaches that inhibit apoptosis, fibrosis, or muscle wasting do not address the primary defect in the basement membrane and are therefore not expected to cure the disease but might offer a supportive therapy. The potential side effects of these drugs must also be considered and therefore it is worth noting that losartan is well tolerated in children (Webb et al., 2014). Additionally, ongoing drug development may lead to new compounds with fewer adverse effects and also more selective agents (e.g., autophagy modulators). But all in all, the risk of side effects must also be balanced with the chance of slowing down the progression of MDC1A.

In order to fully alleviate MDC1A pathology it will probably be necessary to restore the skeletal muscle basement membrane and reinforce skeletal muscle adhesion. This could be achieved with gene therapy, protein (replacement) therapy, or small molecule-based approaches. While gene therapy with large cDNAs (as for laminins) is not feasible at the moment, laminin-111 protein therapy offers a promising approach for MDC1A treatment (Rooney et al., 2012), and a combinatorial treatment strategy may have larger beneficial effects than laminin-111 protein therapy alone. Also, combinatorial approaches directed toward some of the secondary manifestations of MDC1A may be almost as effective as targeting the primary cause (Yamaguchi et al., 2013).

In summary, despite excellent progress in MDC1A-related research, MDC1A remains incurable. Thus, continued intense basic, translational, and clinical research is needed in order to develop an efficient treatment.

ACKNOWLEDGMENTS

I am greatly indebted to Dr Rita Barresi for providing images in Figure 2. I also thank Drs Kinga Gawlik and Johan Holmberg for critical reading of the manuscript.

REFERENCES

Allamand, V., & Guicheney, P. (2002). Merosin-deficient muscular dystrophy, autosomal recessive (MDC1A, MIM#156225, LAMA2 gene coding for α2 chain of laminin). *European Journal of Human Genetics, 10*, 91–94.

Allikian, M. J., Hack, A. A., Mewborn, S., Mayer, U., & McNally, E. M. (2004). Genetic compensation for sarcoglycan loss by integrin α7β1 in muscle. *Journal of Cell Science, 117*, 3821–3830.

Aoki, Y., Nagata, T., Yokota, T., Nakamura, A., Wood, M. J., Partridge, T., et al. (2013). Highly efficient in vivo delivery of PMO into regenerating myotubes and rescue in laminin-α2 chain-null congenital muscular dystrophy mice. *Human Molecular Genetics, 22*, 4914–4928.

Arechavala-Gomeza, V., Anthony, K., Morgan, J., & Muntoni, F. (2012). Antisense oligonucleotide-mediated exon skipping for Duchenne muscular dystrophy: progress and challenges. *Current Gene Therapy, 12*, 152–160.

Aumailley, M., Bruckner-Tuderman, L., Carter, W. G., Deutzmann, R., Edgar, D., Ekblom, P., et al. (2005). A simplified laminin nomenclature. *Matrix Biology, 24*, 326–332.

Bentzinger, C. F., Barzaghi, P., Lin, S., & Rüegg, M. A. (2005). Overexpression of mini-agrin in skeletal muscle increases muscle integrity and regenerative capacity in laminin-α2-deficient mice. *FASEB Journal, 19*, 934–942.

Bönnemann, C. G., Modi, R., Noguchi, S., Mizuno, Y., Yoshida, M., Gussoni, E., et al. (1995). β-sarcoglycan (A3b) mutations cause autosomal recessive muscular dystrophy with loss of the sarcoglycan complex. *Nature Genetics, 11*, 266–273.

Burkin, D. J., Wallace, G. Q., Nicol, K. J., Kaufman, D. J., & Kaufman, S. J. (2001). Enhanced expression of the α7β1 integrin reduces muscular dystrophy and restores viability in dystrophic mice. *Journal of Cell Biology, 152*, 1207–1218.

Cai, D., Frantz, J. D., Tawa, N. E., Jr., Melendez, P. A., Oh, B. C., Lidov, H. G., et al. (2004). IKKβ/NF-κB activation causes severe muscle wasting in mice. *Cell, 119*, 285–298.

Campbell, K. P., & Kahl, S. D. (1989). Association of dystrophin and an integral membrane glycoprotein. *Nature, 338*, 259–262.

Carmignac, V., Quéré, R., & Durbeej, M. (2011). Proteasome inhibition improves the muscle of laminin α2 chain-deficient mice. *Human Molecular Genetics, 20*, 541–552.

Carmignac, V., Svensson, M., Körner, Z., Elowsson, L., Matsumura, C., Gawlik, K. I., et al. (2011). Autophagy is increased in laminin α2 chain-deficient muscle and its inhibition improves muscle morphology in a mouse model of MDC1A. *Human Molecular Genetics, 20*, 4891–4902.

Cohn, R. D., Henry, M. D., Michele, D. E., Barresi, R., Saito, F., Moore, S. A., et al. (2002). Disruption of DAG1 in differentiated skeletal muscle reveals a role for dystroglycan in muscle regeneration. *Cell, 110*, 639–648.

Cohn, R. D., Herrmann, R., Wewer, U. M., & Voit, T. (1997). Changes of laminin β2 chain expression in congenital muscular dystrophy. *Neuromuscular Disorders, 7*, 373–378.

Cohn, R. D., Mayer, U., Saher, G., Herrmann, R., van der Flier, A., Sonnenberg, A., et al. (1999). Secondary reduction of integrin α7B in laminin α2 deficient congenital muscular dystrophy supports an additional transmembrane link in skeletal muscle. *Journal of Neurological Science, 63*, 140–152.

Colognato, H., & Yurchenco, P. D. (1999). The laminin α2 expressed by dystrophic dy(2J) mice is defective in its ability to form polymers. *Current Biology, 22*, 1327–1330.

Connolly, A. M., Keeling, R. M., Streif, E. M., Pestronk, A., & Mehta, S. (2002). Complement 3 deficiency and oral prednisolone improve strength and prolong survival of laminin α2-deficient mice. *Journal of Neuroimmunology, 127*, 80–87.

Crosbie, R. H., Lebakken, C. S., Holt, K. H., Venzke, D. P., Straub, V., Lee, J. C., et al. (1999). Membrane targeting and stabilization of sarcospan is mediated by the sarcoglycan subcomplex. *Journal of Cell Biology, 145*, 153–165.

Darin, N., & Tulinius, M. (2000). Neuromuscular disorders in childhood: a descriptive epidemiological study from western Sweden. *Neuromuscular Disorders, 10*, 1–9.

Deodato, F., Sabatelli, M., Ricci, E., Mercuri, E., Muntoni, F., Sewry, C., et al. (2002). Hypermyelinating neuropathy, mental retardation and epilepsy in a case of merosin deficiency. *Neuromuscular Disorders, 4*, 392–398.

Deutzmann, R., Aumailley, M., Wiedemann, H., Pysny, W., Timpl, R., & Edgar, D. (1990). Cell adhesion, spreading and neurite stimulation by laminin fragment E8 depends on maintenance of secondary and tertiary structure in its rod and globular domain. *European Journal of Biochemistry, 191*, 513–522.

Di Muzio, A., De Angelis, M. V., Di Fulvio, P., Ratti, A., Pizzuti, A., Stuppia, L., et al. (2003). Dysmyelinating sensory-motor neuropathy in merosin-deficient congenital muscular dystrophy. *Muscle & Nerve, 4*, 500–506.

Doe, J. A., Wuebbles, R. D., Allred, E. T., Rooney, J. E., Elorza, M., & Burkin, D. J. (2011). Transgenic overexpression of the α7 integrin reduces muscle pathology and improves viability in the dy(W) mouse model of merosin-deficient congenital muscular dystrophy type 1A. *Journal of Cell Science, 124*, 2287–2297.

Ehrig, K., Leivo, I., Argraves, W. S., Ruoslahti, E., & Engvall, E. (1991). Merosin, a tissue-specific basement membrane protein, is a laminin-like protein. *Proceedings of the National Academy of Sciences of the United States of America, 87*, 3264–3268.

Eisenberg, I., Alexander, M. S., & Kunkel, L. M. (2009). miRNAs in normal and diseased skeletal muscle. *Journal of Cellular and Molecular Medicine, 13*, 2–11.

Elbaz, M., Yanay, N., Aga-Mizrachi, S., Brunschwig, Z., Kassis, I., Ettinger, K., et al. (2012). Losartan, a therapeutic candidate in congenital muscular dystrophy: studies in the dy^{2J}/dy^{2J} Mouse. *Annals of Neurology, 71*, 699–708.

Erb, M., Meinen, S., Barzaghi, P., Sumanovski, L. T., Courdier-Fruh, I., Rüegg, M. A., et al. (2009). Omigapil ameliorates the pathology of muscle dystrophy caused by laminin-α2 deficiency. *Journal of Pharmacology Experimental Therapy, 331*, 787–795.

Ervasti, J. M., & Campbell, K. P. (1993). A role for the dystrophin-glycoprotein complex as a transmembrane linker between laminin and actin. *Journal of Cell Biology, 122*, 809–823.

Fukada, S., Yamamoto, Y., Segawa, M., Sakamoto, K., Nakajima, M., Sato, M., et al. (2007). CD90-positive cells, an additional cell population, produce laminin α2 chain upon transplantation in dy^{3k}/dy^{3k} mice. *Experimental Cell Research, 314*, 193–203.

Gawlik, K. I., Åkerlund, M., Carmignac, V., Elamaa, H., & Durbeej, M. (2010). Distinct roles for laminin globular domains in laminin α1 chain mediated rescue of murine laminin α2 chain deficiency. *PLoS ONE, 5*, e11549.

Gawlik, K. I., & Durbeej, M. (2010). Transgenic overexpression of laminin α1 chain in laminin α2 chain-deficient mice rescues the disease throughout the lifespan. *Muscle & Nerve, 42*, 30–37.

Gawlik, K. I., & Durbeej, M. (2011). Skeletal muscle laminin and MDC1A: pathogenesis and treatment strategies. *Skeletal Muscle, 1*, 9.

Gawlik, K. I., Holmberg, J., & Durbeej, M. (2014). Loss of dystrophin and β-sarcoglycan significantly exacerbates the phenotype of laminin α2 chain-deficient animals. *American Journal of Pathology, 184*, 740–752.

Gawlik, K. I., Li, J.-Y., Petersen, Å., & Durbeej, M. (2006). Laminin α1 chain improves laminin α2 chain deficient neuropathy. *Human Molecular Genetics, 5*, 2690–2700.

Gawlik, K. I., Mayer, U., Blomberg, K., Sonnenberg, A., Ekblom, P., & Durbeej, M. (2006). Laminin α1 chain mediated reduction of laminin α2 chain deficient muscular dystrophy involves integrin α7β1 and dystroglycan. *FEBS Letters, 580*, 1759–1765.

Gawlik, K., Miyagoe-Suzuki, Y., Ekblom, P., Takeda, S., & Durbeej, M. (2004). Laminin α1 chain reduces muscular dystrophy in laminin α2 chain deficient mice. *Human Molecular Genetics, 13*, 1775–1784.

Geranmayeh, F., Clement, E., Feng, L. H., Sewry, C., Pagan, J., Mein, R., et al. (2010). Genotype-phenotype correlation in a large population of muscular dystrophy patients with LAMA2 mutations. *Neuromuscular Disorders, 4*, 241–250.

Girgenrath, M., Beermann, M. L., Vishnudas, V. K., Homma, S., & Miller, J. B. (2009). Pathology is alleviated by doxycycline in a laminin-α2-null model of congenital muscular dystrophy. *Annals of Neurology, 65*, 47–56.

Girgenrath, M., Dominov, J. A., Kostek, C. A., & Miller, J. B. (2004). Inhibition of apoptosis improves outcome in a model of congenital muscular dystrophy. *Journal of Clinical Investigation, 114*, 1635–1639.

Guo, C., Willem, M., Werner, A., Raivich, G., Emerson, M., Neyses, L., et al. (2006). Absence of α7 integrin in dystrophin-deficient mice causes a myopathy similar to Duchenne muscular dystrophy. *Human Molecular Genetics, 15*, 989–998.

Guo, L. T., Shelton, G. D., Wewer, U. M., & Engvall, E. (2005). ADAM12 overexpression does not improve outcome in mice with laminin α2-deficient muscular dystrophy. *Neuromuscular Disorders, 15*, 786–789.

Guo, L. T., Zhang, X. U., Kuang, W., Xu, H., Liu, L. A., Vilquin, J. T., et al. (2003). Laminin α2 deficiency and muscular dystrophy; genotype-phenotype correlation in mutant mice. *Neuromuscular Disorders, 3*, 207–215.

Gupta, V. A., Kawahara, G., Myers, J. A., Chen, A. T., Hall, T. E., Manzini, M. C., et al. (2012). A splice site mutation in laminin-α2 results in a severe muscular dystrophy and growth abnormalities in zebrafish. *PLoS ONE, 8*, e43794.

Häger, M., Bigotti, M. G., Meszaros, R., Carmignac, V., Holmberg, J., Allamand, V., et al. (2008). Cib2 binds integrin α7Bb1D and is reduced in laminin α2 chain-deficient muscular dystrophy. *Journal of Biological Chemistry, 283*, 24760–24769.

Häger, M., Gawlik, K., Nyström, A., Sasaki, T., & Durbeej, M. (2005). Laminin α1 chain corrects male infertility caused by absence of laminin α2 chain. *American Journal of Pathology, 167*, 823–833.

Hagiwara, H., Ohsawa, Y., Asakura, S., Murakami, T., Teshima, T., & Sunada, Y. (2006). Bone marrow transplantation improves outcome in a mouse model of congenital muscular dystrophy. *FEBS Letters, 580*, 4463–4468.

Hall, T. E., Bryson-Richardson, R. J., Berger, S., Jacoby, A. S., Cole, N. J., Hollway, G. E., et al. (2007). The zebrafish candyfloss mutant implies extracellular matrix adhesion failure in laminin α2-deficient congenital muscular dystrophy. *Proceedings of the National Academy of Sciences of the United States of America, 104*, 7093–7097.

Han, R., Kanagawa, M., Yoshida-Moriguchi, T., Rader, E. P., Ng, R. A., Michele, D. E., et al. (2009). Basal lamina strengthens cell membrane integrity via the laminin G domain-binding motif of α-dystroglycan. *Proceedings of the National Academy of Sciences of the United States of America, 106*, 12573–12579.

Hara, Y., Balci-Hayta, B., Yoshida-Moriguchi, T., Kanagawa, M., Beltrán-Valero de Bernabé, D., Gündeşli, H., et al. (2011). A dystroglycan mutation associated with limb-girdle muscular dystrophy. *New England Journal of Medicine, 3*(64), 939–946.

Hayashi, Y. K., Chou, F.-L., Engvall, E., Ogawa, M., Matsuda, C., Hirabayashi, S., et al. (1998). Mutations in the integrin α7 gene cause congenital myopathy. *Nature Genetics, 19*, 94–97.

Hayashi, Y. K., Tezak, Z., Momoi, T., Nonaka, I., Garcia, C. A., Hoffman, E. P., et al. (2001). Massive muscle cell degeneration in the early stage of merosin-deficient congenital muscular dystrophy. *Neuromuscular Disorders, 11*, 350–359.

Helbling-Leclerc, A., Zhang, X., Topaloglu, H., Cruaud, C., Tesson, F., Weissenbach, J., et al. (1995). Mutations in the laminin α2 chain gene (LAMA2) cause merosin-deficient muscular dystrophy. *Nature Genetics, 11*, 216–218.

Hodges, B. S., Hayashi, Y. K., Nonaka, I., Wang, A., Arahata, K., & Kaufman, S. J. (1997). Altered expression of the α7β1 integrin in human and murine muscular dystrophies. *Journal of Cell Science, 110*, 2873–2881.

Hoffman, E. P., Brown, R. J., Jr., & Kunkel, L. M. (1987). Dystrophin: the protein product of the Duchenne muscular dystrophy locus. *Cell, 51*, 919–928.

Holmberg, J., Alajbegovic, A., Gawlik, K. I., Elowsson, L., & Durbeej, M. (2014). Laminin α2 chain-deficiency is associated with microRNA deregulation in skeletal muscle and plasma. *Frontiers in Aging Neuroscience, 6*, 155, 3.

Hopf, M., Göhring, W., Mann, K., & Timpl, R. (2001). Mapping of binding sites for nidogens, fibulin-2, fibronectin and heparin to different IG modules of perlecan. *Journal of Molecular Biology, 311*, 529–541.

Ibraghimov-Beskrovnaya, O., Ervasti, J. M., Leveille, C. J., Slaughter, C. A., Sernett, S. W., & Campbell, K. P. (1992). Primary structure of dystrophin-associated glycoproteins linking dystrophin to the extracellular matrix. *Nature, 355*, 696–702.

Ido, H., Nakamura, A., Kobayashi, R., Ito, S., Li, S., Futaki, S., et al. (2007). The requirement of the glutamic acid residue at the third position from the carboxyl termini of the laminin gamma chains in integrin binding by laminins. *Journal of Biological Chemistry, 282*, 11144–11154.

Jeudy, S., Wardrop, K. E., Alessi, A., & Dominov, J. A. (2011). Bcl-2 inhibits the innate immune response during early pathogenesis of murine congenital muscular dystrophy. *PLoS ONE, 6*, e22369.

Jimenez-Mallebrera, C., Torelli, S., Feng, L., Kim, J., Godfrey, C., Clement, E., et al. (2009). A comparative study of α-dystroglycan glycosylation in dystroglycanopathies suggests that the hypoglycosylation of α-dystroglycan does not consistently correlate with clinical severity. *Brain Pathology, 19*, 596–611.

Kammerer, R. A., Schulthess, T., Landwehr, R., Schumacher, B., Lustig, A., Yurchenco, P. D., et al. (1999). Interaction of agrin with laminin requires a coiled-coil conformation of the agrin-binding site within the laminin gamma1 chain. *EMBO Journal, 23*, 6762–6770.

Katagiri, F., Hara, T., Yamada, Y., Urushibata, S., Hozumi, K., Kikkawa, Y., et al. (2014). Biological activities of the homologous loop regions in the laminin α chain LG modules. *Biochemistry, 22*, 3699–3708.

Körner, Z., Fontes-Oliveira, C. C., Holmberg, J., Carmignac, V., & Durbeej, M. (2014). Bortezomib partially improves laminin α2 chain-deficient muscular dystrophy. *American Journal of Pathology, 184*, 1518–1528.

Kuang, W., Xu, H., Vachon, P. H., & Engvall, E. (1998). Merosin-deficient congenital muscular dystrophy. Partial genetic correction in two mouse models. *Journal of Clinical Investigation, 102*, 844–852.

Kuang, W., Xu, H., Vilquin, J. T., & Engvall, E. (1999). Activation of the lama2 gene in muscle regeneration: abortive regeneration in laminin α2-deficiency. *Laboratory Investigation, 79*, 1601–1613.

Kumar, A., Yamauchi, J., Girgenrath, T., & Girgenrath, M. (2011). Muscle-specific expression of insulin-like growth factor 1 improves outcome in Lama2Dy-w mice, a model for congenital muscular dystrophy type 1A. *Human Molecular Genetics, 20*, 2333–2343.

Langenbach, K. J., & Rando, T. A. (2002). Inhibition of dystroglycan binding to laminin disrupts the PI3K/AKT pathway and survival signaling in muscle cells. *Muscle & Nerve, 26*, 644–653.

Leivo, I., & Engvall, E. (1988). Merosin, a protein specific for basement membranes of Schwann cells, striated muscle, and trophoblast, is expressed late in nerve and muscle development. *Proceedings of the National Academy of Sciences of the United States of America, 85*, 1544–1588.

Li, S., Liquari, P., McKee, K. K., Harrison, D., Patel, R., Lee, S., et al. (2005). Laminin-sulfatide binding initiates basement membrane assembly and enables receptor signaling in Schwann cells and fibroblasts. *Journal of Cell Biology, 169*, 179–189.

Li, Z. F., Shelton, G. D., & Engvall, E. (2005). Elimination of myostatin does not combat muscular dystrophy in dy mice but increases postnatal lethality. *American Journal of Pathology, 166*, 491–497.

Lim, L. E., Duclos, F., Broux, O., Bourg, N., Sunada, Y., Allamand, V., et al. (1995). β-sarcoglycan: characterization and role in limb-girdle muscular dystrophy linked to 4q12. *Nature Genetics, 11*, 257–265.

van Lunteren, E., Moyer, M., & Leahy, P. (2006). Gene expression profiling of diaphragm muscle in α2-laminin (merosin)-deficient *dy/dy* dystrophic mice. *Physiological Genomics, 25*, 85–95.

von der Mark, H., Durr, J., Sonnenberg, A., von der Mark, K., Deutzmann, R., & Goodman, S. L. (1991). Skeletal myoblasts utilize a novel β1-series integrin and not α6β1 for binding to the E8 and T8 fragments of laminin. *Journal of Biological Chemistry, 266*, 23593–23601.

Mayer, U. (2003). Integrins: redundant or important players in skeletal muscle? *Journal of Biological Chemistry, 278*, 14587–14590.

Mayer, U., Saher, G., Fässler, R., Bornemann, A., Echtermeyer, F., von der Mark, H., et al. (1997). Absence of integrin α7 causes a novel form of muscular dystrophy. *Nature Genetics, 7*, 318–723.

Mehuron, T., Kumar, A., Duarte, L., Yamauchi, J., Accorsi, A., & Girgenrath, M. (2014). Dysregulation of matricellular proteins is an early signature of pathology in laminin-deficient muscular dystrophy. *Skeletal Muscle, 4*, 14.

Meinen, S., Barzaghi, P., Lin, S., Lochmuller, H., & Rüegg, M. A. (2007). Linker molecules between laminins and dystroglycan ameliorate laminin-α2-deficient muscular dystrophy at all disease stages. *Journal of Cell Biology, 176*, 979–993.

Meinen, S., Lin, S., & Rüegg, M. A. (2012). Angiotensin II type 1 receptor antagonists alleviate muscle pathology in the mouse model for laminin-α2-deficient congenital muscular dystrophy (MDC1A). *Skeletal Muscle, 2*, 18.

Meinen, S., Lin, S., Thurnherr, R., Erb, M., Meier, T., & Rüegg, M. A. (2011). Apoptosis inhibitors and mini-agrin have additive benefits in congenital muscular dystrophy mice. *EMBO Molecular Medicine, 3*, 465–479.

Millay, D. P., Sargent, M. A., Osinska, H., Baines, C. P., Barton, E. R., Vaugniaux, G., et al. (2008). Genetic and pharmacologic inhibition of mitochondrial-dependent necrosis attenuates muscular dystrophy. *Nature Medicine, 14*, 442–447.

Miyagoe, Y., Hanaoka, K., Nonaka, I., Hayasaka, M., Nabeshima, Y., Arahata, K., et al. (1997). Laminin α2 chain-null mutant mice by targeted disruption of the *Lama2* gene: a new model of merosin (laminin 2)-deficient congenital muscular dystrophy. *FEBS Letters, 415*, 33–39.

Moll, J., Barzaghi, P., Lin, S., Bezakova, G., Lochmuller, H., Engvall, E., et al. (2001). An agrin minigene rescues dystrophic symptoms in a mouse model for congenital muscular dystrophy. *Nature, 413*, 302–307.

Mostacciuolo, M. L., Miorin, M., Martinello, F., Angelini, C., Perini, P., & Trevisan, C. P. (1996). Genetic epidemiology of congenital muscular dystrophy in a sample from north-east Italy. *Human Genetics, 97*, 277–279.

Nevo, Y., Halevy, O., Genin, O., Moshe, I., Turgeman, T., Harel, M., et al. (2010). Fibrosis inhibition and muscle histopathology improvement in laminin-2-deficient mice. *Muscle & Nerve, 42*, 218–229.

Nigro, V., de Sá Moreira, E., Piluso, G., Vainzof, M., Belsito, A., Politano, L., et al. (1996). Autosomal recessive limb-girdle muscular dystrophy, LGMD2F, is caused by a mutation in the δ-sarcoglycan gene. *Nature Genetics, 14*, 195–198.

Noguchi, S., McNally, E. M., Ben Othmane, K., Hagiwara, Y., Mizuno, Y., Yoshida, M., et al. (1995). Mutations in the dystrophin-associated protein γ-sarcoglycan in chromosome 13 muscular dystrophy. *Science, 270*, 819–822.

de Oliveira, B. M., Matsumura, C. Y., Fontes-Oliveira, C. C., Gawlik, K. I., Acosta, H., Wernhoff, P., et al. (2014). Quantitative proteomic analysis reveals metabolic alterations, calcium dysregulation, and increased expression of extracellular matrix proteins in laminin α2 chain-deficient muscle. *Molecular and Cellular Proteomics, 13*, 3001–3013.

Patton, B. L., Miner, J. H., Chiu, A. Y., & Sanes, J. R. (1997). Distribution and functions of laminins in the neuromuscular system of developing, adult, and mutant mice. *Journal of Cell Biology, 139*, 1507–1521.

Patton, B. L., Wang, B., Tarumi, Y. S., Seburn, K. L., & Burgess, R. W. (2008). A single point mutation in the LN domain of LAMA2 causes muscular dystrophy and peripheral amyelination. *Journal of Cell Science, 121*, 1593–1604.

Pegoraro, E., Mancias, P., Swerdlow, S. H., Raikow, R. B., Garcia, C., Marks, H., et al. (1996). Congenital muscular dystrophy with primary laminin α2 (merosin) deficiency presenting as inflammatory myopathy. *Annals of Neurology, 40*, 782–791.

Philpot, J., Bagnall, A., King, C., Dubowitz, V., & Muntoni, F. (1999). Feeding problems in merosin deficient congenital muscular dystrophy. *Archives of Disease in Childhood, 6*, 542–647.

Pillers, D. A., Kempton, J. B., Duncan, N. M., Pang, J., Dwinnel, S. J., & Trune, D. R. (2002). Hearing loss in the laminin-deficient dy mouse model of congenital muscular dystrophy. *Molecular Genetics and Metabolism, 76*, 217–224.

Qiao, C., Li, J., Jiang, J., Zhu, X., Wang, B., Li, J., et al. (2008). Myostatin propeptide gene delivery by adeno-associated virus serotype 8 vectors enhances muscle growth and ameliorates dystrophic phenotypes in mdx mice. *Human Gene Therapy, 19*, 241–254.

Qiao, C., Li, J., Zhu, T., Draviam, R., Watkins, S., Ye, X., et al. (2005). Amelioration of laminin-α2-deficient congenital muscular dystrophy by somatic gene transfer of miniagrin. *Proceedings of the National Academy of Sciences of the United States of America, 102*, 11999–12004.

Quijano-Roy, S., Sparks, S., & Rutkowski, A. (2012). LAMA2-related muscular dystrophy. In R. A. Pagon, M. P. Adam, H. H. Ardinger, T. D. Bird, C. R. Dolan, C. T. Fong, et al. (Eds.), *GeneReviews (Internet)* (pp. 1993–2014). WA: University of Washington. Seattle.

Ringelmann, B., Roder, C., Hallmann, R., Maley, M., Davies, M., Grounds, M., et al. (1999). Expression of laminin α1, α2, α4, and α5 chains, fibronectin, and tenascin-C in skeletal muscle of dystrophic 129ReJ dy/dy mice. *Experimental Cell Research, 246*, 165–182.

Roberds, S. L., Leturcq, F., Allamand, V., Piccolo, F., Jeanpierre, M., Anderson, R. D., et al. (1994). Missense mutations in the adhalin gene linked to autosomal recessive muscular dystrophy. *Cell, 78*, 625–633.

Rooney, J. E., Gurpur, P. B., Yablonka-Reuveni, Z., & Burkin, D. J. (2009). Laminin-111 restores regenerative capacity in a mouse model for α7 integrin congenital myopathy. *American Journal of Pathology, 174*, 256–264.

Rooney, J. E., Knapp, J. R., Hodges, B. L., Wuebbles, R. D., & Burkin, D. J. (2012). Laminin-111 protein therapy reduces muscle pathology and improves viability of a mouse model of merosin-deficient congenital muscular dystrophy. *American Journal of Pathology, 180*, 1593–1602.

Rooney, J. E., Welser, J. V., Dechert, M. A., Flintoff-Dye, N. L., Kaufman, S. J., & Burkin, D. J. (2006). Severe muscular dystrophy in mice that lack dystrophin and α7 integrin. *Journal of Cell Science, 119*, 2185–2195.

Rutkowski, A., Bönnemann, C., Brown, S., Thorsteinsdóttir, S., Dominov, J., Rüegg, M. A., et al. (2013). Report on the Myomatrix Conference April 22–24, 2012, University of Nevada, Reno, Nevada, USA. *Neuromuscular Disorders, 2*, 188–191.

Sandri, M. (2010). Autophagy in skeletal muscle. *FEBS Letters, 584*, 1411–1416.

Sanes, J. R. (2003). The basement membrane/basal lamina of skeletal muscle. *Journal of Biological Chemistry, 278*, 12601–12604.

Sasaki, T., Giltay, R., Talts, U., Timpl, R., & Talts, J. F. (2002). Expression and distribution of laminin α1 and α2 chains in embryonic and adult mouse tissues: an immunochemical approach. *Experimental Cell Research, 275*, 185–199.

Smirnov, S. P., McDearmon, E. L., Li, S., Ervasti, J. M., Tryggvason, K., & Yurchenco, P. D. (2002). Contributions of the LG modules and furin processing to laminin-2 functions. *Journal of Biological Chemistry, 277*, 18928–18937.

Song, W. K., Wang, W., Foster, R. F., Bielser, D. A., & Kaufman, S. J. (1992). H36-α7 is a novel integrin α chain that is developmentally regulated during skeletal myogenesis. *Journal of Cell Biology, 117*, 643–657.

Sunada, Y., Bernier, S. M., Kozak, C. A., Yamada, Y., & Campbell, K. P. (1994). Deficiency of merosin in dystrophic dy mice and genetic linkage of laminin M chain to dy locus. *Journal of Biological Chemistry, 269*, 13279–13732.

Sunada, Y., Bernier, S. M., Utani, A., Yamada, Y., & Campbell, K. P. (1995). Identification of a novel mutant transcript of laminin α2 chain gene responsible for muscular dystrophy and dysmyelination in dy2J mice. *Human Molecular Genetics, 4*, 1055–1061.

Sunada, Y., Edgar, T. S., Lotz, B. P., Rust, R. S., & Campbell, K. P. (1995). Merosin-negative congenital muscular dystrophy associated with extensive brain abnormalities. *Neurology, 11*, 2084–2089.
Sztal, T. E., Sonntag, C., Hall, T. E., & Currie, P. D. (2012). Epistatic dissection of laminin-receptor interactions in dystrophic zebrafish muscle. *Human Molecular Genetics, 21*, 4718–4731.
Talts, J. F., Andac, Z., Gohring, W., Brancaccio, A., & Timpl, R. (1999). Binding of the G domains of laminin α1 and α2 chains and perlecan to heparin, sulfatides, α-dystroglycan and several extracellular matrix proteins. *EMBO Journal, 18*, 863–870.
Talts, J. F., Mann, K., Yamada, Y., & Timpl, R. (1998). Structural analysis and proteolytic processing of recombinant G domain of mouse laminin α2 chain. *FEBS Letters, 426*, 71–76.
Talts, J. F., Sasaki, T., Miosge, N., Gohring, W., Mann, K., Mayne, R., et al. (2000). Structural and functional analyses of the recombinant G domain of the laminin α4 chain and its proteolytic processing in tissues. *Journal of Biological Chemistry, 275*, 35192–35199.
Taniguchi, M., Kurahashi, H., Noguchi, S., Sese, J., Okinaga, T., Tsukahara, T., et al. (2006). Expression profiling of muscles from Fukuyama-type congenital muscular dystrophy and laminin-α2 deficient congenital muscular dystrophy; is congenital muscular dystrophy a primary fibrotic disease? *Biochemical and Biophysical Research Communications, 342*, 489–502.
Tisi, D., Talts, J. F., Timpl, R., & Hohenester, E. (2000). Structure of the C-terminal laminin G-like domain pair of the laminin α2 chain harbouring binding sites for α-dystroglycan and heparin. *EMBO Journal, 19*, 1432–1440.
Tomé, F. M., Evangelista, T., Leclerc, A., Sunada, Y., Manole, E., Estournet, B., et al. (1994). Congenital muscular dystrophy with merosin deficiency. *Comptes Rendus de l'Académie des Sciences – Series III, 314*, 351–357.
Tomomura, M., Fujii, T., Sakagami, H., & Tomomura, A. (2011). Serum calcium-decreasing factor, caldecrin ameliorates muscular dystrophy in dy/dy mice. *In Vivo, 25*, 157–164.
Vachon, P. H., Xu, H., Liu, L., Loechel, F., Hayashi, Y., Arahata, K., et al. (1997). Integrins (α7β1) in muscle function and survival. Disrupted expression in merosin-deficient congenital muscular dystrophy. *Journal of Clinical Investigation, 10*, 1870–1881.
Van Ry, P. M., Minogue, P., Hodges, B. L., & Burkin, D. J. (2014). Laminin-111 improves muscle repair in a mouse model of merosin-deficient congenital muscular dystrophy. *Human Molecular Genetics, 23*, 383–396.
Vilquin, J.-T., Guerette, B., Puymirat, J., Yaffe, D., Tomé, F. M. S., Fardeau, M., et al. (1999). Myoblast transplantations lead to the expression of the laminin α2 chain in normal and dystrophic (dy/dy) mouse muscles. *Gene Therapy, 6*, 792–800.
Voit, T., & Tomé, F. S. (2004). The congenital muscular dystrophies. In A. Engel, & C. Franzini-Armstrong (Eds.), *Myology* (pp. 1203–1238). New York: McGraw-Hill.
Vuolteenaho, R., Nissinen, M., Sainio, K., Byers, M., Eddy, R., Hirvonen, H., et al. (1994). Human laminin M chain (merosin): complete primary structure, chromosomal assignment, and expression of the M and A chain in human fetal tissues. *Journal of Cell Biology, 124*, 381–394.
Wagner, W. J., Chang, A. C., Owens, J., Hong, M. J., Brooks, A., & Coligan, J. E. (2000). Aberrant development of thymocytes in mice lacking laminin-2. *Developmental Immunology, 7*, 179–193.
Wallace, G. Q., & McNally, E. M. (2009). Mechanisms of muscle degeneration, regeneration, and repair in the muscular dystrophies. *Annual Review of Physiology, 71*, 37–57.
Wang, C. H., Bönnemann, C. G., Rutkowski, A., Sejersen, T., Bellini, J., Battista, V., et al. (2010). Consensus statement on standard care for congenital muscular dystrophies. *Journal of Child Neurology, 25*, 1559–1581.

Wardrop, K. E., & Dominov, J. A. (2011). Proinflammatory signals and the loss of lymphatic vessel hyaluronan receptor-1 (LYVE-1) in the early pathogenesis of laminin α2-deficient skeletal muscle. *Journal of Histochemistry and Cytochemistry, 59*, 167–179.

Webb, N. J., Wells, T. G., Shahinfar, S., Massaad, R., Dankner, W. M., Lam, C., et al. (2014). A randomized, open-label, dose-response study of losartan in hypertensive children. *Clinical Journal of the American Society of Nephrology, 9*, 1441–1448.

Williams, D. A., Head, S. I., Bakker, A. J., & Stephenson, D. G. (1990). Resting calcium concentrations in isolated skeletal muscle fibers of dystrophic mice. *Journal of Physiology, 428*, 243–256.

Wizemann, H., Garbe, J. H., Friedrich, M. V., Timpl, R., Sasaki, T., & Hohenester, E. (2003). Distinct requirements for heparin and alpha-dystroglycan binding revealed by structure-based mutagenesis of the laminin α2 LG4-LG5 domain pair. *Journal of Molecular Biology, 332*, 635–642.

Xiong, H., Tan, D., Wang, S., Song, S., Yang, H., Gao, K., et al. (2015). Genotype/phenotype analysis in Chinese laminin-α2 deficient congenital muscular dystrophy patients. *Clinical Genetics*. http://dx.doi.org/10.111/cge12366.

Xu, R., Chandrasekharan, K., Yoon, J. H., Camboni, M., & Martin, P. T. (2007). Overexpression of the cytotoxic T cell (CT) carbohydrate inhibits muscular dystrophy in the dy^W mouse model of congenital muscular dystrophy 1A. *American Journal of Pathology, 171*, 181–199.

Xu, H., Christmas, P., Wu, X. R., Wewer, U. M., & Engvall, E. (1994). Defective muscle basement membrane and lack of M-laminin in the dystrophic dy/dy mouse. *Proceedings of the National Academy of Sciences of the United States of America, 91*, 5572–5576.

Xu, H., Wu, X. R., Wewer, U. M., & Engvall, E. (1994). Murine muscular dystrophy caused by a mutation in the laminin α2 (Lama2) gene. *Nature Genetics, 8*, 297–302.

Yamauchi, J., Kumar, A., Duarte, L., Mehuron, T., & Girgenrath, M. (2013). Triggering regeneration and tackling apoptosis: a combinatorial approach to treating congenital muscular dystrophy type 1 A. *Human Molecular Genetics, 22*, 4306–4317.

Yoon, S., Stadler, G., Beermann, M. L., Schmidt, E. V., Windelborn, J. A., Schneiderat, P., et al. (2013). Immortalized myogenic cells from congenital muscular dystrophy type1A patients recapitulate aberrant caspase activation in pathogenesis: a new tool for MDC1A research. *Skeletal Muscle, 3*, 28.

Yoshida-Moriguchi, T., Yu, L., Stalnaker, S. H., Davis, S., Kunz, S., Madson, M., et al. (2010). O-mannosyl phosphorylation of α-dystroglycan is required for laminin binding. *Science, 327*, 88–92.

Yu, Q., Sali, A., Van der Meulen, J., Creeden, B. K., Gordish-Dressman, H., Rutkowski, A., et al. (2013). Omigapil treatment decreases fibrosis and improves respiratory rate in dy(2J) mouse model of congenital muscular dystrophy. *PLoS ONE, 8*, e65468.

Yuasa, K., Fukumoto, S., Kamasaki, Y., Yamada, A., Fukumoto, E., Kanaoka, K., et al. (2004). Laminin α2 is essential for odontoblast differentiation regulating dentin sialoprotein expression. *Journal of Biological Chemistry, 279*, 10286–10292.

Yurchenco, P. D. (2011). Basement membranes: cell scaffoldings and signaling platforms. *Cold Spring Harbor Perspectives in Biology, 1*, 3.

Yurchenco, P. D., Quan, Y., Colognato, H., Mathus, T., Harrison, D., Yamada, Y., et al. (1997). The α chain of laminin-1 is independently secreted and drives secretion of its β- and γ-chain partners. *Proceedings of the National Academy of Sciences of the United States of America, 4*, 10189–10194.

Zhang, Y., Li, H., Lian, Z., & Li, N. (2010). Myofibroblasts protect myoblasts from intrinsic apoptosis associated with differentiation via b1 integrin-PI3K/Akt pathway. *Development Growth and Differentiation, 52*, 725–733.

CHAPTER THREE

Type IV Collagens and Basement Membrane Diseases: Cell Biology and Pathogenic Mechanisms

Mao Mao, Marcel V. Alavi, Cassandre Labelle-Dumais and Douglas B. Gould*
Departments of Ophthalmology and Anatomy, Institute for Human Genetics, UCSF School of Medicine, San Francisco, CA, USA
*Corresponding author: E-mail: gouldd@vision.ucsf.edu

Contents

Current Topics in Membranes, Volume 76
ISSN 1063-5823
http://dx.doi.org/10.1016/bs.ctm.2015.09.002

Abstract

Basement membranes are highly specialized extracellular matrices. Once considered inert scaffolds, basement membranes are now viewed as dynamic and versatile environments that modulate cellular behaviors to regulate tissue development, function, and repair. Increasing evidence suggests that, in addition to providing structural support to neighboring cells, basement membranes serve as reservoirs of growth factors that direct and fine-tune cellular functions. Type IV collagens are a major component of all basement membranes. They evolved along with the earliest multicellular organisms and have been integrated into diverse fundamental biological processes as time and evolution shaped the animal kingdom. The roles of basement membranes in humans are as complex and diverse as their distributions and molecular composition. As a result, basement membrane defects result in multisystem disorders with ambiguous and overlapping boundaries that likely reflect the simultaneous interplay and integration of multiple cellular pathways and processes. Consequently, there will be no single treatment for basement membrane disorders, and therapies are likely to be as varied as the phenotypes. Understanding tissue-specific pathology and the underlying molecular mechanism is the present challenge; personalized medicine will rely upon understanding how a given mutation impacts diverse cellular functions.

1. GENOMIC ORGANIZATION AND PROTEIN STRUCTURE OF TYPE IV COLLAGENS

1.1 Introduction and history

Basement membrane proteins are usually large and insoluble, and early structural and molecular studies were hampered by the limited availability of isolated basement membrane components. Nevertheless, elegant biochemical and electron microscopic studies were fundamental to the current understanding of the molecular nature of type IV collagens. The discovery of type IV collagen was made by Dr Nicholas Kefalides at the University of Chicago while studying proteins extracted from glomerular basement membranes (GBMs) of dogs (Kefalides, 1966). Dr Kefalides described a glycoprotein that accounted for 30% of the basement membrane by weight and whose glycine content was approximately one-third of all amino acids, suggesting that it was a type of collagen. In contrast to collagens isolated from Achilles tendon, this novel type of collagen had abnormally high levels of hydroxyproline and hydroxylysine. Type IV collagens were eventually recognized as a distinct form of collagen in that they have frequent imperfections or interruptions in their triple helical domain and are heavily cross-linked by disulfide- and lysine-derived bonds (Kefalides, 1973). Moreover, unlike fibrillar collagens in which the amino and carboxyl termini are cleaved after being secreted into the extracellular matrix, type IV collagens exist as protomers with intact globular ends (Kefalides, 1973; Minor et al., 1976; Olsen, Alper, & Kefalides, 1973). Rotary shadowing studies revealed that type IV collagens have rod-like structures 380–390 nm in length with a terminal globular domain 8–12 nm in diameter (Timpl, Wiedemann, van Delden, Furthmayr, & Kuhn, 1981). Initially thought to be trimers made up of three identical alpha chains, biosynthetic and protease digestion analyses demonstrated that distinct chains, which were later designated as α1(IV) and α2(IV), exist in a 2:1 ratio in the basement membrane (Crouch, Sage, & Bornstein, 1980; Mayne & Zettergren, 1980; Tryggvason, Robey, & Martin, 1980). Additional alpha chains were later discovered in basement membranes from other tissues (Fagg et al., 1990; Hostikka et al., 1990; Pihlajaniemi, Pohjolainen, & Myers, 1990; Zhou, Ding, Zhao, & Reeders, 1994). In mammals, six distinct but related type IV collagen alpha chains (α1(IV) to α6(IV) encoded by *COL4A1* to *COL4A6* genes, respectively) have been described. Based on similar exon–intron organization, exon sizes, sequence similarities, and shared features of their encoded proteins, *COL4A1*, *COL4A3*, and *COL4A5* belong to the α1-like group, and *COL4A2*, *COL4A4*, and *COL4A6* belong to the α2-like group (Netzer, Suzuki, Itoh, Hudson, & Khalifah, 1998). The α1(IV) chain (or COL4A1) and α2(IV) chain (or COL4A2) are considered the classical type IV collagen alpha chains, as they are present in nearly all basement membranes and have been the most extensively studied (Timpl, 1989). The other four alpha chains have more restricted distributions. For example, type IV collagen networks containing the α3(IV), α4(IV), and α5(IV) chains are present in the inner ear, testis, lung, and glomerular and tubular basement membranes of the kidney, whereas networks composed of the α5(IV) and a6(IV) chains are found in basement membranes of the skin, esophagus, smooth muscle cells, and synovia and in Bowman's capsule in the kidney (Kruegel & Miosge, 2010; Mariyama, Leinonen, Mochizuki,

Tryggvason, & Reeders, 1994; Ninomiya et al., 1995; Sanes, Engvall, Butkowski, & Hunter, 1990; Yoshioka et al., 1994). Moreover, in several tissues there is a developmental switch in type IV collagen network composition whereby the α1(IV) and α2(IV) chains are expressed during development while other chains are acquired later during organogenesis to coexist with or replace the α1(IV) and α2(IV) network (Gunwar et al., 1998; Kalluri, Shield, Todd, Hudson, & Neilson, 1997; Kelley, Sado, & Duncan, 2002). This chapter will primarily focus on COL4A1 and COL4A2, although the general role of type IV collagens will be discussed and specific differences highlighted where appropriate.

1.2 Genomic structure

Type IV collagens are major constituents of basement membranes and have been conserved since the emergence of metazoans over half a billion years ago (Boute et al., 1996; Fidler et al., 2014). The six genes exist as pairs organized in a head-to-head orientation on three different chromosomes where the genes within a pair are transcribed from opposite strands (Momota et al., 1998; Poschl, Pollner, & Kuhn, 1988; Sugimoto, Oohashi, & Ninomiya, 1994). In humans, *COL4A1* and *COL4A2* are located on chromosome 13, *COL4A3* and *COL4A4* on chromosome 2, and *COL4A5* and *COL4A6* on the X chromosome (Figure 1). The corresponding mouse

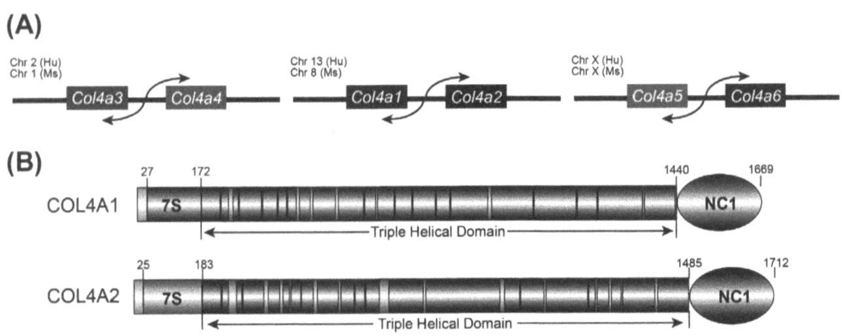

Figure 1 ***Chromosomal arrangements for type IV collagens and domain structures for COL4A1 and COL4A2.*** (A) Human (Hu) and mouse (Ms) type IV collagens are located on three distinct chromosomes as three pairs of genes transcribed from shared bidirectional promoters. (B) Type IV collagens have three functional domains. Following the signal peptide (yellow box), type IV collagens contain a 7S domain at the N-terminus, a triple helical domain and an NC1 domain at the C-terminus. Numbers above the schematics indicate amino acids in human COL4A1 or COL4A2. Gray boxes indicate repeat interruptions in the triple helical domain. Chr, chromosome. The program DOG v2.0 was used to draw the protein structure (Ren et al., 2009). (See color plate)

genes are located on chromosomes 8, 1, and X, respectively. In humans, genes encoding type IV collagens comprise 48–58 exons and span 150,000–290,000 base pairs (bp). Based upon sequence alignments, it is proposed that three independent duplication events facilitated the present genomic organization. Duplication and inversion of a single ancestral gene resulted in the formation of the first head-to-head pair that subsequently diverged. A second duplication event encompassing the entire locus created a second pair. The *COL4A3/COL4A4* gene pair is more divergent from the other two gene pairs suggesting that *COL4A3/COL4A4* was the product of the second duplication event. A third and final duplication later separated the more closely related *COL4A1/COL4A2* and *COL4A5/COL4A6* pairs (Zhou et al., 1994). This genomic head-to-head arrangement of genes that are transcribed in opposite directions is also conserved for the *Col4a1* and *Col4a2* orthologs (*Cg25c* and *viking*) in Drosophila (Yasothornsrikul, Davis, Cramer, Kimbrell, & Dearolf, 1997) and is distinct from the genomic organization of fibrillar collagens, which are dispersed throughout the genome (Myers & Emanuel, 1987).

The paired genes also share a common bidirectional promoter, which ensures the coordinated expression of type IV collagen alpha chains that will form trimeric proteins (Miner & Sanes, 1994; Peissel et al., 1995; Schmidt, Pollner, Poschl, & Kuhn, 1992; Timpl, 1989). In the case of human *COL4A1/COL4A2*, the transcription start sites are separated by a 127 bp promoter region that has a palindromic sequence structure (Poschl et al., 1988; Soininen, Huotari, Hostikka, Prockop, & Tryggvason, 1988). The promoter does not have a canonical TATA box which usually ensures directional transcription (Breathnach & Chambon, 1981). Instead, it contains an A/T-rich region approximately 30 bp upstream of the transcription start sites and three elements, a GC box, a CCAAT box, and a CTC box, that are also found in several other basement membrane proteins, which are binding sites for three distinct transcription factors (Sp1, a CCAAT binding protein, and CTCBF, respectively) (Fischer et al., 1993; Genersch et al., 1995; Schmidt et al., 1993). Additional regulatory elements including enhancers located in the first intron of both *COL4A1* and *COL4A2* are required for transcriptional activity of the bidirectional promoter (Fischer et al., 1993; Pollner, Fischer, Poschl, & Kuhn, 1990), and a downstream silencer in *COL4A2* has been reported (Haniel, Welge-Lussen, Kuhn, & Poschl, 1995). Interactions between *cis*-regulatory elements are proposed to regulate the transcription of human *COL4A1* and *COL4A2* genes (Pollner, Schmidt, Fischer, Kuhn, & Poschl, 1997). The *COL4A3/COL4A4* pair shares many regulatory elements with the *COL4A1/COL4A2* pair (Mariyama, Zheng, Yang-Feng, & Reeders, 1992; Momota et al., 1998). In the case of the *COL4A5/COL4A6* pair, however, an additional promoter region for *COL4A6* has been described (Segal, Zhuang, Rondeau, Sraer, & Zhou, 2001). Unlike other gene pairs, expression of *COL4A5* and *COL4A6* does not always colocalize (Ninomiya et al., 1995). Accordingly, *COL4A6* is transcribed from two distinct promoters in a tissue-specific manner, resulting in transcripts that differ at their amino termini encoding two different signal peptides. Differential promoter usage in a tissue-specific manner accounts, at least partially, for the different expression patterns of *COL4A5 and COL4A6* (Segal et al., 2001; Sugimoto et al., 1994; Sund, Maeshima, & Kalluri, 2005).

1.3 Protein domain structure

The amino acid sequences of the mouse and human COL4A1 and COL4A2 orthologs are highly conserved and the protein structure is shared between paralogs (Brazel et al., 1987; Brazel, Pollner, Oberbaumer, & Kuhn, 1988; Hostikka & Tryggvason, 1988; Muthukumaran, Blumberg, & Kurkinen, 1989; Saus et al., 1989; Soininen, Haka-Risku, Prockop, & Tryggvason, 1987). In addition to an amino terminal signal peptide, type IV collagens contain three major structural domains: the 7S, the triple helical (collagenous) and the globular noncollagenous 1 (NC1) domains (Figure 1). A number of functional subdomains and putative binding sites for potential interacting proteins have also been mapped (Parkin et al., 2011). In humans (UniProt ID P02462) and mice (UniProt ID P02463), COL4A1 is composed of 1669 amino acids with the signal peptide and the 7S, triple helical, and NC1 domains being 27, 145, 1272, and 225 amino acids, respectively. COL4A2 (UniProt ID P08572) in humans comprises 1712 amino acids with the signal peptide and the 7S, triple helical, and NC1 domains being 25, 158, 1302, and 227 amino acids, respectively. In the mouse, the COL4A2 protein (UniProt ID P08122) is slightly shorter owing to five fewer amino acids in the triple helical domain.

During translation, the signal peptide directs the translocation of type IV collagen alpha chains to the endoplasmic reticulum (ER) where it is removed. Following translation, type IV collagens assemble into three types of heterotrimers (called protomers) in the ER (Figure 2). Protomer formation initiates at the carboxyl terminal NC1 domain and proceeds toward the amino terminal 7S domain in a zipper-like fashion. After a series of posttranslational modifications (see Section 2), the type IV collagen

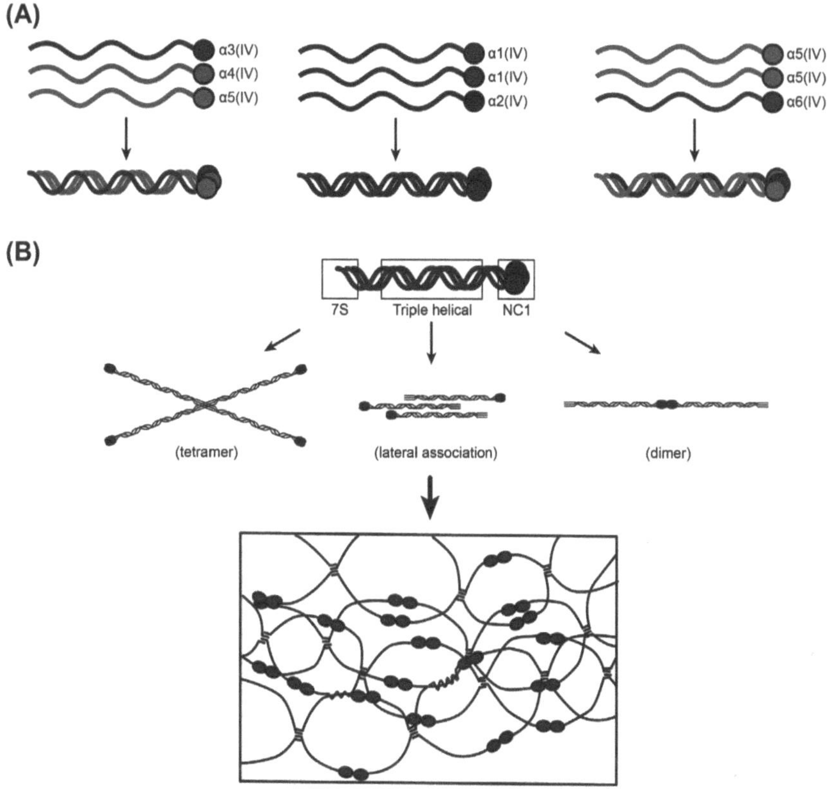

Figure 2 ***Type IV collagen heterotrimer and network formation.*** (A) Among 56 possible combinations, only α1α1α2, α3α4α5, and α5α5α6 heterotrimers are formed. (B) In the basement membrane, type IV collagens form an irregular and complex polygonal network. The 7S domains of collagen molecules interact to form tetramers and the NC1 domains interact to form dimers. Lateral associations along the triple helical domain allow branching and further strengthening of the type IV collagen network. (See color plate)

protomer is secreted into the extracellular space where it self-assembles into an intricate and complex supramolecular network resembling a spider web or chicken wire mesh (Figure 2). The network is formed when NC1 domains from two protomers interact in a head-to-head orientation. At the other end of each protomer, the 7S domains form tetrameric, antiparallel lateral interactions with three other protomers (Timpl et al., 1981). In addition to interactions at the carboxyl and amino termini, lateral associations along the triple helical domain allow branching to occur leading to the formation of irregular and complex polygonal networks (Yurchenco & Ruben, 1987).

1.3.1 7S domain

The 7S domain was first isolated as a large molecular complex resistant to bacterial collagenase digestion (Risteli, Bachinger, Engel, Furthmayr, & Timpl, 1980; Timpl, Risteli, & Bachinger, 1979). The 7S domain was named so because it has a sedimentation coefficient of approximately 7.2 Svedbergs (S) when subjected to ultracentrifugation. Depending on the digestion conditions (37 °C or 20 °C), the 7S domain can appear as a short (225,000 Da at 37 °C) or a long (360,000 Da at 20 °C) form. The long form was later shown to contain a part of the triple helical collagenous domain. Under rotary shadowing electron microscopy, the short form appears as a compact, rectangular, rod-like structure with a size of 30 nm. The long form shares the rectangular structure but has four thinner, 28 nm arms sticking out from the center in a symmetric fashion. Interestingly, the polymeric form of type IV collagen isolated with limited pepsin digestion has a similar organization, with four 328 nm long threads connected at one end to a central structure morphologically similar to the core 7S domain (Kuhn et al., 1981). These observations provided the first evidence that type IV collagen protomers form tetramers by association through their amino terminal 7S domains (Risteli et al., 1980).

The amino acid sequence of the 7S domains of type IV collagens revealed more molecular details (Glanville, Qian, Siebold, Risteli, & Kuhn, 1985; Siebold et al., 1987). The 7S domain starts with a region of approximately 20 amino acids that is enriched in cysteine and lysine residues, followed by a 100-amino acid region that consists of the Gly—Xaa—Yaa triplets typical of a collagenous domain. The amino terminal noncollagenous region in all type IV collagen alpha chains contains four conserved cysteine residues that form intra- and intermolecular disulfide cross-links. In addition to disulfide bonds, cross-links can also form between lysine and hydroxylysine residues. The collagenous region of the 7S domain comprises the antiparallel, lateral overlapping regions of the four aggregating type IV collagen molecules. With the exception of the α4(IV) chain (Leinonen, Mariyama, Mochizuki, Tryggvason, & Reeders, 1994), there is also a fifth cysteine residue in the X position of a Gly—Xaa—Yaa triplet in the collagenous region. The presence of the cysteine residue in a Gly—Xaa—Yaa triplet is extremely rare, and it was proposed to form an intermolecular disulfide bond with one of the four cysteine residues at the amino terminus of an adjacent molecule (Glanville et al., 1985). Isolated type IV collagens can spontaneously oligomerize in vitro through hydrophobic associations of 7S domains, which are eventually stabilized into a tetramer by intermolecular covalent cross-links

(Bachinger, Fessler, & Fessler, 1982; Duncan, Fessler, Bachinger, & Fessler, 1983). In addition to cross-linking and glycosylation sites of the hydroxylysine residues, there is one asparagine residue within the 7S domain that carries an N-linked heteropolysaccharide. The extensive cross-linking and glycosylation are responsible for conferring resistance to bacterial collagenase digestion. Following the 7S domain, there is a short noncollagenous region 5 to 12 residues in length that is thought to provide flexibility during network formation.

1.3.2 Triple helical domain

The triple helical domain comprises the majority of the type IV collagens. This domain constitutes the signature feature of all collagens and consists of Gly—Xaa—Yaa repeats. There is a requirement for glycine at every third amino acid, as the absence of a side chain allows glycine residues to fit into the core of the triple helix (Ramachandran & Kartha, 1955). Xaa and Yaa are often proline and hydroxyproline (Shoulders & Raines, 2009). The triple helical domains of type IV collagens are approximately 1300 residues (1272 and 1302 in human COL4A1 and COL4A2, respectively), slightly larger than the triple helical domain of fibrillar collagens. A notable feature of the triple helical domain of type IV collagens is the presence of short but frequent interruptions of the Gly—Xaa—Yaa triplet repeats. Unlike fibrillar collagens that are highly resistant to proteolytic digestion, type IV collagens isolated from various sources can be digested into fragments of different lengths, suggesting the presence of interruptions within the triple helical domain (Schuppan, Timpl, & Glanville, 1980). The first evidence for the presence of interruptions came from peptide end sequencing of a large fragment of type IV collagen isolated from mouse tumors, in which the eight amino acids at the amino terminus were found not to follow the Gly—Xaa—Yaa pattern (Timpl, Bruckner, & Fietzek, 1979). Subsequent amino acid sequencing analyses confirmed the presence of multiple interruptions (Babel & Glanville, 1984; Brazel et al., 1987; Schuppan, Glanville, & Timpl, 1982; Schuppan, Glanville, Timpl, Dixit, & Kang, 1984; Schuppan et al., 1980). The number of interruptions varies from 21 to 26 between alpha chains (Brazel et al., 1987, 1988; Hostikka & Tryggvason, 1988; Leinonen et al., 1994; Mariyama et al., 1994; Zhou et al., 1994; Zhou, Hertz, Leinonen, & Tryggvason, 1992). Most of the interruptions occur at similar positions, suggesting their functional importance (Leinonen et al., 1994). COL4A1 has 21 interruptions whereas COL4A2 has 23 interruptions, 18 of which are position matched with interruptions in COL4A1. Interruptions vary in length from 1 to 24 residues, and most of the large interruptions occur nearer the amino terminus of the protein and are believed to confer flexibility to a structure that would otherwise be rigid. The triple helical domains of COL4A1 and COL4A2 are devoid of cysteine residues except in interruptions. Three cysteine residues are found in two interruptions of the COL4A1 triple helical domain and four cysteine residues in two interruptions of the COL4A2 triple helical domain. The interruptions are speculated to facilitate lateral associations during type IV collagen network assembly, and the presence of cysteine residues in those interruptions are thought to mediate the formation of interchain cross-linking bridges and strengthen lateral association between triple helical domains (Yurchenco & Furthmayr, 1984; Yurchenco & Ruben, 1987, 1988). Furthermore, some interruptions were shown to serve as cell-binding sites (Vandenberg et al., 1991). Collectively, these findings demonstrate a critical role for repeat interruptions in type IV collagen's supramolecular network organization.

1.3.3 NC1 domain

The NC1 domain is a 12.8 nm globular domain that is located at the carboxyl terminus of type IV collagens (Timpl et al., 1981). NC1 domains have relatively high sequence similarities among all chains (52–69% identity) and for each alpha chain the sequence is highly conserved among orthologs (e.g., 96.9% identity in human vs mouse for COL4A1) (Leinonen et al., 1994; Oberbaumer et al., 1985; Pihlajaniemi et al., 1985; Schwarz-Magdolen, Oberbaumer, & Kuhn, 1986). The NC1 domains can be divided into two homologous halves. Each half contains six conserved cysteine residues in corresponding positions within a highly conserved region, forming three sets of intrachain disulfide bridges within each subdomain (Siebold, Deutzmann, & Kuhn, 1988). NC1 domains are often studied as hexamers, as they can be easily purified using bacterial collagenase digestion of native type IV collagens isolated from basement membranes.

NC1 domains serve multiple critical functions. NC1 domains were implicated as the sites of nucleation for heterotrimer formation by directing heterotrimer formation during reassembly of heat-denatured type IV collagen (Dolz, Engel, & Kuhn, 1988). Removing the NC1 domain by pepsin digestion or disrupting the hexametric structure of NC1 domains with acetic acid severely affected the proper reassembly in vitro. Moreover, the NC1 domains were shown to be responsible for the chain selectivity and molecular stoichiometry of type IV collagen heterotrimers. With six different type IV collagen alpha chains, 56 different combinations of trimeric protomers are theoretically possible. However, only three heterotrimers exist: α1α1α2, α3α4α5, and α5α5α6. The suggestion that NC1 domains may be responsible for chain selectivity came from the observation that dissociated NC1 monomers reassociate in vitro into NC1 hexamers comparable to their native forms, and purified NC1 monomers from α1 to α5 chains mixed in equal moles form only two types of hexamers in vitro (Borza et al., 2001; Boutaud et al., 2000). Crystallography of the NC1 hexamer of the α1α1α2 protomer revealed the structural basis for this interaction (Sundaramoorthy, Meiyappan, Todd, & Hudson, 2002; Than et al., 2002), and structural comparison of the NC1 domains from all six alpha chains across species suggests that the NC1 domains contain the codes for selective chain assembly (Khoshnoodi, Sigmundsson, et al., 2006). This was tested in a subsequent study using mutant NC1 domains to determine the in vitro assembly of the α3α4α5 heterotrimer, in which the 40 residues at the carboxyl terminus of the α5(IV) chain were found to selectively bind to the α3(IV) chain, whereas the 58 residues at the amino terminus of α3(IV) chain are necessary to bind to the α5(IV) chain (Kang et al., 2008). Furthermore, kinetic analyses demonstrated that the NC1 domain of the α2(IV) chain has a higher affinity to the NC1 domain of the α1(IV) chain than to the NC1 domain of the α2(IV) chain (Khoshnoodi, Sigmundsson, et al., 2006). Since the α2(IV), α4(IV), and α6(IV) chains only occur once in their corresponding heterotrimers, it was proposed that the α2(IV)-like chains play a major regulatory role in determining the molecular stoichiometry of the type IV collagen trimers (Khoshnoodi, Cartailler, Alvares, Veis, & Hudson, 2006).

Within the basement membrane, the NC1 domain plays a critical role for network formation and stabilization. Crystal structural analysis suggested that NC1 hexamers are stabilized via an unusual type of covalent cross-link between adjoining heterotrimers (Than et al., 2002). Mass spectrometry confirmed a cross-link between a methionine (Met^{1553} in COL4A1) and a hydroxylysine (Hyl^{1651} in COL4A1) residue of opposing protomers (Vanacore, Friedman, Ham, Sundaramoorthy, & Hudson, 2005; Vanacore et al., 2004). A novel sulfilimine bond (−S=N−) was discovered to cross-link the Met^{1553} residue and the Hyl^{1651} residue (Vanacore et al., 2009). Investigation of the occurrence of the sulfilimine bond in 31 species spanning 11 major phyla revealed that this bond appeared at the time of the divergence of sponge and cnidarian, suggesting its importance in organogenesis (Fidler et al., 2014; Vanacore et al., 2009). Peroxidasin, a heme peroxidase in basement membranes, was later discovered as the enzyme that catalyzes sulfilimine bond formation (Bhave et al., 2012). Like type IV collagens, Peroxidasin also exists since the emergence of metazoans (Ero-Tolliver, Hudson, & Bhave, 2015; Fidler et al., 2014).

2. TYPE IV COLLAGEN BIOSYNTHESIS

Type IV collagen biosynthesis is a complex multistep process that relies on the concerted action of multiple proteins and cofactors (Figure 3). Although the series of biosynthetic events underlying type I collagen maturation and secretion has been studied in more details, much remains to be learned about the events and enzymes controlling type IV collagen synthesis, maturation, and secretion. The following section reviews the general understanding of collagen biosynthesis and discusses how it might relate to type IV collagen.

2.1 Heat shock protein 47

Type IV collagen alpha chains are cotranslationally translocated into the ER where they assemble into defined trimers before reaching the extracellular matrix via the secretory pathway. Multiple folding enzymes and molecular chaperones are required for the successful assembly and secretion of collagens.

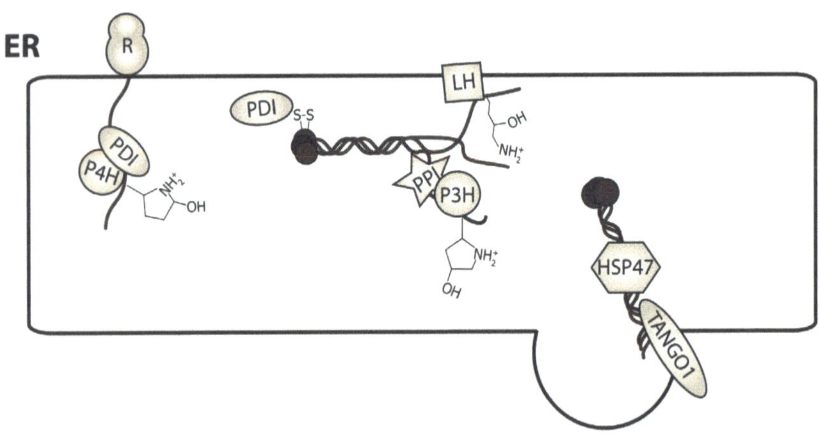

Figure 3 ***Type IV collagen biosynthetic pathway.*** Various enzymes posttranslationally modify nascent type IV collagens, and chemical chaperones prevent their aggregation in the ER. After heterotrimer formation, type IV collagen protomers are packed into specialized cargo vesicles to be transported via the Golgi to the extracellular matrix. HSP47, heat shock protein 47; LH, lysyl hydroxylase; PDI, protein disulfide isomerase; PPI, peptidylprolyl isomerase; P3H, prolyl 3-hydroxylase; P4H, prolyl 4-hydroxylase; R, ribosome; TANGO1, transport and Golgi organization 1. (See color plate)

Among them is heat shock protein 47 (HSP47), which preferentially binds and stabilizes the triple helical region of collagens on their passage from the ER to the Golgi (Koide, Aso, Yorihuzi, & Nagata, 2000; Nagata, 1996; Ono, Miyazaki, Ishida, Uehata, & Nagata, 2012; Tasab, Batten, & Bulleid, 2000). HSP47 binds to type I, II, III, IV, and V collagens in a pH-dependent manner (Natsume, Koide, Yokota, Hirayoshi, & Nagata, 1994; Saga, Nagata, Chen, & Yamada, 1987). The arginine within the Gly–Xaa–Arg sequence and the Yaa residue of the preceding Gly–Xaa–Yaa motif are thought to be required for HSP47 recognition (Koide et al., 2006; Koide, Takahara, Asada, & Nagata, 2002; Tasab, Jenkinson, & Bulleid, 2002). As collagens move into the more acidic Golgi, HSP47 dissociates and is recycled back to the ER (Saga et al., 1987). Recessive mutations in *SERPINH1* (the gene encoding HSP47) cause osteogenesis imperfecta (OI) in patients (Christiansen et al., 2010) and dogs (Lindert et al., 2015). OI is a connective tissue disorder characterized by brittle bones that are prone to fracture and is caused in the majority of cases by dominant mutations in *COL1A1* or *COL1A2* (Barsh, Roush, Bonadio, Byers, & Gelinas, 1985; Chu et al., 1983; Pihlajaniemi et al., 1984). *Hsp47* knockout mice die by embryonic day (E) 11.5 and exhibit reduced secretion of processed type IV and type I collagens and abnormal basement membranes (Nagai et al., 2000). This embryonic phenotype is reminiscent of that observed in mice homozygous for null alleles of *Col4a1* and *Col4a2*, which also die around E11.5 and exhibit basement membrane defects (Harbers, Kuehn, Delius, & Jaenisch, 1984; Lohler, Timpl, & Jaenisch, 1984; Nagai et al., 2000; Poschl et al., 2004). Accordingly, in *Hsp47* knockout embryos, type IV collagen accumulated in the ER and was absent from the basement membrane that was marked by focal disruptions (Marutani, Yamamoto, Nagai, Kubota, & Nagata, 2004). *Hsp47*-deficient cells had a significantly reduced rate of type IV collagen secretion, and the heterotrimers that were successfully secreted were more sensitive to protease digestion, supporting the existence of quantitative and qualitative abnormalities in type IV collagen in the absence of HSP47 (Marutani et al., 2004; Matsuoka et al., 2004).

2.2 Protein disulfide isomerase

Proper trimer formation and secretion of collagens requires posttranslational modifications that result from the coordinated action of multiple enzymes along the secretory pathway. Nascent type IV collagens interact with ER resident proteins to ensure proper assembly, folding, and trafficking. NC1 domains are stabilized by intramolecular cross-links formed by protein disulfide isomerase (PDI) before initiating trimer formation (Doyle & Smith, 1998; Koivu, 1987; Lim, Doyle, Balian, & Smith, 1998). PDI represents one of the most abundant ER resident proteins and is an oxidoreductase of the thioredoxin superfamily with multiple functions. In addition to catalyzing disulfide bond formation and isomerization, it acts as a molecular chaperone. Both functions are essential for proper collagen maturation (Wilkinson & Gilbert, 2004). In *C. elegans*, *PDI* mutations lead to aberrant collagen deposition, severe morphological defects and death (Winter, McCormack, & Page, 2007). Patients with a dominant negative mutation in PDI were reported to have a severe subtype of OI, and their skin fibroblasts had pathological amounts of disulfide bridges between PDI and their substrates, which were associated with ER stress (Rauch et al., 2015). PDI also forms complexes with other collagen-modifying enzymes including prolyl 4-hydroxylases (P4Hs) (see below).

2.3 Peptidylprolyl isomerases

Following folding of the NC1 domain, heterotrimer assembly is initiated and proceeds by the progressive winding of the triple helical domains in a carboxyl-to-amino terminal direction. Next to glycine, proline is the most abundant residue in the triple helical domain of type IV collagens. COL4A1 and COL4A2 have 325 and 286 proline residues, respectively. Proline exists as either a cis or trans isomer in nascent collagen propeptides, and peptidylprolyl isomerases (PPIases) catalyze the conformational change from cis to trans proline, a crucial step in collagen triple helix formation (Bachinger, 1987; Bachinger, Bruckner, Timpl, & Engel, 1978; Bachinger, Bruckner, Timpl, Prockop, & Engel, 1980; Bachinger, Morris, & Davis, 1993; Bruckner & Eikenberry, 1984; Steinmann, Bruckner, & Superti-Furga, 1991). Cyclophilins, FK506 binding proteins (FKBP), and parvulins are the three major PPIase families (reviewed in Schmidpeter & Schmid, 2015). Mutations in cyclophilin and FKBP family members cause recessive forms of OI, highlighting their importance in collagen maturation (Alanay et al., 2010; Barnes et al., 2010; Pyott et al., 2011; van Dijk et al., 2009). Interestingly, PPIase deficiency results in type I collagen overmodification likely because hindered triple helix formation allows more time for other posttranslational modifications to take place (Choi et al., 2009; Morello et al., 2006; Vranka et al., 2010).

2.4 Prolyl 4-hydroxylases

Before the triple helix forms, nascent type IV collagens undergo several posttranslational modifications. Proline residues in the triple helical domain can be hydroxylated at the fourth carbon of the proline ring by P4Hs or at the third carbon by prolyl 3-hydroxylases (P3Hs). Prolyl 4-hydroxylation occurs at the Yaa position of the Gly—Xaa—Yaa sequence motif in collagen and other proteins containing collagen-like domains (Kivirikko & Myllyharju, 1998). Most of the prolines at the Yaa position are hydroxylated (Myllyharju & Kivirikko, 2004), and the proportions of 4-hydroxyprolines (4Hyps) are consistent between different collagen types (Kivirikko, Myllyla, & Pihlajaniemi, 1991; Kivirikko & Pihlajaniemi, 1998). 4Hyps promote electrostatic interactions between collagen chains (reviewed in Shoulders & Raines, 2009), thereby providing thermal stability to the triple helix and allowing collagens to persist at physiological temperatures (Berg & Prockop, 1973; Jimenez, Harsch, & Rosenbloom, 1973; Rosenbloom, Harsch, & Jimenez, 1973). Collagen prolyl 4-hydroxylation is accomplished in the ER lumen by a tetrameric protein complex composed of two α- and two β-subunits. PDI comprises the β-subunits while the α-subunit can vary (Myllyharju, 2008). In *C. elegans* homozygous mutations for either *phy-1* or *phy-2*, encoding two P4H α-subunits, resulted in reduced growth while *phy-1/phy-2* double mutants were embryonic lethal (Friedman et al., 2000). This suggests partial functional redundancy of P4H α-subunits in worms. Mammals have three isoforms for the α-subunit called P4HA1, P4HA2, and P4HA3. P4HA1 is the predominant P4H in most human cell types, while P4HA2 dominates in chondrocytes and capillary endothelial cells (Annunen, Autio-Harmainen, & Kivirikko, 1998; Nissi, Autio-Harmainen, Marttila, Sormunen, & Kivirikko, 2001). Mice heterozygous for a *P4ha1* null allele appeared to be normal, while homozygous mutants had abnormal assembly of type IV collagen and died at E10.5 (Holster et al., 2007). P4HA2-deficient mice had no obvious phenotype (Aro et al., 2015); however, when but in the context of heterozygosity for *P4ha1*, the double mutant mice had severe extracellular matrix abnormalities and chondrodysplasia, supporting a functional redundancy between different P4H isoenzymes (Aro et al., 2015). Less is known about P4HA3; no mutations have been reported in patients, and animal models have not been described.

2.5 Prolyl 3-hydroxylases

Prolyl 3-hydroxylation occurs after prolyl 4-hydroxylation in the Xaa position of a Gly—Xaa—4Hyp sequence motif in the triple helical domain (Gryder, Lamon, & Adams, 1975; Kefalides, 1975; Kresina & Miller, 1979). Prolyl 3-hydroxylation depends on prior prolyl 4-hydroxylation and on the surrounding amino acid context, which limits the number of potential prolyl 3-hydroxylation sites (Tiainen, Pasanen, Sormunen, & Myllyharju, 2008). In general, collagens have far fewer 3-hydroxyprolines (3Hyp) compared to 4Hyps, and the number of 3Hyps varies between tissues and types of collagens (Hudson & Eyre, 2013). Type IV collagens have relatively high amounts of 3Hyps compared to other collagens, with about 6–16 3Hyps per 1000 amino acids in bovine GBMs and other tissues from various species (Dean, Barr, Freytag, & Hudson, 1983; Pokidysheva et al., 2014; Risteli et al., 1980). 3Hyps generate regions of lower stability in the triple helix and may be involved in the binding of other extracellular matrix molecules (Mizuno, Hayashi, Peyton, & Bachinger, 2004). Like P4H, mammals have three P3H isoforms (P3H1–P3H3). P3H1 is part of a multiprotein complex with cartilage-associated protein (CRTAP) and cyclophilin B (CypB), and mutations in all three genes lead to recessive forms of OI (Byers & Pyott, 2012), suggesting that type I collagen is an important P3H1 substrate. P3H2 is strongly expressed in tissues where type IV collagen is abundant and hydroxylates type IV collagen-derived peptides more effectively than type I collagen-derived peptides in vitro (Tiainen et al., 2008). *P3h2* null mice had no obvious phenotypic abnormalities despite a reduction in prolyl 3-hydroxylation levels of type I and type IV collagens in various ocular tissues and tendon (Hudson et al., 2015). Patients with mutations in the *LEPREL1* gene, which encodes P3H2, had increased ocular growth resulting in myopia (Guo et al., 2014; Jiang et al., 2015). The absence of obvious phenotypes in *P3h2* null mice could possibly be explained by potential functional redundancy with P3H3 during development and in specific cell types, as the expression pattern of P3H3 overlaps with those of P3H1 and P3H2 (Vranka, Stadler, & Bachinger, 2009). To date, the precise role of P3H3 remains elusive, as no animal models or human mutations have been reported.

2.6 Lysyl hydroxylases

Lysyl hydroxylation occurs at lysine residues in Gly–Xaa–Lys sequence motifs in the triple helical domain (Yamauchi & Sricholpech, 2012). Hydroxylated lysine residues provide sites for intermolecular cross-links and carbohydrate attachments (Kivirikko & Pihlajaniemi, 1998). The extent of lysyl hydroxylation is highly variable, depends on the type of collagen and is age and tissue-specific (Miller & Gay, 1982). Lysine residues of type IV collagens are highly hydroxylated compared to other types of collagens (Miller & Gay, 1982). *C. elegans* mutant for lysyl hydroxylase showed disrupted processing and secretion of type IV collagen. These worms had contraction-induced body wall detachment similar to that observed in worms with type IV collagen mutations, suggesting that lysyl hydroxylation is important for proper type IV collagen secretion (Norman & Moerman, 2000). Mammals have three lysyl hydroxylases (LH—LH3) encoded by the genes procollagen-lysine 1, 2-oxoglutarate 5-dioxygenase (*PLOD*) 1 to 3 (Yamauchi & Sricholpech, 2012), which are differentially expressed during development (Salo et al., 2006). LH1 deficiency causes Ehlers—Danlos syndrome (Hautala, Heikkinen, Kivirikko, & Myllyla, 1993; Pinnell, Krane, Kenzora, & Glimcher, 1972), and LH2 deficiency causes Bruck syndrome (van der Slot et al., 2003), two connective tissue disorders resembling diseases associated with type III and type I collagen mutations, respectively. LH3 deficiency in a patient resulted in a complex connective tissue disorder with features that overlap with a number of known collagen disorders (Salo et al., 2008). Consistent with findings in *C. elegans*, LH3-deficient mice die around E9.5 and show disrupted basement membranes associated with abnormal type IV collagen processing (Rautavuoma et al., 2004). LH3 differs from LH1 and LH2 in that it not only catalyzes hydroxylation of lysine residues but also subsequent glycosylation of the hydroxylysine to either galactosylhydroxylysyl or glucosylgalactosylhydroxylysyl residues, a process important for type IV collagen secretion and basement membrane formation (Ruotsalainen et al., 2006; Sipila et al., 2007). Interestingly, investigations in distinct *Lh3* mouse mutant lines have demonstrated that the galactosylhydroxylysyl glucosyltransferase (GGT) activity but not the lysine hydroxylase activity of LH3 was essential for the formation of the basement membrane (Ruotsalainen et al., 2006). Mice with a point mutation that blocked the lysine hydroxylase activity but retained most of the GGT activity of LH3 developed normally and had only subtle extracellular matrix defects. In contrast, a hypomorphic *Lh3* mouse mutant line showed disrupted basement membrane formation and embryonic lethality, and the survival rate of mutant embryos was correlated with the GGT activity (Ruotsalainen et al., 2006). These findings were further supported by studies using primary fibroblasts isolated from *Lh3* mutant mice or patients, demonstrating that deficiency in LH3-mediated GGT correlated with abnormal extracellular matrix deposition (Risteli et al., 2009). Of interest, it was recently reported that type IV collagen glycosylation can modulate its interactions with members of the integrin family of cell surface receptors in the extracellular matrix (Stawikowski, Aukszi, Stawikowska, Cudic, & Fields, 2014), which raises the possibility that glycosylation might influence type IV collagen-mediated signaling to regulate cell function and behaviors.

2.7 Transport and Golgi organization 1

Collagen constitute exceptionally large cargo and require specialized trafficking vesicles for subsequent transport to the extracellular space via the Golgi apparatus. Hetero oligomers of TANGO1 (*transport and Golgi organization 1*) and cTAGE5 (*cutaneous T-cell lymphoma-associated antigen 5*) are critical components for the formation of trafficking vesicles (Malhotra & Erlmann, 2011; Malhotra, Erlmann, & Nogueira, 2015). TANGO1 binds cargoes directly or indirectly via its luminal SH3 domains, while its cytoplasmic domain recruits other proteins in order to form extended COPII vesicles for transport of large extracellular matrix molecules including collagens (Saito et al., 2009, 2011). Accordingly, TANGO1 knockout mice showed impaired type I, II, III, IV, VII, and IX collagen secretion, while other extracellular matrix proteins were found to be secreted into the extracellular space (Wilson et al., 2011). As a consequence, collagens accumulated in the ER, leading to the activation of the unfolded protein response pathway (Wilson et al., 2011).

3. TYPE IV COLLAGEN-RELATED PATHOLOGY

3.1 COL4A3–A6-associated pathology

A role for type IV collagen in acquired and inherited human diseases was originally discovered after its implication in Goodpasture disease and Alport syndrome (Hudson, 2004). The involvement of type IV collagen in these two prototypical basement membrane diseases has been recognized for many years and is the subject of excellent reviews (Cosgrove, 2012; Hudson, 2004; Hudson, Tryggvason, Sundaramoorthy, & Neilson, 2003; Kashtan, 1999; Thorner, 2007). Goodpasture disease and Alport syndrome are two distinct disorders that primarily affect the kidney GBM (Hudson, 2004). The GBM is an essential component of the glomerular filtration barrier, and its disruption or dysfunction can lead to loss of renal function and eventually kidney failure.

3.1.1 Goodpasture disease

Goodpasture syndrome is an acquired autoimmune condition first defined in the 1950s by Stanton and Tange to describe pathophysiological features of patients originally reported by Goodpasture in 1919 (Stanton & Tange, 1958). The classic clinical presentation of Goodpasture syndrome is lung hemorrhage associated with rapidly progressive glomerulonephritis that was later demonstrated to be mediated by autoantibodies against the GBM (Lerner, Glassock, & Dixon, 1967; Wilson, Borza, & Hudson, 2002). The term Goodpasture *syndrome* is now used to describe the clinical constellation of glomerulonephritis and pulmonary hemorrhage, irrespective of the underlying cause (Cui & Zhao, 2011), while Goodpasture disease (or anti-GBM disease) is used to describe an organ-specific autoimmune disorder characterized by rapidly progressive glomerulonephritis and pulmonary hemorrhage caused by antibodies against the glomerular and alveolar basement membranes (Cui & Zhao, 2011; Peto & Salama, 2011; Salama, Levy, Lightstone, & Pusey, 2001). The pathogenesis of Goodpasture disease is specifically attributed to the production of antibodies against the NC1 domain of the α3 chain, and to a lesser extent α5 chain of type IV collagen (Kalluri, Sun, Hudson, & Neilson, 1996; Leinonen, Netzer, Boutaud, Gunwar, & Hudson, 1999; Merkel et al., 1996; Pedchenko et al., 2010; Saus, Wieslander, Langeveld, Quinones, & Hudson, 1988; Wieslander et al., 1984), although anti-GBM antibodies could potentially recognize other alpha chains (Pedchenko et al., 2010; Zhao et al., 2009). The specific epitopes of the NC1 domain targeted by autoantibodies are inaccessible in their native hexamer conformation, and it was suggested that environmental insults are required to expose the cryptic epitopes and elicit an immune response triggering disease (Borza et al., 2000; Wieslander et al., 1985).

3.1.2 Alport syndrome

A role for type IV collagens in an inherited genetic disease was subsequently discovered when mutations in *COL4A5*, and later *COL4A3* and *COL4A4*, were found to underlie X-linked and autosomal recessive forms of Alport syndrome, respectively (Barker et al., 1990; Hostikka et al., 1990; Lemmink et al., 1994; Mochizuki et al., 1994). Alport syndrome is characterized by hereditary sensorineural deafness, ocular abnormalities, and progressive glomerulonephritis primarily affecting males (Alport, 1927; Hudson et al., 2003; Kashtan, 1999; Kruegel, Rubel, & Gross, 2013). Progressive hearing loss is a highly penetrant feature of Alport syndrome and usually develops by late childhood or early adolescence (Jais et al., 2003). Ophthalmologic findings include anterior lenticonus characterized by a thin, fragile lens capsule (Choi, Na, Bae, & Roh, 2005; Citirik, Batman, Men, Tuncel, & Zileliogl u, 2007), dot-and-fleck retinopathy (Savige et al., 2010), and temporal retinal thinning (Kruegel et al., 2013; Savige et al., 2015). The presence of ocular abnormalities was found to have prognostic value, as they positively correlate with the development of renal failure before the age of 30 in Alport syndrome patients (Savige et al., 2015; Zhang et al., 2008). The renal manifestations observed in Alport syndrome typically include hematuria, proteinuria, and hypertension. The ultrastructural and histological features of glomerular pathology observed in patients with Alport syndrome include splitting and progressive changes of thickness of the GBM that eventually culminates in end-stage kidney disease (Cosgrove, 2012).

Approximately 85% of Alport syndrome cases are caused by mutations in *COL4A5* (Hudson et al., 2003). Because it is located on the X chromosome, *COL4A5* mutations lead to a highly penetrant disease in hemizygous males while random X-inactivation results in variable disease outcomes in heterozygous females ranging from no disease to deafness and end-stage renal disease (Rheault, 2012). The remaining 15% of Alport cases are caused by mutations in genes coding for COL4A3 and COL4A4 (COL4A5-binding partners) and are autosomal recessive. Heterozygous *COL4A3* or *COL4A4* mutations can also cause autosomal dominant thin basement membrane nephropathy and benign familial hematuria (Kashtan, 1998, 2004; Tryggvason & Patrakka, 2006). The similarities and selectivity of the organs affected in Alport syndrome and Goodpasture disease are consistent with the tissue distributions of the collagen type IV alpha chains underlying these diseases (Kalluri, Gattone, & Hudson, 1998; Kruegel & Miosge, 2010; Ninomiya et al., 1995). During normal development, the $\alpha1\alpha1\alpha2$ network in the GBM is gradually replaced by the $\alpha3\alpha4\alpha5$ network (Hudson et al., 2003; Miner & Sanes, 1994). In Alport syndrome, there is absence of the $\alpha3\alpha4\alpha5$ network and compensatory persistence of the embryonic $\alpha1\alpha1\alpha2$ network. This network is more susceptible to proteolytic degradation compared to the more resistant and heavily cross-linked $\alpha3\alpha4\alpha5$ network, leading to basement membrane damage and renal failure (Cosgrove, 2012; Kruegel et al., 2013). The absence of obvious pathology in the lungs of patients with *COL4A3*, *COL4A4*, and *COL4A5* mutations, an organ severely affected in Goodpasture disease, could be explained by functional redundancy with the $\alpha1\alpha1\alpha2$ type IV collagen network present in the lungs (Gunwar et al., 1991). Mutations in genes coding for COL4A3, COL4A4, and COL4A5 also cause glomerular nephropathy in mice (Cosgrove et al., 1996; Korstanje et al., 2014; Lu et al., 1999; Miner & Sanes, 1996; Rheault et al., 2004) and recapitulate many of the pathophysiological hallmarks of Alport syndrome. While the roles of *COL4A3*, *COL4A4*, and *COL4A5* mutations in human disease are well established, evidence for the contribution of *COL4A6* mutations is lacking except for the observation that large deletions involving both *COL4A5* and *COL4A6* genes are present in rare cases of diffuse leiomyomatosis associated with Alport syndrome (Anker et al., 2003; Garcia-Torres, Cruz, Orozco, Heidet, & Gubler, 2000; Hudson et al., 2003; Thielen et al., 2003; Uliana et al., 2011).

3.2 COL4A1/COL4A2-associated pathology

The first report of what is now known to be a *Col4a1* mutation was the description of a mutant mouse strain called *bruised* (*Bru*) that was identified from an N-ethyl-N-nitrosourea mutagenesis screen (Lyon, Glenister, & West, 1984). While homozygosity for the *Bru* mutation was embryonically lethal, heterozygous mice were smaller than their control littermates and had reduced viability. Those that survived had ocular abnormalities, cerebral hemorrhages, and apparent body bruising. Although initially attributed to a deletion on chromosome 8 (Cattanach, Burtenshaw, Rasberry, & Evans, 1993), *Bru* was later found to be a missense mutation of a conserved glycine residue in the triple helical domain of COL4A1 (p.G627W) (Van Agtmael et al., 2005). Taking advantage of the close proximity and head-to-head arrangement of *Col4a1* and *Col4a2*, a targeted mutagenesis approach was used to inactivate both genes simultaneously and address their functions (Poschl et al., 2004). The targeted mutation deleted exon 1 of *Col4a1* and exons 1−3 of *Col4a2*, generating null alleles for both genes. Mice heterozygous for the *Col4a1* and *Col4a2* null alleles were viable and fertile without any obvious phenotype. Homozygous mutant mice, however, did not survive beyond E12. At E11.5, bleeding in the pericardium, blood vessel dilation, and neuronal ectopia were observed in mutant embryos, implicating defects of the vascular and pial basement membranes, respectively. Furthermore, *Col4a1/Col4a2*-deficient embryos exhibited abnormal vascular development marked by reduced capillary plexus density in the vicinity of the pial basement membrane and fewer and disorganized capillaries invading the neuroectoderm. Although embryonic basement membrane alterations were clearly evident in *Col4a1/Col4a2*-deficient embryos, the most obvious defects were detected in Reichert's membrane, resulting in excessive amounts of maternal blood in the yolk sac cavity. The presence of basement membranes in *Col4a1/Col4a2*-deficient embryos indicates that COL4A1 and COL4A2 are dispensable for the initiation of basement membrane formation but are required for viability (Poschl et al., 2004).

3.2.1 Ocular dysgenesis

Concurrently, independent groups at MRC Harwell, GSF Research Center and The Jackson Laboratory identified *Col4a1* mutations through random chemical mutagenesis (Favor et al., 2007; Gould, Marchant, Savinova, Smith, & John, 2007; Gould et al., 2005; Thaung et al., 2002; Van Agtmael et al., 2005). In all cases, heterozygous mutant mice were identified by virtue of having ocular anterior segment dysgenesis and cataracts. Subsequent evaluations demonstrated that some mutant mice had optic nerve hypoplasia (Gould et al., 2007) and that ocular dysgenesis was associated with elevated intraocular pressures and progressive loss of retinal ganglion cells, modeling glaucoma (Mao et al., 2015; Van Agtmael et al., 2005). Consistent with these observations, patients with *COL4A1* mutations have been reported to have various ocular defects that include cataracts, anterior segment dysgenesis, microphthalmia, optic nerve hypoplasia, and glaucoma (Colin et al., 2014; Coupry et al., 2010; Deml et al., 2014; Livingston et al., 2011; Rodahl et al., 2013; Shah et al., 2012; Sibon et al., 2007; Slavotinek et al., 2014; Tonduti et al., 2012; Xia et al., 2014; Yoneda et al., 2013).

3.2.2 Porencephaly

Although *Col4a1* mutant mice were originally discovered because of ocular anterior segment dysgenesis, subsequent analyses have revealed pathology in multiple organs. The past decade of research has demonstrated that heterozygous, semidominant mutations in genes coding for COL4A1 or COL4A2 can cause a broad spectrum of multisystem disorders in mice and humans. Perhaps the most serious consequences of *COL4A1* and *COL4A2* mutations arise from their role in cerebrovascular disease. Accordingly, the first human disease reported to result from *COL4A1* mutations was porencephaly (Gould et al., 2005). Porencephaly is a rare disease characterized by cerebral white matter lesions and cystic cerebral cavities that often communicate with the lateral ventricles. Mice heterozygous for a semidominant *Col4a1* mutation were shown to develop porencephaly and perinatal brain hemorrhages (Gould et al., 2005), and although the disease is most commonly sporadic, *COL4A1* mutations were found in patients with an apparent autosomal dominant form of familial porencephaly (Breedveld et al., 2006; Gould et al., 2005). Subsequently, a number of de novo and inherited *COL4A1* mutations have been reported in patients with porencephaly (Aguglia et al., 2004; Bilguvar et al., 2009; Breedveld et al., 2006; Colin et al., 2014; Lemmens et al., 2013; Lichtenbelt, Pistorius, De Tollenaer, Mancini, & De Vries, 2012; Livingston et al., 2011; Meuwissen et al., 2011; Niwa et al., 2015; Shah et al., 2012, 2010; Sibon et al., 2007; Takenouchi et al., 2015; Tonduti et al., 2012; Vahedi, Boukobza, et al., 2007; Vahedi, Kubis, et al., 2007; Vermeulen et al., 2011; de Vries et al.,

2009; Yoneda et al., 2013). Although less frequent, mutations in the gene coding for the COL4A1 obligate trimeric partner, COL4A2, were also found to underlie sporadic and inherited porencephaly in patients (Verbeek et al., 2012; Yoneda et al., 2013). Porencephaly is generally attributed to embryonic germinal matrix hemorrhages, and *Col4a1* mutant mice were found to develop intracerebral hemorrhages (ICHs) that were detectable as early as E10.5 and persisted throughout life (Favor et al., 2007; Gould et al., 2005, 2006; Jeanne, Jorgensen, & Gould, 2015). Concomitantly, *Col4a1* mutant mice exhibit cerebrovascular developmental defects characterized by distorted and enlarged blood vessels as well as increased vascular tortuosity and density that preceded subcutaneous hematomas and ICHs that are readily visible at birth. Thus, although a distinct clinical entity, porencephaly likely represents the severe end of the cerebrovascular disease continuum caused by *COL4A1* and *COL4A2* mutations.

3.2.3 Small vessel disease

In addition to porencephaly, fetal ICHs, and aberrant vascular development, *Col4a1* and *Col4a2* mutant mice exhibit highly penetrant multifocal and recurrent ICHs that are consistent with cerebral small vessel disease (Gould et al., 2005, 2006; Jeanne et al., 2015; Van Agtmael et al., 2010). Although multifocal hemorrhages are present in the cerebral cortices of young mice, by 1–3 months of age the lesions are predominantly observed in the basal ganglia. Transmission electron microscopy of cerebral blood vessels also revealed ultrastructural defects including disruptions, splitting, herniation, and focal variations in the thickness of vascular basement membranes (Gould et al., 2006). Furthermore, mice aged for over 8 months developed age-related macroangiopathic lesions that appeared as very large caliber vessels with fibrotic walls that were associated with thrombi and parenchymal bleeding (Jeanne et al., 2015). Reduction in red blood cell number and hemoglobin level leading to anemia has also been reported in *Col4a1* mutant mice (Favor et al., 2007; Jeanne et al., 2015; Van Agtmael et al., 2010). Although anemia could be a direct consequence of cerebral or systemic hemorrhages, other explanations have not been ruled out. Vascular defects in the central nervous system are not restricted to the brain and typically affect the retina, presenting as retinal vascular tortuosity and arteriolar silvering (Gould et al., 2006; Jeanne et al., 2015; Van Agtmael et al., 2010).

Over the past 10 years, numerous patients have been reported with *COL4A1* or *COL4A2* mutations. While the phenotypic spectrum is broad, *COL4A1* and *COL4A2* mutations are most often identified in patients with familial or sporadic forms of small vessel disease with cerebral involvement (Choi, 2015; Falcone, Malik, Dichgans, & Rosand, 2014; Gould et al., 2006; Joutel & Faraci, 2014; Joutel, Haddad, Ratelade, & Nelson, 2015; Kuo, Labelle-Dumais, & Gould, 2012; Yamamoto, Craggs, Baumann, Kalimo, & Kalaria, 2011). Notably, de novo and inherited mutations in *COL4A1* and *COL4A2* cause multifocal and recurrent ICHs in young and old patients (Corlobe et al., 2013; Gunda et al., 2014; Jeanne et al., 2012; Kuo et al., 2012; Vahedi, Kubis, et al., 2007; de Vries & Mancini, 2012; Weng et al., 2012). Furthermore, large-scale genetic studies found positive or suggestive associations for *COL4A1* mutations with a spectrum of defects associated with small vessel disease including arterial calcification (Livingston et al., 2011; O'Donnell et al., 2011), arterial stiffness (Adi et al., 2015; Tarasov et al., 2009), deep ICH (Rannikmae et al., 2015), lacunar ischemic stroke (Rannikmae et al., 2015), reduced white matter volume (Rannikmae et al., 2015), and vascular leukoencephalopathy (Ayrignac et al., 2015; Di Donato, Banchi, Federico, & Dotti, 2014). In one retrospective study of 52 patients with *COL4A1* mutations, stroke occurred in 17.3% of subjects with a mean age at onset of 36 years (Lanfranconi & Markus, 2010). One-third of these subjects had lacunar ischemic strokes and two-thirds had hemorrhagic strokes. Imaging of all subjects showed leukoaraiosis (63.5%), subcortical microbleeds (52.9%), porencephaly (46%), symptomatic intracranial aneurysms (44.4%), enlarged perivascular spaces (19.2%), and lacunar infarctions (13.5%) (Lanfranconi & Markus, 2010). Collectively, these studies have defined the cerebrovascular manifestations observed in patients with *COL4A1* or *COL4A2* mutations and validated these mutations as bona fide causes of cerebral small vessel disease in humans.

3.2.4 Cerebral cortical lamination defects

In addition to and independent from the vascular defects observed in the central nervous system, *Col4a1* mutant mice exhibit structural cerebral cortical malformations and neuronal localization defects (Labelle-Dumais et al., 2011). *Col4a1* mutant mice displayed variable but consistent cerebral cortex lamination defects ranging from mild distortions and ectopia to widespread heterotopia and regions devoid of obvious lamination (Labelle-Dumais et al., 2011; Kuo et al., 2014). Ectopia and disorganized lamination of the *Col4a1* mutant cerebral cortex arose from developmental neuronal migration defects associated with breaches in the pial basement membrane. This finding is in agreement with the presence of neuronal ectopia reported in mice homozygous for the *Col4a1* and *Col4a2* null alleles and points to a role for *Col4a1* in cerebral cortical development (Poschl et al., 2004).

3.2.5 Myopathy

Ocular dysgenesis and cerebral cortical lamination defects, features consistently observed in *Col4a1* mutant mice, represent two of the three pathophysiological hallmarks of a subgroup of congenital muscular dystrophy that includes muscle–eye–brain disease and Walker–Warburg syndrome. Consistent with a role in this class of diseases, *Col4a1* mutant mice have myopathy characterized by elevated serum creatine kinase levels, reduced grip force, and increased numbers of nonperipheral nuclei that are indicative of degenerating and regenerating myofibers (Labelle-Dumais et al., 2011). Two putative *COL4A1* mutations were identified in patients diagnosed with muscle–eye–brain disease/Walker–Warburg syndrome, underscoring a role for type IV collagen in muscle biology and disease. Muscle function depends on the concerted action of myofibers, peripheral nerves, and blood vessels. While their role in the vasculature is well established, COL4A1 and COL4A2 are also present in neural and sarcolemmal basement membranes (Fox et al., 2007; Labelle-Dumais et al., 2011; Ninomiya et al., 1995), but the relative contributions of each of these basement membranes to myopathy remain to be determined. Supporting a role for COL4A1 and COL4A2 in neural basement membranes, the NC1 domains of the α1α1α2 heterotrimer are involved in synaptogenesis at the neuromuscular junction, and *Col4a1* mutant mice exhibit transient synaptic maturation defects in the early postnatal period (Fox et al., 2007). In support of a role for COL4A1 and COL4A2 in muscle myofiber basement membranes, myopathy resulting from *Col4a1* and *Col4a2* mutations has been reported in invertebrates. For instance, in *C. elegans*, type IV collagen homologues emb-9 and let-2 are required for muscle integrity, maintenance and function, and mutations result in contraction-induced muscle fiber ruptures and embryonic lethality (Gupta, Graham, & Kramer, 1997). In addition, reduced expression of the collagen IV-encoding gene *Cg25C* in *Drosophila* led to impaired muscle attachment (Borchiellini, Coulon, & Le Parco, 1996), and *Col4a1* mutant flies showed aberrant organization of larval body wall muscles and centronuclear myopathy of the oviduct muscles, resulting in the gradual development of female infertility (Kelemen-Valkony et al., 2012).

3.2.6 HANAC syndrome and nephropathy

Further supporting a role for COL4A1 in muscle development and disease, six families with *COL4A1* mutations that clustered within a 31-amino acid region of the COL4A1 triple helical domain were reported with a clinical diagnosis of HANAC syndrome (hereditary angiopathy with nephropathy, aneurysms, and muscle cramps) (Alamowitch et al., 2009; Plaisier et al.,

2005, 2010, 2007). In addition to having cerebrovascular defects reminiscent of small vessel disease and retinal tortuosity, HANAC patients typically present with muscle cramps and elevated creatine kinase levels indicative of myopathy. Another cardinal feature of HANAC syndrome is the occurrence of nephropathy. HANAC patients develop renal dysfunction characterized by the presence of multiple cysts and chronic kidney failure with or without hematuria. Consistent with these findings, *Col4a1* mutant mice were found to have renal defects including delayed glomerulogenesis, glomerular cysts in adulthood, as well as periglomerular and perivascular inflammation (Chen et al., 2015; Gould et al., 2006; Van Agtmael et al., 2005). *Col4a1* mutant mice also exhibit impaired renal function characterized by highly penetrant microalbuminuria and hematuria (Chen et al., 2015; Gould et al., 2006). In addition, transmission electron microscopy revealed focal disruptions of the GBM; occasional morphological abnormalities of the glomerular parietal epithelial cells; and focal thickening, splitting, and multilamination of Bowman's capsule's basement membrane (Chen et al., 2015; Gould et al., 2006). In contrast to what is observed in Alport syndrome in which there is a persistence of the $\alpha1\alpha1\alpha2$ network, no changes in the expression and distribution pattern of $\alpha3\alpha4\alpha5$ and $\alpha5\alpha5\alpha6$ networks occurred to compensate for the effects of the mutant $\alpha1\alpha1\alpha2$ network in *Col4a1* mutant mice (Chen et al., 2015; Van Agtmael et al., 2005). Together, these findings indicate that in addition to the $\alpha3\alpha4\alpha5$ and $\alpha5\alpha5\alpha6$ networks, the $\alpha1\alpha1\alpha2$ network is also required for proper renal function.

4. MECHANISMS FOR TYPE IV COLLAGEN-RELATED PATHOLOGY

4.1 Overview

As a consequence of both the abundance and functional importance of glycine residues in the triple helical domain, glycine missense mutations constitute the "signature" collagen mutations. These mutations, or mutations in genes encoding proteins required for trimer biosynthesis, can cause intracellular trimer accumulation and delayed or failed secretion. If accumulated proteins are not efficiently removed by ER-associated degradation or the autophagy–lysosomal pathway, they can lead to activation of the unfolded protein response, ER stress, and cellular dysfunction or death (Bateman, Boot-Handford, & Lamande, 2009; Lamande et al., 1995). Irrespective of whether the accumulated proteins trigger ER stress or are efficiently degraded, failed secretion can contribute to an extracellular collagen deficiency that can alter the structure and function of the extracellular matrix. Alternatively, mutant trimers may be secreted and can have deleterious effects (Bateman et al., 2009; Byers, Wallis, & Willing, 1991; Marini et al., 2007). Thus, the potential pathogenic mechanisms underlying collagen-related disease can be considered broadly in terms of proximal (intracellular) and distal (extracellular) insults. Proximal insults are those related to intracellular protein accumulation, while distal mechanisms comprise both extracellular deficiency and the presence of mutant proteins in the basement membrane. The quantitative or qualitative extracellular defects can have repercussions including perturbations of growth factor signaling and/or altered binding to extracellular matrix components and cell surface receptors. Thus, the potential pathogenic mechanisms are diverse and not mutually exclusive, as there could be a complex interplay between proximal and distal insults taking place at different stages of pathogenesis or in a tissue-specific manner.

4.2 Dominant negative effects of mutant proteins

Understanding the relative roles and potential diversity of proximal and distal insults will dictate therapeutic approaches for patients with *COL4A1* and *COL4A2* mutations. The observation that mice heterozygous for *Col4a1* or *Col4a2* point mutations had multisystem disorders (Chen et al., 2015; Jeanne et al., 2015; Kuo et al., 2012; Van Agtmael et al., 2010, 2005), whereas mice heterozygous for null alleles of both *Col4a1* and *Col4a2* did not have obvious abnormalities (Poschl et al., 2004), suggests that the presence of mutant proteins is required for pathology. While this could be held as support for the pathogenicity of intracellular or extracellular mutant heterotrimers, this observation does not rule out the potential importance of extracellular deficiency in mice with *Col4a1* or *Col4a2* point mutations. It is possible that the intracellular accumulation is not itself toxic but that mutant proteins titrate the levels of extracellular collagen below a pathogenic threshold that is not achieved in mice heterozygous for null mutations. Complementation experiments in *Drosophila* support a mixed hypomorph (deficiency)—antimorph (intracellular toxicity or extracellular disruption) mechanism, as pathology in flies heterozygous for mutations in the *Col4a1* ortholog could be partially suppressed by increasing the dosage of the transgenic wild-type gene (Kelemen-Valkony et al., 2012).

Assuming that COL4A1 and COL4A2 monomers assort randomly in the ER, heterozygous *Col4a1* mutant animals should form at least three different species of heterotrimers. The NC1 domain of COL4A2 may initiate assembly with two, one, or no mutant COL4A1 monomers (designated as $\alpha1^{\star}\alpha1^{\star}\alpha2$, $\alpha1^{\star}\alpha1\alpha2$, and $\alpha1\alpha1\alpha2$), and the relative proportions of these three heterotrimers should be 25%, 50%, and 25%, respectively. Heterozygous *Col4a2* mutant animals should form only two species of heterotrimers ($\alpha1\alpha1\alpha2^{\star}$ and $\alpha1\alpha1\alpha2$) in equal proportions. The potential for mutant proteins to be toxic or disruptive depends on the fates of the mutant heterotrimers. In contrast to their control littermates, E9.5 embryos that were homozygous for a *Col4a1* mutation showed intense intracellular COL4A1 immunolabeling, but little or no COL4A1 was detected in Reichert's membrane (Gould et al., 2005). These data suggest that $\alpha1^{\star}\alpha1^{\star}\alpha2$ heterotrimers (the only possibility in homozygous mutants) are not secreted at levels detectable by immunolabeling. Heterozygous mutant littermates show both intracellular and extracellular labeling. Because these signals can be attributed to $\alpha1^{\star}\alpha1^{\star}\alpha2$ and $\alpha1\alpha1\alpha2$ heterotrimers, respectively, the fate of $\alpha1^{\star}\alpha1\alpha2$ heterotrimers, which constitute up to half of all heterotrimers in heterozygous animals, remains unknown. Together, these data support the potential pathogenicity of intracellular accumulation and extracellular deficiency and leave open the possibility for an extracellular effect of mutant heterotrimers.

4.3 Potential role of ER stress

Elevated intracellular COL4A1 and COL4A2 levels resulting from *COL4A1* and *COL4A2* mutations have been documented in multiple cell types in vitro and in vivo (Firtina et al., 2009; Jeanne et al., 2015, 2012; Kuo et al., 2014; Labelle-Dumais et al., 2011; Murray et al., 2014). However, the extent to which intracellular accumulation of mutant type IV collagen represents a toxic insult contributing to pathology is not clear. In lens epithelial cells, the increased intracellular COL4A1 signal colocalized with ER resident proteins and activated the unfolded protein response (Firtina et al., 2009; Gould et al., 2007). Similar responses have also been detected in the vasculature of *Col4a1* mutant mice (Van Agtmael et al., 2010) and in primary skin fibroblasts from a patient with hemorrhagic stroke and a *COL4A2* mutation (Murray et al., 2014). Moreover, reduced proliferation and increased apoptosis was detected in the patient's fibroblasts. While mutant collagen accumulates and can elicit an ER stress response under some conditions, it was undetectable in other paradigms (Jeanne et al., 2012; Kuo et al., 2014). Thus, the role of ER stress and the unfolded protein response in pathogenesis remains an open question.

4.4 Cell autonomous and noncell autonomous mechanisms

A conditional *Col4a1* mutation that expresses mutant protein in the presence of Cre recombinase was recently generated (Jeanne et al., 2015). In addition to its utility to define the spatial and temporal parameters of *Col4a1*-related pathology, it has the potential to address the relative importance of intracellular and extracellular insults in disease. In the context of conditional mutant protein expression one would expect intracellular insults to behave cell autonomously and extracellular insults to behave noncell autonomously. Vascular endothelial cells, pericytes, and astrocytes contribute to a shared cerebrovascular basement membrane, and the conditional *Col4a1* mutant mouse line was used to test the relative role of each of these cell types in cerebrovascular disease (Jeanne et al., 2015). While astrocytes contributed little to the phenotype, conditional expression of the *Col4a1* mutation in both pericytes and vascular endothelial cells led to ICHs; however, neither cell type alone was able to recapitulate the full phenotype resulting from the equivalent germ line mutation. One interpretation of these data is that there is a cell autonomous effect but that the full phenotype requires simultaneous insults in vascular endothelial cells and pericytes. An alternative conclusion is that an extracellular insult is being partially complemented by normal $\alpha1\alpha1\alpha2$ heterotrimers contributed by the other cell types.

4.5 Genetic background effects suggest mechanistic heterogeneity

Studies addressing the effects of the genetic context on *Col4a1*-related pathology raised important considerations for understanding the relative contributions of proximal and distal insults. Ocular dysgenesis, myopathy, and ICH are all more severe in *Col4a1* mutant mice maintained on a pure C57BL/6J (B6) genetic background than they are in *Col4a1* mutant mice that have been crossed to the CAST/EiJ (CAST) inbred strain for a single generation (called CASTB6F1) (Gould et al., 2007; Jeanne et al., 2015; Labelle-Dumais et al., 2011). These data imply that the CAST background has one or more loci that can suppress pathology caused by *Col4a1* mutations. Two independent genetic screens for modifier loci identified a single interval on CAST chromosome 1 that suppresses ocular dysgenesis (Gould et al., 2007) and myopathy (Mao, Jeanne, and Gould, unpublished). Surprisingly, this locus does not appear to be responsible for ICH suppression by the CAST background (Mao, Jeanne, and Gould, unpublished). The observation that the chromosome 1 locus suppresses ocular dysgenesis and myopathy, but does not suppress ICH, suggests that there may be tissue-specific pathogenic mechanisms and that while ocular dysgenesis and myopathy are likely mechanistically linked, ICH is distinct.

A study using primary fibroblasts from *Col4a1* mutant mice found that B6, but not CASTB6F1, mutant cells had significantly increased intracellular COL4A1 levels (Jeanne et al., 2015). Interestingly, mutant cells from both genetic backgrounds had similar levels of extracellular COL4A1 that were significantly lower than those of control cells. This difference was also observed in vivo in the retinal vasculature of B6 and CASTB6F1 *Col4a1* mutant mice. The ability of the CASTB6F1 background to alleviate intracellular accumulation without changing the extracellular levels points to a role of intracellular toxicity. Together these observations support a model whereby ocular dysgenesis and myopathy may share a pathogenic mechanism that is distinct from that underlying cerebrovascular disease in which proximal insults may be relatively more important than distal insults. However, until the mechanism(s) underlying ICH suppression is identified, it remains possible that the modification of cerebrovascular disease by the CAST background is unrelated to this observation.

4.6 Evidence for allelic heterogeneity and mechanistic heterogeneity

An independent line of investigation that compared the cellular and phenotypic consequences of different mutations in an allelic series extends the contention of tissue-specific mechanistic heterogeneity and supports a conclusion that secreted mutant heterotrimers can be pathogenic (Jeanne et al., 2015; Kuo et al., 2014). Characterization of nine different mutations (seven missense mutations of glycine residues in the triple helical domain—six in COL4A1, one in COL4A2; one missense mutation in the NC1 domain of COL4A1; and the *Col4a1*$^{\Delta ex41}$ allele (Gould et al., 2005) caused by a splice site mutation that skips exon 41 but maintains the open reading frame) demonstrated potential domain- and position-dependent effects on heterotrimer biosynthesis (Figure 4). Intracellular COL4A1 and COL4A2 levels were concordant for each of the alleles with the exception of the *Col4a1*^{S1582P} mutation, which had disproportionately low levels of intracellular COL4A2 (Kuo et al., 2014). Because this mutation is in the NC1 domain of COL4A1, it is likely that the mutant proteins do not bind and sequester COL4A2. In contrast, proteins with mutations in the triple helical domain are expected to be incorporated into heterotrimers and, for those mutations, intracellular COL4A1 levels tended to be higher for mutations

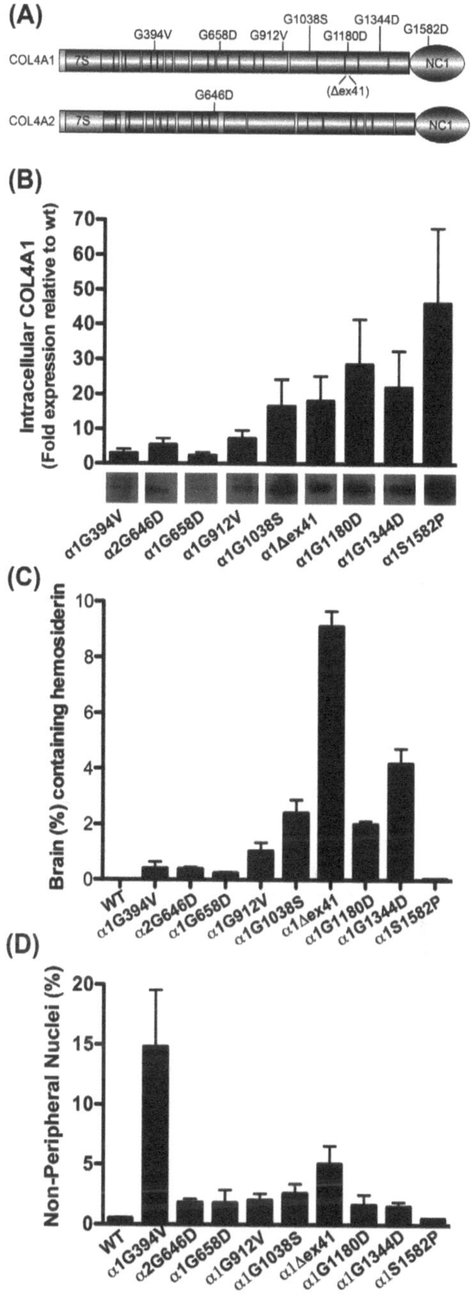

Figure 4 ***Col4a1 and Col4a2 allelic heterogeneity and tissue-specific mechanistic heterogeneity.*** (A) Diagram illustrating the mutations reported in the allelic series studies. (B) Mutations nearer the NC1 domain had the greatest intracellular COL4A1 accumulation. (C) Quantification for intracerebral hemorrhages revealed that the *Col4a1*$^{+/\Delta ex41}$ mutation leads to the most severe phenotype and that point mutations in the triple helix-forming domain nearer the carboxyl terminus tended to cause more hemorrhages. (D) Quantification of nonperipheral nuclei revealed that the *Col4a1*^{G394V} mutation, which is in an integrin-binding domain, causes the most severe myopathy. *Figures modified from Jeanne et al. (2015), Kuo et al. (2014).*

nearer the carboxyl termini (*Col4a1*^{G1038S}, *Col4a1*$^{\Delta ex41}$, *Col4a1*^{G1180D}, and *Col4a1*^{G1344D}) compared to mutations nearer the amino termini (*Col4a1*^{G394V}, *Col4a2*^{G646D}, *Col4a1*^{G658D}, and *Col4a1*^{G912V}). If one assumes that α1α1α2 heterotrimers are uniformly produced and secreted across all mutations, then the allelic differences in intracellular and extracellular COL4A1 levels between mutations are explained by the relative success with which α1*α1α2 and α1*α1*α2 heterotrimers are secreted, implying that mutant heterotrimers can be secreted and may have pathogenic implications. Definitive evidence for the secretion of mutant heterotrimers was reported recently when mice homozygous for a *Col4a1*^{G498V} mutation were shown to be viable and to have secreted mutant COL4A1 in basement membranes (Chen et al., 2015).

Comparing the severity of ICHs in aged mice in this allelic series confirmed the impact of allelic heterogeneity and extended the genotype/phenotype correlations (Jeanne et al., 2015) (Figure 4). First, the NC1 domain mutation (*Col4a1*^{S1582P}) caused less severe cerebrovascular disease than did the triple helical domain mutations, supporting the differential effect of mutations in distinct domains. Second, for point mutations within the triple helical domain, there was a position effect whereby mutations nearer the carboxyl termini caused more severe ICH than mutations nearer the amino termini. In this regard, this class of mutations behaves like a graded series in which ICH severity is correlated with levels of COL4A1 intracellular accumulation. Third, there appears to be a "class effect" whereby the *Col4a1*$^{\Delta ex41}$ mutation that skips 17 amino acids from the triple helical domain is more severe than missense mutations. Notably, this disproportionate effect includes the *Col4a1*^{G1180D} mutation, which is located within exon 41 and had similar levels of intracellular accumulation. Similar genotype/phenotype correlations have been described previously with other types of collagens and can even extend further to include the type of the amino acid that replaces glycines, with amino acids with charged or branched side chains being more disruptive to the trimer assembly process (Bateman et al., 2009; Byers et al., 1991; Kuivaniemi, Tromp, & Prockop, 1991; Marini et al., 2007). Another study of an allelic series of *Col4a1* mutations suggested that pathology may also be milder for mutations in amino acids occurring in Xaa or Yaa positions (Van Agtmael et al., 2005).

An interesting discrepancy arose when the effect of allelic heterogeneity on the severity of myopathy was evaluated (Kuo et al., 2014). Similar to the effect of allelic heterogeneity on ICH, domain and class effects were observed. Mice with a mutation in the NC1 domain (*Col4a1*^{S1582P}) were indistinguishable from control animals, while myopathy resulting from the *Col4a1*$^{\Delta ex41}$ mutation was more severe than that resulting from the positionally matched point mutation (*Col4a1*^{G1180D}). However, in contrast to the trends observed in ICH quantification, there was no apparent position effect for glycine missense mutations within the triple helical domain (Figure 4). Most surprisingly, *Col4a1*^{G394V}, which was among the mutations with the least intracellular accumulation and mildest ICH, caused the most severe myopathy (Kuo et al., 2014). This clearly shows that myopathy severity does not correlate with intracellular COL4A1 or COL4A2 levels; however, this finding has clinical relevance, as this mutation occurs adjacent to putative integrin-binding domains present in COL4A1. The mutations that cause HANAC syndrome, which typically include myopathy, also cluster within a 31-amino acid region of the COL4A1 triple helical domain that encompasses nearby putative integrin-binding sites (Plaisier et al., 2010). Together, these observations support the existence of one or more functional subdomains near the amino terminus of the triple helical domain that are disproportionately important for myopathy but not for ICH. The clear discordance for the *Col4a1*^{G394V} mutation on ICH and myopathy underscores mechanistic heterogeneity for *Col4a1*-related diseases. The fact that this mutant is efficiently secreted and that mice that are homozygous for a nearby mutation are viable indicates that these mutations act by an extracellular mechanism that may involve cell—matrix interactions.

4.7 Development of mechanism-based therapies

Much still remains to be discovered about the relative contributions and diversity of proximal and distal mechanisms underlying multisystem disorders caused by *Col4a1* and *Col4a2* mutations. Identifying and understanding the nature, role, and relative importance of these insults in diseases is critical for the development of targeted therapeutic interventions in patients with *COL4A1* and *COL4A2* mutations. The evidence to date supports roles for both proximal and distal insults in COL4A1-related pathology. Providing further support for an important role of distal insults, mutations in genes coding for laminins, another major class of basement membrane proteins, can cause diverse pathologies that overlap with COL4A1- and COL4A2-related diseases (Barak et al., 2011; Chen et al., 2013; Gawlik et al., 2006; Helbling-Leclerc et al., 1995; Menezes et al., 2014; Miyagoe et al., 1997; Radner et al., 2013; Willem et al., 2002; Xu, Christmas, Wu, Wewer, & Engvall, 1994; Yao, Chen, Norris, & Strickland, 2014; Zenker et al., 2004). In addition, there is a significant overlap in the pathologies described in *Col4a1* and *Col4a2* mutant mice and those reported with mutations in collagen network-forming enzymes (Khan et al., 2011; Yan et al., 2014), other basement membrane collagens (Marneros & Olsen, 2005), nonbasement membrane collagens (Aikio et al., 2013; Ylikarppa et al., 2003), growth factors, and cell surface receptors (Beggs et al., 2003; Cohn et al., 2002; Ervasti & Campbell, 1993; Hayashi et al., 1998; Luo et al., 2011; Moore et al., 2002; Niewmierzycka, Mills, St-Arnaud, Dedhar, & Reichardt, 2005; Rooney et al., 2006; Schmid & Anton, 2003). Not only does this vast spectrum of matrix-associated diseases support the importance of extracellular insults, it further emphasizes the potential diversity of pathogenic mechanisms that can result from distal insults. Determining the identities of the suppressor genes will also be important for understanding these pathogenic mechanisms and will provide guidance as how to circumvent or overcome their detrimental effects therapeutically.

A critical observation emerged from experiments conducted in *C. elegans* that may foreshadow translational benefits for patients. Mutations in the *Col4a1* and *Col4a2* orthologs in worms caused intracellular accumulation of the proteins at the expense of their secretion, just as they do in mammals (Guo, Johnson, & Kramer, 1991; Gupta et al., 1997; Sibley, Graham, von Mende, & Kramer, 1994). The consequence was contraction-induced detachment of the body wall muscles leading to larval death. A key experiment demonstrated that rearing the animals in conditions that promote protein folding was sufficient to decrease intracellular accumulation, restore secretion of mutant collagen, and rescue muscle integrity and viability of mutant animals that would have otherwise died (Guo et al., 1991; Gupta et al., 1997; Sibley et al., 1994). The significance of this finding is that if mutant proteins are folded and secreted, muscle pathology and death are prevented. 4-phenylbutyrate (4PBA) is an FDA-approved drug that can prevent aggregation of misfolded proteins associated with human diseases (de Almeida et al., 2007; Bonapace, Waheed, Shah, & Sly, 2004; Iannitti & Palmieri, 2011; Ozcan et al., 2006; Perlmutter, 2002; Welch & Brown, 1996; Zode et al., 2011). When applied to mutant mouse or patient cells in vitro, 4PBA decreased intracellular and increased extracellular COL4A1 and COL4A2 levels in mutant cells compared to their untreated counterparts (Kuo et al., 2014; Jeanne et al., 2015; Murray et al., 2014). Moreover, 4PBA improved COL4A1 secretion and reduced ICH in vivo in mice that were treated from birth to 1 month of age (Jeanne et al., 2015). 4PBA-treated *Col4a1*$^{+/\Delta ex41}$ mice had significantly milder ICH compared to untreated *Col4a1*$^{+/\Delta ex41}$ littermates. Collectively, these data are consistent with the allelic series and 4PBA treatment acting as genetic and pharmacologic rheostats controlling heterotrimer biosynthesis efficiency and ICH severity. While chemical chaperones are promising and may prove to be effective in alleviating proximal insults and quantitative distal insults, they are not expected to be efficacious—and may even be harmful—for alleles or phenotypes that result from qualitative distal insults. Thus, before targeted therapeutics can be developed to treat patients with *COL4A1* and *COL4A2* mutations, there is a critical need for continued exploration and comprehensive understanding of the complex interplay of pathogenic mechanisms underlying each component of these multisystem disorders.

REFERENCES

Adi, D., Xie, X., Xiang, Y., Ma, Y. T., Yang, Y. N., Fu, Z. Y., et al. (2015). Polymorphisms of COL4A1 gene are associated with arterial pulse wave velocity in healthy Han Chinese and Uygur subjects. *International Journal of Clinical and Experimental Medicine, 8*(2), 2693–2701.

Aguglia, U., Gambardella, A., Breedveld, G. J., Oliveri, R. L., Le Piane, E., Messina, D., et al. (2004). Suggestive evidence for linkage to chromosome 13qter for autosomal dominant type 1 porencephaly. *Neurology, 62*(9), 1613–1615.

Aikio, M., Hurskainen, M., Brideau, G., Hagg, P., Sormunen, R., Heljasvaara, R., et al. (2013). Collagen XVIII short isoform is critical for retinal vascularization, and overexpression of the Tsp-1 domain affects eye growth and cataract formation. *Investigative Ophthalmology & Visual Science, 54*(12), 7450–7462. http://dx.doi.org/10.1167/iovs.13-13039.

Alamowitch, S., Plaisier, E., Favrole, P., Prost, C., Chen, Z., Van Agtmael, T., et al. (2009). Cerebrovascular disease related to COL4A1 mutations in HANAC syndrome. *Neurology, 73*(22), 1873–1882. http://dx.doi.org/10.1212/WNL.0b013e3181c3fd12.

Alanay, Y., Avaygan, H., Camacho, N., Utine, G. E., Boduroglu, K., Aktas, D., et al. (2010). Mutations in the gene encoding the RER protein FKBP65 cause autosomal-recessive osteogenesis imperfecta. *American Journal of Human Genetics, 86*(4), 551–559. http://dx.doi.org/10.1016/j.ajhg.2010.02.022.

de Almeida, S. F., Picarote, G., Fleming, J. V., Carmo-Fonseca, M., Azevedo, J. E., & de Sousa, M. (2007). Chemical chaperones reduce endoplasmic reticulum stress and prevent mutant HFE aggregate formation. *Journal of Biological Chemistry, 282*(38), 27905–27912. http://dx.doi.org/10.1074/jbc.M702672200.

Alport, A. C. (1927). Hereditary familial congenital haemorrhagic nephritis. *British Medical Journal, 1*(3454), 504–506.

Anker, M. C., Arnemann, J., Neumann, K., Ahrens, P., Schmidt, H., & Konig, R. (2003). Alport syndrome with diffuse leiomyomatosis. *American Journal of Medical Genetics, Part A, 119A*(3), 381–385. http://dx.doi.org/10.1002/ajmg.a.20019.

Annunen, P., Autio-Harmainen, H., & Kivirikko, K. I. (1998). The novel type II prolyl 4-hydroxylase is the main enzyme form in chondrocytes and capillary endothelial cells, whereas the type I enzyme predominates in most cells. *Journal of Biological Chemistry, 273*(11), 5989–5992.

Aro, E., Salo, A. M., Khatri, R., Finnila, M., Miinalainen, I., Sormunen, R., et al. (2015). Severe extracellular matrix abnormalities and chondrodysplasia in mice lacking collagen prolyl 4-hydroxylase isoenzyme II in combination with a reduced amount of isoenzyme I. *Journal of Biological Chemistry, 290*(27), 16964–16978. http://dx.doi.org/10.1074/jbc.M115.662635.

Ayrignac, X., Carra-Dalliere, C., Menjot de Champfleur, N., Denier, C., Aubourg, P., Bellesme, C., et al. (2015). Adult-onset genetic leukoencephalopathies: a MRI pattern-based approach in a comprehensive study of 154 patients. *Brain, 138*(Pt 2), 284–292. http://dx.doi.org/10.1093/brain/awu353.

Babel, W., & Glanville, R. W. (1984). Structure of human-basement-membrane (type IV) collagen. Complete amino-acid sequence of a 914-residue-long pepsin fragment from the alpha 1(IV) chain. *European Journal of Biochemistry, 143*(3), 545–556.

Bachinger, H. P. (1987). The influence of peptidyl-prolyl cis-trans isomerase on the in vitro folding of type III collagen. *Journal of Biological Chemistry, 262*(35), 17144–17148.

Bachinger, H. P., Bruckner, P., Timpl, R., & Engel, J. (1978). The role of cis-trans isomerization of peptide bonds in the coil leads to and comes from triple helix conversion of collagen. *European Journal of Biochemistry, 90*(3), 605–613.

Bachinger, H. P., Bruckner, P., Timpl, R., Prockop, D. J., & Engel, J. (1980). Folding mechanism of the triple helix in type-III collagen and type-III pN-collagen. Role of disulfide bridges and peptide bond isomerization. *European Journal of Biochemistry, 106*(2), 619–632.

Bachinger, H. P., Fessler, L. I., & Fessler, J. H. (1982). Mouse procollagen IV. Characterization and supramolecular association. *Journal of Biological Chemistry, 257*(16), 9796–9803.

Bachinger, H. P., Morris, N. P., & Davis, J. M. (1993). Thermal stability and folding of the collagen triple helix and the effects of mutations in osteogenesis imperfecta on the triple helix of type I collagen. *American Journal of Medical Genetics, 45*(2), 152–162. http://dx.doi.org/10.1002/ajmg.1320450204.

Barak, T., Kwan, K. Y., Louvi, A., Demirbilek, V., Saygi, S., Tuysuz, B., et al. (2011). Recessive LAMC3 mutations cause malformations of occipital cortical development. *Nature Genetics, 43*(6), 590–594. http://dx.doi.org/10.1038/ng.836.

Barker, D. F., Hostikka, S. L., Zhou, J., Chow, L. T., Oliphant, A. R., Gerken, S. C., et al. (1990). Identification of mutations in the COL4A5 collagen gene in Alport syndrome. *Science, 248*(4960), 1224–1227.

Barnes, A. M., Carter, E. M., Cabral, W. A., Weis, M., Chang, W., Makareeva, E., et al. (2010). Lack of cyclophilin B in osteogenesis imperfecta with normal collagen folding. *New England Journal of Medicine, 362*(6), 521–528. http://dx.doi.org/10.1056/NEJMoa0907705.

Barsh, G. S., Roush, C. L., Bonadio, J., Byers, P. H., & Gelinas, R. E. (1985). Intron-mediated recombination may cause a deletion in an alpha 1 type I collagen chain in a lethal form of osteogenesis imperfecta. *Proceedings of the National Academy of Sciences of the United States of America, 82*(9), 2870–2874.

Bateman, J. F., Boot-Handford, R. P., & Lamande, S. R. (2009). Genetic diseases of connective tissues: cellular and extracellular effects of ECM mutations. *Nature Reviews Genetics, 10*(3), 173–183. http://dx.doi.org/10.1038/nrg2520.

Beggs, H. E., Schahin-Reed, D., Zang, K., Goebbels, S., Nave, K. A., Gorski, J., et al. (2003). FAK deficiency in cells contributing to the basal lamina results in cortical abnormalities resembling congenital muscular dystrophies. *Neuron, 40*(3), 501–514.

Berg, R. A., & Prockop, D. J. (1973). The thermal transition of a non-hydroxylated form of collagen. Evidence for a role for hydroxyproline in stabilizing the triple-helix of collagen. *Biochemical and Biophysical Research Communications, 52*(1), 115–120.

Bhave, G., Cummings, C. F., Vanacore, R. M., Kumagai-Cresse, C., Ero-Tolliver, I. A., Rafi, M., et al. (2012). Peroxidasin forms sulfilimine chemical bonds using hypohalous acids in tissue genesis. *Nature Chemical Biology, 8*(9), 784–790. http://dx.doi.org/10.1038/nchembio.1038.

Bilguvar, K., DiLuna, M. L., Bizzarro, M. J., Bayri, Y., Schneider, K. C., Lifton, R. P., &, Breastfeeding Trial, G. (2009). COL4A1 mutation in preterm intraventricular hemorrhage. *Journal of Pediatrics, 155*(5), 743–745. http://dx.doi.org/10.1016/j.jpeds.2009.04.014.

Bonapace, G., Waheed, A., Shah, G. N., & Sly, W. S. (2004). Chemical chaperones protect from effects of apoptosis-inducing mutation in carbonic anhydrase IV identified in retinitis pigmentosa 17. *Proceedings of the National Academy of Sciences of the United States of America, 101*(33), 12300–12305. http://dx.doi.org/10.1073/pnas.0404764101.

Borchiellini, C., Coulon, J., & Le Parco, Y. (1996). The function of type IV collagen during *Drosophila* muscle development. *Mechanisms of Development, 58*(1–2), 179–191.

Borza, D. B., Bondar, O., Ninomiya, Y., Sado, Y., Naito, I., Todd, P., et al. (2001). The NC1 domain of collagen IV encodes a novel network composed of the alpha 1, alpha 2, alpha 5, and alpha 6 chains in smooth muscle basement membranes. *Journal of Biological Chemistry, 276*(30), 28532–28540. http://dx.doi.org/10.1074/jbc.M103690200.

Borza, D. B., Netzer, K. O., Leinonen, A., Todd, P., Cervera, J., Saus, J., et al. (2000). The goodpasture autoantigen. Identification of multiple cryptic epitopes on the NC1 domain of the alpha3(IV) collagen chain. *Journal of Biological Chemistry, 275*(8), 6030–6037.

Boutaud, A., Borza, D. B., Bondar, O., Gunwar, S., Netzer, K. O., Singh, N., et al. (2000). Type IV collagen of the glomerular basement membrane. Evidence that the chain specificity of network assembly is encoded by the noncollagenous NC1 domains. *Journal of Biological Chemistry, 275*(39), 30716–30724. http://dx.doi.org/10.1074/jbc.M004569200.

Boute, N., Exposito, J. Y., Boury-Esnault, N., Vacelet, J., Noro, N., et al. (1996). Type IV collagen in sponges, the missing link in basement membrane ubiquity. *Biological Cell, 88*(1–2), 37–44.

Brazel, D., Oberbaumer, I., Dieringer, H., Babel, W., Glanville, R. W., Deutzmann, R., et al. (1987). Completion of the amino acid sequence of the alpha 1 chain of human basement membrane collagen (type IV) reveals 21 non-triplet interruptions located within the collagenous domain. *European Journal of Biochemistry, 168*(3), 529–536.

Brazel, D., Pollner, R., Oberbaumer, I., & Kuhn, K. (1988). Human basement membrane collagen (type IV). The amino acid sequence of the alpha 2(IV) chain and its comparison with the alpha 1(IV) chain reveals deletions in the alpha 1(IV) chain. *European Journal of Biochemistry, 172*(1), 35–42.

Breathnach, R., & Chambon, P. (1981). Organization and expression of eucaryotic split genes coding for proteins. *Annual Review of Biochemistry, 50,* 349–383. http://dx.doi.org/10.1146/annurev.bi.50.070181.002025.

Breedveld, G., de Coo, I. F., Lequin, M. H., Arts, W. F., Heutink, P., Gould, D. B., et al. (2006). Novel mutations in three families confirm a major role of COL4A1 in hereditary porencephaly. *Journal of Medical Genetics, 43*(6), 490–495. http://dx.doi.org/10.1136/jmg.2005.035584.

Bruckner, P., & Eikenberry, E. F. (1984). Formation of the triple helix of type I procollagen in cellulo. Temperature-dependent kinetics support a model based on cis in equilibrium trans isomerization of peptide bonds. *European Journal of Biochemistry, 140*(2), 391–395.

Byers, P. H., & Pyott, S. M. (2012). Recessively inherited forms of osteogenesis imperfecta. *Annual Review of Genetics, 46,* 475–497. http://dx.doi.org/10.1146/annurev-genet-110711-155608.

Byers, P. H., Wallis, G. A., & Willing, M. C. (1991). Osteogenesis imperfecta: translation of mutation to phenotype. *Journal of Medical Genetics, 28*(7), 433–442.

Cattanach, B. M., Burtenshaw, M. D., Rasberry, C., & Evans, E. P. (1993). Large deletions and other gross forms of chromosome imbalance compatible with viability and fertility in the mouse. *Nature Genetics, 3*(1), 56–61. http://dx.doi.org/10.1038/ng0193-56.

Chen, Z., Migeon, T., Verpont, M. C., Zaidan, M., Sado, Y., Kerjaschki, D., et al. (2015). HANAC syndrome Col4a1 mutation causes neonate glomerular hyperpermeability and adult glomerulocystic kidney disease. *Journal of the American Society of Nephrology*. http://dx.doi.org/10.1681/ASN.2014121217.

Chen, Z. L., Yao, Y., Norris, E. H., Kruyer, A., Jno-Charles, O., Akhmerov, A., et al. (2013). Ablation of astrocytic laminin impairs vascular smooth muscle cell function and leads to hemorrhagic stroke. *Journal of Cell Biology, 202*(2), 381–395. http://dx.doi.org/10.1083/jcb.201212032.

Choi, J., Na, K., Bae, S., & Roh, G. (2005). Anterior lens capsule abnormalities in Alport syndrome. *Korean Journal of Ophthalmology, 19*(1), 84–89.

Choi, J. C. (2015). Genetics of cerebral small vessel disease. *Journal of Stroke, 17*(1), 7–16. http://dx.doi.org/10.5853/jos.2015.17.1.7.

Choi, J. W., Sutor, S. L., Lindquist, L., Evans, G. L., Madden, B. J., Bergen, H. R., 3rd, et al. (2009). Severe osteogenesis imperfecta in cyclophilin B-deficient mice. *PLoS Genetics, 5*(12), e1000750. http://dx.doi.org/10.1371/journal.pgen.1000750.

Christiansen, H. E., Schwarze, U., Pyott, S. M., AlSwaid, A., Al Balwi, M., Alrasheed, S., et al. (2010). Homozygosity for a missense mutation in SERPINH1, which encodes the collagen chaperone protein HSP47, results in severe recessive osteogenesis imperfecta. *American Journal of Human Genetics, 86*(3), 389–398. http://dx.doi.org/10.1016/j.ajhg.2010.01.034.

Chu, M. L., Williams, C. J., Pepe, G., Hirsch, J. L., Prockop, D. J., & Ramirez, F. (1983). Internal deletion in a collagen gene in a perinatal lethal form of osteogenesis imperfecta. *Nature, 304*(5921), 78–80.

Citirik, M., Batman, C., Men, G., Tuncel, M., & Zilelioglu, O. (2007). Electron microscopic examination of the anterior lens capsule in a case of Alport's syndrome. *Clinical & Experimental Optometry, 90*(5), 367–370. http://dx.doi.org/10.1111/j.1444-0938.2007.00134.x.

Cohn, R. D., Henry, M. D., Michele, D. E., Barresi, R., Saito, F., Moore, S. A., et al. (2002). Disruption of DAG1 in differentiated skeletal muscle reveals a role for dystroglycan in muscle regeneration. *Cell, 110*(5), 639–648.

Colin, E., Sentilhes, L., Sarfati, A., Mine, M., Guichet, A., Ploton, C., et al. (2014). Fetal intracerebral hemorrhage and cataract: think COL4A1. *Journal of Perinatology, 34*(1), 75–77. http://dx.doi.org/10.1038/jp.2013.135.

Corlobe, A., Tournier-Lasserve, E., Mine, M., Menjot de Champfleur, N., Carra Dalliere, C., et al. (2013). COL4A1 mutation revealed by an isolated brain hemorrhage. *Cerebrovascular Diseases, 35*(6), 593–594. http://dx.doi.org/10.1159/000351520.

Cosgrove, D. (2012). Glomerular pathology in Alport syndrome: a molecular perspective. *Pediatric Nephrology, 27*(6), 885–890. http://dx.doi.org/10.1007/s00467-011-1868-z.

Cosgrove, D., Meehan, D. T., Grunkemeyer, J. A., Kornak, J. M., Sayers, R., Hunter, W. J., et al. (1996). Collagen COL4A3 knockout: a mouse model for autosomal Alport syndrome. *Genes & Development, 10*(23), 2981–2992.

Coupry, I., Sibon, I., Mortemousque, B., Rouanet, F., Mine, M., & Goizet, C. (2010). Ophthalmological features associated with COL4A1 mutations. *Archives of Ophthalmology, 128*(4), 483–489. http://dx.doi.org/10.1001/archophthalmol.2010.42.

Crouch, E., Sage, H., & Bornstein, P. (1980). Structural basis for apparent heterogeneity of collagens in human basement membranes: type IV procollagen contains two distinct chains. *Proceedings of the National Academy of Sciences of the United States of America, 77*(2), 745–749.

Cui, Z., & Zhao, M. H. (2011). Advances in human antiglomerular basement membrane disease. *Nature Reviews Nephrology*, 7(12), 697–705. http://dx.doi.org/10.1038/nrneph.2011.89.

Dean, D. C., Barr, J. F., Freytag, J. W., & Hudson, B. G. (1983). Isolation of type IV procollagen-like polypeptides from glomerular basement membrane. Characterization of pro-alpha 1(IV). *Journal of Biological Chemistry, 258*(1), 590–596.

Deml, B., Reis, L. M., Maheshwari, M., Griffis, C., Bick, D., & Semina, E. V. (2014). Whole exome analysis identifies dominant COL4A1 mutations in patients with complex ocular phenotypes involving microphthalmia. *Clinical Genetics, 86*(5), 475–481. http://dx.doi.org/10.1111/cge.12379.

Di Donato, I., Banchi, S., Federico, A., & Dotti, M. T. (2014). Adult-onset genetic leukoencephalopathies. Focus on the more recently defined forms. *Current Molecular Medicine.*
van Dijk, F. S., Nesbitt, I. M., Zwikstra, E. H., Nikkels, P. G., Piersma, S. R., Fratantoni, S. A., et al. (2009). PPIB mutations cause severe osteogenesis imperfecta. *American Journal of Human Genetics, 85*(4), 521–527. http://dx.doi.org/10.1016/j.ajhg.2009.09.001.
Dolz, R., Engel, J., & Kuhn, K. (1988). Folding of collagen IV. *European Journal of Biochemistry, 178*(2), 357–366.
Doyle, S. A., & Smith, B. D. (1998). Role of the pro-alpha2(I) COOH-terminal region in assembly of type I collagen: disruption of two intramolecular disulfide bonds in pro-alpha2(I) blocks assembly of type I collagen. *Journal of Cellular Biochemistry, 71*(2), 233–242.
Duncan, K. G., Fessler, L. I., Bachinger, H. P., & Fessler, J. H. (1983). Procollagen IV. Association to tetramers. *Journal of Biological Chemistry, 258*(9), 5869–5877.
Ero-Tolliver, I. A., Hudson, B. G., & Bhave, G. (2015). The ancient immunoglobulin domains of peroxidasin are required to form sulfilimine cross-links in collagen IV. *Journal of Biological Chemistry, 290*(35), 21741–21748. http://dx.doi.org/10.1074/jbc.M115.673996.
Ervasti, J. M., & Campbell, K. P. (1993). A role for the dystrophin-glycoprotein complex as a transmembrane linker between laminin and actin. *Journal of Cell Biology, 122*(4), 809–823.
Fagg, W. R., Timoneda, J., Schwartz, C. E., Langeveld, J. P., Noelken, M. E., & Hudson, B. G. (1990). Glomerular basement membrane: evidence for collagenous domain of the alpha 3 and alpha 4 chains of collagen IV. *Biochemical and Biophysical Research Communications, 170*(1), 322–327.
Falcone, G. J., Malik, R., Dichgans, M., & Rosand, J. (2014). Current concepts and clinical applications of stroke genetics. *Lancet Neurology, 13*(4), 405–418. http://dx.doi.org/10.1016/S1474-4422(14)70029-8.
Favor, J., Gloeckner, C. J., Janik, D., Klempt, M., Neuhauser-Klaus, A., Pretsch, W., et al. (2007). Type IV procollagen missense mutations associated with defects of the eye, vascular stability, the brain, kidney function and embryonic or postnatal viability in the mouse, *Mus musculus*: an extension of the Col4a1 allelic series and the identification of the first two Col4a2 mutant alleles. *Genetics, 175*(2), 725–736. http://dx.doi.org/10.1534/genetics.106.064733.
Fidler, A. L., Vanacore, R. M., Chetyrkin, S. V., Pedchenko, V. K., Bhave, G., Yin, V. P., et al. (2014). A unique covalent bond in basement membrane is a primordial innovation for tissue evolution. *Proceedings of the National Academy of Sciences of the United States of America, 111*(1), 331–336. http://dx.doi.org/10.1073/pnas.1318499111.
Firtina, Z., Danysh, B. P., Bai, X., Gould, D. B., Kobayashi, T., & Duncan, M. K. (2009). Abnormal expression of collagen IV in lens activates unfolded protein response resulting in cataract. *Journal of Biological Chemistry, 284*(51), 35872–35884. http://dx.doi.org/10.1074/jbc.M109.060384.
Fischer, G., Schmidt, C., Opitz, J., Cully, Z., Kuhn, K., & Poschl, E. (1993). Identification of a novel sequence element in the common promoter region of human collagen type IV genes, involved in the regulation of divergent transcription. *Biochemical Journal, 292*(Pt 3), 687–695.
Fox, M. A., Sanes, J. R., Borza, D. B., Eswarakumar, V. P., Fassler, R., Hudson, B. G., et al. (2007). Distinct target-derived signals organize formation, maturation, and maintenance of motor nerve terminals. *Cell, 129*(1), 179–193. http://dx.doi.org/10.1016/j.cell.2007.02.035.
Friedman, L., Higgin, J. J., Moulder, G., Barstead, R., Raines, R. T., & Kimble, J. (2000). Prolyl 4-hydroxylase is required for viability and morphogenesis in *Caenorhabditis elegans*. *Proceedings of the National Academy of Sciences of the United States of America, 97*(9), 4736–4741.

Garcia-Torres, R., Cruz, D., Orozco, L., Heidet, L., & Gubler, M. C. (2000). Alport syndrome and diffuse leiomyomatosis. Clinical aspects, pathology, molecular biology and extracellular matrix studies. A synthesis. *Nephrologie, 21*(1), 9–12.

Gawlik, K. I., Mayer, U., Blomberg, K., Sonnenberg, A., Ekblom, P., & Durbeej, M. (2006). Laminin alpha1 chain mediated reduction of laminin alpha2 chain deficient muscular dystrophy involves integrin alpha7beta1 and dystroglycan. *FEBS Letters, 580*(7), 1759–1765. http://dx.doi.org/10.1016/j.febslet.2006.02.027.

Genersch, E., Eckerskorn, C., Lottspeich, F., Herzog, C., Kuhn, K., & Poschl, E. (1995). Purification of the sequence-specific transcription factor CTCBF, involved in the control of human collagen IV genes: subunits with homology to Ku antigen. *EMBO Journal, 14*(4), 791–800.

Glanville, R. W., Qian, R. Q., Siebold, B., Risteli, J., & Kuhn, K. (1985). Amino acid sequence of the N-terminal aggregation and cross-linking region (7S domain) of the alpha 1 (IV) chain of human basement membrane collagen. *European Journal of Biochemistry, 152*(1), 213–219.

Gould, D. B., Marchant, J. K., Savinova, O. V., Smith, R. S., & John, S. W. (2007). Col4a1 mutation causes endoplasmic reticulum stress and genetically modifiable ocular dysgenesis. *Human Molecular Genetics, 16*(7), 798–807. http://dx.doi.org/10.1093/hmg/ddm024.

Gould, D. B., Phalan, F. C., Breedveld, G. J., van Mil, S. E., Smith, R. S., Schimenti, J. C., et al. (2005). Mutations in Col4a1 cause perinatal cerebral hemorrhage and porencephaly. *Science, 308*(5725), 1167–1171. http://dx.doi.org/10.1126/science.1109418.

Gould, D. B., Phalan, F. C., van Mil, S. E., Sundberg, J. P., Vahedi, K., Massin, P., et al. (2006). Role of COL4A1 in small-vessel disease and hemorrhagic stroke. *New England Journal of Medicine, 354*(14), 1489–1496. http://dx.doi.org/10.1056/NEJMoa053727.

Gryder, R. M., Lamon, M., & Adams, E. (1975). Sequence position of 3-hydroxyproline in basement membrane collagen. Isolation of glycyl-3-hydroxyprolyl-4-hydroxyproline from swine kidney. *Journal of Biological Chemistry, 250*(7), 2470–2474.

Gunda, B., Mine, M., Kovacs, T., Hornyak, C., Bereczki, D., Varallyay, G., et al. (2014). COL4A2 mutation causing adult onset recurrent intracerebral hemorrhage and leukoencephalopathy. *Journal of Neurology, 261*(3), 500–503. http://dx.doi.org/10.1007/s00415-013-7224-4.

Gunwar, S., Ballester, F., Noelken, M. E., Sado, Y., Ninomiya, Y., & Hudson, B. G. (1998). Glomerular basement membrane. Identification of a novel disulfide-cross-linked network of alpha3, alpha4, and alpha5 chains of type IV collagen and its implications for the pathogenesis of Alport syndrome. *Journal of Biological Chemistry, 273*(15), 8767–8775.

Gunwar, S., Bejarano, P. A., Kalluri, R., Langeveld, J. P., Wisdom, B. J., Jr., Noelken, M. E., et al. (1991). Alveolar basement membrane: molecular properties of the noncollagenous domain (hexamer) of collagen IV and its reactivity with Goodpasture autoantibodies. *American Journal of Respiratory Cell and Molecular Biology, 5*(2), 107–112.

Guo, H., Tong, P., Peng, Y., Wang, T., Liu, Y., Chen, J., et al. (2014). Homozygous loss-of-function mutation of the LEPREL1 gene causes severe non-syndromic high myopia with early-onset cataract. *Clinical Genetics, 86*(6), 575–579. http://dx.doi.org/10.1111/cge.12309.

Guo, X. D., Johnson, J. J., & Kramer, J. M. (1991). Embryonic lethality caused by mutations in basement membrane collagen of *C. elegans*. *Nature, 349*(6311), 707–709. http://dx.doi.org/10.1038/349707a0.

Gupta, M. C., Graham, P. L., & Kramer, J. M. (1997). Characterization of alpha1(IV) collagen mutations in *Caenorhabditis elegans* and the effects of alpha1 and alpha2(IV) mutations on type IV collagen distribution. *Journal of Cell Biology, 137*(5), 1185–1196.

Haniel, A., Welge-Lussen, U., Kuhn, K., & Poschl, E. (1995). Identification and characterization of a novel transcriptional silencer in the human collagen type IV gene COL4A2. *Journal of Biological Chemistry, 270*(19), 11209–11215.

Harbers, K., Kuehn, M., Delius, H., & Jaenisch, R. (1984). Insertion of retrovirus into the first intron of alpha 1(I) collagen gene to embryonic lethal mutation in mice. *Proceedings of the National Academy of Sciences of the United States of America, 81*(5), 1504–1508.

Hautala, T., Heikkinen, J., Kivirikko, K. I., & Myllyla, R. (1993). A large duplication in the gene for lysyl hydroxylase accounts for the type VI variant of Ehlers–Danlos syndrome in two siblings. *Genomics, 15*(2), 399–404. http://dx.doi.org/10.1006/geno.1993.1074.

Hayashi, Y. K., Chou, F. L., Engvall, E., Ogawa, M., Matsuda, C., Hirabayashi, S., et al. (1998). Mutations in the integrin alpha7 gene cause congenital myopathy. *Nature Genetics, 19*(1), 94–97. http://dx.doi.org/10.1038/ng0598-94.

Helbling-Leclerc, A., Zhang, X., Topaloglu, H., Cruaud, C., Tesson, F., Weissenbach, J., et al. (1995). Mutations in the laminin alpha 2-chain gene (LAMA2) cause merosin-deficient congenital muscular dystrophy. *Nature Genetics, 11*(2), 216–218. http://dx.doi.org/10.1038/ng1095-216.

Holster, T., Pakkanen, O., Soininen, R., Sormunen, R., Nokelainen, M., Kivirikko, K. I., et al. (2007). Loss of assembly of the main basement membrane collagen, type IV, but not fibril-forming collagens and embryonic death in collagen prolyl 4-hydroxylase I null mice. *Journal of Biological Chemistry, 282*(4), 2512–2519. http://dx.doi.org/10.1074/jbc.M606608200.

Hostikka, S. L., Eddy, R. L., Byers, M. G., Hoyhtya, M., Shows, T. B., & Tryggvason, K. (1990). Identification of a distinct type IV collagen alpha chain with restricted kidney distribution and assignment of its gene to the locus of X chromosome-linked Alport syndrome. *Proceedings of the National Academy of Sciences of the United States of America, 87*(4), 1606–1610.

Hostikka, S. L., & Tryggvason, K. (1988). The complete primary structure of the alpha 2 chain of human type IV collagen and comparison with the alpha 1(IV) chain. *Journal of Biological Chemistry, 263*(36), 19488–19493.

Hudson, B. G. (2004). The molecular basis of Goodpasture and Alport syndromes: beacons for the discovery of the collagen IV family. *Journal of the American Society of Nephrology, 15*(10), 2514–2527. http://dx.doi.org/10.1097/01.ASN.0000141462.00630.76.

Hudson, B. G., Tryggvason, K., Sundaramoorthy, M., & Neilson, E. G. (2003). Alport's syndrome, Goodpasture's syndrome, and type IV collagen. *New England Journal of Medicine, 348*(25), 2543–2556. http://dx.doi.org/10.1056/NEJMra022296.

Hudson, D. M., & Eyre, D. R. (2013). Collagen prolyl 3-hydroxylation: a major role for a minor post-translational modification? *Connective Tissue Research, 54*(4–5), 245–251. http://dx.doi.org/10.3109/03008207.2013.800867.

Hudson, D. M., Joeng, K. S., Werther, R., Rajagopal, A., Weis, M., Lee, B. H., et al. (2015). Post-translationally abnormal collagens of prolyl 3-hydroxylase-2 null mice offer a pathobiological mechanism for the high myopia linked to human LEPREL1 mutations. *Journal of Biological Chemistry, 290*(13), 8613–8622. http://dx.doi.org/10.1074/jbc.M114.634915.

Iannitti, T., & Palmieri, B. (2011). Clinical and experimental applications of sodium phenylbutyrate. *Drugs in R&D, 11*(3), 227–249. http://dx.doi.org/10.2165/11591280-000000000-00000.

Jais, J. P., Knebelmann, B., Giatras, I., De Marchi, M., Rizzoni, G., Renieri, A., et al. (2003). X-linked Alport syndrome: natural history and genotype-phenotype correlations in girls and women belonging to 195 families: a "European Community Alport Syndrome Concerted Action" study. *Journal of the American Society of Nephrology, 14*(10), 2603–2610.

Jeanne, M., Jorgensen, J., & Gould, D. B. (2015). Molecular and genetic analyses of collagen type IV mutant mouse models of spontaneous intracerebral hemorrhage identify mechanisms for stroke prevention. *Circulation, 131*(18), 1555–1565. http://dx.doi.org/10.1161/CIRCULATIONAHA.114.013395.

Jeanne, M., Labelle-Dumais, C., Jorgensen, J., Kauffman, W. B., Mancini, G. M., Favor, J., et al. (2012). COL4A2 mutations impair COL4A1 and COL4A2 secretion and cause hemorrhagic stroke. *American Journal of Human Genetics, 90*(1), 91–101. http://dx.doi.org/10.1016/j.ajhg.2011.11.022.

Jiang, D., Li, J., Xiao, X., Li, S., Jia, X., Sun, W., et al. (2015). Detection of mutations in LRPAP1, CTSH, LEPREL1, ZNF644, SLC39A5, and SCO2 in 298 families with early-onset high myopia by exome sequencing. *Investigative Ophthalmology & Visual Science, 56*(1), 339–345. http://dx.doi.org/10.1167/iovs.14-14850.

Jimenez, S., Harsch, M., & Rosenbloom, J. (1973). Hydroxyproline stabilizes the triple helix of chick tendon collagen. *Biochemical and Biophysical Research Communications, 52*(1), 106–114.

Joutel, A., & Faraci, F. M. (2014). Cerebral small vessel disease: insights and opportunities from mouse models of collagen IV-related small vessel disease and cerebral autosomal dominant arteriopathy with subcortical infarcts and leukoencephalopathy. *Stroke, 45*(4), 1215–1221. http://dx.doi.org/10.1161/STROKEAHA.113.002878.

Joutel, A., Haddad, I., Ratelade, J., & Nelson, M. T. (2015). Perturbations of the cerebrovascular matrisome: a convergent mechanism in small vessel disease of the brain? *Journal of Cerebral Blood Flow & Metabolism.* http://dx.doi.org/10.1038/jcbfm.2015.62.

Kalluri, R., Gattone, V. H., 2nd, & Hudson, B. G. (1998). Identification and localization of type IV collagen chains in the inner ear cochlea. *Connective Tissue Research, 37*(1–2), 143–150.

Kalluri, R., Shield, C. F., Todd, P., Hudson, B. G., & Neilson, E. G. (1997). Isoform switching of type IV collagen is developmentally arrested in X-linked Alport syndrome leading to increased susceptibility of renal basement membranes to endoproteolysis. *Journal of Clinical Investigation, 99*(10), 2470–2478. http://dx.doi.org/10.1172/JCI119431.

Kalluri, R., Sun, M. J., Hudson, B. G., & Neilson, E. G. (1996). The Goodpasture autoantigen. Structural delineation of two immunologically privileged epitopes on alpha3(IV) chain of type IV collagen. *Journal of Biological Chemistry, 271*(15), 9062–9068.

Kang, J. S., Colon, S., Hellmark, T., Sado, Y., Hudson, B. G., & Borza, D. B. (2008). Identification of noncollagenous sites encoding specific interactions and quaternary assembly of alpha 3 alpha 4 alpha 5(IV) collagen: implications for Alport gene therapy. *Journal of Biological Chemistry, 283*(50), 35070–35077. http://dx.doi.org/10.1074/jbc.M806396200.

Kashtan, C. E. (1998). Alport syndrome and thin glomerular basement membrane disease. *Journal of the American Society of Nephrology, 9*(9), 1736–1750.

Kashtan, C. E. (1999). Alport syndrome. An inherited disorder of renal, ocular, and cochlear basement membranes. *Medicine (Baltimore), 78*(5), 338–360.

Kashtan, C. E. (2004). Familial hematuria due to type IV collagen mutations: Alport syndrome and thin basement membrane nephropathy. *Current Opinion in Pediatrics, 16*(2), 177–181.

Kefalides, N. A. (1966). A collagen of unusual composition and a glycoprotein isolated from canine glomerular basement membrane. *Biochemical and Biophysical Research Communications, 22*(1), 26–32.

Kefalides, N. A. (1973). Structure and biosynthesis of basement membranes. *International Review of Connective Tissue Research, 6*, 63–104.

Kefalides, N. A. (1975). Basement membranes: structural and biosynthetic considerations. *Journal of Investigative Dermatology, 65*(1), 85–92.

Kelemen-Valkony, I., Kiss, M., Csiha, J., Kiss, A., Bircher, U., et al. (2012). *Drosophila* basement membrane collagen Col4a1 mutations cause severe myopathy. *Matrix Biology, 31*(1), 29–37. http://dx.doi.org/10.1016/j.matbio.2011.09.004.

Kelley, P. B., Sado, Y., & Duncan, M. K. (2002). Collagen IV in the developing lens capsule. *Matrix Biology, 21*(5), 415–423.

Khan, K., Rudkin, A., Parry, D. A., Burdon, K. P., McKibbin, M., Logan, C. V., et al. (2011). Homozygous mutations in PXDN cause congenital cataract, corneal opacity,

and developmental glaucoma. *American Journal of Human Genetics, 89*(3), 464–473. http://dx.doi.org/10.1016/j.ajhg.2011.08.005.
Khoshnoodi, J., Cartailler, J. P., Alvares, K., Veis, A., & Hudson, B. G. (2006). Molecular recognition in the assembly of collagens: terminal noncollagenous domains are key recognition modules in the formation of triple helical protomers. *Journal of Biological Chemistry, 281*(50), 38117–38121. http://dx.doi.org/10.1074/jbc.R600025200.
Khoshnoodi, J., Sigmundsson, K., Cartailler, J. P., Bondar, O., Sundaramoorthy, M., & Hudson, B. G. (2006). Mechanism of chain selection in the assembly of collagen IV: a prominent role for the alpha2 chain. *Journal of Biological Chemistry, 281*(9), 6058–6069. http://dx.doi.org/10.1074/jbc.M506555200.
Kivirikko, K. I., & Myllyharju, J. (1998). Prolyl 4-hydroxylases and their protein disulfide isomerase subunit. *Matrix Biology, 16*(7), 357–368.
Kivirikko, K. I., Myllyla, R., & Pihlajaniemi, T. (1991). Hydroxylation of proline and lysine residues in collagens and other animal and plant proteins. In J. J. Harding, & M. J. C. Crabbe (Eds.), *Post-Translational modifications of proteins* (1st ed.). Boca Raton: CRC Press.
Kivirikko, K. I., & Pihlajaniemi, T. (1998). Collagen hydroxylases and the protein disulfide isomerase subunit of prolyl 4-hydroxylases. *Advances In Enzymology and Related Areas of Molecular Biology, 72*, 325–398.
Koide, T., Asada, S., Takahara, Y., Nishikawa, Y., Nagata, K., & Kitagawa, K. (2006). Specific recognition of the collagen triple helix by chaperone HSP47: minimal structural requirement and spatial molecular orientation. *Journal of Biological Chemistry, 281*(6), 3432–3438. http://dx.doi.org/10.1074/jbc.M509707200.
Koide, T., Aso, A., Yorihuzi, T., & Nagata, K. (2000). Conformational requirements of collagenous peptides for recognition by the chaperone protein HSP47. *Journal of Biological Chemistry, 275*(36), 27957–27963. http://dx.doi.org/10.1074/jbc.M003026200.
Koide, T., Takahara, Y., Asada, S., & Nagata, K. (2002). Xaa-Arg-Gly triplets in the collagen triple helix are dominant binding sites for the molecular chaperone HSP47. *Journal of Biological Chemistry, 277*(8), 6178–6182. http://dx.doi.org/10.1074/jbc.M106497200.
Koivu, J. (1987). Disulfide bonding as a determinant of the molecular composition of types I, II and III procollagen. *FEBS Letters, 217*(2), 216–220.
Korstanje, R., Caputo, C. R., Doty, R. A., Cook, S. A., Bronson, R. T., Davisson, M. T., et al. (2014). A mouse Col4a4 mutation causing Alport glomerulosclerosis with abnormal collagen alpha3alpha4alpha5(IV) trimers. *Kidney International, 85*(6), 1461–1468. http://dx.doi.org/10.1038/ki.2013.493.
Kresina, T. F., & Miller, E. J. (1979). Isolation and characterization of basement membrane collagen from human placental tissue. Evidence for the presence of two genetically distinct collagen chains. *Biochemistry, 18*(14), 3089–3097.
Kruegel, J., & Miosge, N. (2010). Basement membrane components are key players in specialized extracellular matrices. *Cellular and Molecular Life Sciences, 67*(17), 2879–2895. http://dx.doi.org/10.1007/s00018-010-0367-x.
Kruegel, J., Rubel, D., & Gross, O. (2013). Alport syndrome—insights from basic and clinical research. *Nature Reviews Nephrology, 9*(3), 170–178. http://dx.doi.org/10.1038/nrneph.2012.259.
Kuhn, K., Wiedemann, H., Timpl, R., Risteli, J., Dieringer, H., Voss, T., et al. (1981). Macromolecular structure of basement membrane collagens. *FEBS Letters, 125*(1), 123–128.
Kuivaniemi, H., Tromp, G., & Prockop, D. J. (1991). Mutations in collagen genes: causes of rare and some common diseases in humans. *FASEB Journal, 5*(7), 2052–2060.
Kuo, D. S., Labelle-Dumais, C., & Gould, D. B. (2012). COL4A1 and COL4A2 mutations and disease: insights into pathogenic mechanisms and potential therapeutic targets. *Human Molecular Genetics, 21*(R1), R97–R110. http://dx.doi.org/10.1093/hmg/dds346.

Kuo, D. S., Labelle-Dumais, C., Mao, M., Jeanne, M., Kauffman, W. B., Allen, J., et al. (2014). Allelic heterogeneity contributes to variability in ocular dysgenesis, myopathy and brain malformations caused by Col4a1 and Col4a2 mutations. *Human Molecular Genetics, 23*(7), 1709–1722. http://dx.doi.org/10.1093/hmg/ddt560.

Labelle-Dumais, C., Dilworth, D. J., Harrington, E. P., de Leau, M., Lyons, D., Kabaeva, Z., et al. (2011). COL4A1 mutations cause ocular dysgenesis, neuronal localization defects, and myopathy in mice and Walker–Warburg syndrome in humans. *PLoS Genetics,* 7(5), e1002062. http://dx.doi.org/10.1371/journal.pgen.1002062.

Lamande, S. R., Chessler, S. D., Golub, S. B., Byers, P. H., Chan, D., Cole, W. G., et al. (1995). Endoplasmic reticulum-mediated quality control of type I collagen production by cells from osteogenesis imperfecta patients with mutations in the pro alpha 1 (I) chain carboxyl-terminal propeptide which impair subunit assembly. *Journal of Biological Chemistry, 270*(15), 8642–8649.

Lanfranconi, S., & Markus, H. S. (2010). COL4A1 mutations as a monogenic cause of cerebral small vessel disease: a systematic review. *Stroke, 41*(8), e513–518. http://dx.doi.org/10.1161/STROKEAHA.110.581918.

Leinonen, A., Mariyama, M., Mochizuki, T., Tryggvason, K., & Reeders, S. T. (1994). Complete primary structure of the human type IV collagen alpha 4(IV) chain. Comparison with structure and expression of the other alpha (IV) chains. *Journal of Biological Chemistry, 269*(42), 26172–26177.

Leinonen, A., Netzer, K. O., Boutaud, A., Gunwar, S., & Hudson, B. G. (1999). Goodpasture antigen: expression of the full-length alpha3(IV) chain of collagen IV and localization of epitopes exclusively to the noncollagenous domain. *Kidney International, 55*(3), 926–935. http://dx.doi.org/10.1046/j.1523-1755.1999.055003926.x.

Lemmens, R., Maugeri, A., Niessen, H. W., Goris, A., Tousseyn, T., Demaerel, P., et al. (2013). Novel COL4A1 mutations cause cerebral small vessel disease by haploinsufficiency. *Human Molecular Genetics, 22*(2), 391–397. http://dx.doi.org/10.1093/hmg/dds436.

Lemmink, H. H., Mochizuki, T., van den Heuvel, L. P., Schroder, C. H., Barrientos, A., Monnens, L. A., et al. (1994). Mutations in the type IV collagen alpha 3 (COL4A3) gene in autosomal recessive Alport syndrome. *Human Molecular Genetics, 3*(8), 1269–1273.

Lerner, R. A., Glassock, R. J., & Dixon, F. J. (1967). The role of anti-glomerular basement membrane antibody in the pathogenesis of human glomerulonephritis. *Journal of Experimental Medicine, 126*(6), 989–1004.

Lichtenbelt, K. D., Pistorius, L. R., De Tollenaer, S. M., Mancini, G. M., & De Vries, L. S. (2012). Prenatal genetic confirmation of a COL4A1 mutation presenting with sonographic fetal intracranial hemorrhage. *Ultrasound in Obstetrics & Gynecology, 39*(6), 726–727. http://dx.doi.org/10.1002/uog.11070.

Lim, A. L., Doyle, S. A., Balian, G., & Smith, B. D. (1998). Role of the pro-alpha2(I) COOH-terminal region in assembly of type I collagen: truncation of the last 10 amino acid residues of pro-alpha2(I) chain prevents assembly of type I collagen heterotrimer. *Journal of Cellular Biochemistry, 71*(2), 216–232.

Lindert, U., Weis, M. A., Rai, J., Seeliger, F., Hausser, I., Leeb, T., et al. (2015). Molecular consequences of the SERPINH1/HSP47 mutation in the Dachshund natural model of osteogenesis imperfecta. *Journal of Biological Chemistry, 290*(29), 17679–17689. http://dx.doi.org/10.1074/jbc.M115.661025.

Livingston, J., Doherty, D., Orcesi, S., Tonduti, D., Piechiecchio, A., La Piana, R., et al. (2011). COL4A1 mutations associated with a characteristic pattern of intracranial calcification. *Neuropediatrics, 42*(6), 227–233. http://dx.doi.org/10.1055/s-0031-1295493.

Lohler, J., Timpl, R., & Jaenisch, R. (1984). Embryonic lethal mutation in mouse collagen I gene causes rupture of blood vessels and is associated with erythropoietic and mesenchymal cell death. *Cell, 38*(2), 597–607.

Lu, W., Phillips, C. L., Killen, P. D., Hlaing, T., Harrison, W. R., Elder, F. F., et al. (1999). Insertional mutation of the collagen genes Col4a3 and Col4a4 in a mouse model of Alport syndrome. *Genomics, 61*(2), 113–124. http://dx.doi.org/10.1006/geno.1999.5943.

Luo, R., Jeong, S. J., Jin, Z., Strokes, N., Li, S., & Piao, X. (2011). G protein-coupled receptor 56 and collagen III, a receptor-ligand pair, regulates cortical development and lamination. *Proceedings of the National Academy of Sciences of the United States of America, 108*(31), 12925–12930. http://dx.doi.org/10.1073/pnas.1104821108.

Lyon, M. F., Glenister, P. H., & West, J. D. (1984). Bruised (Bru). *Mouse News Letter, 71*, 1.

Malhotra, V., & Erlmann, P. (2011). Protein export at the ER: loading big collagens into COPII carriers. *EMBO Journal, 30*(17), 3475–3480. http://dx.doi.org/10.1038/emboj.2011.255.

Malhotra, V., Erlmann, P., & Nogueira, C. (2015). Procollagen export from the endoplasmic reticulum. *Biochemical Society Transactions, 43*(1), 104–107. http://dx.doi.org/10.1042/BST20140286.

Mao, M., Smith, R. S., Alavi, M. V., Marchant, J. K., Cosma, M., Libby, R. T., John, S. W. M., & Gould, D. B. (2015). Strain dependent anterior segment dysgenesis and progression to glaucoma in Col4a1 mutant mice. *Investigative Ophthalmology & Visual Science, 56*. http://dx.doi.org/10.1167/iovs.15-17527.

Marini, J. C., Forlino, A., Cabral, W. A., Barnes, A. M., San Antonio, J. D., Milgrom, S., et al. (2007). Consortium for osteogenesis imperfecta mutations in the helical domain of type I collagen: regions rich in lethal mutations align with collagen binding sites for integrins and proteoglycans. *Human Mutation, 28*(3), 209–221. http://dx.doi.org/10.1002/humu.20429.

Mariyama, M., Leinonen, A., Mochizuki, T., Tryggvason, K., & Reeders, S. T. (1994). Complete primary structure of the human alpha 3(IV) collagen chain. Coexpression of the alpha 3(IV) and alpha 4(IV) collagen chains in human tissues. *Journal of Biological Chemistry, 269*(37), 23013–23017.

Mariyama, M., Zheng, K., Yang-Feng, T. L., & Reeders, S. T. (1992). Colocalization of the genes for the alpha 3(IV) and alpha 4(IV) chains of type IV collagen to chromosome 2 bands q35–q37. *Genomics, 13*(3), 809–813.

Marneros, A. G., & Olsen, B. R. (2005). Physiological role of collagen XVIII and endostatin. *FASEB Journal, 19*(7), 716–728. http://dx.doi.org/10.1096/fj.04-2134rev.

Marutani, T., Yamamoto, A., Nagai, N., Kubota, H., & Nagata, K. (2004). Accumulation of type IV collagen in dilated ER leads to apoptosis in Hsp47-knockout mouse embryos via induction of CHOP. *Journal of Cell Science, 117*(Pt 24), 5913–5922. http://dx.doi.org/10.1242/jcs.01514.

Matsuoka, Y., Kubota, H., Adachi, E., Nagai, N., Marutani, T., Hosokawa, N., et al. (2004). Insufficient folding of type IV collagen and formation of abnormal basement membrane-like structure in embryoid bodies derived from Hsp47-null embryonic stem cells. *Molecular Biology of the Cell, 15*(10), 4467–4475. http://dx.doi.org/10.1091/mbc.E04-01-0050.

Mayne, R., & Zettergren, J. G. (1980). Type IV collagen from chicken muscular tissues. Isolation and characterization of the pepsin-resistant fragments. *Biochemistry, 19*(17), 4065–4072.

Menezes, M. J., McClenahan, F. K., Leiton, C. V., Aranmolate, A., Shan, X., & Colognato, H. (2014). The extracellular matrix protein laminin alpha2 regulates the maturation and function of the blood–brain barrier. *Journal of Neuroscience, 34*(46), 15260–15280. http://dx.doi.org/10.1523/JNEUROSCI.3678-13.2014.

Merkel, F., Kalluri, R., Marx, M., Enders, U., Stevanovic, S., Giegerich, G., et al. (1996). Autoreactive T-cells in Goodpasture's syndrome recognize the N-terminal NC1 domain on alpha 3 type IV collagen. *Kidney International, 49*(4), 1127–1133.

Meuwissen, M. E., de Vries, L. S., Verbeek, H. A., Lequin, M. H., Govaert, P. P., Schot, R., et al. (2011). Sporadic COL4A1 mutations with extensive prenatal porencephaly resembling hydranencephaly. *Neurology, 76*(9), 844–846. http://dx.doi.org/10.1212/WNL.0b013e31820e7751.

Miller, E. J., & Gay, S. (1982). Collagen: an overview. *Methods in Enzymology, 82*(Pt A), 3–32.

Miner, J. H., & Sanes, J. R. (1994). Collagen IV alpha 3, alpha 4, and alpha 5 chains in rodent basal laminae: sequence, distribution, association with laminins, and developmental switches. *Journal of Cell Biology, 127*(3), 879–891.

Miner, J. H., & Sanes, J. R. (1996). Molecular and functional defects in kidneys of mice lacking collagen alpha 3(IV): implications for Alport syndrome. *Journal of Cell Biology, 135*(5), 1403–1413.

Minor, R. R., Clark, C. C., Strause, E. L., Koszalka, T. R., Brent, R. L., & Kefalides, N. A. (1976). Basement membrane procollagen is not converted to collagen in organ cultures of parietal yolk sac endoderm. *Journal of Biological Chemistry, 251*(6), 1789–1794.

Miyagoe, Y., Hanaoka, K., Nonaka, I., Hayasaka, M., Nabeshima, Y., Arahata, K., et al. (1997). Laminin alpha2 chain-null mutant mice by targeted disruption of the Lama2 gene: a new model of merosin (laminin 2)-deficient congenital muscular dystrophy. *FEBS Letters, 415*(1), 33–39.

Mizuno, K., Hayashi, T., Peyton, D. H., & Bachinger, H. P. (2004). The peptides acetyl-(Gly-3(S)Hyp-4(R)Hyp)10-NH2 and acetyl-(Gly-Pro-3(S)Hyp)10-NH2 do not form a collagen triple helix. *Journal of Biological Chemistry, 279*(1), 282–287. http://dx.doi.org/10.1074/jbc.M308181200.

Mochizuki, T., Lemmink, H. H., Mariyama, M., Antignac, C., Gubler, M. C., Pirson, Y., et al. (1994). Identification of mutations in the alpha 3(IV) and alpha 4(IV) collagen genes in autosomal recessive Alport syndrome. *Nature Genetics, 8*(1), 77–81. http://dx.doi.org/10.1038/ng0994-77.

Momota, R., Sugimoto, M., Oohashi, T., Kigasawa, K., Yoshioka, H., & Ninomiya, Y. (1998). Two genes, COL4A3 and COL4A4 coding for the human alpha3(IV) and alpha4(IV) collagen chains are arranged head-to-head on chromosome 2q36. *FEBS Letters, 424*(1–2), 11–16.

Moore, S. A., Saito, F., Chen, J., Michele, D. E., Henry, M. D., Messing, A., et al. (2002). Deletion of brain dystroglycan recapitulates aspects of congenital muscular dystrophy. *Nature, 418*(6896), 422–425. http://dx.doi.org/10.1038/nature00838.

Morello, R., Bertin, T. K., Chen, Y., Hicks, J., Tonachini, L., Monticone, M., et al. (2006). CRTAP is required for prolyl 3-hydroxylation and mutations cause recessive osteogenesis imperfecta. *Cell, 127*(2), 291–304. http://dx.doi.org/10.1016/j.cell.2006.08.039.

Murray, L. S., Lu, Y., Taggart, A., Van Regemorter, N., Vilain, C., Abramowicz, M., et al. (2014). Chemical chaperone treatment reduces intracellular accumulation of mutant collagen IV and ameliorates the cellular phenotype of a COL4A2 mutation that causes haemorrhagic stroke. *Human Molecular Genetics, 23*(2), 283–292. http://dx.doi.org/10.1093/hmg/ddt418.

Muthukumaran, G., Blumberg, B., & Kurkinen, M. (1989). The complete primary structure for the alpha 1-chain of mouse collagen IV. Differential evolution of collagen IV domains. *Journal of Biological Chemistry, 264*(11), 6310–6317.

Myers, J. C., & Emanuel, B. S. (1987). Chromosomal localization of human collagen genes. *Collagen and Related Research, 7*(2), 149–159.

Myllyharju, J. (2008). Prolyl 4-hydroxylases, key enzymes in the synthesis of collagens and regulation of the response to hypoxia, and their roles as treatment targets. *Annals of Medicine, 40*(6), 402–417. http://dx.doi.org/10.1080/07853890801986594.

Myllyharju, J., & Kivirikko, K. I. (2004). Collagens, modifying enzymes and their mutations in humans, flies and worms. *Trends in Genetics, 20*(1), 33–43. http://dx.doi.org/10.1016/j.tig.2003.11.004.

Nagai, N., Hosokawa, M., Itohara, S., Adachi, E., Matsushita, T., Hosokawa, N., et al. (2000). Embryonic lethality of molecular chaperone hsp47 knockout mice is associated with defects in collagen biosynthesis. *Journal of Cell Biology, 150*(6), 1499–1506.

Nagata, K. (1996). Hsp47: a collagen-specific molecular chaperone. *Trends in Biochemical Sciences, 21*(1), 22–26.

Natsume, T., Koide, T., Yokota, S., Hirayoshi, K., & Nagata, K. (1994). Interactions between collagen-binding stress protein HSP47 and collagen. Analysis of kinetic parameters by surface plasmon resonance biosensor. *Journal of Biological Chemistry, 269*(49), 31224–31228.

Netzer, K. O., Suzuki, K., Itoh, Y., Hudson, B. G., & Khalifah, R. G. (1998). Comparative analysis of the noncollagenous NC1 domain of type IV collagen: identification of structural features important for assembly, function, and pathogenesis. *Protein Science,* 7(6), 1340–1351. http://dx.doi.org/10.1002/pro.5560070610.

Niewmierzycka, A., Mills, J., St-Arnaud, R., Dedhar, S., & Reichardt, L. F. (2005). Integrin-linked kinase deletion from mouse cortex results in cortical lamination defects resembling cobblestone lissencephaly. *Journal of Neuroscience, 25*(30), 7022–7031. http://dx.doi.org/10.1523/JNEUROSCI.1695-05.2005.

Ninomiya, Y., Kagawa, M., Iyama, K., Naito, I., Kishiro, Y., Seyer, J. M., et al. (1995). Differential expression of two basement membrane collagen genes, COL4A6 and COL4A5, demonstrated by immunofluorescence staining using peptide-specific monoclonal antibodies. *Journal of Cell Biology, 130*(5), 1219–1229.

Nissi, R., Autio-Harmainen, H., Marttila, P., Sormunen, R., & Kivirikko, K. I. (2001). Prolyl 4-hydroxylase isoenzymes I and II have different expression patterns in several human tissues. *Journal of Histochemistry and Cytochemistry, 49*(9), 1143–1153.

Niwa, T., Aida, N., Osaka, H., Wada, T., Saitsu, H., & Imai, Y. (2015). Intracranial hemorrhage and tortuosity of veins detected on susceptibility-weighted imaging of a child with a type IV collagen alpha1 mutation and schizencephaly. *Magnetic Resonance in Medical Sciences, 14*(3), 223–226. http://dx.doi.org/10.2463/mrms.2014-0060.

Norman, K. R., & Moerman, D. G. (2000). The let-268 locus of *Caenorhabditis elegans* encodes a procollagen lysyl hydroxylase that is essential for type IV collagen secretion. *Developmental Biology, 227*(2), 690–705. http://dx.doi.org/10.1006/dbio.2000.9897.

Oberbaumer, I., Laurent, M., Schwarz, U., Sakurai, Y., Yamada, Y., Vogeli, G., et al. (1985). Amino acid sequence of the non-collagenous globular domain (NC1) of the alpha 1(IV) chain of basement membrane collagen as derived from complementary DNA. *European Journal of Biochemistry, 147*(2), 217–224.

Olsen, B. R., Alper, R., & Kefalides, N. A. (1973). Structural characterization of a soluble fraction from lens-capsule basement membrane. *European Journal of Biochemistry, 38*(2), 220–228.

Ono, T., Miyazaki, T., Ishida, Y., Uehata, M., & Nagata, K. (2012). Direct in vitro and in vivo evidence for interaction between Hsp47 protein and collagen triple helix. *Journal of Biological Chemistry, 287*(9), 6810–6818. http://dx.doi.org/10.1074/jbc.M111.280248.

Ozcan, U., Yilmaz, E., Ozcan, L., Furuhashi, M., Vaillancourt, E., Smith, R. O., et al. (2006). Chemical chaperones reduce ER stress and restore glucose homeostasis in a mouse model of type 2 diabetes. *Science, 313*(5790), 1137–1140. http://dx.doi.org/10.1126/science.1128294.

O'Donnell, C. J., Kavousi, M., Smith, A. V., Kardia, S. L., Feitosa, M. F., Hwang, S. J., et al. (2011). Genome-wide association study for coronary artery calcification with follow-up in myocardial infarction. *Circulation, 124*(25), 2855–2864. http://dx.doi.org/10.1161/CIRCULATIONAHA.110.974899.

Parkin, J. D., San Antonio, J. D., Pedchenko, V., Hudson, B., Jensen, S. T., & Savige, J. (2011). Mapping structural landmarks, ligand binding sites, and missense mutations to the collagen IV heterotrimers predicts major functional domains, novel interactions, and variation in phenotypes in inherited diseases affecting basement membranes. *Human Mutation, 32*(2), 127–143. http://dx.doi.org/10.1002/humu.21401.

Pedchenko, V., Bondar, O., Fogo, A. B., Vanacore, R., Voziyan, P., Kitching, A. R., et al. (2010). Molecular architecture of the Goodpasture autoantigen in anti-GBM nephritis.

New England Journal of Medicine, 363(4), 343–354. http://dx.doi.org/10.1056/NEJMoa0910500.

Peissel, B., Geng, L., Kalluri, R., Kashtan, C., Rennke, H. G., Gallo, G. R., et al. (1995). Comparative distribution of the alpha 1(IV), alpha 5(IV), and alpha 6(IV) collagen chains in normal human adult and fetal tissues and in kidneys from X-linked Alport syndrome patients. *Journal of Clinical Investigation, 96*(4), 1948–1957. http://dx.doi.org/10.1172/JCI118241.

Perlmutter, D. H. (2002). Chemical chaperones: a pharmacological strategy for disorders of protein folding and trafficking. *Pediatric Research, 52*(6), 832–836. http://dx.doi.org/10.1203/00006450-200212000-00004.

Peto, P., & Salama, A. D. (2011). Update on antiglomerular basement membrane disease. *Current Opinion in Rheumatology, 23*(1), 32–37. http://dx.doi.org/10.1097/BOR.0b013e328341009f.

Pihlajaniemi, T., Dickson, L. A., Pope, F. M., Korhonen, V. R., Nicholls, A., Prockop, D. J., et al. (1984). Osteogenesis imperfecta: cloning of a pro-alpha 2(I) collagen gene with a frameshift mutation. *Journal of Biological Chemistry, 259*(21), 12941–12944.

Pihlajaniemi, T., Pohjolainen, E. R., & Myers, J. C. (1990). Complete primary structure of the triple-helical region and the carboxyl-terminal domain of a new type IV collagen chain, alpha 5(IV). *Journal of Biological Chemistry, 265*(23), 13758–13766.

Pihlajaniemi, T., Tryggvason, K., Myers, J. C., Kurkinen, M., Lebo, R., Cheung, M. C., et al. (1985). cDNA clones coding for the pro-alpha1(IV) chain of human type IV procollagen reveal an unusual homology of amino acid sequences in two halves of the carboxyl-terminal domain. *Journal of Biological Chemistry, 260*(12), 7681–7687.

Pinnell, S. R., Krane, S. M., Kenzora, J. E., & Glimcher, M. J. (1972). A heritable disorder of connective tissue. Hydroxylysine-deficient collagen disease. *New England Journal of Medicine, 286*(19), 1013–1020. http://dx.doi.org/10.1056/NEJM197205112861901.

Plaisier, E., Alamowitch, S., Gribouval, O., Mougenot, B., Gaudric, A., Antignac, C., et al. (2005). Autosomal-dominant familial hematuria with retinal arteriolar tortuosity and contractures: a novel syndrome. *Kidney International, 67*(6), 2354–2360. http://dx.doi.org/10.1111/j.1523-1755.2005.00341.x.

Plaisier, E., Chen, Z., Gekeler, F., Benhassine, S., Dahan, K., Marro, B., et al. (2010). Novel COL4A1 mutations associated with HANAC syndrome: a role for the triple helical CB3 [IV] domain. *American Journal of Medical Genetics, Part A, 152A*(10), 2550–2555. http://dx.doi.org/10.1002/ajmg.a.33659.

Plaisier, E., Gribouval, O., Alamowitch, S., Mougenot, B., Prost, C., Verpont, M. C., et al. (2007). COL4A1 mutations and hereditary angiopathy, nephropathy, aneurysms, and muscle cramps. *New England Journal of Medicine, 357*(26), 2687–2695. http://dx.doi.org/10.1056/NEJMoa071906.

Pokidysheva, E., Boudko, S., Vranka, J., Zientek, K., Maddox, K., Moser, M., et al. (2014). Biological role of prolyl 3-hydroxylation in type IV collagen. *Proceedings of the National Academy of Sciences of the United States of America, 111*(1), 161–166. http://dx.doi.org/10.1073/pnas.1307597111.

Pollner, R., Fischer, G., Poschl, E., & Kuhn, K. (1990). Regulation of divergent transcription of the genes coding for basement membrane type IV collagen. *Annals of the New York Academy of Sciences, 580*, 44–54.

Pollner, R., Schmidt, C., Fischer, G., Kuhn, K., & Poschl, E. (1997). Cooperative and competitive interactions of regulatory elements are involved in the control of divergent transcription of human Col4A1 and Col4A2 genes. *FEBS Letters, 405*(1), 31–36.

Poschl, E., Pollner, R., & Kuhn, K. (1988). The genes for the alpha 1(IV) and alpha 2(IV) chains of human basement membrane collagen type IV are arranged head-to-head and separated by a bidirectional promoter of unique structure. *EMBO Journal, 7*(9), 2687–2695.

Poschl, E., Schlotzer-Schrehardt, U., Brachvogel, B., Saito, K., Ninomiya, Y., & Mayer, U. (2004). Collagen IV is essential for basement membrane stability but dispensable for initiation of its assembly during early development. *Development, 131*(7), 1619–1628. http://dx.doi.org/10.1242/dev.01037.

Pyott, S. M., Schwarze, U., Christiansen, H. E., Pepin, M. G., Leistritz, D. F., Dineen, R., et al. (2011). Mutations in PPIB (cyclophilin B) delay type I procollagen chain association and result in perinatal lethal to moderate osteogenesis imperfecta phenotypes. *Human Molecular Genetics, 20*(8), 1595–1609. http://dx.doi.org/10.1093/hmg/ddr037.

Radner, S., Banos, C., Bachay, G., Li, Y. N., Hunter, D. D., Brunken, W. J., et al. (2013). beta2 and gamma3 laminins are critical cortical basement membrane components: ablation of Lamb2 and Lamc3 genes disrupts cortical lamination and produces dysplasia. *Developmental Neurobiology, 73*(3), 209–229. http://dx.doi.org/10.1002/dneu.22057.

Ramachandran, G. N., & Kartha, G. (1955). Structure of collagen. *Nature, 176*(4482), 593–595.

Rannikmae, K., Davies, G., Thomson, P. A., Bevan, S., Devan, W. J., Falcone, G. J., &, International Stroke Genetics, C. (2015). Common variation in COL4A1/COL4A2 is associated with sporadic cerebral small vessel disease. *Neurology, 84*(9), 918–926. http://dx.doi.org/10.1212/WNL.0000000000001309.

Rauch, F., Fahiminiya, S., Majewski, J., Carrot-Zhang, J., Boudko, S., Glorieux, F., et al. (2015). Cole-Carpenter syndrome is caused by a heterozygous missense mutation in P4HB. *American Journal of Human Genetics, 96*(3), 425–431. http://dx.doi.org/10.1016/j.ajhg.2014.12.027.

Rautavuoma, K., Takaluoma, K., Sormunen, R., Myllyharju, J., Kivirikko, K. I., & Soininen, R. (2004). Premature aggregation of type IV collagen and early lethality in lysyl hydroxylase 3 null mice. *Proceedings of the National Academy of Sciences of the United States of America, 101*(39), 14120–14125. http://dx.doi.org/10.1073/pnas.0404966101.

Ren, J., Wen, L., Gao, X., Jin, C., Xue, Y., & Yao, X. (2009). DOG 1.0: illustrator of protein domain structures. *Cell Research, 19*(2), 271–273. http://dx.doi.org/10.1038/cr.2009.6.

Rheault, M. N. (2012). Women and Alport syndrome. *Pediatric Nephrology, 27*(1), 41–46. http://dx.doi.org/10.1007/s00467-011-1836-7.

Rheault, M. N., Kren, S. M., Thielen, B. K., Mesa, H. A., Crosson, J. T., Thomas, W., et al. (2004). Mouse model of X-linked Alport syndrome. *Journal of the American Society of Nephrology, 15*(6), 1466–1474.

Risteli, J., Bachinger, H. P., Engel, J., Furthmayr, H., & Timpl, R. (1980). 7-S collagen: characterization of an unusual basement membrane structure. *European Journal of Biochemistry, 108*(1), 239–250.

Risteli, M., Ruotsalainen, H., Salo, A. M., Sormunen, R., Sipila, L., Baker, N. L., et al. (2009). Reduction of lysyl hydroxylase 3 causes deleterious changes in the deposition and organization of extracellular matrix. *Journal of Biological Chemistry, 284*(41), 28204–28211. http://dx.doi.org/10.1074/jbc.M109.038190.

Rodahl, E., Knappskog, P. M., Majewski, J., Johansson, S., Telstad, W., Krakenes, J., et al. (2013). Variants of anterior segment dysgenesis and cerebral involvement in a large family with a novel COL4A1 mutation. *American Journal of Ophthalmology, 155*(5), 946–953. http://dx.doi.org/10.1016/j.ajo.2012.11.028.

Rooney, J. E., Welser, J. V., Dechert, M. A., Flintoff-Dye, N. L., Kaufman, S. J., & Burkin, D. J. (2006). Severe muscular dystrophy in mice that lack dystrophin and alpha7 integrin. *Journal of Cell Science, 119*(Pt 11), 2185–2195. http://dx.doi.org/10.1242/jcs.02952.

Rosenbloom, J., Harsch, M., & Jimenez, S. (1973). Hydroxyproline content determines the denaturation temperature of chick tendon collagen. *Archives of Biochemistry and Biophysics, 158*(2), 478–484.

Ruotsalainen, H., Sipila, L., Vapola, M., Sormunen, R., Salo, A. M., Uitto, L., et al. (2006). Glycosylation catalyzed by lysyl hydroxylase 3 is essential for basement membranes. *Journal of Cell Science, 119*(Pt 4), 625–635. http://dx.doi.org/10.1242/jcs.02780.

Saga, S., Nagata, K., Chen, W. T., & Yamada, K. M. (1987). pH-dependent function, purification, and intracellular location of a major collagen-binding glycoprotein. *Journal of Cell Biology, 105*(1), 517–527.

Saito, K., Chen, M., Bard, F., Chen, S., Zhou, H., Woodley, D., et al. (2009). TANGO1 facilitates cargo loading at endoplasmic reticulum exit sites. *Cell, 136*(5), 891–902. http://dx.doi.org/10.1016/j.cell.2008.12.025.

Saito, K., Yamashiro, K., Ichikawa, Y., Erlmann, P., Kontani, K., Malhotra, V., et al. (2011). cTAGE5 mediates collagen secretion through interaction with TANGO1 at endoplasmic reticulum exit sites. *Molecular Biology of the Cell, 22*(13), 2301–2308. http://dx.doi.org/10.1091/mbc.E11-02-0143.

Salama, A. D., Levy, J. B., Lightstone, L., & Pusey, C. D. (2001). Goodpasture's disease. *Lancet, 358*(9285), 917–920. http://dx.doi.org/10.1016/S0140-6736(01)06077-9.

Salo, A. M., Cox, H., Farndon, P., Moss, C., Grindulis, H., Risteli, M., et al. (2008). A connective tissue disorder caused by mutations of the lysyl hydroxylase 3 gene. *American Journal of Human Genetics, 83*(4), 495–503. http://dx.doi.org/10.1016/j.ajhg.2008.09.004.

Salo, A. M., Sipila, L., Sormunen, R., Ruotsalainen, H., Vainio, S., & Myllyla, R. (2006). The lysyl hydroxylase isoforms are widely expressed during mouse embryogenesis, but obtain tissue- and cell-specific patterns in the adult. *Matrix Biology, 25*(8), 475–483. http://dx.doi.org/10.1016/j.matbio.2006.08.260.

Sanes, J. R., Engvall, E., Butkowski, R., & Hunter, D. D. (1990). Molecular heterogeneity of basal laminae: isoforms of laminin and collagen IV at the neuromuscular junction and elsewhere. *Journal of Cell Biology, 111*(4), 1685–1699.

Saus, J., Quinones, S., MacKrell, A., Blumberg, B., Muthukumaran, G., Pihlajaniemi, T., et al. (1989). The complete primary structure of mouse alpha 2(IV) collagen. Alignment with mouse alpha 1(IV) collagen. *Journal of Biological Chemistry, 264*(11), 6318–6324.

Saus, J., Wieslander, J., Langeveld, J. P., Quinones, S., & Hudson, B. G. (1988). Identification of the Goodpasture antigen as the alpha 3(IV) chain of collagen IV. *Journal of Biological Chemistry, 263*(26), 13374–13380.

Savige, J., Liu, J., DeBuc, D. C., Handa, J. T., Hageman, G. S., Wang, Y. Y., et al. (2010). Retinal basement membrane abnormalities and the retinopathy of Alport syndrome. *Investigative Ophthalmology & Visual Science, 51*(3), 1621–1627. http://dx.doi.org/10.1167/iovs.08-3323.

Savige, J., Sheth, S., Leys, A., Nicholson, A., Mack, H. G., & Colville, D. (2015). Ocular features in Alport syndrome: pathogenesis and clinical significance. *Clinical Journal of the American Society of Nephrology, 10*(4), 703–709. http://dx.doi.org/10.2215/CJN.10581014.

Schmid, R. S., & Anton, E. S. (2003). Role of integrins in the development of the cerebral cortex. *Cerebral Cortex, 13*(3), 219–224.

Schmidpeter, P. A., & Schmid, F. X. (2015). Prolyl isomerization and its catalysis in protein folding and protein function. *Journal of Molecular Biology, 427*(7), 1609–1631. http://dx.doi.org/10.1016/j.jmb.2015.01.023.

Schmidt, C., Fischer, G., Kadner, H., Genersch, E., Kuhn, K., & Poschl, E. (1993). Differential effects of DNA-binding proteins on bidirectional transcription from the common promoter region of human collagen type IV genes COL4A1 and COL4A2. *Biochimica et Biophysica Acta, 1174*(1), 1–10.

Schmidt, C., Pollner, R., Poschl, E., & Kuhn, K. (1992). Expression of human collagen type IV genes is regulated by transcriptional and post-transcriptional mechanisms. *FEBS Letters, 312*(2–3), 174–178.

Schuppan, D., Glanville, R. W., & Timpl, R. (1982). Covalent structure of mouse type-IV collagen. Isolation, order and partial amino-acid sequence of cyanogen-bromide and

tryptic peptides of pepsin fragment P1 from the alpha 1(IV) chain. *European Journal of Biochemistry, 123*(3), 505–512.
Schuppan, D., Glanville, R. W., Timpl, R., Dixit, S. N., & Kang, A. H. (1984). Sequence comparison of pepsin-resistant segments of basement-membrane collagen alpha 1(IV) chains from bovine lens capsule and mouse tumour. *Biochemical Journal, 220*(1), 227–233.
Schuppan, D., Timpl, R., & Glanville, R. W. (1980). Discontinuities in the triple helical sequence Gly-X-Y of basement membrane (type IV) collagen. *FEBS Letters, 115*(2), 297–300.
Schwarz-Magdolen, U., Oberbaumer, I., & Kuhn, K. (1986). cDNA and protein sequence of the NC1 domain of the alpha 2-chain of collagen IV and its comparison with alpha 1(IV). *FEBS Letters, 208*(2), 203–207.
Segal, Y., Zhuang, L., Rondeau, E., Sraer, J. D., & Zhou, J. (2001). Regulation of the paired type IV collagen genes COL4A5 and COL4A6. Role of the proximal promoter region. *Journal of Biological Chemistry, 276*(15), 11791–11797. http://dx.doi.org/10.1074/jbc.M007477200.
Shah, S., Ellard, S., Kneen, R., Lim, M., Osborne, N., Rankin, J., et al. (2012). Childhood presentation of COL4A1 mutations. *Developmental Medicine and Child Neurology, 54*(6), 569–574. http://dx.doi.org/10.1111/j.1469-8749.2011.04198.x.
Shah, S., Kumar, Y., McLean, B., Churchill, A., Stoodley, N., Rankin, J., et al. (2010). A dominantly inherited mutation in collagen IV A1 (COL4A1) causing childhood onset stroke without porencephaly. *European Journal of Paediatric Neurology, 14*(2), 182–187. http://dx.doi.org/10.1016/j.ejpn.2009.04.010.
Shoulders, M. D., & Raines, R. T. (2009). Collagen structure and stability. *Annual Review of Biochemistry, 78*, 929–958. http://dx.doi.org/10.1146/annurev.biochem.77.032207.120833.
Sibley, M. H., Graham, P. L., von Mende, N., & Kramer, J. M. (1994). Mutations in the alpha 2(IV) basement membrane collagen gene of *Caenorhabditis elegans* produce phenotypes of differing severities. *EMBO Journal, 13*(14), 3278–3285.
Sibon, I., Coupry, I., Menegon, P., Bouchet, J. P., Gorry, P., Burgelin, I., et al. (2007). COL4A1 mutation in Axenfeld-Rieger anomaly with leukoencephalopathy and stroke. *Annals of Neurology, 62*(2), 177–184. http://dx.doi.org/10.1002/ana.21191.
Siebold, B., Deutzmann, R., & Kuhn, K. (1988). The arrangement of intra- and intermolecular disulfide bonds in the carboxyterminal, non-collagenous aggregation and cross-linking domain of basement-membrane type IV collagen. *European Journal of Biochemistry, 176*(3), 617–624.
Siebold, B., Qian, R. A., Glanville, R. W., Hofmann, H., Deutzmann, R., & Kuhn, K. (1987). Construction of a model for the aggregation and cross-linking region (7S domain) of type IV collagen based upon an evaluation of the primary structure of the alpha 1 and alpha 2 chains in this region. *European Journal of Biochemistry, 168*(3), 569–575.
Sipila, L., Ruotsalainen, H., Sormunen, R., Baker, N. L., Lamande, S. R., Vapola, M., et al. (2007). Secretion and assembly of type IV and VI collagens depend on glycosylation of hydroxylysines. *Journal of Biological Chemistry, 282*(46), 33381–33388. http://dx.doi.org/10.1074/jbc.M704198200.
Slavotinek, A. M., Garcia, S. T., Chandratillake, G., Bardakjian, T., Ullah, E., Wu, D., et al. (2014). Exome sequencing in 32 patients with anophthalmia/microphthalmia and developmental eye defects. *Clinical Genetics*. http://dx.doi.org/10.1111/cge.12543.
van der Slot, A. J., Zuurmond, A. M., Bardoel, A. F., Wijmenga, C., Pruijs, H. E., Sillence, D. O., et al. (2003). Identification of PLOD2 as telopeptide lysyl hydroxylase, an important enzyme in fibrosis. *Journal of Biological Chemistry, 278*(42), 40967–40972. http://dx.doi.org/10.1074/jbc.M307380200.
Soininen, R., Haka-Risku, T., Prockop, D. J., & Tryggvason, K. (1987). Complete primary structure of the alpha 1-chain of human basement membrane (type IV) collagen. *FEBS Letters, 225*(1–2), 188–194.

Soininen, R., Huotari, M., Hostikka, S. L., Prockop, D. J., & Tryggvason, K. (1988). The structural genes for alpha 1 and alpha 2 chains of human type IV collagen are divergently encoded on opposite DNA strands and have an overlapping promoter region. *Journal of Biological Chemistry, 263*(33), 17217–17220.

Stanton, M. C., & Tange, J. D. (1958). Goodpasture's syndrome (pulmonary haemorrhage associated with glomerulonephritis). *Australasian Annals of Medicine, 7*(2), 132–144.

Stawikowski, M. J., Aukszi, B., Stawikowska, R., Cudic, M., & Fields, G. B. (2014). Glycosylation modulates melanoma cell alpha2beta1 and alpha3beta1 integrin interactions with type IV collagen. *Journal of Biological Chemistry, 289*(31), 21591–21604. http://dx.doi.org/10.1074/jbc.M114.572073.

Steinmann, B., Bruckner, P., & Superti-Furga, A. (1991). Cyclosporin A slows collagen triple-helix formation in vivo: indirect evidence for a physiologic role of peptidyl-prolyl cis-trans-isomerase. *Journal of Biological Chemistry, 266*(2), 1299–1303.

Sugimoto, M., Oohashi, T., & Ninomiya, Y. (1994). The genes COL4A5 and COL4A6, coding for basement membrane collagen chains alpha 5(IV) and alpha 6(IV), are located head-to-head in close proximity on human chromosome Xq22 and COL4A6 is transcribed from two alternative promoters. *Proceedings of the National Academy of Sciences of the United States of America, 91*(24), 11679–11683.

Sund, M., Maeshima, Y., & Kalluri, R. (2005). Bifunctional promoter of type IV collagen COL4A5 and COL4A6 genes regulates the expression of alpha5 and alpha6 chains in a distinct cell-specific fashion. *Biochemical Journal, 387*(Pt 3), 755–761. http://dx.doi.org/10.1042/BJ20041870.

Sundaramoorthy, M., Meiyappan, M., Todd, P., & Hudson, B. G. (2002). Crystal structure of NC1 domains. Structural basis for type IV collagen assembly in basement membranes. *Journal of Biological Chemistry, 277*(34), 31142–31153. http://dx.doi.org/10.1074/jbc.M201740200.

Takenouchi, T., Ohyagi, M., Torii, C., Kosaki, R., Takahashi, T., & Kosaki, K. (2015). Porencephaly in a fetus and HANAC in her father: variable expression of COL4A1 mutation. *American Journal of Medical Genetics, Part A, 167A*(1), 156–158. http://dx.doi.org/10.1002/ajmg.a.36823.

Tarasov, K. V., Sanna, S., Scuteri, A., Strait, J. B., Orru, M., Parsa, A., et al. (2009). COL4A1 is associated with arterial stiffness by genome-wide association scan. *Circulation: Cardiovascular Genetics, 2*(2), 151–158. http://dx.doi.org/10.1161/CIRCGENETICS.108.823245.

Tasab, M., Batten, M. R., & Bulleid, N. J. (2000). Hsp47: a molecular chaperone that interacts with and stabilizes correctly-folded procollagen. *EMBO Journal, 19*(10), 2204–2211. http://dx.doi.org/10.1093/emboj/19.10.2204.

Tasab, M., Jenkinson, L., & Bulleid, N. J. (2002). Sequence-specific recognition of collagen triple helices by the collagen-specific molecular chaperone HSP47. *Journal of Biological Chemistry, 277*(38), 35007–35012. http://dx.doi.org/10.1074/jbc.M202782200.

Than, M. E., Henrich, S., Huber, R., Ries, A., Mann, K., Kuhn, K., et al. (2002). The 1.9-A crystal structure of the noncollagenous (NC1) domain of human placenta collagen IV shows stabilization via a novel type of covalent Met-Lys cross-link. *Proceedings of the National Academy of Sciences of the United States of America, 99*(10), 6607–6612. http://dx.doi.org/10.1073/pnas.062183499.

Thaung, C., West, K., Clark, B. J., McKie, L., Morgan, J. E., Arnold, K., et al. (2002). Novel ENU-induced eye mutations in the mouse: models for human eye disease. *Human Molecular Genetics, 11*(7), 755–767.

Thielen, B. K., Barker, D. F., Nelson, R. D., Zhou, J., Kren, S. M., & Segal, Y. (2003). Deletion mapping in Alport syndrome and Alport syndrome-diffuse leiomyomatosis reveals potential mechanisms of visceral smooth muscle overgrowth. *Human Mutation, 22*(5), 419. http://dx.doi.org/10.1002/humu.9191.

Thorner, P. S. (2007). Alport syndrome and thin basement membrane nephropathy. *Nephron Clinical Practice, 106*(2), c82–88. http://dx.doi.org/10.1159/000101802.

Tiainen, P., Pasanen, A., Sormunen, R., & Myllyharju, J. (2008). Characterization of recombinant human prolyl 3-hydroxylase isoenzyme 2, an enzyme modifying the basement membrane collagen IV. *Journal of Biological Chemistry, 283*(28), 19432–19439. http://dx.doi.org/10.1074/jbc.M802973200.

Timpl, R. (1989). Structure and biological activity of basement membrane proteins. *European Journal of Biochemistry, 180*(3), 487–502.

Timpl, R., Bruckner, P., & Fietzek, P. (1979). Characterization of pepsin fragments of basement membrane collagen obtained from a mouse tumor. *European Journal of Biochemistry, 95*(2), 255–263.

Timpl, R., Risteli, J., & Bachinger, H. P. (1979). Identification of a new basement membrane collagen by the aid of a large fragment resistant to bacterial collagenase. *FEBS Letters, 101*(2), 265–268.

Timpl, R., Wiedemann, H., van Delden, V., Furthmayr, H., & Kuhn, K. (1981). A network model for the organization of type IV collagen molecules in basement membranes. *European Journal of Biochemistry, 120*(2), 203–211.

Tonduti, D., Pichiecchio, A., La Piana, R., Livingston, J. H., Doherty, D. A., Majumdar, A., et al. (2012). COL4A1-related disease: raised creatine kinase and cerebral calcification as useful pointers. *Neuropediatrics, 43*(5), 283–288. http://dx.doi.org/10.1055/s-0032-1325116.

Tryggvason, K., & Patrakka, J. (2006). Thin basement membrane nephropathy. *Journal of the American Society of Nephrology, 17*(3), 813–822. http://dx.doi.org/10.1681/ASN.2005070737.

Tryggvason, K., Robey, P. G., & Martin, G. R. (1980). Biosynthesis of type IV procollagens. *Biochemistry, 19*(7), 1284–1289.

Uliana, V., Marcocci, E., Mucciolo, M., Meloni, I., Izzi, C., Manno, C., et al. (2011). Alport syndrome and leiomyomatosis: the first deletion extending beyond COL4A6 intron 2. *Pediatric Nephrology, 26*(5), 717–724. http://dx.doi.org/10.1007/s00467-010-1693-9.

Vahedi, K., Boukobza, M., Massin, P., Gould, D. B., Tournier-Lasserve, E., & Bousser, M. G. (2007). Clinical and brain MRI follow-up study of a family with COL4A1 mutation. *Neurology, 69*(16), 1564–1568. http://dx.doi.org/10.1212/01.wnl.0000295994.46586.e7.

Vahedi, K., Kubis, N., Boukobza, M., Arnoult, M., Massin, P., Tournier-Lasserve, E., et al. (2007). COL4A1 mutation in a patient with sporadic, recurrent intracerebral hemorrhage. *Stroke, 38*(5), 1461–1464. http://dx.doi.org/10.1161/STROKEAHA.106.475194.

Van Agtmael, T., Bailey, M. A., Schlotzer-Schrehardt, U., Craigie, E., Jackson, I. J., Brownstein, D. G., et al. (2010). Col4a1 mutation in mice causes defects in vascular function and low blood pressure associated with reduced red blood cell volume. *Human Molecular Genetics, 19*(6), 1119–1128. http://dx.doi.org/10.1093/hmg/ddp584.

Van Agtmael, T., Schlotzer-Schrehardt, U., McKie, L., Brownstein, D. G., Lee, A. W., Cross, S. H., et al. (2005). Dominant mutations of Col4a1 result in basement membrane defects which lead to anterior segment dysgenesis and glomerulopathy. *Human Molecular Genetics, 14*(21), 3161–3168. http://dx.doi.org/10.1093/hmg/ddi348.

Vanacore, R., Ham, A. J., Voehler, M., Sanders, C. R., Conrads, T. P., Veenstra, T. D., et al. (2009). A sulfilimine bond identified in collagen IV. *Science, 325*(5945), 1230–1234. http://dx.doi.org/10.1126/science.1176811.

Vanacore, R. M., Friedman, D. B., Ham, A. J., Sundaramoorthy, M., & Hudson, B. G. (2005). Identification of S-hydroxylysyl-methionine as the covalent cross-link of the noncollagenous (NC1) hexamer of the alpha1alpha1alpha2 collagen IV network: a role for the post-translational modification of lysine 211 to hydroxylysine 211 in hexamer assembly. *Journal of Biological Chemistry, 280*(32), 29300–29310. http://dx.doi.org/10.1074/jbc.M502752200.

Vanacore, R. M., Shanmugasundararaj, S., Friedman, D. B., Bondar, O., Hudson, B. G., & Sundaramoorthy, M. (2004). The alpha1.alpha2 network of collagen IV. Reinforced stabilization of the noncollagenous domain-1 by noncovalent forces and the absence of Met-Lys cross-links. *Journal of Biological Chemistry, 279*(43), 44723–44730. http://dx.doi.org/10.1074/jbc.M406344200.

Vandenberg, P., Kern, A., Ries, A., Luckenbill-Edds, L., Mann, K., & Kuhn, K. (1991). Characterization of a type IV collagen major cell binding site with affinity to the alpha 1 beta 1 and the alpha 2 beta 1 integrins. *Journal of Cell Biology, 113*(6), 1475–1483.

Verbeek, E., Meuwissen, M. E., Verheijen, F. W., Govaert, P. P., Licht, D. J., Kuo, D. S., et al. (2012). COL4A2 mutation associated with familial porencephaly and small-vessel disease. *European Journal of Human Genetics, 20*(8), 844–851. http://dx.doi.org/10.1038/ejhg.2012.20.

Vermeulen, R. J., Peeters-Scholte, C., Van Vugt, J. J., Barkhof, F., Rizzu, P., van der Schoor, S. R., et al. (2011). Fetal origin of brain damage in 2 infants with a COL4A1 mutation: fetal and neonatal MRI. *Neuropediatrics, 42*(1), 1–3. http://dx.doi.org/10.1055/s-0031-1275343.

Vranka, J., Stadler, H. S., & Bachinger, H. P. (2009). Expression of prolyl 3-hydroxylase genes in embryonic and adult mouse tissues. *Cell Structure and Function, 34*(2), 97–104.

Vranka, J. A., Pokidysheva, E., Hayashi, L., Zientek, K., Mizuno, K., Ishikawa, Y., et al. (2010). Prolyl 3-hydroxylase 1 null mice display abnormalities in fibrillar collagen-rich tissues such as tendons, skin, and bones. *Journal of Biological Chemistry, 285*(22), 17253–17262. http://dx.doi.org/10.1074/jbc.M110.102228.

de Vries, L. S., Koopman, C., Groenendaal, F., Van Schooneveld, M., Verheijen, F. W., Verbeek, E., et al. (2009). COL4A1 mutation in two preterm siblings with antenatal onset of parenchymal hemorrhage. *Annals of Neurology, 65*(1), 12–18. http://dx.doi.org/10.1002/ana.21525.

de Vries, L. S., & Mancini, G. M. (2012). Intracerebral hemorrhage and COL4A1 and COL4A2 mutations, from fetal life into adulthood. *Annals of Neurology, 71*(4), 439–441. http://dx.doi.org/10.1002/ana.23544.

Welch, W. J., & Brown, C. R. (1996). Influence of molecular and chemical chaperones on protein folding. *Cell Stress & Chaperones, 1*(2), 109–115.

Weng, Y. C., Sonni, A., Labelle-Dumais, C., de Leau, M., Kauffman, W. B., Jeanne, M., et al. (2012). COL4A1 mutations in patients with sporadic late-onset intracerebral hemorrhage. *Annals of Neurology, 71*(4), 470–477. http://dx.doi.org/10.1002/ana.22682.

Wieslander, J., Barr, J. F., Butkowski, R. J., Edwards, S. J., Bygren, P., Heinegard, D., et al. (1984). Goodpasture antigen of the glomerular basement membrane: localization to noncollagenous regions of type IV collagen. *Proceedings of the National Academy of Sciences of the United States of America, 81*(12), 3838–3842.

Wieslander, J., Langeveld, J., Butkowski, R., Jodlowski, M., Noelken, M., & Hudson, B. G. (1985). Physical and immunochemical studies of the globular domain of type IV collagen. Cryptic properties of the Goodpasture antigen. *Journal of Biological Chemistry, 260*(14), 8564–8570.

Wilkinson, B., & Gilbert, H. F. (2004). Protein disulfide isomerase. *Biochimica et Biophysica Acta, 1699*(1–2), 35–44. http://dx.doi.org/10.1016/j.bbapap.2004.02.017.

Willem, M., Miosge, N., Halfter, W., Smyth, N., Jannetti, I., Burghart, E., et al. (2002). Specific ablation of the nidogen-binding site in the laminin gamma1 chain interferes with kidney and lung development. *Development, 129*(11), 2711–2722.

Wilson, C., Borza, D., & Hudson, B. (2002). Autoimmune renal disease involving renal basement membrane antigen. In A. N. Theofilopoulos, & A. C. Bona (Eds.), *The molecular pathology of autoimmune diseases* (2nd ed., pp. 981–1010). Newark: Gordon&Breach Science/Harwood Academic.

Wilson, D. G., Phamluong, K., Li, L., Sun, M., Cao, T. C., Liu, P. S., et al. (2011). Global defects in collagen secretion in a Mia3/TANGO1 knockout mouse. *Journal of Cell Biology, 193*(5), 935–951. http://dx.doi.org/10.1083/jcb.201007162.

Winter, A. D., McCormack, G., & Page, A. P. (2007). Protein disulfide isomerase activity is essential for viability and extracellular matrix formation in the nematode *Caenorhabditis elegans*. *Developmental Biology, 308*(2), 449–461. http://dx.doi.org/10.1016/j.ydbio.2007.05.041.

Xia, X. Y., Li, N., Cao, X., Wu, Q. Y., Li, T. F., Zhang, C., et al. (2014). A novel COL4A1 gene mutation results in autosomal dominant non-syndromic congenital cataract in a Chinese family. *BMC Medical Genetics, 15*, 97. http://dx.doi.org/10.1186/s12881-014-0097-2.

Xu, H., Christmas, P., Wu, X. R., Wewer, U. M., & Engvall, E. (1994). Defective muscle basement membrane and lack of M-laminin in the dystrophic dy/dy mouse. *Proceedings of the National Academy of Sciences of the United States of America, 91*(12), 5572–5576.

Yamamoto, Y., Craggs, L., Baumann, M., Kalimo, H., & Kalaria, R. N. (2011). Review: molecular genetics and pathology of hereditary small vessel diseases of the brain. *Neuropathology and Applied Neurobiology, 37*(1), 94–113. http://dx.doi.org/10.1111/j.1365-2990.2010.01147.x.

Yamauchi, M., & Sricholpech, M. (2012). Lysine post-translational modifications of collagen. *Essays in Biochemistry, 52*, 113–133. http://dx.doi.org/10.1042/bse0520113.

Yan, X., Sabrautzki, S., Horsch, M., Fuchs, H., Gailus-Durner, V., Beckers, J., et al. (2014). Peroxidasin is essential for eye development in the mouse. *Human Molecular Genetics, 23*(21), 5597–5614. http://dx.doi.org/10.1093/hmg/ddu274.

Yao, Y., Chen, Z. L., Norris, E. H., & Strickland, S. (2014). Astrocytic laminin regulates pericyte differentiation and maintains blood–brain barrier integrity. *Nature Communications, 5*, 3413. http://dx.doi.org/10.1038/ncomms4413.

Yasothornsrikul, S., Davis, W. J., Cramer, G., Kimbrell, D. A., & Dearolf, C. R. (1997). viking: identification and characterization of a second type IV collagen in *Drosophila*. *Gene, 198*(1–2), 17–25.

Ylikarppa, R., Eklund, L., Sormunen, R., Kontiola, A. I., Utriainen, A., Maatta, M., et al. (2003). Lack of type XVIII collagen results in anterior ocular defects. *FASEB Journal, 17*(15), 2257–2259. http://dx.doi.org/10.1096/fj.02-1001fje.

Yoneda, Y., Haginoya, K., Kato, M., Osaka, H., Yokochi, K., Arai, H., et al. (2013). Phenotypic spectrum of COL4A1 mutations: porencephaly to schizencephaly. *Annals of Neurology, 73*(1), 48–57. http://dx.doi.org/10.1002/ana.23736.

Yoshioka, K., Hino, S., Takemura, T., Maki, S., Wieslander, J., Takekoshi, Y., et al. (1994). Type IV collagen alpha 5 chain. Normal distribution and abnormalities in X-linked Alport syndrome revealed by monoclonal antibody. *American Journal of Pathology, 144*(5), 986–996.

Yurchenco, P. D., & Furthmayr, H. (1984). Self-assembly of basement membrane collagen. *Biochemistry, 23*(8), 1839–1850.

Yurchenco, P. D., & Ruben, G. C. (1987). Basement membrane structure in situ: evidence for lateral associations in the type IV collagen network. *Journal of Cell Biology, 105*(6 Pt 1), 2559–2568.

Yurchenco, P. D., & Ruben, G. C. (1988). Type IV collagen lateral associations in the EHS tumor matrix. Comparison with amniotic and in vitro networks. *American Journal of Pathology, 132*(2), 278–291.

Zenker, M., Aigner, T., Wendler, O., Tralau, T., Muntefering, H., Fenski, R., et al. (2004). Human laminin beta2 deficiency causes congenital nephrosis with mesangial sclerosis and distinct eye abnormalities. *Human Molecular Genetics, 13*(21), 2625–2632. http://dx.doi.org/10.1093/hmg/ddh284.

Zhang, K. W., Colville, D., Tan, R., Jones, C., Alexander, S. I., Fletcher, J., et al. (2008). The use of ocular abnormalities to diagnose X-linked Alport syndrome in children. *Pediatric Nephrology, 23*(8), 1245–1250. http://dx.doi.org/10.1007/s00467-008-0759-4.

Zhao, J., Cui, Z., Yang, R., Jia, X. Y., Zhang, Y., & Zhao, M. H. (2009). Anti-glomerular basement membrane autoantibodies against different target antigens are associated with disease severity. *Kidney International, 76*(10), 1108–1115. http://dx.doi.org/10.1038/ki.2009.348.

Zhou, J., Ding, M., Zhao, Z., & Reeders, S. T. (1994). Complete primary structure of the sixth chain of human basement membrane collagen, alpha 6(IV). Isolation of the cDNAs for alpha 6(IV) and comparison with five other type IV collagen chains. *Journal of Biological Chemistry, 269*(18), 13193–13199.

Zhou, J., Hertz, J. M., Leinonen, A., & Tryggvason, K. (1992). Complete amino acid sequence of the human alpha 5 (IV) collagen chain and identification of a single-base mutation in exon 23 converting glycine 521 in the collagenous domain to cysteine in an Alport syndrome patient. *Journal of Biological Chemistry, 267*(18), 12475–12481.

Zode, G. S., Kuehn, M. H., Nishimura, D. Y., Searby, C. C., Mohan, K., Grozdanic, S. D., et al. (2011). Reduction of ER stress via a chemical chaperone prevents disease phenotypes in a mouse model of primary open angle glaucoma. *Journal of Clinical Investigation, 121*(9), 3542–3553. http://dx.doi.org/10.1172/JCI58183.

CHAPTER FOUR

Epidermal Basement Membrane in Health and Disease

Cristina Has* and Alexander Nyström*
Department of Dermatology, University Medical Center Freiburg, Freiburg, Germany
*Corresponding authors: E-mails: cristina.has@uniklinik-freiburg.de; alexander.nystroem@uniklinik-freiburg.de

Contents

Abstract

Skin, as the organ protecting the individual from environmental aggressions, constantly meets external insults and is dependent on mechanical toughness for its preserved function. Accordingly, the epidermal basement membrane (BM) zone has adapted to enforce tissue integrity. It harbors anchoring structures created through unique organization of common BM components and expression of proteins exclusive to the epidermal BM zone. Evidence for the importance of its correct assembly and the non-redundancy of its components for skin integrity is apparent from the multiple skin

Current Topics in Membranes, Volume 76
ISSN 1063-5823
http://dx.doi.org/10.1016/bs.ctm.2015.05.003

blistering disorders caused by mutations in genes coding for proteins associated with the epidermal BM and from autoimmune disorders in which autoantibodies target these molecules. However, it has become clear that these proteins not only provide mechanical support but are also critically involved in tissue homeostasis, repair, and regeneration. In this chapter, we provide an overview of the unique organization and components of the epidermal BM. A special focus will be given to its function during regeneration, and in inherited and acquired diseases.

Abbreviations

ADAM A disintegrin and metalloproteinase
BM Basement membrane
BPAG1 Bullous pemphigoid antigen 1
BPAG2 Bullous pemphigoid antigen 2
BP180 The 180-kD bullous pemphigoid antigen
DEJ Dermal–epidermal junction
EB Epidermolysis bullosa
ECM Extracellular matrix
EGF Epidermal growth factor
FERM F for 4.1 protein, E for ezrin, R for radixin, and M for moesin
FGF Fibroblast growth factor
Flii Flightless I
IF Intermediate filament
K5 Keratin 5
K14 Keratin 14
LAD Linear IgA dermatosis
LE Laminin epidermal growth factor-like
LN Laminin N-terminal
NC Noncollagenous
TGFβ Transforming growth factor-β

1. INTRODUCTION

Skin creates the physical boundary between self and the outside world. It constantly meets external challenges in the form of physical, immunological, and chemical insults, and acts as a first line of defense to combat these challenges. Thus, the skin is an organ that both defines and preserves the individual. Structurally it is composed of an outer keratinized stratified epithelium, the epidermis, connected to an inner connective tissue, the dermis. The epidermis is avascular and mainly populated by epidermal keratinocytes, with minor contributions of melanocytes, mechanosensitive Merkel cells, antigen presenting Langerhans cells, and other immune cells. Through intricate differentiation, production of lipids, extensive cross-linking, and highly evolved cell–cell contacts, keratinocytes establish a barrier that protects from penetration of chemical, physical, and biological agents and loss of fluids. Some mechanical support is also provided by the epidermal keratin cytoskeleton linking neighboring cells through specialized adhesion junctions. Still, most strength comes from the dermal extracellular matrix (ECM) and here primarily from fibrillar collagens.

The dermis is highly vascularized. Dermal fibroblasts are the major cell types in the dermis and the prime producers of the ECM. The dermis is divided into two layers, the superficial papillary matrix that is thin and extends toward the epidermis, and the deep reticular dermis composed of dense ECM.

In the interface between the epidermis and the dermis lies the epidermal basement membrane (BM), which is one of the largest uninterrupted BMs in the body. BMs in their classical configuration have limited strength. However, given the high physical demands of the skin, the epidermal BM has evolved to contain additional supportive structures that ensure firm adhesion of the epidermis to the dermis and protect against shearing forces. This zone is collectively known as the dermal–epidermal junction (DEJ). The DEJ is a functional unit composed of both cellular and extracellular components: the plasma membrane of the basal keratinocyte with its hemidesmosomes, a lamina lucida, lamina densa, and a sublamina densa fibrous zone or reticular layer (Figure 1); it is established by input from both epidermal keratinocytes and dermal fibroblasts. In addition to creating mechanical support, the DEJ is vital in regulating communication between the epidermis and dermis and actively contributes to tissue reconstruction and repair.

Like most other BMs, the epidermal BM contains all four principal components of BMs, laminins, type IV collagens, perlecan, and nidogens, but both it and the extended DEJ also harbor multiple additional specialized proteins. It is primarily these proteins that provide the DEJ with its unique structure and abilities. In the following sections, we will review the molecular composition of the DEJ, then examine its function during regeneration, and lastly discuss genetic and acquired disorders connected to the DEJ. As a general structure we have chosen to start with the epidermal BM and then subsequently expand to its associated epidermal and dermal adhesive structures.

2. COMPOSITION AND ARCHITECTURE OF THE EPIDERMAL BM AND ASSOCIATED ADHESION STRUCTURES

2.1 Epidermal BM

2.1.1 Laminins in the epidermal BM

The laminin network in the epidermal BM of mature skin is primarily composed of epithelial-specific laminin-α3A-containing laminins in the

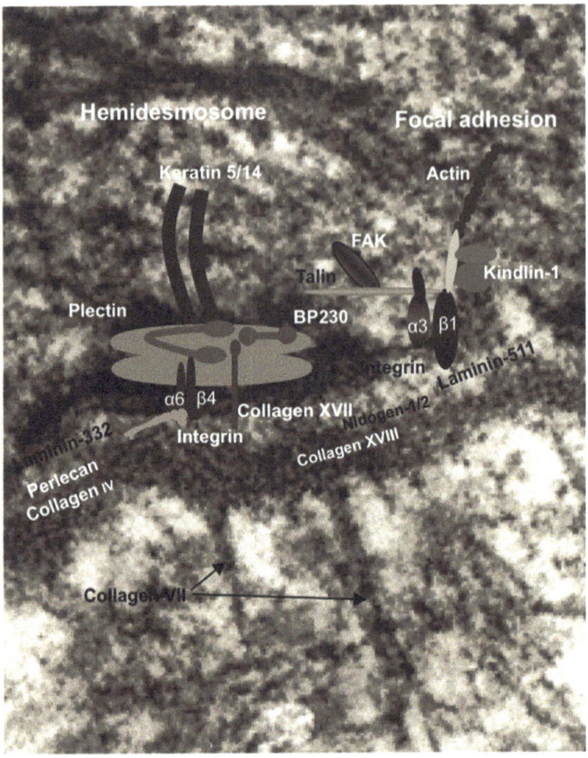

Figure 1 ***Schematic representation of the major molecules of the epidermal basement membrane and associated structures.*** An electron microscopic picture representing the dermal–epidermal junction serves as background. (See color plate)

form of laminin-3A32 (here referred to as 332) and -3A11, as well as the ubiquitous laminin-511 (Ryan et al., 1996; Sugawara, Tsuruta, Ishii, Jones, & Kobayashi, 2008). However, there is additional minor contribution from several other laminins, although their specific function in skin is not well known. Laminin-α2 shows a weak linear expression at the epidermal BM and around hair follicles (Sewry et al., 1996); in fact analysis of laminin-α2 expression in skin biopsies has been used to diagnose congenital muscular dystrophy caused by laminin-α2 deficiency (Allamand & Guicheney, 2002). At sites where nerves cross the DEJ, expression of a laminin-γ3 chain-containing laminin, likely laminin-423 or -523, occurs (Koch et al., 1999; Libby et al., 2000). There are dynamic changes in expression of laminins during skin morphogenesis and during aging. The largely embryonic laminin, laminin-111, is transiently expressed during skin development (Fleischmajer et al., 2000; Li et al., 2003) as is laminin-β2, which is found in fetal human skin but only weakly if at all in adult skin, indicating that laminin-3A21 or -521 is present during embryonic skin development (Champliaud et al., 1996; Iivanainen et al., 1995; Sollberg, Peltonen, & Uitto, 1992). Laminin-511 seems to appear later than laminin-332 (Fleischmajer et al., 2000; Lu, Miyazaki, Mizushima, & Nemoto, 2001; Roediger, Miosge, & Gersdorff, 2010), and its expression is also suggested to be reduced with age (Pouliot, Saunders, & Kaur, 2002).

Laminin-332 is the most abundant laminin and ECM protein expressed by keratinocytes (Botta et al., 2012) but is also present in multiple epithelial BMs outside the skin including those of kidney, lung, thymus, brain, gastrointestinal tract, and lung. Studies with transgenic mice suggest that it fills its most vital function in skin as laminin-332 expression in keratinocytes rescues laminin-γ2 knockout mice from neonatal lethality (Adair-Kirk et al., 2012; Kuster, Guarnieri, Ault, Flaherty, & Swiatek, 1997; Meng et al., 2003; Ryan, Lee, Miyashita, & Carter, 1999). Laminin-332 is not evenly distributed throughout the epidermal BM; it condenses in hemidesmosomes (see section below) and is downregulated in hair follicles during hair morphogenesis (Nanba, Hieda, & Nakanishi, 2000). Laminin-332 is solely contributed by keratinocytes, but as for the other laminins in the epidermal BM, epithelial–mesenchymal interactions are crucially involved in promoting expression (Fleischmajer et al., 2000). To some extent laminin-332 shares similar functions with laminin-511; both laminins support keratinocyte adhesion (Pouliot et al., 2002) and are important in maintaining structural integrity of the BM. Analyses of laminin-α5-deficient mice have shown that the epidermal BM is separated in the lamina lucida at some exposed sites (Miner, Cunningham, & Sanes, 1998). The epidermal BM is split at a similar level in laminin-332 deficiency, leading to junctional epidermolysis bullosa (EB) (see Section 4.1.1). There are also some apparent functional differences. These differences are most well characterized in terms of hair growth. In general, laminin-511 seems to support hair growth, whereas laminin-332 can largely be said to inhibit it (Gao et al., 2008; Imanishi et al., 2010; Li et al., 2003).

Structural analysis of purified laminin network from epidermal BM reveals that it is organized into honeycomb-like structures (Behrens et al., 2012). This is intriguing, as the predominant laminin-332 only carries one laminin N-terminal (LN) domain available for building laminin networks, since both the laminin-α3A and γ2 chains are truncated. Consequently, the epidermal laminin network is created through heterotypic lateral aggregation of laminins, likely with key contributions from both laminin-511 and

-311, although laminin-311 also has limitations to its networking ability (Behrens et al., 2012; Odenthal et al., 2004). Interactions between the laminin-α3A laminin epidermal growth factor-like (LE) domain and the laminin-β3 LN domain have been suggested to be an additional mechanism of polymerization of laminin-α3 networks (Rousselle & Beck, 2013).

On average, the long axis of individual laminin-332 molecules are not positioned perpendicular to the keratinocyte plasma membrane but lie slantwise in an approximately 30° angle with their γ2 chain facing furthest away (McMillan, Akiyama, & Shimizu, 2003). The positioning of the γ2 chain is interesting as it contains many high-affinity interaction sites for other BM proteins (Sasaki et al., 2001). Intriguingly, it frequently undergoes maturation through proteolytic removal of the short-arm domain by astacin-like proteases (Amano et al., 2000), implying that interaction with other BM ligands are needed for correct deposition in the BM but not for function of the mature BM. The α3A chain also matures through proteolysis. Under pathological conditions the LG4 domain can also be released by MT1-MMP (Koshikawa, Minegishi, Sharabi, Quaranta, & Seiki, 2005). Both N- and C-terminal processing of the LE domain and in the linker between LG3 and 4, respectively, may occur through astacin-like proteases (Amano et al., 2000). The C-terminus has additionally been shown to be cleaved in vitro by plasmin, thrombin, MMP-2, and MT1-MMP (Rousselle & Beck, 2013). The function of N-terminal processing is not clear but it may play a role in regenerative processes, since patients with mutations in this domain suffer from increased formation of vascular granulation tissue (McLean et al., 2003). Removal of the LG4–5 domain has implications for integration into the BM and receptor interaction and consequently keratinocyte behavior. The LG4–5 domain through its high heparin affinity likely promotes incorporation of laminin-332 into the BM (Sigle et al., 2004; Tran et al., 2008; Tsubota et al., 2005), although one study did not find a clear correlation of LG4–5 retention and deposition (Baudoin, Fantin, & Meneguzzi, 2005).

In vitro analyses indicate that both laminin-332 and -511 are able to strongly interact with integrin α3β1 and α6β4 through their LG1–3 domain, with crucial contribution of the β and γ chains (Nishiuchi et al., 2006). Investigation of laminin-α3-deficient mice suggests that integrin α6β4 is the major receptor for mature laminin-332 and integrin α3β1 the major laminin-511 receptor (Ryan et al., 1999). In addition to integrin receptors, laminin-332 interacts with multiple proteins. Laminin-332 interacts with collagen XVII (Nishie, Kiritsi, Nystrom, Hofmann, & Bruckner-Tuderman, 2011), and strong heparin affinity of the α3 LG4–5 domain enables binding to syndecan 1 and 4 (Rousselle & Beck, 2013). The γ2 short arm binds perlecan, nidogen 1, fibulin 2 (Sasaki et al., 2001) and additionally, weakly collagen VII (Chen et al., 1999), whereas the β3 short arm has strong affinity for collagen VII (Chen et al., 1999). It is important to note that the laminin-β3 short arm, in contrast to the other two short arms, is not proteolytically matured, although cleavage by MMP-7 may occur during certain pathological events (Remy, Trespeuch, Bachy, Scoazec, & Rousselle, 2006). The lack of processing of the β3 chain under physiological conditions emphasizes the importance of the maintenance of collagen VII—laminin-332 interaction for skin stability (see below).

2.1.2 Collagen IV in the epidermal BM

Collagen IV arranges into a second network below the laminin network. Six genetically distinct α chains make up three heterodimers (see Chapter 4), which show tissue distinct expression (Yurchenco, 2011). Collagen $[\alpha 1(IV)]_2\alpha 2(IV)$ is the main form in the epidermal BM with some additional contribution from $[\alpha 5(IV)]_2\alpha 6(IV)$ (Hasegawa et al., 2007). A complete absence of collagen $[\alpha 1(IV)]_2\alpha 2(IV)$ is embryonically lethal, but reduced functionality and pathologies stemming from point mutations have been described and involve vascular destabilization (Gould et al., 2005, 2006). Mouse embryos lacking collagen $[\alpha 1(IV)]_2\alpha 2(IV)$ show an irregular and partially absent epidermal BM, suggesting that collagen IV is vital for maintaining epidermal BM integrity (Poschl et al., 2004). So far no skin phenotype has been described in patients with syndromic *COL4A1 and COL4A2* mutations, but this may have gone unnoticed due focus on other aspects of these disorders.

2.1.3 Adhesion of the laminin and collagen IV networks in the epidermal BM

For BM stability the laminin and collagen IV networks need to be connected. Nidogens were viewed as candidates to perform this task in the epidermal BM. Two forms exist, nidogen 1 and nidogen 2. Both forms are present in the epidermal BM and contributed by mesenchymal cells (Nischt et al., 2007). Genetic studies have shown that nidogen 1 and 2 are not needed for mediating adhesion of the two networks (Mokkapati et al., 2008). Mice lacking both forms of nidogen do not display obvious dermal blistering. At the molecular level absence of nidogen does not interfere or alter incorporation of other epidermal BM components. As analyzed by electron microscopy, the epidermal BM appears ultrastructurally normal, although there are tendencies to minor localized separations between the lamina densa and papillary matrix (Mokkapati et al., 2008). These results have clearly indicated that nidogens are dispensable for connecting the laminin and collagen IV networks.

The BM component perlecan is an additional possible linker, but it is not needed for formation of the epidermal BM (Costell et al., 1999). Perlecan is expressed by both epidermal keratinocytes and dermal fibroblasts (Sher et al., 2006). Fibrillin 1, which forms microfibrils participating in anchorage of the epidermal BM to the papillary matrix, interacts with perlecan, and loss of perlecan reduces the number of microfibrils (Tiedemann et al., 2005). Perlecan has, through its heparan sulfate chains, affinity for multiple BM components including, laminin and collagen IV (Battaglia, Mayer, Aumailley, & Timpl, 1992). Analyses of purified epidermal BM superstructures have indicated that the two networks are indeed brought together by perlecan. Perlecan interaction with the networks was not continuous but occurred at distinct sites. When glycanated perlecan was present, laminin-332 antibodies precipitated collagen IV networks, and vice versa. After digestion of the heparan sulfate chain the interaction between the collagen IV and laminin networks was lost (Behrens et al., 2012). The studies further suggested that perlecan binds collagen IV through heparan sulfate chains and laminin-332 by its protein core (Behrens et al., 2012). Although important in this setting, evidence also suggests that perlecan may not be needed for preserving epidermal BM integrity in vivo. First, in patients and mice hypomorphic for perlecan expression, no skin involvement has been reported (Rodgers, Sasaki, Aszodi, & Jacenko, 2007; Stum et al., 2006), nor has it been observed in mice with perlecan expression primarily restricted to cartilage (Inomata et al., 2012). Further, mice only expressing perlecan lacking heparan sulfate chains do not display an obvious skin phenotype (Rossi et al., 2003). One possibility of absence of phenotype is compensation from the other heparan sulfate proteoglycans expressed in skin, such as agrin and collagen XVIII (Jung et al., 2012; Saarela, Rehn, Oikarinen, Autio-Harmainen, & Pihlajaniemi, 1998), although lack of both collagen XVIII and heparan sulfate-decorated perlecan does also not give an obvious skin phenotype (Rossi et al., 2003). Not much is known about agrin in skin, except that it is expressed by both epidermal keratinocytes and dermal fibroblasts (Shin, Oh, Kim, Jung, & Chung, 2011; Sprenger et al., 2013). Finally, although the laminin and collagen IV networks in purified epidermal BM are critically linked by perlecan (Behrens et al., 2012) a completely developed DEJ in vivo may not be as dependent on perlecan due to additional supportive structures spanning the epidermal BM.

Perlecan also has nonstructural functions in skin. Its protein core interacts with the cell surface matrix receptors integrin $\alpha2\beta1$ and α-dystroglycan, which are expressed on keratinocytes (Herzog et al., 2004; Iozzo, Zoeller, & Nystrom, 2009; Woodall et al., 2008). Studies on skin equivalents deficient in perlecan suggest that it supports keratinocyte survival and terminal differentiation through sequestering of growth factors, such as fibroblast growth factor 7 (FGF7) (Sher et al., 2006) and also possibly through direct signaling of its C-terminal domain V, also known as endorepellin (Iozzo et al., 2009). The studies are supported by the finding of impaired corneal epithelium caused by reduced proliferation and differentiation in mice lacking perlecan expression in cornea (Inomata et al., 2012).

2.1.4 Additional important components

Fibulins are a family of calcium-binding ECM proteins with elongated structure. Multiple epidermal growth factor (EGF)-like repeats provide the calcium-binding properties (Timpl, Sasaki, Kostka, & Chu, 2003). Fibulins 1 and 2 are present in multiple BMs (de Vega, Iwamoto, & Yamada, 2009), including the epidermal BM (Miosge et al., 1996). They bind several BM components such as the laminin-$\gamma2$ short arm and nidogens; additionally, fibulin 2 binds the perlecan protein core (Hopf, Gohring, Kohfeldt, Yamada, & Timpl, 1999). Fibulins take part in assembly and stabilization of ECM complexes. Fibulin 1- and 2-deficient mice do not display obvious BM abnormalities (de Vega et al., 2009; Kostka et al., 2001). However, fibulin 2-deficient mice show localized, mild, and transient dermal–epidermal separation, potentially owing to fibulin 2 playing a role in strengthening epidermal anchoring by laminin-332 (de Vega et al., 2009; Longmate et al., 2014).

Collagen XVIII is, as discussed above, present in the epidermal BM. It is a hybrid collagen/proteoglycan contributed by keratinocytes (Saarela et al., 1998). Mice lacking collagen XVIII have a broadened epidermal BM suggesting a structural role in maintenance of epidermal BM integrity (Utriainen et al., 2004).

FRAS1 (Fraser ECM complex subunit 1), FREM1 (FRAS1-related ECM one), and FREM2 (FRAS1-related ECM 2), are members of a new family of ECM proteins associated with the human genetic disorder Fraser syndrome (MIM#219000) or related FRAS–FREM complex diseases (Short, Wiradjaja, & Smyth, 2007). They form a ternary complex known as the Fraser complex which is associated with the epidermal BM. AMACO (VWA2, von Willebrand factor A domain containing 2) is an additional component of this complex (Gebauer et al., 2009; Richardson et al.,

2014). Also mutations affecting the cytoplasmic Post synaptic density protein (PSD95), Drosophila disc large tumor suppressor (Dlg1), and zonula occludens-1 protein (PDZ) domain-containing protein GRIP1 Fraser's syndrome and so by affecting intracellular trafficking and deposition of Fraser complex components (Vogel et al., 2012). This complex is located in the sublamina densa region and is important for embryonic epidermal BM stabilization, since loss of function of any of the proteins, except AMACO, leads to transient embryonic skin blistering (Short et al., 2007). The transient detachment, which is at the sublamina densa level, has suggested that the proteins help mediate skin stability during development until another stabilizer of skin integrity is introduced. Sublamina densa separation and temporal expression data have indicated that collagen VII (see below) is this stabilizer (Short et al., 2007). Recent studies implicate decreased expression of nephronectin as a cause of reduced epidermal–dermal attachment upon absence of FREM1 (Kiyozumi et al., 2012). Nephronectin is transiently expressed throughout the epidermal BM during development (Fujiwara et al., 2011; Kiyozumi et al., 2012). It would in its turn mediate embryonic epidermal–dermal stabilization by interaction with integrin $\alpha8\beta1$, which is expressed by superficial dermal fibroblasts in a similar tempospatial distribution as nephronectin (Fujiwara et al., 2011).

2.1.5 Attachment of the epidermal BM to the papillary matrix

Skin, as discussed, needs highly developed anchoring structures to withstand frictional damage. Collagen VII has evolved as a specialized nonredundant component of the DEJ ensuring firm attachment of the epidermal BM to the papillary matrix. It is expressed by both epidermal keratinocytes and dermal fibroblasts (Nystrom, Bruckner-Tuderman, & Kern, 2013). Collagen VII is a homotrimer composed of three identical $\alpha1$ chains of over more than 2900 amino acids. The α chain comprises two noncollagenous (NC) domains flanking a central collagenous domain with multiple imperfections (Chung & Uitto, 2010). Most ligand-binding activity is localized to the N-terminal NC1 domain. It makes up about half of the α chain, and the collagenous domain—the primary structural element—most of the other half. The C-terminal NC2 domain is small but critical for intermolecular organization (Bruckner-Tuderman et al., 1995). Association of the $\alpha1$ chains and folding of the collagen VII homotrimer occur intracellularly. Fully folded, the collagen VII triple helix spans 420 nm, making it the longest collagen triple helix present in vertebrates (Bachinger et al., 1990; Hulmes, 2008). Although some interruptions occur in the collagen VII triple helix it is in general a rigid structure, excluding use of conventional intracellular protein transport and secretion systems. For its transport and secretion the cells use a specialized system, which includes the endoplasmic reticulum membrane-spanning cargo loading guide TANGO1 (MIA3, melanoma inhibitory activity family, member 3) (Saito et al., 2009).

Once secreted collagen VII undergoes proteolytic maturation, which allows formation of adhesive superstructures termed anchoring fibrils. Removal of the most C-terminal part of the NC2 domain converts procollagen VII to mature collagen VII that undergoes spontaneous fibril formation. After secretion, the procollagen VII molecules arrange as antiparallel dimers with a 60 nm overlap. Collagen VII maturation, which can be mediated by proteinases of the bone morphogenetic protein-1 family (Rattenholl et al., 2002), promotes antiparallel dimer formation (Colombo et al., 2003), and the dimers are stabilized by C-terminal intermolecular disulfide bonds (Burgeson, 1993; Colombo et al., 2003). Studies suggest that the dimers are further stabilized by intermolecular cross-linking of C-terminal glutamine residues by tissue transglutaminase 2 (Kuttner, Mack, Gretzmeier, Bruckner-Tuderman, & Dengjel, 2014; Raghunath et al., 1996). The dimers condense laterally to form the anchoring fibrils (Burgeson, 1993). The end-to-end arrangement allows the ligand-binding NC1 domain to be present at both ends of the fibril. Laminin-332, primarily via its β3 short arm (Chen et al., 1999), and collagen IV, interact with the collagen VII NC1 domain with high affinity (Brittingham, Uitto, & Fertala, 2006). Also fibronectin and collagen I bind the NC1 domain (Lapiere et al., 1994; Wegener, Leineweber, & Seeger, 2013). Each NC1 domain, which is approximately 50 nm in length (Brittingham et al., 2006), of the homotrimer may interact separately with ligands, thus maximizing interactions with epidermal BM ligands and also potentially connecting the laminin and collagen IV networks. The collagenous domain spreads into the papillary matrix where it entraps collagen fibrils of other entities. This firmly anchors the epidermal BM to the papillary dermal matrix. Loss of function of anchoring fibrils has dire consequences both for homeostasis and regeneration (see Section 4.1.3).

2.2 Hemidesmosomes

The main ultrastructurally recognizable structures that mediate stable adhesion between the basal keratinocytes and the BM are the hemidesmosomes. Together, the hemidesmosome-anchoring filament complex forms a continuous structural link between the basal keratinocyte keratin intermediate filaments (IFs) and the underlying BM and dermal components. During wound healing, hemidesmome disassembly enables keratinocyte migration and proliferation. Hemidesmosomes are placed on the plasma membrane of basal keratinocytes that face the lamina lucida, and ultrastructurally comprise an inner and an outer electron dense plaque (Figure 1). At the molecular level, they represent supramolecular assemblies of distinct intracellular and transmembrane proteins (Figure 1). The functional relevance of these protein complexes in cutaneous integrity is emphasized by their association with human genetic or autoimmune disorders (Table 1 and Section 4). In stratified epithelia, the inner plaque of the hemidesmosomes (type 1) consists of two plakin homologues, BPAG1 (bullous pemphigoid antigen 1, BPAG1e, also known as BP230 or the 230-kD bullous pemphigoid antigen (BPAG)) and plectin. Plakins are a family of cytolinkers with a multimodular structure; they connect the microfilament, microtubule, and IF systems with each other and with junctional complexes in membranes (Bouameur, Favre, & Borradori, 2014). BPAG1 and plectin have a similar overall tripartite structure consisting of a central rod domain, which mediates self-association, flanked by N- and C-terminal domains that harbor binding sites for other proteins. Epithelial BPAG1 interacts via its N-terminal domains with the hemidesmosomal transmembrane proteins $\alpha6\beta4$ integrin and collagen XVII (reviewed in (Bouameur et al., 2014)) and via its C-terminal tail to the epidermis-specific IF network consisting of keratin 5 and 14 (K5, K14). In a similar manner, plectin interacts with the K5/K14 IF and connects them to the membrane and further to the ECM, by the interaction with $\alpha6\beta4$ integrin. The plectin N-terminal actin-binding domain binds to actin and $\alpha6\beta4$ integrin, while the C-terminal plakin domains interact with IF and $\alpha6\beta4$ integrin. Several plectin isoforms exist that differ from each other in their N-termini; plectin 1a is considered the major variant in hemidesmosomes, while plectin 1c is also able to localize either in or close to hemidesmosomes (Walko et al., 2011). Plectin can also interact with other IF proteins, such as simple epithelial keratins, the muscle-specific desmin, and IF proteins expressed in the nervous system. Mice with conditionally deleted plectin in stratified epithelia die early after birth, showing signs of starvation and growth retardation. Blistering was observed on their extremities and on the oral epithelium after initial nursing, impairing food uptake. Knockout epidermis was very fragile and showed focal epidermal barrier defects caused by the presence of small skin lesions. Stratification, proliferation, and differentiation of knockout skin seemed unaffected by epidermis-restricted plectin deficiency (Ackerl et al., 2007). In an alternative mouse model, tamoxifen-induced mutation of the plectin

Table 1 Epidermal basement membrane (BM): human disorders and mouse models

Protein	Human genetic disorders	Human autoimmune disorders	Mouse models[a]	Wound healing phenotype in mouse model	Other features
BPAG1e	EBS	Bullous pemphigoid Mucous membrane pemphigoid Pemphigoid gestationis Lichen planus pemphigoides Linear IgA disease Brunsting–Perry pemphigoid	EBS-like phenotype (Guo et al., 1995)	Not reported	
Plectin	EBS w. muscular dystrophy EBS w. pyloric atresia EBS Ogna	Paraneoplastic pemphigus Antiplectin pemphigoid	EBS-like phenotype (Andra et al., 1997; Walko et al., 2011)	Mice expressing plectin isoform 1 only show reduced leukocyte infiltration of wounds (Abrahamsberg et al., 2005)	Knockout mice neonatal lethal
Collagen XVII	JEB generalized intermediate JEB localized JEB late onset	Bullous pemphigoid Pemphigoid gestationis	JEB-like phenotype (Nishie et al., 2007; Hurskainen et al., 2015)	Not reported	
LAD-1, LABD-97		Linear IgA disease			

(Continued)

Table 1 Epidermal basement membrane (BM): human disorders and mouse models—cont'd

Protein	Human genetic disorders	Human autoimmune disorders	Mouse models[a]	Wound healing phenotype in mouse model	Other features
Integrin $\alpha6\beta4$	JEB w. pyloric atresia JEB localized	Mucous membrane pemphigoid	JEB-like phenotype (Dowling, Yu, & Fuchs, 1996; van der Neut, Krimpenfort, Calafat, Niessen, & Sonnenberg, 1996; Georges-Labouesse et al., 1996; Nikolopoulos et al., 2005)	Mice lacking the integrin $\beta4$ cytoplasmic tail show delayed reepithelialization wounds (Nikolopoulos et al., 2005)	Knockout mice neonatal lethal
CD151	Pretibial EB w. deafness and nephropathy	Not known	Renal disease, no skin phenotype	Wound healing is defective in mice lacking tetraspanin CD151 (Cowin et al., 2006)	
Kindlin-1	Kindler syndrome	Not known	Conventional knockout animals are perinataly lethal because of intestinal epithelial dysfunction (Ussar et al., 2008); loss of expression in keratinocytes only mirrors Kindler syndrome (Rognoni et al., 2014)	Not known	
Kindlin-2	Not known	Not known	Deletion causes lethality at the periimplantation stage (Montanez et al., 2008)	Not known	

Integrin α2	Not known	Not known	Knockout mice appear normal and are fertile (Chen, Diacovo, Grenache, Santoro, & Zutter, 2002; Holtkotter et al., 2002).	Normal reepithelialization but increased wound angiogenesis (Grenache et al., 2007; Zweers et al., 2007)
Integrin α3	ILNEB	Not known	ILNEB-like phenotype, early lethal (Kreidberg et al., 1996)	Faster wound reepithelialization in mice lacking expression of integrin α3 on keratinocytes (Margadant et al., 2009; Mitchell et al., 2009)
Integrin β1	Not known	Not known	Conventional knockout mice are early embryonic lethal (Stephens et al., 1995). Keratinocyte-restricted deletion lead to changes in the epidermal BM, blistering, and dermal fibrosis (Brakebusch et al., 2000)	Mice with keratinocyte-specific deletion of integrin β1 display delayed reepithelialization and abnormal epidermis after wound closure (Grose et al., 2002). Loss of integrin β1 expression on dermal fibroblast causes delayed dermal healing (Liu et al., 2010)

(Continued)

Table 1 Epidermal basement membrane (BM): human disorders and mouse models—cont'd

Protein	Human genetic disorders	Human autoimmune disorders	Mouse models[a]	Wound healing phenotype in mouse model	Other features
Laminin-332	JEB severe generalized JEB intermediate JEB inversa	Mucous membrane pemphigoid	JEB-like phenotype (Bubier et al., 2010; Kuster et al., 1997; Meng et al., 2003; Muhle et al., 2006; Ryan et al., 1999)	Not reported	
Laminin-511	Not known	Antilaminin-γ1 pemphigoid	Mice lacking laminin-α5 die at midgestation embryogenesis and display multiple organ defects (Miner et al., 1998). Knockout of laminin-β1 and -γ1 is early embryonic lethal (Miner, Li, Mudd, Go, & Sutherland, 2004; Smyth et al., 1998)		Indications of destabilization of epidermal BM when laminin-α5 is absent (Miner et al., 1998)
Nidogen1/2	Not known	Not known	Nidogen-1 or -2-deficient mice appear normal and fertile; compound-deficient mice are perinatal lethal due to lung and heart defects (Bader et al., 2005; Murshed et al., 2000; Schymeinsky et al., 2002)	Nidogen-1 deficient mice heal skin wounds with delayed epidermal maturation and granulation tissue defects (Baranowsky et al., 2010)	

Perlecan	Schwartz–Jampel syndrome Silver–Handmaker syndrome	Not known	Knockout mice are embryonic and neonatal lethal with vascular and cerebral defects and chondrodysplasia (Costell et al., 1999; Arikawa-Hirasawa, Watanabe, Takami, Hassell, & Yamada, 1999; Rodgers et al., 2007)	Mice only expressing the perlecan protein core have delayed wound healing caused by impaired angiogenesis and granulation tissue formation (Zhou et al., 2004)	No overt cutaneous phenotype in mice lacking perlecan expression in skin (Xu et al., 2010)
Collagen IV	Perinatal cerebral hemorrhage, sporadic late-onset intracerebral hemorrhage and porencephaly	Goodpasture syndrome with pulmonary hemorrhage and glomerulonephritis but no cutaneous manifestations	Mice lacking collagen IV alpha1(IV)(2)alpha2(IV) die at midgestation with structural weakness of multiple BMs (Poschl et al., 2004)		
Collagen VII	DEB dominant localized or generalized DEB recessive localized, generalized or severe generalized	EB acquisita Bullous systemic lupus erythematosus	Recessive DEB-like (Fritsch et al., 2008; Heinonen et al., 1999; Ito et al., 2009) Tamoxifen-inducible knockout (Nystrom, Velati, et al., 2013)	Delayed reepithelialization and granulation tissue formation (Nystrom, Velati, et al., 2013)	

BM, basement membrane; EB, epidermolysis bullosa; EBS, EB simplex; JEB, junctional EB; DEB, dystrophic EB; ILNEB, interstitial lung disease, nephrotic syndrome, EB.
[a]Mouse models for autoimmune blistering disorders are not included in this table.

gene led to mice with a mosaic plectin deletion pattern in adult epidermis, combined with microblister formation and epidermal barrier defects (Ackerl et al., 2007).

The two main transmembrane components of hemidesmosomes are collagen XVII (also known as BPAG2, bullous pemphigoid antigen 2 or BP180, the 180-kD bullous pemphigoid antigen) and α6β4 integrin. Collagen XVII is a type II transmembrane protein and a homotrimer consisting of three 180 kDa alpha-1 (XVII) chains, with an extracellular collagenous domain (Franzke, Bruckner, & Bruckner-Tuderman, 2005). Collagen XVII exists as a full-length transmembrane protein and a soluble 120 kDa ectodomain that is proteolytically released from the cell surface by sheddases of a disintegrin and metalloproteinase (ADAM) family, ADAM-10, -9, and -17 (Franzke, Bruckner-Tuderman, & Blobel, 2009). This shedding seems to be dependent on the conformation of the NC16A domain and the steric availability of the cleavage site (Franzke, Tasanen, Borradori, Huotari, & Bruckner-Tuderman, 2004). Regulation of cleavage is complex: it is dependent on plasma membrane lipid organization (Zimina, Bruckner-Tuderman, & Franzke, 2005), on extracellular phosphorylation of collagen XVII by ecto-casein kinase 2 (Zimina et al., 2007), and on the coiled-coil structure (Nishie, Jackow, Hofmann, Franzke, & Bruckner-Tuderman, 2012). The cleavage occurs within the NC16A domain, and various N-termini have been identified (Nishie et al., 2010). Notably, ectodomain shedding generates neoepitopes on collagen XVII, which may serve as a target of blister-inducing autoantibodies (Nishie et al., 2010) (see Section 4.2). Part of the 120 kDa ectodomain is further processed to a 97 kDa fragment (LABD97), an autoantigen in linear IgA dermatosis (LAD) (Hofmann et al., 2009). In vitro, plasmin is able to generate a stable 97 kDa fragment that is also targeted by LAD sera. The precise role of collagen XVII shedding is elusive, but it may include regulation of cell migration and differentiation. The intracellular ligands of collagen XVII are BPAG1 and β4 integrin, and its extracellular ligands are α6 integrin and laminin-332. Recently, collagen XVII was reported to be highly expressed in hair follicle stem cells. Analysis of *Col17a1*-deficient mice, which shows premature hair graying and hair loss, suggested that collagen XVII is critical for the self-renewal of both hair follicle stem cells and melanocytes (Tanimura et al., 2011). In addition, Col17-knockout mice demonstrate phenotypic features closely resembling the human disease junctional EB caused by null mutations in the collagen XVII gene, *COL17A1* (Nishie et al., 2007) (see Section 4.1.1).

α6β4 integrin is a noncovalent heterodimer composed of two type I transmembrane subunits. The extracellular moiety of α6β4 binds to laminins, preferentially to laminin-332 (reviewed in (de Pereda, Ortega, Alonso-Garcia, Gomez-Hernandez, Sonnenberg, 2009)). The intracellular region of α6β4 consists of the short tail of the α6A isoform and a large β4 cytodomain, which is responsible for the interactions with plectin and collagen XVII. Binding of plectin to the β4 subunit is critical for the formation and the stability of hemidesmosomes, and it is likely to also be an initial step in their assembly. The primary plectin interaction site is formed by the first pair of fibronectin type III domains and a small part of the connecting segment of β4 that binds to the actin-binding domain of plectin (Geerts et al., 1999; de Pereda, Lillo, Sonnenberg, 2009). The secondary binding site is formed between the carboxy terminus of the connecting segment of β4 and the plakin domain of plectin (Koster, van Wilpe, Kuikman, Litjens, & Sonnenberg, 2004; Rezniczek, de Pereda, Reipert, & Wiche, 1998). Phosphorylation on the C-tail of β4 augments the disassembly of hemidesmosomes. Several growth factors have been implicated in regulating hemidesmosomal disassembly, such as EGF, hepatocyte growth factor, and macrophage-stimulating protein (Barrientos, Stojadinovic, Golinko, Brem, & Tomic-Canic, 2008; Litjens, de Pereda, & Sonnenberg, 2006; Santoro, Gaudino, & Marchisio, 2003). These factors collectively induce the Ras/MAPK-signaling pathway and stimulate β4 phosphorylation on serine residues in the connecting segment (Rabinovitz, Tsomo, & Mercurio, 2004; Wilhelmsen et al., 2007). Phosphorylation on T1736 by PKD1 (protein kinase D1) or other Ca2+/calmodulin-dependent kinase (CAMK)-related protein kinases results in the dissociation of the C-tail and the plakin domain of plectin, whereas phosphorylation of β4 at S1356 and S1364 by ERK1/2 and p90RSK1/2 causes the loss of the plectin–actin-binding domain binding (Frijns et al., 2012; Frijns, Sachs, Kreft, Wilhelmsen, & Sonnenberg, 2010). Syndecans engage the distal region of the β4 cytoplasmic domain and bring it to the plasma membrane, where it can be acted upon by Src family kinases (Wang, Leavitt, Ramaswamy, & Rapraeger, 2010).

The ligands of α6 integrin include CD151, collagen XVII, and laminin -332. The cytoplasmic tail of the α6A subunit, one of the two splice variants (A and B) of the α6 subunit that is predominantly expressed in the epidermis, has a membrane proximal GFFKR sequence recognized by calreticulin, Rab21, Mss4, BIN1, and other proteins, and contains a binding site for the PDZ domain of TIP-2/GIPC at its C-terminus (reviewed in (de Pereda, Ortega, et al., 2009)). Moreover, there is a cross talk between integrin α6β4 and insulin-like growth factor-1 receptor (IGF1R) through direct α6β4 binding to IGF1 and subsequent α6β4–IGF1–IGF1R ternary complex formation in anchorage-independent conditions (Fujita et al., 2012).

Ablation in mice of either α6β4 or α3β1, through null mutation of the gene encoding the, α6, β4, or α3 integrin subunit, results in epidermal blistering of varying severity. While α6β4 and α3β1 are both required for integrity of the DEJ, neither one is essential for epidermal morphogenesis during skin development (DiPersio et al., 2000). Conditional knockout mice, in which the gene encoding the β4 integrin subunit was inactivated only in small stretches of the skin, are viable. Despite the complete loss of hemidesmosomes in regions lacking α6β4 integrin, the distribution of a range of adhesion receptors and BM proteins was unaltered. Moreover, loss of α6β4 did not affect differentiation, proliferation, or survival, except for areas in which keratinocytes had detached from the BM (Raymond, Kreft, Janssen, Calafat, & Sonnenberg, 2005). Mice carrying a targeted deletion of the signaling domain of β4 develop normally and do not display signs of skin fragility. The epidermis of these mice contains well-structured hemidesmosomes and adheres stably to the BM. However, it is hypoplastic due to reduced proliferation of basal keratinocytes and undergoes wound repair at a reduced rate. Keratinocytes from β4 mutant mice undergo extensive spreading but fail to proliferate and migrate in response to EGF on laminin-332. Inhibitor studies indicate that α6β4 promotes keratinocyte proliferation and migration through its effect on NF-kappaB and P-JNK (Nikolopoulos et al., 2005). Mice lacking integrin α6 die at birth with severe skin blistering (Georges-Labouesse et al., 1996). In a mouse line with epidermal deletion of integrin α6 controlled by tamoxifen injection, the deletion was mosaic, but hairless skin such as ears, tails, and paws were affected and showed chronic inflammation associated with hyperproliferation, and expression of laminin-111 (Niculescu et al., 2011). Loss of α6 integrins in keratinocytes leads to an increase in transforming growth factor-β (TGFβ) and AP1 signaling and in expression of differentiation genes (Rodius, Indra, Thibault, Pfister, & Georges-Labouesse, 2007).

CD151 (also known as Raph blood group or TSPAN24), a member of the tetraspanin family of cell surface proteins, acts as a stabilizer of integrin functions. It forms complexes with α3β1 and α6β4 integrins at the basolateral surface of basal keratinocytes (Zoller, 2009). This molecule is expressed in epithelia, endothelia, muscle cells, renal glomerular podocytes, Schwann and dendritic cells, but also in platelets and megakaryocytes

(Sachs et al., 2006). CD151 is considered to play a role in the organization and stability of hemidesmosomes by facilitating the formation of stable laminin-binding complexes with integrin α6β4, as well as being involved in cellular signaling (Sterk et al., 2000, 2002). In one report, CD151-null mice were normal, healthy, and fertile with no skin or hemidesmosome pathology (Wright et al., 2004), whereas another research group reported CD151-null mice that had substantial renal disease, including focal glomerulosclerosis, disorganization of the glomerular membrane, and tubular cystic dilatation. Their skin and hearing apparatus were not involved (Sachs et al., 2006).

CD151 can support de novo carcinogenesis. During two-stage mouse skin chemical carcinogenesis, CD151 reduces tumor lag time and increases incidence, multiplicity, size, and progression to malignant squamous cell carcinoma, while supporting both cell survival during tumor initiation and cell proliferation during the promotion phase. CD151 support of keratinocyte survival and proliferation may depend on activation of transcription factor STAT3 (signal transducer and activators of transcription 3), a regulator of cell proliferation and apoptosis. CD151 also supports protein kinase C (PKC)α-α6β4 integrin association and PKC-dependent β4 S1424 phosphorylation, while regulating α6β4 distribution. CD151 ablation, while minimally affecting normal cell and normal mouse functions, markedly sensitized mouse skin and epidermoid cells to chemicals/drugs, as well as to agents targeting EGF receptor, PKC, Jak2/Tyk2, and STAT3 (Li et al., 2013).

Other receptors that bind laminin-332 are the syndecan family of cell surface heparan sulfate proteoglycans. This family of receptors is composed of syndecans 1–4, of which syndecan 1 is abundantly expressed on epithelial cells (Bernfield et al., 1992; Rapraeger, Jalkanen, & Bernfield, 1986). The syndecans, by virtue of their heparan sulfate chains binding most or all matrix components, may serve as organizers of matrix adhesion and signaling by recruiting integrins and growth factor receptors and activating them at matrix adhesion sites.

Beside structural components, other proteins are involved in the regulation of hemidesmosomes. For example, the mouse mutant ragged-opossum (Ra(op)) expresses a dominant-negative form of the SOX18 transcription factor that interferes with the function of wild-type SOX18 and of the related SOXF-subgroup proteins SOX7 and -17. Skin and oral mucosa in homozygous Ra(op) mice display extensive detachment of epithelium from the underlying mesenchymal tissue, caused by tearing of epithelial cells just above the plasma membrane due to hemidesmosome disruption.

In addition, several hemidesmosome proteins expression were found to be dysregulated in the Ra(op) mice, suggesting that SOXF transcription factors play a role in regulating formation of cytoplasmic plaque protein assembly, and that disrupted SOXF function results in EB-like skin phenotypes (Oommen et al., 2012).

Flightless I (Flii), a highly conserved member of the gelsolin family of actin-remodeling proteins associates with actin structures and is involved in cellular motility and adhesion. Flii affects hemidesmosome formation and integrin-mediated keratinocyte adhesion and migration (Kopecki, Arkell, Powell, & Cowin, 2009). Impaired hemidesmosome formation and sparse arrangements of keratin IF and actin cytoskeleton were observed in Flii transgenic mice with their skin being significantly more fragile than wild-type mice. Decreased expression of CD151 and laminin-binding integrins α3, β1, α6, and β4 were observed in Flii overexpressing wounds, which could contribute to the impaired wound reepithelialization observed in these mice. Flii interacts with proteins directly linked to the cytoplasmic domain of integrin receptors, suggesting that it may be a mechanical link between ligand-bound integrin receptors and the actin cytoskeleton driving adhesion-signaling pathways (Kopecki et al., 2009).

2.3 Focal adhesions

Next to the hemidesmosomes, integrin β1 adhesion complexes provide additional cohesion and signaling hubs by connecting the actin cytoskeleton to the basal cell plasma membrane and, subsequently, to the ECM. These multimolecular complexes at the basolateral plasma membrane of basal keratinocytes include proteins found in focal adhesions in vitro, e.g., integrin α2β1, α3β1, transmembrane collagen XIII, and kindlins (also known as fermitin family homologs) (Winograd-Katz, Fassler, Geiger, & Legate, 2014). The intracellular β1 integrin tails interact with a multitude of proteins with ubiquitous expression in cells, such as talin, integrin-linked kinase, and focal adhesion kinase (Winograd-Katz et al., 2014).

The α2β1 integrin is the collagen and perlecan domain V/endorepellin receptor on keratinocytes, while α3β1 is receptor for the major laminin isoforms of the DEJ. The fact that neither α2β1 nor α3β1 can compensate for loss of α6β4 indicates that αβ1 integrins should be important for tasks other than adhesion (Wen et al., 2010).

Integrin α3β1 is a receptor for laminin-511, -521, and -332, the major components of epithelial BM (Nishiuchi et al., 2006) and promotes epidermal adhesion primarily by maintaining BM integrity (Longmate et al., 2014). The integrin α3 subunit is a type I transmembrane protein with a large extracellular region, a single transmembrane helix domain, and a short unstructured cytoplasmic tail (Campbell & Humphries, 2011). The extracellular region is composed of a seven-bladed β-propeller, a thigh, and two calf domains with flexible linkers. Both α3 and β1 integrin subunits are synthesized as precursors. After N-linked glycosylation, folding and association of the α and β subunits in the endoplasmic reticulum, the heterodimer is transported to and through the Golgi network, in which the N-linked high-mannose oligosaccharides are processed into complex oligosaccharides. Subsequently, the α3 precursor is cleaved at the calf-2 into the heavy and light chains, which are held together by a disulfide bond to generate the mature form that is expressed at the plasma membrane (Lissitzky et al., 2000). Here, the laminin-α3 and -α5 chains are the main extracellular ligands which bind to α3β1 integrin through their C-terminal globular domains (reviewed in (Sugawara et al., 2008). CD151 forms a stoichiometric complex with integrin α3β1 and regulates its endocytosis and posttranslation modification (Baldwin et al., 2008; Yauch, Berditchevski, Harler, Reichner, & Hemler, 1998). Integrin α3-null mice die within 24 h after birth due to defects in kidney and/or lung organogenesis (Kreidberg et al., 1996). In the skin of these mice, loss of integrin α3 affects proper organization and stabilization of the epidermal BM (DiPersio, Hodivala-Dilke, Jaenisch, Kreidberg, & Hynes, 1997; Margadant et al., 2009), but not epidermal morphogenesis (Conti, Rudling, Robson, & Hodivala-Dilke, 2003; DiPersio et al., 2000). The epidermis-specific integrin α3 knockout mice are viable and exhibit skin inflammation and hair loss. Microblisters and duplication of the cutaneous BM are observed, but hemidesmosome assembly and keratinocyte differentiation are not impaired (Margadant et al., 2009). Interestingly, reepithelialization of the wounds was faster than in wild-type mice (Margadant et al., 2009; Mitchell et al., 2009). This was due to the effect of α3β1 integrin on directional migration of keratinocytes (Margadant et al., 2009) and on wound angiogenesis and keratinocyte-to-endothelial cell cross-talk (Mitchell et al., 2009).

Kindlin proteins represent a novel family of evolutionarily conserved FERM (4.1 protein, ezrin, radixin, and moesin) domain-containing proteins and are members of B4.1 superfamily. The FERM domain is bipartite, being interrupted by a pleckstrin homology domain. The kindlin family consists of three conserved protein homologs in vertebrates, of which kindlin-1 and -2 are expressed in epidermal keratinocytes (Bandyopadhyay, Rothschild, Kim, Calderwood, & Raghavan, 2012; He, Esser, Heinemann,

Bruckner-Tuderman, & Has, 2011). Both are associated with focal adhesions and are involved in integrin activation by interacting with the second NPxY motif on the β1 intracellular tail (Goult et al., 2009; Harburger, Bouaouina, & Calderwood, 2009; Montanez et al., 2008; Ussar et al., 2008). Kindlin-1 controls keratinocyte adhesion through β1-class integrins and proliferation and differentiation of cutaneous epithelial stem cells by promoting αvβ6 integrin-mediated TGFβ activation and inhibiting Wnt-β-catenin signaling through integrin-independent regulation of Wnt ligand expression (Rognoni et al., 2014). A new function of kindlin-1 in regulating spindle assembly emerged recently. Kindlin-1 localizes to centrosomes, its concentration peaking during G2/M, where it associates with various pericentriolar material proteins including polo-like kinase 1. Short interfering RNA-mediated depletion of kindlin-1 increases formation of abnormal mitotic spindles and decreases cellular survival. This effect is dependent not only on the ability of kindlin-1 to bind integrins but also on polo-like kinase 1-mediated kindlin-1 phosphorylation (Patel et al., 2013).

While, keratinocytes express both kindlin-1 and kindlin-2, in the absence of kindlin-1 in Kindler syndrome, kindlin-2 cannot fully compensate and rescue the defect. Knockdown of either of the kindlins affects adhesion, survival, and migration of the keratinocytes and integrin activation, suggesting that kindlins have overlapping functions. Nevertheless, the presence of at least one kindlin partially assures keratinocyte functions, whereas loss of both kindlins has a cumulative impact on cell adhesion, survival, and migration, as well as on integrin activation (He et al., 2011). Notably, although both kindlins localize to focal adhesions in wild-type cells, only kindlin-1 localizes to the integrin-β6-rich adhesions of integrin-β1-null cells (Bandyopadhyay et al., 2012). Beyond that, kindlin-2 is expressed in all epidermal layers and is part of both with β1 and E-cadherin adhesomes (He et al., 2014).

The phenotype of kindlin-1 knockout mice recapitulates only in part the corresponding human disorder, the Kindler syndrome (Ussar et al., 2008). Animals appear normal at birth, but 2 days postnatal they become dehydrated, fail to thrive, and die between postnatal day 3 and 5. The skin is atrophic with reduced keratinocyte proliferation, but adhesion of keratinocytes to the BM, differentiation, and epidermal barrier are not altered. Perinatal lethality is due to severe progressive intestinal dysfunction, with severe inflammation and more than 80% of the colonic epithelium detached by postnatal day 3 (Ussar et al., 2008). Intriguingly, colitis is a rare feature in patients with Kindler syndrome (Kern et al., 2007). Loss of kindlin-1 in mouse keratinocytes produces enlarged and hyperactive stem cell compartments, which lead to a hyperthickened epidermis, ectopic hair follicle development, and increased skin tumor susceptibility (Rognoni et al., 2014). These findings are only partially recapitulated by the human disorder, which is characterized by profound epidermal atrophy (Herz et al., 2006; Piccinni et al., 2013).

3. THE ROLE OF COMPONENTS OF THE EPIDERMAL BM IN CUTANEOUS REGENERATION

Proteins of the DEJ actively participate in tissue regeneration after wounding. Individual proteins fill different functions during the healing process as reflected by their distinct temporal/spatial patterns of expression. Some components of the epidermal BM support migration and are needed for closure of the wound, whereas others promote permanent adhesion and differentiation and are involved in remodeling of the closed wound to mature and functional skin. Figure 2 shows an immunohistological and schematic summary of the below discussed appearance and function of individual DEJ members during wound healing.

The ECM protein fibronectin supports initial keratinocyte migration. Fibronectin, which in normal skin shows a broad distribution, is deposited under migrating keratinocytes and its receptor integrin $\alpha5\beta1$, otherwise practically undetectable, is upregulated in migrating keratinocytes (Larjava, Salo, Haapasalmi, Kramer, & Heino, 1993). Laminin-332 is essential for proper wound reepithelialization, as clearly indicated by frequent occurrence of chronic wounds in patients with impaired laminin-332 function (Yuen, Huizinga, & Jonkman, 2013). Its role in wound healing is further supported by altered deposition under epidermal tongues in acquired chronic wounds (Hunefeld et al., 2013; Usui, Mansbridge, Carter, Fujita, & Olerud, 2008). Very early after wounding, laminin-332 expression is increased in wound keratinocytes with subsequent deposition under the epidermal wound front (Ryan, Tizard, VanDevanter, & Carter, 1994). The expression of laminin-332 is partially regulated by microRNAs; miR-198 inhibits expression of laminin-$\gamma2$. TGFβ signaling, increased in wounded tissue, turns off miR-198 which allows laminin-$\gamma2$ expression (Sundaram et al., 2013). Laminin-332 has been described as both a substrate for migration and a protein supporting stasis and adhesion (Goldfinger, Stack, & Jones, 1998; O'Toole, Marinkovich, Hoeffler, Furthmayr, & Woodley, 1997). Much

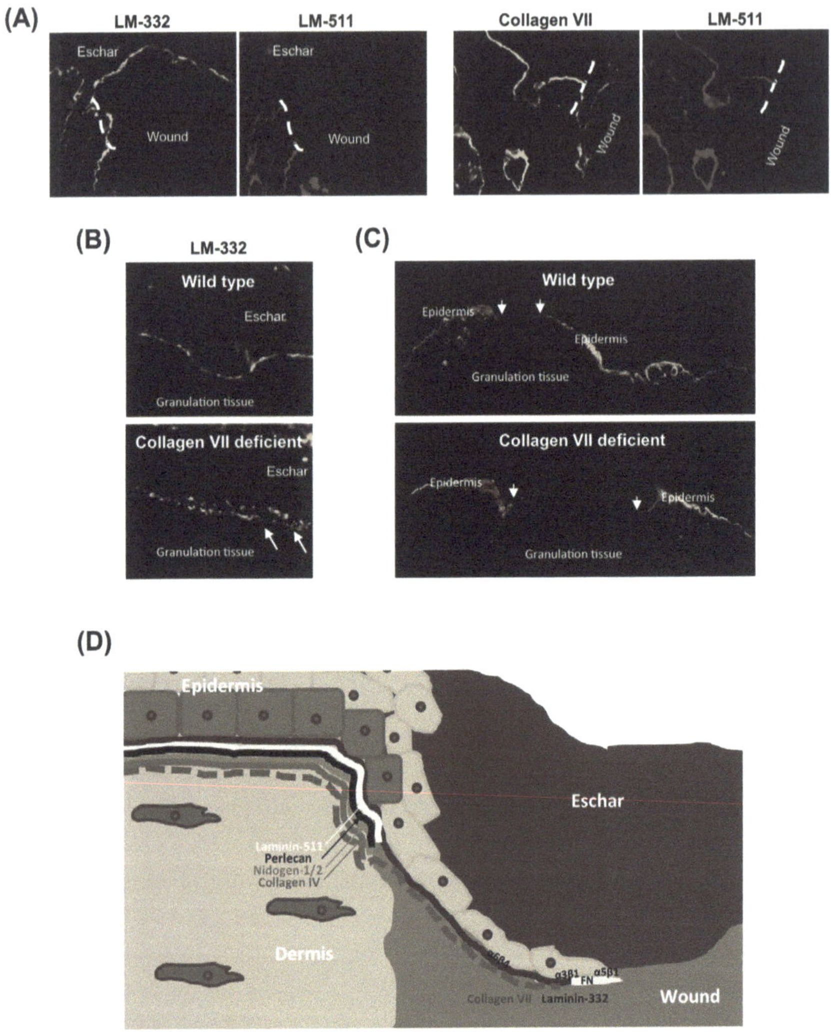

Figure 2 ***Epidermal basement membrane components in skin wound closure.*** (A) Three-day-old full-thickness mouse wounds stained for laminin-332, laminin-511, and collagen VII as indicated. (B) Laminin-332 deposition in wild-type- and collagen VII-deficient seven-day-old mouse wounds. Note the altered irregular laminin-332 deposition in absence of collagen VII. (C) Wounds as in (B) stained for keratin 1 to visualize the delayed reepithelialization and subsequent epithelial maturation of collagen VII-deficient wounds. (D) Schematic representation of expression of key epidermal basement membrane components during wound healing. (See color plate)

data suggest that proteolytic processing regulate the pro- and antimigratory properties of laminin-332 (Hintermann & Quaranta, 2004). However, the exact relationship between specific laminin chain processing and functional behavior is not completely understood. The intricate proteolytic maturation of laminin-3A32 has implications for wound healing events. Cleavage of the laminin-γ2 short arm, which occurs naturally during BM maturation, seems to turn laminin-332 into a protein supporting migration (Gagnoux-Palacios et al., 2001; Ogawa, Tsubota, Maeda, Kariya, & Miyazaki, 2004). Thus, the laminin-332 in an intact epidermal BM might already be in a promigratory form. Analysis of reepithelializing wounds indicates that prolaminin-332 is present in the provisional BM under migratory cells in the leading epidermal edge, whereas mature laminin-332 is present further back under the static epidermis (Goldfinger et al., 1999). Further, integrin α3β1 is basally localized in leading edge keratinocytes where it likely interacts with prolaminin-332 retaining the LG4–5 domain, since the other major integrin α3β1 ligand—laminin-511—is not expressed in reepithelializing wounds (Larjava et al., 1993; Nystrom, Velati, et al., 2013). Colocalization of integrin α6β4 and mature laminin-332 is seen under static keratinocytes further back in the wound (Goldfinger et al., 1999). These observations, together with in vitro studies showing that prolaminin-332 promotes migration and loss of the α3 LG4–5 domain promotes stasis and adhesion (Goldfinger et al., 1998), have led to the hypothesis that loss of the LG4–5 domain switches laminin-332 from a molecule promoting migration to a molecule promoting hemidesmosome formation. It has been proposed that laminin-332 retaining the LG4–5 domain primarily interacts with α3β1, whereas loss of LG4–5 results in increased binding to integrin α6β4, which promotes hemidesmosome formation. However, the situation is likely more complex, as conflicting data exist. Other studies show no correlation between retention or loss of LG4–5 and migration (Baudoin et al., 2005) or even indicate that loss of LG4–5 promotes migration (Tsubota et al., 2005). In support of the latter is the fact that loss of integrin α3β1 in keratinocytes accelerates wound reepithelialization (Margadant et al., 2009), although the altered migration may in fact be a consequence of changed ECM deposition (deHart, Healy, & Jones, 2003). There are indications that laminin-332 is not only important for reepithelialization but also for epidermal–dermal communication to regulate aspects of dermal healing. In healing full-thickness wounds, laminin-332 deposits can be found in the superficial dermis (Nystrom, Velati, et al., 2013) and patients with mutations in the N-terminus of the laminin-α3 chain develop excessive vascular granulation tissue at selected areas (McLean et al., 2003).

Laminin-511 is not deposited under migrating keratinocytes (Figure 2(A) and (Botta et al., 2012)). After reepithelialization its amount increases and deposition is seen in the maturing epidermal BM (Botta et al., 2012; Nystrom, Velati, et al., 2013). Integrin α3β1 is believed to be the main keratinocyte integrin receptor for laminin-511 (Nishiuchi et al., 2003). The appearance of localized skin blisters when this receptor is absent (Has et al., 2012), together with indications of migration restrictive functions of integrin α3β1 (Margadant et al., 2009), suggest that laminin-511—integrin α3β1 interaction limits migration and stabilizes the epidermis in maturing skin after wounding. Lutheran blood group glycoprotein, or basal cell adhesion molecule, is an additional laminin-511 receptor. It is barely, if at all, expressed in healthy epidermis (Schon et al., 2000). Its expression during wound healing has not been studied, but it is present on basal keratinocytes in pathological events such as cancer and inflammation, implying important functional interactions with laminin-511 under these conditions (Drewniok et al., 2004).

Even though collagen IV supports keratinocyte migration in vitro (Putnins, Firth, Lohachitranont, Uitto, & Larjava, 1999) and is present together with fibronectin and laminin-332 in keratinocyte migration tracks in vitro (Kirfel, Rigort, Borm, Schulte, & Herzog, 2003), it is not a substrate used for migration during in vivo skin wound closure. Collagen IV is not highly deposited in the provisional BM under migrating keratinocytes. Rather, collagen IV accumulates further back from the wound edge in the maturing epidermal BM in the wound margin (Betz et al., 1992; Larjava et al., 1993), likely indicating BM-stabilizing properties.

Both nidogens and perlecan show similar expression in healing wounds. They are not deposited into the provisional epidermal BM used for keratinocyte migration but become part of the maturing epidermal BM present in the wound margins (Andriessen et al., 1997; Botta et al., 2012; Oksala et al., 1995). As for laminin-511 and collagen IV, this suggests functions related to tissue maturation rather that for supporting migration of keratinocytes during wound healing. Mice lacking nidogen 1 close skin wounds with speeds similar to controls. However, the wounds show delay in epidermal and granulation tissue maturation (Baranowsky et al., 2010), which points to nonstructural functions of nidogens during skin wound healing.

Perlecan affects epidermal function and epidermal–dermal interactions, as discussed in Section 2.1.3, through multiple mechanisms. First, it fills a stabilizing structural function by linking the laminin and collagen IV networks and organizing microfibrils (Behrens et al., 2012; Tiedemann et al., 2005). Second, it regulates growth factor presentation by sequestering growth factors via its protein core and heparan sulfate chains (Iozzo et al., 2009). Third, it influences cells by direct engagement of cell surface receptors through its protein core (Iozzo et al., 2009). All of these functions are potentially important for skin restoration after wounding. Perlecan deposition promotes epidermal adhesion and reduces cell migration (Botta et al., 2012). Importantly, interaction of perlecan's protein core with laminin-332 seems to be crucial for perlecan deposition into regenerating epidermal BMs, since limited perlecan deposition occurs when keratinocytes are devoid of laminin-332 expression (Behrens et al., 2012; Botta et al., 2012; Urbano et al., 2009). The increased presence of perlecan in the epidermal BM at wound margins can be accounted for in that establishment of keratinocyte adherens junctions promotes perlecan expression (Botta et al., 2012). Adherens junctions increase activin A, which triggers expression of perlecan. The late appearance of perlecan in the epidermal BM during healing is in line with the concept that apart from inhibiting migration, likely via integrin $\alpha2\beta1$ (Woodall et al., 2008), it also promotes epidermal proliferation and differentiation via signaling and growth factor presentation of such factors as FGF7 (Sher et al., 2006). That loss of perlecan does not impact early stages during skin regeneration is supported by the lack of reepithelialization phenotype in mice lacking the cognate perlecan domain V/endorepellin receptor integrin $\alpha2\beta1$ (Grenache et al., 2007; Zweers et al., 2007). Further support is found from wound healing studies on mice lacking glycanated perlecan. Mice only expressing perlecan without heparan sulfate chains show no overt abnormality in wound reepithelialization compared to wild-type mice (Zhou et al., 2004). The mice, however, display delayed granulation tissue formation and reduced wound angiogenesis (Zhou et al., 2004), highlighting the importance of presentation of growth factors such as VEGFs and FGFs for completion of these events.

Loss of collagen XVIII does not greatly influence the reepithelialization rate of skin wounds (Seppinen et al., 2008). However, gross wound healing is accelerated by increased myofibroblast formation and angiogenesis (Seppinen et al., 2008). Both observations are in line with the fact that collagen XVIII's NC1 domain/endostatin has strong antiangiogenic and antiscarring properties (O'Reilly et al., 1997; Yamaguchi et al., 2012) through integrin $\alpha5\beta1$, VEGFR2, and potentially an additional unidentified receptor (Kim et al., 2002; Sudhakar et al., 2003). The absence of functional full-length collagen XVIII may delay maturation of the epidermal BM during wound healing, which manifests as a transiently increased fragility of reepithelialized wounds (Seppinen et al., 2008).

Although highly specialized for ensuring firm attachment of the epidermal BM to the papillary ECM, collagen VII has multiple roles in wound healing. Initial studies suggested that collagen VII was only expressed in the wound margins and not present under migrating keratinocytes (Larjava et al., 1993). However, our analysis showed that collagen VII is expressed under epidermal tongues, almost as far into the wound as laminin-332 (Figure 2(A) and (Nystrom, Velati, et al., 2013)). The discrepancies are likely due to improved antibodies and a more controlled experimental setup. Alternatively, there may be intrinsic differences between dermal and mucosal wound healing. Our studies revealed that collagen VII is involved in both epidermal as well as dermal wound closure of full-thickness wounds (Nystrom, Velati, et al., 2013). The absence of collagen VII leads to abnormalities in both dermal and epidermal wound healing. Through its high affinity for laminin-332, collagen VII organizes laminin-332 deposition in the provisional epidermal BM (Figure 2(B)). When collagen VII is absent laminin-332 deposition is disorganized which in its turn alters integrin $\alpha6\beta4$ downstream signaling ultimately impacting keratinocyte migration and delaying reepithelialization (Figure 2(C) and (Nystrom, Velati, et al., 2013)). As discussed above evidence of the importance of correct laminin-332 deposition for skin restoration can be found in the occurrence of chronic wounds in patients with reduced functional laminin-332 and altered laminin-332 deposition in diabetic ulcers and chronic venous ulcers (Nystrom, Velati, et al., 2013; Usui et al., 2008; Yuen, Huizinga, et al., 2013). Interestingly, patients with collagen VII deficiency also develop chronic wounds (see Section 4.1.3) and additionally, loss of collagen VII deposition in the epidermal BM is seen in acquired chronic wounds (Hopkinson, Anglin, Evans, & Harding, 1997; Nystrom, Velati, et al., 2013). Collagen VII also promotes dermal healing (Nystrom, Velati, et al., 2013). Absence of collagen VII leads to delayed maturation of granulation tissue and lingering inflammation. The effects can be accounted for by collagen VII's role in promoting fibroblast migration and regulating expression of growth factors important for dermal wound closure. Collagen VII-deficient fibroblasts cultured under woundlike conditions show reduced expression of FGF2 but increased expression of TGFβ (Kuttner et al., 2013; Ng et al., 2012; Nystrom, Velati, et al., 2013; Odorisio et al., 2014). The altered growth factor expression is dependent on the absence of collagen VII, as the expression can be normalized by supplementation with recombinant collagen VII (Nystrom, Velati, et al., 2013). Unrestrained TGFβ activity is the likely basis for the strong scarring phenotype that is connected with reduced collagen VII functionality (see Section 4.1.3). Lastly, the significance of collagen VII for skin regeneration can be seen from promotion of both reepithelialization as well as scar-free dermal healing, due to decreased profibrotic TGFβ2 and increased TGFβ3 expression (which may under certain circumstances be antifibrotic (Occleston, Laverty, O'Kane, & Ferguson, 2008)) after topical application of collagen VII (Wang et al., 2013).

4. DISORDERS OF THE EPIDERMAL BM

4.1 Genetic disorders

Inherited EB is a heterogeneous group of mechanobullous disorders resulting from mutations in genes encoding adhesion proteins of the cutaneous DEJ (Bruckner-Tuderman & Has, 2014). Extensive genetic studies and characterization of genotype–phenotype correlations in patients with EB has contributed substantially to our understanding of how these proteins function and interact. EB is classified in four types, according to the level of skin cleavage which corresponds to the defective adhesion structure and molecule: EB simplex with intraepidermal split, junctional EB with lamina lucida split, dystrophic EB (DEB) with dermal split, and Kindler syndrome with mixed levels of skin cleavage (Figure 3) (Fine et al., 2014). The genetic disorders associated with the epidermal BM discussed in detail in this chapter are junctional EB, DEB, the Kindler syndrome, and the newly identified interstitial lung disease, nephrotic syndrome, and EB (ILNEB) syndrome. Mutations in the genes encoding intracellular hemidesmosomal components plectin and BPAG1 are the cause of rare subtypes of EB simplex.

4.1.1 Disorders with defects of proteins of the hemidesmosomes and anchoring filaments

Junctional EB comprises clinically and genetically heterogeneous autosomal recessive disorders characterized by tissue separation along the lamina lucida within the BM zone, and alterations of the hemidesmosome-anchoring filament complex. The affected proteins include collagen XVII, integrin α6β4, and laminin-332. Junctional EB covers a broad spectrum of clinical severity, which mainly depends on the affected molecule and the type of mutation (reviewed in (Bruckner-Tuderman & Has, 2014)).

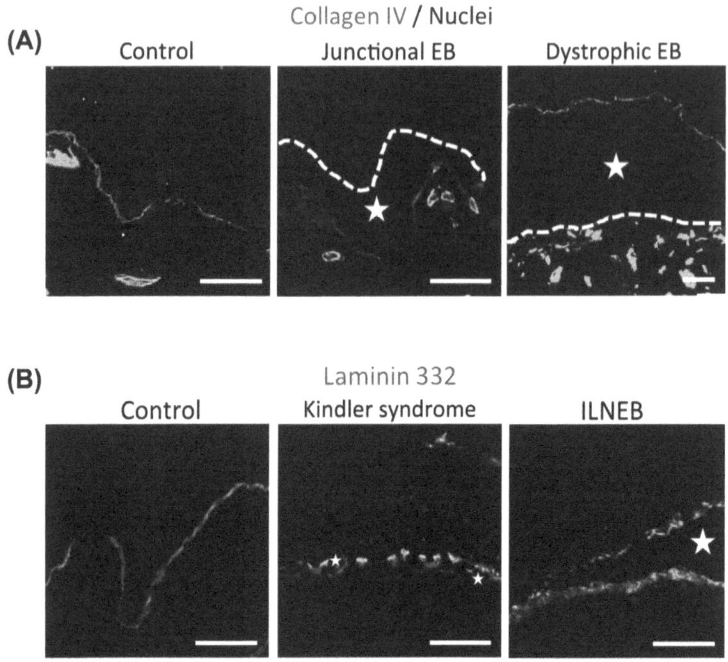

Figure 3 ***Skin cleavage and basement membrane anomalies in epidermolysis bullosa (EB).*** (A) Immunofluorescence staining with antibodies to collagen IV and (B) laminin-332 on skin sections of control individuals and patients with different types of EB. The blisters are indicated by asterisk. Scale bars = 100 μm. ILNEB, interstitial lung disease, nephrotic syndrome, and EB.

Lack of laminin-332 leads to generalized severe junctional EB (also known as Herlitz type, MIM#226700), with a lethal course due to extreme fragility of the skin and mucous membranes. Missense, splicing or in-frame mutations resulting in residual laminin-332 expression and function are associated with milder yet variable degrees of chronic skin fragility, but essentially a normal life span. Here, skin blistering may be generalized or localized and accompanied by mucosal involvement, nail dystrophy, and amelogenesis imperfecta. Importantly, these patients experience chronic nonhealing wounds on sites of mechanical skin exposure. In the vast majority of cases mutations reside in the *LAMB3* gene, followed by *LAMA3* and *LAMC2* (Kiritsi, Has, & Bruckner-Tuderman, 2013). The laryngo-onycho-cutaneous syndrome (MIM#245660) is a subtype of junctional EB caused by specific mutations in the *LAMA3* gene (McLean et al., 2003). Thus far, the disorder is confined to patients from the Punjabi Muslim population or from Iran. It is characterized by formation of extensive granulation tissue in skin areas with repeated trauma and in the laryngeal and conjunctival mucosa. The onset is at birth with a hoarse cry; later, chronic wounds and granulation tissue develop. The ocular lesions can lead to blindness (Kadyan et al., 2010); the larynx involvement may require tracheostomy. Mutations that just affect the laminin-α3a *LAMA3A* transcript are associated with the LOC syndrome (Barzegar et al., 2013; McLean et al., 2003). In two kindreds with autosomal-dominant amelogenesis imperfecta characterized by generalized severe enamel hypoplasia with deep linear grooves and pits, whole-exome sequencing identified heterozygous mutations in the last exon of *LAMB3* that likely truncate the protein (Kim et al., 2013).

Most mutations affecting the collagen XVII gene, *COL17A1*, result in lack of protein expression and generalized intermediate junctional EB (MIM#226650). The full blown clinical picture caused by loss of collagen XVII comprises congenital generalized blistering, and other progressive signs which develop with advancing age, such as skin atrophy and dyspigmentation, dystrophy and loss of nails and alopecia (Figure 4(A)). Mucosal involvement is mild and may be oral, ocular, nasal, or genitourinary. Teeth are always affected by amelogenesis imperfecta, manifesting as enamel pits and by increased incidence of caries. Heterozygous carriers of *COL17A1* mutations may have enamel defects, demonstrating that reduced amounts or mutated collagen XVII may be sufficient for dermal–epidermal stability, but not for proper dental development (Asaka et al., 2009). Mutations leading to expression of residual amounts or mutant collagen XVII result in milder, sometimes localized or late onset, skin fragility (Has & Kern, 2010; Has et al., 2014; Kiritsi et al., 2011; Tasanen et al., 2000). Fingernails may be normal or dystrophic, but toenails are mostly dystrophic; teeth are usually affected, but hair and mucous membranes are not. In the context of a *Lamc2* hypomorphic mouse model with low levels of laminin-332 expression (Bubier et al., 2010), *Col17a1* emerged as a strong genetic modifier of the phenotype (Sproule et al., 2014). Whether this mechanism holds true in the human disorder remains to be investigated.

Integrin α6β4 mutations are associated with variable degrees of skin fragility, pyloric atresia, and urogenital involvement (MIM#226730, 226650, 131800). The clinical spectrum ranges from nonlethal forms with very mild skin features, with or without pyloric atresia and/or genitourinary involvement, to severe lethal phenotypes (Diociaiuti et al., 2013; Schumann et al., 2013; Yuen, Sinke, & Jonkman, 2013). Most mutations affect the integrin β4 gene (*ITGB4*); genotype–phenotype correlations mostly depend

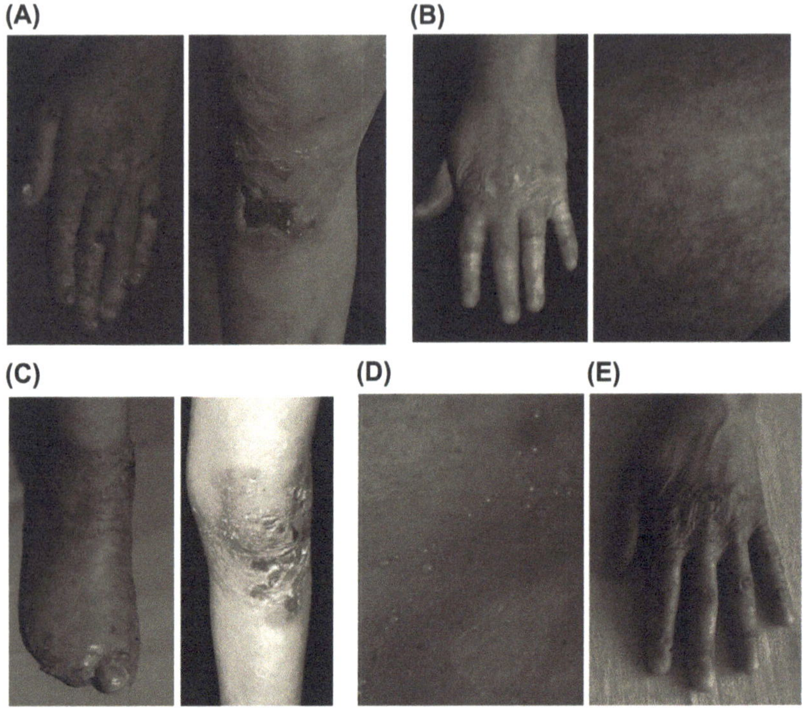

Figure 4 ***Clinical spectrum of genetic and acquired disorders of the epidermal basement membrane and associated structures.*** (A) Five-year-old boy with junctional epidermolysis bullosa (EB) due to a homozygous *COL17A1* splice site mutation. Note blisters, erosions, crusts, atrophic scarring dyspigmentation, and nail dystrophy. (B) 44-year-old patient with Kindler syndrome. Note skin atrophy, fibrotic changes of the fingers and poikiloderma on the limbs. (C) Five-year-old boy with dystrophic EB due to a homozygous *COL7A1* frameshift mutation. Note wounds, crusts, scarring, and mutilating deformity of the foot. (D) Erythema and tens blisters in a patient with bullous pemphigoid. (E) Scarring and nail loss in a patient with EB acquisita (Otten et al., 2014). (See color plate)

on the type of mutation, but also on unknown modifying factors (Bruckner-Tuderman, 2008; Dang et al., 2008).

In humans, a homozygous nonsense mutation in CD151 in two siblings resulted in hereditary nephropathy, sensorineural deafness, and pretibial EB (Karamatic Crew et al., 2004). This single report awaits confirmation before it can be accepted as a separate EB subtype (Fine et al., 2014).

4.1.2 Disorders with defects of focal adhesion proteins

Although the presence of focal adhesions in tissue, in situ, is a matter of debate, the genetic skin fragility disorders associated with alterations in focal adhesion proteins clearly support their contribution to skin integrity in vivo. Mutations in the *FERMT1* gene encoding kindlin-1 are the cause of the Kindler syndrome (MIM#173650), whereas mutations in the gene *ITGA3* encoding the integrin α3 subunit result in ILNEB (MIM#614748). Both disorders manifest with rather mild skin fragility. Intriguingly kindlin-1 and integrin α3 deficiency lead to significant alterations of the epidermal BM (Figure 3) (Bruckner-Tuderman & Has, 2014).

Kindler syndrome is considered to be a distinct EB subtype because, in contrast to the other major types of EB, there are typically multiple cleavage planes—intraepidermal, junctional, or sublamina densa (Fine et al., 2008). In addition to skin fragility, patients with Kindler syndrome suffer from photosensitivity, mucosal involvement, progressive scarring, and early aggressive mucocutaneous squamous cell carcinomas (Has et al., 2011) (Figure 4(B)). Most of the about 60 mutations reported so far are predicted to lead to premature termination of translation and to loss of the kindlin-1 protein or of its function (https://grenada.lumc.nl/LOVD2/mendelian_genes/home.php?select_db=FERMT1).

The clinical features of ILNEB are based on the findings in four patients reported so far in the literature (Has et al., 2012; Nicolaou et al., 2012). The disease first manifests with respiratory distress, as early as the first day of life. All patients were oxygen-dependent and exhibited recurrent aspiration pneumonias and lung infections. Chest radiographs and high resolution computed tomographies showed the picture of interstitial lung disease. All cases also had nephrotic syndrome and renal failure. Skin blistering started between 2 and 4 months of age and was rather mild, or completely unnoticed. Mucous membranes were not affected. Posttraumatic nail dystrophy and sparse hair were observed. All patients survived the neonatal period but died of multiorgan failure before the age of 19 months. Although the respiratory and renal features dominated the clinical picture, it was the investigation of the skin fragility that gave the clue to the diagnosis (Has et al., 2012). The *ITGA3* mutations disclosed so far lead to premature termination codons and lack of protein expression, or to amino acid substitutions which alter posttranslational modification of integrin α3 and prevent its membrane incorporation and function (Has et al., 2012; Nicolaou et al., 2012; Yamada & Sekiguchi, 2013).

4.1.3 Disorders with defects of proteins of the papillary matrix

In DEB (MIM#131750, 226600, 604129, 607523, 132000), blistering takes place at the level of the anchoring fibrils in the uppermost dermis.

The anchoring fibrils can be either completely absent, reduced in number or, if present, ultrastructurally altered. Mutations in the gene for collagen VII, *COL7A1*, underlie both dominant and recessive forms of DEB, which display a broad range of clinical severity (Figure 4(C)). Along with skin and mucosal blistering resulting in scarring, nail, and teeth involvement is common, as is scarring alopecia. The most severe subtype evolves with chronic wounds, mutilating scarring, multiorgan involvement, and early metastasizing cutaneous squamous cell carcinomas which restrict life expectancy to 30–40 years. At the other end of the phenotypic spectrum, patients may exhibit minor acral skin fragility and nail dystrophy. Although all subtypes of DEB are caused by mutations in *COL7A1* the molecular pathology remains challenging because of the large number of more than 650 mutations, and their variety of combinations (see Human Gene Mutation Database; http://www.hgmd.org). Glycine substitutions in the triple helix of collagen VII have a particular significance because they can disrupt the local helix stability and the salt bridges that function to stabilize subregions of the triple helix. Depending on the localization of the mutated glycine within the large triple helix, altered protein folding renders the collagen VII molecule more or less unstable and sensitive to degradation. Glycine substitution mutations in collagen VII may be inherited in a dominant or recessive manner, and genotype–phenotype correlations are difficult to predict (Has & Bruckner-Tuderman, 2014). Specific arginine and glycine substitutions in the collagenous subdomains cause the rare subtype of DEB inversa which affects mainly the great folds (van den Akker et al., 2011). Although the mechanisms underlying this predilection of blistering in the folds remain unknown, the hypothesis was that glycine substitutions located near the borders of collagenous subdomains exert milder changes than those located in the centers of collagenous subdomains (Chiaverini et al., 2010) and that the pathophysiology is temperature dependent (van den Akker et al., 2011). More than 100 *COL7A1* splicing mutations are known at the canonical splice sites as well as within regulatory regions in introns or exons. Most of them are recessive, but some cause in-frame skipping of exons and may act in a dominant manner. There may be some degree of clinical variability between DEB cases with the same mutation, or within the same family due to genetic or epigenetic modifying factors (reviewed in (Has & Bruckner-Tuderman, 2014)). This is further illustrated by large phenotypic variation seen in inbred rats with spontaneous DEB (Nystrom, Buttgereit, et al., 2013).

4.2 Acquired disorders

Circulating autoantibodies targeting structural constituents of the DEJ are the hallmark of autoimmune blistering disorders, in particular of pemphigoid diseases and EB acquisita (Table 1). The clinical picture consists of tense blisters and mucocutaneous erosions as a result of autoantibody binding to structural components of the hemidesmosomes and the papillary dermis (Figure 4(D)). Our understanding about the etiologic factors that induce the production of autoantibodies is limited. The pathogenesis of autoimmune blistering disorders is highly complex, and its detailed discussion is beyond the scope of this chapter. Animal and experimental models contributed strongly to the elucidation of the mechanisms of these diseases (reviewed in (Bieber et al., 2010; Chiriac et al., 2013; Nishie, 2013; Sesarman & Sitaru, 2013; Sitaru, 2007)).

The pemphigoid group is heterogeneous, both from clinical and molecular points of view. It includes bullous pemphigoid, mucous membrane pemphigoid, pemphigoid gestationis, LAD, lichen planus pemphigoides, and antilaminin-γ1 pemphigoid (syn. anti-p200 pemphigoid) with distinct molecular targets of the autoantibodies (Table 1) (Kasperkiewicz, Zillikens, & Schmidt, 2012). Bullous pemphigoid is the most common autoimmune blistering skin disorder that mostly affects the elderly (Figure 4(D)). Autoantibodies from bullous pemphigoid patients react with two hemidesmosomal components, collagen XVII and BPAG1. Almost all patients with bullous pemphigoid have autoantibodies binding to an immunodominant region of collagen XVII, the NC16A domain located extracellularly close to the transmembrane region (see Section 2.2). Additional antigenic sites exist on both the extracellular and intracellular domains of collagen XVII (Di Zenzo et al., 2004). The binding of autoantibodies to collagen XVII following the activation of complement and inflammatory pathways leads to the degradation of collagen XVII. Recent investigations have suggested other pathways, including a complement-independent pathway and a pathway involving IgE-autoantibodies (reviewed in (Nishie, 2014; Schmidt & Zillikens, 2013)).

EB acquisita is a chronic bullous disease with clinical features similar to DEB (Figure 4(E)), or manifesting as an inflammatory, widespread bullous disease. EB acquisita is characterized by the presence of autoantibodies against collagen VII (reviewed in (Kim & Kim, 2013; Ludwig & Zillikens, 2011)). The NC1 domain of collagen VII contains the major antigenic epitopes for EB acquisita (Gupta, Woodley, & Chen, 2012). In vivo functional epitope mapping of pathogenic autoantibodies in experimental EB acquisita demonstrated that multiple, distinct epitopes distributed over the entire NC1 (but not NC2) domain of collagen VII induce blistering skin disease in vivo (Csorba et al., 2014).

Both pemphigoid diseases and EB acquisita are characterized by a subepidermal cleavage plane. While histopathology identifies the type of inflammatory infiltrate and the rough cleavage level, direct immunofluorescence studies detect tissue bound autoantibodies, and indirect immunofluorescence studies disclose the antigen-specific autoantibodies in patients' sera. Immunoblot and immunoprecipitation studies of keratinocyte extracts show the presence of autoantibodies binding to specific proteins. Enzyme-linked immunosorbent assays utilizing recombinant proteins have been found to be highly specific and sensitive (Otten, Hashimoto, Hertl, Payne, & Sitaru, 2014). The precise diagnosis is critical for the prognosis and treatment (Otten et al., 2014).

5. CONCLUSIONS AND FUTURE PROSPECTS

Much progress has been made in understanding the composition and architecture of the epidermal BM in health and disease, mainly by studying genetically manipulated mouse models and naturally occurring mutations in humans. Although designed to assure stability, the epidermal BM and associated structures undergo coordinated dynamic changes, which assure cutaneous homeostasis and regeneration. These processes are regulated in complex manners through epithelial–mesenchymal interactions, and become deregulated in normal physiological processes such as aging, and in pathologic conditions such as chronic wounds and cancer. The lessons to be learned are how to control and manipulate individual factors or their interactions in order to restore/maintain proper BM function, promote wound healing, and prevent/treat neoplasia.

ACKNOWLEDGMENTS

This work was supported by grants from DEBRA international, Else Kröner-Fresenius-Foundation and Fritz Thyssen Foundation (C.H.), and from the German Federal Ministry for Education and Research BMBF, under the frame of E-Rare-2, the ERA-Net for Research Programmes on Rare Diseases (A.N).

REFERENCES

Abrahamsberg, C., Fuchs, P., Osmanagic-Myers, S., Fischer, I., Propst, F., Elbe-Burger, A., et al. (2005). Targeted ablation of plectin isoform 1 uncovers role of cytolinker proteins in leukocyte recruitment. *Proceedings of the National Academy of Sciences of the United States of America, 102*(51), 18449–18454.

Ackerl, R., Walko, G., Fuchs, P., Fischer, I., Schmuth, M., & Wiche, G. (2007). Conditional targeting of plectin in prenatal and adult mouse stratified epithelia causes keratinocyte fragility and lesional epidermal barrier defects. *Journal of Cell Science, 120*(Pt 14), 2435–2443.

Adair-Kirk, T. L., Griffin, G. L., Meyer, M. J., Kelley, D. G., Miner, J. H., Keene, D. R., et al. (2012). Keratinocyte-targeted expression of human laminin gamma2 rescues skin blistering and early lethality of laminin gamma2 deficient mice. *PLoS One, 7*(9), e45546.

van den Akker, P. C., Mellerio, J. E., Martinez, A. E., Liu, L., Meijer, R., Dopping-Hepenstal, P. J., et al. (2011). The inversa type of recessive dystrophic epidermolysis bullosa is caused by specific arginine and glycine substitutions in type VII collagen. *Journal of Medical Genetics, 48*(3), 160–167.

Allamand, V., & Guicheney, P. (2002). Merosin-deficient congenital muscular dystrophy, autosomal recessive (MDC1A, MIM#156225, LAMA2 gene coding for alpha2 chain of laminin). *European Journal of Human Genetics, 10*(2), 91–94.

Amano, S., Scott, I. C., Takahara, K., Koch, M., Champliaud, M. F., Gerecke, D. R., et al. (2000). Bone morphogenetic protein 1 is an extracellular processing enzyme of the laminin 5 gamma 2 chain. *The Journal of Biological Chemistry, 275*(30), 22728–22735.

Andra, K., Lassmann, H., Bittner, R., Shorny, S., Fassler, R., Propst, F., et al. (1997). Targeted inactivation of plectin reveals essential function in maintaining the integrity of skin, muscle, and heart cytoarchitecture. *Genes and Development, 11*(23), 3143–3156.

Andriessen, M. P., van den Born, J., Latijnhouwers, M. A., Bergers, M., van de Kerkhof, P. C., & Schalkwijk, J. (1997). Basal membrane heparan sulphate proteoglycan expression during wound healing in human skin. *The Journal of Pathology, 183*(3), 264–271.

Arikawa-Hirasawa, E., Watanabe, H., Takami, H., Hassell, J. R., & Yamada, Y. (1999). Perlecan is essential for cartilage and cephalic development. *Nature Genetics, 23*(3), 354–358.

Asaka, T., Akiyama, M., Domon, T., Nishie, W., Natsuga, K., Fujita, Y., et al. (2009). Type XVII collagen is a key player in tooth enamel formation. *American Journal of Pathology, 174*(1), 91–100.

Bachinger, H. P., Morris, N. P., Lunstrum, G. P., Keene, D. R., Rosenbaum, L. M., Compton, L. A., et al. (1990). The relationship of the biophysical and biochemical characteristics of type VII collagen to the function of anchoring fibrils. *The Journal of Biological Chemistry, 265*(17), 10095–10101.

Bader, B. L., Smyth, N., Nedbal, S., Miosge, N., Baranowsky, A., Mokkapati, S., et al. (2005). Compound genetic ablation of nidogen 1 and 2 causes basement membrane defects and perinatal lethality in mice. *Molecular and Cellular Biology, 25*(15), 6846–6856.

Baldwin, G., Novitskaya, V., Sadej, R., Pochec, E., Litynska, A., Hartmann, C., et al. (2008). Tetraspanin CD151 regulates glycosylation of (alpha)3(beta)1 integrin. *The Journal of Biological Chemistry, 283*(51), 35445–35454.

Bandyopadhyay, A., Rothschild, G., Kim, S., Calderwood, D. A., & Raghavan, S. (2012). Functional differences between kindlin-1 and kindlin-2 in keratinocytes. *Journal of Cell Science, 125*(Pt 9), 2172–2184.

Baranowsky, A., Mokkapati, S., Bechtel, M., Krugel, J., Miosge, N., Wickenhauser, C., et al. (2010). Impaired wound healing in mice lacking the basement membrane protein nidogen 1. *Matrix Biology, 29*(1), 15–21.

Barrientos, S., Stojadinovic, O., Golinko, M. S., Brem, H., & Tomic-Canic, M. (2008). Growth factors and cytokines in wound healing. *Wound Repair and Regeneration, 16*(5), 585–601.

Barzegar, M., Mozafari, N., Kariminejad, A., Asadikani, Z., Ozoemena, L., & McGrath, J. A. (2013). A new homozygous nonsense mutation in LAMA3A underlying laryngo-onycho-cutaneous syndrome. *British Journal of Dermatology, 169*(6), 1353–1356.

Battaglia, C., Mayer, U., Aumailley, M., & Timpl, R. (1992). Basement-membrane heparan sulfate proteoglycan binds to laminin by its heparan sulfate chains and to nidogen by sites in the protein core. *European Journal of Biochemistry, 208*(2), 359–366.

Baudoin, C., Fantin, L., & Meneguzzi, G. (2005). Proteolytic processing of the laminin alpha3 G domain mediates assembly of hemidesmosomes but has no role on keratinocyte migration. *Journal of Investigative Dermatology, 125*(5), 883–888.

Behrens, D. T., Villone, D., Koch, M., Brunner, G., Sorokin, L., Robenek, H., et al. (2012). The epidermal basement membrane is a composite of separate laminin- or collagen IV-containing networks connected by aggregated perlecan, but not by nidogens. *The Journal of Biological Chemistry, 287*(22), 18700–18709.

Bernfield, M., Kokenyesi, R., Kato, M., Hinkes, M. T., Spring, J., Gallo, R. L., et al. (1992). Biology of the syndecans: a family of transmembrane heparan sulfate proteoglycans. *Annual Review of Cell Biology, 8*, 365–393.

Betz, P., Nerlich, A., Wilske, J., Tubel, J., Wiest, I., Penning, R., et al. (1992). The time-dependent rearrangement of the epithelial basement membrane in human skin wounds–immunohistochemical localization of collagen IV and VII. *International Journal of Legal Medicine, 105*(2), 93–97.

Bieber, K., Sun, S., Ishii, N., Kasperkiewicz, M., Schmidt, E., Hirose, M., et al. (2010). Animal models for autoimmune bullous dermatoses. *Experimental Dermatology, 19*(1), 2–11.

Botta, A., Delteil, F., Mettouchi, A., Vieira, A., Estrach, S., Negroni, L., et al. (2012). Confluence switch signaling regulates ECM composition and the plasmin proteolytic cascade in keratinocytes. *Journal of Cell Science, 125*(Pt 18), 4241–4252.

Bouameur, J. E., Favre, B., & Borradori, L. (2014). Plakins, a versatile family of cytolinkers: roles in skin integrity and in human diseases. *Journal of Investigative Dermatology, 134*(4), 885–894.

Brakebusch, C., Grose, R., Quondamatteo, F., Ramirez, A., Jorcano, J. L., Pirro, A., et al. (2000). Skin and hair follicle integrity is crucially dependent on beta 1 integrin expression on keratinocytes. *The EMBO Journal, 19*(15), 3990–4003.

Brittingham, R., Uitto, J., & Fertala, A. (2006). High-affinity binding of the NC1 domain of collagen VII to laminin 5 and collagen IV. *Biochemical and Biophysical Research, 343*(3), 692–699.

Bruckner-Tuderman, L. (2008). Secondary modifiers and the phenotypic variability of junctional epidermolysis bullosa. *Acta Dermato-Venereologica, 88*(5), 436.

Bruckner-Tuderman, L., & Has, C. (2014). Disorders of the cutaneous basement membrane zone–the paradigm of epidermolysis bullosa. *Matrix Biology, 33*, 29–34.

Bruckner-Tuderman, L., Nilssen, O., Zimmermann, D. R., Dours-Zimmermann, M. T., Kalinke, D. U., Gedde-Dahl, T., Jr., et al. (1995). Immunohistochemical and mutation analyses demonstrate that procollagen VII is processed to collagen VII through removal of the NC-2 domain. *The Journal of Cell Biology, 131*(2), 551–559.

Bubier, J. A., Sproule, T. J., Alley, L. M., Webb, C. M., Fine, J. D., Roopenian, D. C., et al. (2010). A mouse model of generalized non-Herlitz junctional epidermolysis bullosa. *Journal of Investigative Dermatology, 130*(7), 1819–1828.

Burgeson, R. E. (1993). Type VII collagen, anchoring fibrils, and epidermolysis bullosa. *Journal of Investigative Dermatology, 101*(3), 252–255.

Campbell, I. D., & Humphries, M. J. (2011). Integrin structure, activation, and interactions. *Cold Spring Harbor Perspectives in Biology, 3*(3).

Champliaud, M. F., Lunstrum, G. P., Rousselle, P., Nishiyama, T., Keene, D. R., & Burgeson, R. E. (1996). Human amnion contains a novel laminin variant, laminin 7, which like laminin 6, covalently associates with laminin 5 to promote stable epithelial-stromal attachment. *The Journal of Cell Biology, 132*(6), 1189–1198.

Chen, J., Diacovo, T. G., Grenache, D. G., Santoro, S. A., & Zutter, M. M. (2002). The alpha(2) integrin subunit-deficient mouse: a multifaceted phenotype including defects of branching morphogenesis and hemostasis. *American Journal of Pathology, 161*(1), 337–344.

Chen, M., Marinkovich, M. P., Jones, J. C., O'Toole, E. A., Li, Y. Y., & Woodley, D. T. (1999). NC1 domain of type VII collagen binds to the beta3 chain of laminin 5 via a unique subdomain within the fibronectin-like repeats. *Journal of Investigative Dermatology, 112*(2), 177–183.

Chiaverini, C., Charlesworth, A. V., Youssef, M., Cuny, J. F., Rabia, S. H., Lacour, J. P., et al. (2010). Inversa dystrophic epidermolysis bullosa is caused by missense mutations at specific positions of the collagenic domain of collagen type VII. *Journal of Investigative Dermatology, 130*(10), 2508–2511.

Chiriac, M. T., Licarete, E., Sas, A. G., Rados, A. M., Lupan, I., Chiriac, A. M., et al. (2013). Passive transfer of collagen XVII-specific antibodies induces sustained blistering disease in adult mice. *Orphanet Journal of Rare Diseases, 8*, 17.

Chung, H. J., & Uitto, J. (2010). Type VII collagen: the anchoring fibril protein at fault in dystrophic epidermolysis bullosa. *Dermatologic Clinics, 28*(1), 93–105.

Colombo, M., Brittingham, R. J., Klement, J. F., Majsterek, I., Birk, D. E., Uitto, J., et al. (2003). Procollagen VII self-assembly depends on site-specific interactions and is promoted by cleavage of the NC2 domain with procollagen C-proteinase. *Biochemistry, 42*(39), 11434–11442.

Conti, F. J., Rudling, R. J., Robson, A., & Hodivala-Dilke, K. M. (2003). alpha3beta1-integrin regulates hair follicle but not interfollicular morphogenesis in adult epidermis. *Journal of Cell Science, 116*(Pt 13), 2737–2747.

Costell, M., Gustafsson, E., Aszodi, A., Morgelin, M., Bloch, W., Hunziker, E., et al. (1999). Perlecan maintains the integrity of cartilage and some basement membranes. *The Journal of Cell Biology, 147*(5), 1109–1122.

Cowin, A. J., Adams, D., Geary, S. M., Wright, M. D., Jones, J. C., & Ashman, L. K. (2006). Wound healing is defective in mice lacking tetraspanin CD151. *Journal of Investigative Dermatology, 126*(3), 680–689.

Csorba, K., Chiriac, M. T., Florea, F., Ghinia, M. G., Licarete, E., Rados, A., et al. (2014). Blister-inducing antibodies target multiple epitopes on collagen VII in mice. *Journal of Cellular and Molecular Medicine, 18*(9), 1727–1739.

Dang, N., Klingberg, S., Rubin, A. I., Edwards, M., Borelli, S., Relic, J., et al. (2008). Differential expression of pyloric atresia in junctional epidermolysis bullosa with ITGB4 mutations suggests that pyloric atresia is due to factors other than the mutations and not predictive of a poor outcome: three novel mutations and a review of the literature. *Acta Dermato-Venereologica, 88*(5), 438–448.

Di Zenzo, G., Grosso, F., Terracina, M., Mariotti, F., De Pita, O., Owaribe, K., et al. (2004). Characterization of the anti-BP180 autoantibody reactivity profile and epitope mapping in bullous pemphigoid patients. *Journal of Investigative Dermatology, 122*(1), 103–110.

Diociaiuti, A., Castiglia, D., Morini, F., Boldrini, R., Fortugno, P., Zambruno, G., et al. (2013). Long-term follow-up of a spontaneously improving patient with junctional epidermolysis bullosa associated with ITGB4 c.3977-19T>A splicing mutation. *Acta Dermato-Venereologica, 93*(1), 116–118.

DiPersio, C. M., Hodivala-Dilke, K. M., Jaenisch, R., Kreidberg, J. A., & Hynes, R. O. (1997). alpha3beta1 Integrin is required for normal development of the epidermal basement membrane. *The Journal of Cell Biology, 137*(3), 729–742.

DiPersio, C. M., van der Neut, R., Georges-Labouesse, E., Kreidberg, J. A., Sonnenberg, A., & Hynes, R. O. (2000). alpha3beta1 and alpha6beta4 integrin receptors for laminin-5 are not essential for epidermal morphogenesis and homeostasis during skin development. *Journal of Cell Science, 113*(Pt 17), 3051–3062.

Dowling, J., Yu, Q. C., & Fuchs, E. (1996). Beta4 integrin is required for hemidesmosome formation, cell adhesion and cell survival. *The Journal of Cell Biology, 134*(2), 559–572.

Drewniok, C., Wienrich, B. G., Schon, M., Ulrich, J., Zen, Q., Telen, M. J., et al. (2004). Molecular interactions of B-CAM (basal-cell adhesion molecule) and laminin in epithelial skin cancer. *Archives of Dermatological Research, 296*(2), 59–66.
Fine, J. D., Bruckner-Tuderman, L., Eady, R. A., Bauer, E. A., Bauer, J. W., Has, C., et al. (2014). Inherited epidermolysis bullosa: updated recommendations on diagnosis and classification. *Journal of the American Academy of Dermatology, 70*(6), 1103–1126.
Fine, J. D., Eady, R. A., Bauer, E. A., Bauer, J. W., Bruckner-Tuderman, L., Heagerty, A., et al. (2008). The classification of inherited epidermolysis bullosa (EB): Report of the Third International Consensus Meeting on Diagnosis and Classification of EB. *Journal of the American Academy of Dermatology, 58*(6), 931–950.
Fleischmajer, R., Kuroda, K., Utani, A., Douglas MacDonald, E., Perlish, J. S., Arikawa-Hirasawa, E., et al. (2000). Differential expression of laminin alpha chains during proliferative and differentiation stages in a model for skin morphogenesis. *Matrix Biology, 19*(7), 637–647.
Franzke, C. W., Bruckner-Tuderman, L., & Blobel, C. P. (2009). Shedding of collagen XVII/BP180 in skin depends on both ADAM10 and ADAM9. *The Journal of Biological Chemistry, 284*(35), 23386–23396.
Franzke, C. W., Bruckner, P., & Bruckner-Tuderman, L. (2005). Collagenous transmembrane proteins: recent insights into biology and pathology. *The Journal of Biological Chemistry, 280*(6), 4005–4008.
Franzke, C. W., Tasanen, K., Borradori, L., Huotari, V., & Bruckner-Tuderman, L. (2004). Shedding of collagen XVII/BP180: structural motifs influence cleavage from cell surface. *The Journal of Biological Chemistry, 279*(23), 24521–24529.
Frijns, E., Kuikman, I., Litjens, S., Raspe, M., Jalink, K., Ports, M., et al. (2012). Phosphorylation of threonine 1736 in the C-terminal tail of integrin beta4 contributes to hemidesmosome disassembly. *Molecular Biology of the Cell, 23*(8), 1475–1485.
Frijns, E., Sachs, N., Kreft, M., Wilhelmsen, K., & Sonnenberg, A. (2010). EGF-induced MAPK signaling inhibits hemidesmosome formation through phosphorylation of the integrin {beta}4. *The Journal of Biological Chemistry, 285*(48), 37650–37662.
Fritsch, A., Loeckermann, S., Kern, J. S., Braun, A., Bosl, M. R., Bley, T. A., et al. (2008). A hypomorphic mouse model of dystrophic epidermolysis bullosa reveals mechanisms of disease and response to fibroblast therapy. *Journal of Clinical Investigation, 118*(5), 1669–1679.
Fujita, M., Ieguchi, K., Davari, P., Yamaji, S., Taniguchi, Y., Sekiguchi, K., et al. (2012). Cross-talk between integrin alpha6beta4 and insulin-like growth factor-1 receptor (IGF1R) through direct alpha6beta4 binding to IGF1 and subsequent alpha6beta4-IGF1-IGF1R ternary complex formation in anchorage-independent conditions. *The Journal of Biological Chemistry, 287*(15), 12491–12500.
Fujiwara, H., Ferreira, M., Donati, G., Marciano, D. K., Linton, J. M., Sato, Y., et al. (2011). The basement membrane of hair follicle stem cells is a muscle cell niche. *Cell, 144*(4), 577–589.
Gagnoux-Palacios, L., Allegra, M., Spirito, F., Pommeret, O., Romero, C., Ortonne, J. P., et al. (2001). The short arm of the laminin gamma2 chain plays a pivotal role in the incorporation of laminin 5 into the extracellular matrix and in cell adhesion. *The Journal of Cell Biology, 153*(4), 835–850.
Gao, J., DeRouen, M. C., Chen, C. H., Nguyen, M., Nguyen, N. T., Ido, H., et al. (2008). Laminin-511 is an epithelial message promoting dermal papilla development and function during early hair morphogenesis. *Genes and Development, 22*(15), 2111–2124.
Gebauer, J. M., Keene, D. R., Olsen, B. R., Sorokin, L. M., Paulsson, M., & Wagener, R. (2009). Mouse AMACO, a kidney and skin basement membrane associated molecule that mediates RGD-dependent cell attachment. *Matrix Biology, 28*(8), 456–462.
Geerts, D., Fontao, L., Nievers, M. G., Schaapveld, R. Q., Purkis, P. E., Wheeler, G. N., et al. (1999). Binding of integrin alpha6beta4 to plectin prevents plectin association

with F-actin but does not interfere with intermediate filament binding. *The Journal of Cell Biology, 147*(2), 417–434.
Georges-Labouesse, E., Messaddeq, N., Yehia, G., Cadalbert, L., Dierich, A., & Le Meur, M. (1996). Absence of integrin alpha 6 leads to epidermolysis bullosa and neonatal death in mice. *Nature Genetics, 13*(3), 370–373.
Goldfinger, L. E., Hopkinson, S. B., deHart, G. W., Collawn, S., Couchman, J. R., & Jones, J. C. (1999). The alpha3 laminin subunit, alpha6beta4 and alpha3beta1 integrin coordinately regulate wound healing in cultured epithelial cells and in the skin. *Journal of Cell Science, 112*(Pt 16), 2615–2629.
Goldfinger, L. E., Stack, M. S., & Jones, J. C. (1998). Processing of laminin-5 and its functional consequences: role of plasmin and tissue-type plasminogen activator. *The Journal of Cell Biology, 141*(1), 255–265.
Gould, D. B., Phalan, F. C., Breedveld, G. J., van Mil, S. E., Smith, R. S., Schimenti, J. C., et al. (2005). Mutations in Col4a1 cause perinatal cerebral hemorrhage and porencephaly. *Science, 308*(5725), 1167–1171.
Gould, D. B., Phalan, F. C., van Mil, S. E., Sundberg, J. P., Vahedi, K., Massin, P., et al. (2006). Role of COL4A1 in small-vessel disease and hemorrhagic stroke. *New England Journal of Medicine, 354*(14), 1489–1496.
Goult, B. T., Bouaouina, M., Harburger, D. S., Bate, N., Patel, B., Anthis, N. J., et al. (2009). The structure of the N-terminus of kindlin-1: a domain important for alphaiibbeta3 integrin activation. *Journal of Molecular Biology, 394*(5), 944–956.
Grenache, D. G., Zhang, Z., Wells, L. E., Santoro, S. A., Davidson, J. M., & Zutter, M. M. (2007). Wound healing in the alpha2beta1 integrin-deficient mouse: altered keratinocyte biology and dysregulated matrix metalloproteinase expression. *Journal of Investigative Dermatology, 127*(2), 455–466.
Grose, R., Hutter, C., Bloch, W., Thorey, I., Watt, F. M., Fassler, R., et al. (2002). A crucial role of beta 1 integrins for keratinocyte migration in vitro and during cutaneous wound repair. *Development, 129*(9), 2303–2315.
Guo, L., Degenstein, L., Dowling, J., Yu, Q. C., Wollmann, R., Perman, B., et al. (1995). Gene targeting of BPAG1: abnormalities in mechanical strength and cell migration in stratified epithelia and neurologic degeneration. *Cell, 81*(2), 233–243.
Gupta, R., Woodley, D. T., & Chen, M. (2012). Epidermolysis bullosa acquisita. *Clinics in Dermatology, 30*(1), 60–69.
Harburger, D. S., Bouaouina, M., & Calderwood, D. A. (2009). Kindlin-1 and -2 directly bind the C-terminal region of beta integrin cytoplasmic tails and exert integrin-specific activation effects. *The Journal of Biological Chemistry, 284*(17), 11485–11497.
deHart, G. W., Healy, K. E., & Jones, J. C. (2003). The role of alpha3beta1 integrin in determining the supramolecular organization of laminin-5 in the extracellular matrix of keratinocytes. *Experimental Cell Research, 283*(1), 67–79.
Has, C., & Bruckner-Tuderman, L. (2014). The genetics of skin fragility. *Annuual Review of Genomics and Human Genetics, 15*, 245–268.
Has, C., Castiglia, D., del Rio, M., Diez, M. G., Piccinni, E., Kiritsi, D., et al. (2011). Kindler syndrome: extension of FERMT1 mutational spectrum and natural history. *Human Mutation, 32*(11), 1204–1212.
Has, C., & Kern, J. S. (2010). Collagen XVII. *Dermatologic Clinics, 28*(1), 61–66.
Has, C., Kiritsi, D., Mellerio, J. E., Franzke, C. W., Wedgeworth, E., Tantcheva-Poor, I., et al. (2014). The missense mutation p.R1303Q in type XVII collagen underlies junctional epidermolysis bullosa resembling Kindler syndrome. *Journal of Investigative Dermatology, 134*(3), 845–849.
Has, C., Sparta, G., Kiritsi, D., Weibel, L., Moeller, A., Vega-Warner, V., et al. (2012). Integrin alpha3 mutations with kidney, lung, and skin disease. *New England Journal of Medicine, 366*(16), 1508–1514.

Hasegawa, H., Naito, I., Nakano, K., Momota, R., Nishida, K., Taguchi, T., et al. (2007). The distributions of type IV collagen alpha chains in basement membranes of human epidermis and skin appendages. *Archives of Histology and Cytology, 70*(4), 255–265.

He, Y., Esser, P., Heinemann, A., Bruckner-Tuderman, L., & Has, C. (2011). Kindlin-1 and -2 have overlapping functions in epithelial cells implications for phenotype modification. *American Journal of Pathology, 178*(3), 975–982.

He, Y., Sonnenwald, T., Sprenger, A., Hansen, U., Dengjel, J., Bruckner-Tuderman, L., et al. (2014). RhoA activation by CNFy restores cell-cell adhesion in kindlin-2-deficient keratinocytes. *The Journal of Pathology, 233*(3), 269–280.

Heinonen, S., Mannikko, M., Klement, J. F., Whitaker-Menezes, D., Murphy, G. F., & Uitto, J. (1999). Targeted inactivation of the type VII collagen gene (Col7a1) in mice results in severe blistering phenotype: a model for recessive dystrophic epidermolysis bullosa. *Journal of Cell Science, 112*(Pt 21), 3641–3648.

Herz, C., Aumailley, M., Schulte, C., Schlotzer-Schrehardt, U., Bruckner-Tuderman, L., & Has, C. (2006). Kindlin-1 is a phosphoprotein involved in regulation of polarity, proliferation, and motility of epidermal keratinocytes. *The Journal of Biological Chemistry, 281*(47), 36082–36090.

Herzog, C., Has, C., Franzke, C. W., Echtermeyer, F. G., Schlotzer-Schrehardt, U., Kroger, S., et al. (2004). Dystroglycan in skin and cutaneous cells: beta-subunit is shed from the cell surface. *Journal of Investigative Dermatology, 122*(6), 1372–1380.

Hintermann, E., & Quaranta, V. (2004). Epithelial cell motility on laminin-5: regulation by matrix assembly, proteolysis, integrins and erbB receptors. *Matrix Biology, 23*(2), 75–85.

Hofmann, S. C., Voith, U., Schonau, V., Sorokin, L., Bruckner-Tuderman, L., & Franzke, C. W. (2009). Plasmin plays a role in the in vitro generation of the linear IgA dermatosis antigen LADB97. *Journal of Investigative Dermatology, 129*(7), 1730–1739.

Holtkotter, O., Nieswandt, B., Smyth, N., Muller, W., Hafner, M., Schulte, V., et al. (2002). Integrin alpha 2-deficient mice develop normally, are fertile, but display partially defective platelet interaction with collagen. *The Journal of Biological Chemistry, 277*(13), 10789–10794.

Hopf, M., Gohring, W., Kohfeldt, E., Yamada, Y., & Timpl, R. (1999). Recombinant domain IV of perlecan binds to nidogens, laminin-nidogen complex, fibronectin, fibulin-2 and heparin. *European Journal of Biochemistry, 259*(3), 917–925.

Hopkinson, I., Anglin, I. E., Evans, D. L., & Harding, K. G. (1997). Collagen VII expression in human chronic wounds and scars. *The Journal of Pathology, 182*(2), 192–196.

Hulmes, D. J. S. (2008). Collagen diversity, synthesis and assembly. In P. Fratzl (Ed.), *Collagen: Structure and mechanics*. New York: Springer Science+Business Media, LLC.

Hunefeld, C., Mezger, M., Kern, J. S., Nystrom, A., Bruckner-Tuderman, L., Handgretinger, R., et al. (2013). One goal, different strategies—molecular and cellular approaches for the treatment of inherited skin fragility disorders. *Experimental Dermatology, 22*(3), 162–167.

Hurskainen, T., Kokkonen, N., Sormunen, R., Jackow, J., Loffek, S., Soininen, R., et al. (2015). Deletion of the major bullous pemphigoid epitope region of collagen XVII induces blistering, autoimmunization, and itching in mice. *Journal of Investigative Dermatology, 135*(5), 1303–1310.

Iivanainen, A., Vuolteenaho, R., Sainio, K., Eddy, R., Shows, T. B., Sariola, H., et al. (1995). The human laminin beta 2 chain (S-laminin): structure, expression in fetal tissues and chromosomal assignment of the LAMB2 gene. *Matrix Biology, 14*(6), 489–497.

Imanishi, H., Tsuruta, D., Tateishi, C., Sugawara, K., Paus, R., Tsuji, T., et al. (2010). Laminin-511, inducer of hair growth, is down-regulated and its suppressor in hair growth, laminin-332 up-regulated in chemotherapy-induced alopecia. *Journal of Dermatological Science, 58*(1), 43–54.

Inomata, T., Ebihara, N., Funaki, T., Matsuda, A., Watanabe, Y., Ning, L., et al. (2012). Perlecan-deficient mutation impairs corneal epithelial structure. *Investigative Opthalmology and Visual Science, 53*(3), 1277–1284.

Iozzo, R. V., Zoeller, J. J., & Nystrom, A. (2009). Basement membrane proteoglycans: modulators Par Excellence of cancer growth and angiogenesis. *Molecules and Cells, 27*(5), 503–513.

Ito, K., Sawamura, D., Goto, M., Nakamura, H., Nishie, W., Sakai, K., et al. (2009). Keratinocyte-/fibroblast-targeted rescue of Col7a1-disrupted mice and generation of an exact dystrophic epidermolysis bullosa model using a human COL7A1 mutation. *American Journal of Pathology, 175*(6), 2508–2517.

Jung, J. Y., Oh, J. H., Kim, Y. K., Shin, M. H., Lee, D., & Chung, J. H. (2012). Acute UV irradiation increases heparan sulfate proteoglycan levels in human skin. *Journal of Korean Medical Science, 27*(3), 300–306.

Kadyan, A., Aralikatti, A., Shah, S., Jewell, R., Paul, L., Darling, J., et al. (2010). Laryngo-onycho-cutaneous syndrome. *Ophthalmology, 117*(5), 1056. e1052.

Karamatic Crew, V., Burton, N., Kagan, A., Green, C. A., Levene, C., Flinter, F., et al. (2004). CD151, the first member of the tetraspanin (TM4) superfamily detected on erythrocytes, is essential for the correct assembly of human basement membranes in kidney and skin. *Blood, 104*(8), 2217–2223.

Kasperkiewicz, M., Zillikens, D., & Schmidt, E. (2012). Pemphigoid diseases: pathogenesis, diagnosis, and treatment. *Autoimmunity, 45*(1), 55–70.

Kern, J. S., Herz, C., Haan, E., Moore, D., Nottelmann, S., von Lilien, T., et al. (2007). Chronic colitis due to an epithelial barrier defect: the role of kindlin-1 isoforms. *The Journal of Pathology, 213*(4), 462–470.

Kim, J. H., & Kim, S. C. (2013). Epidermolysis bullosa acquisita. *Journal of the European Academy of Dermatology and Venereology, 27*(10), 1204–1213.

Kim, J. W., Seymen, F., Lee, K. E., Ko, J., Yildirim, M., Tuna, E. B., et al. (2013). LAMB3 mutations causing autosomal-dominant amelogenesis imperfecta. *Journal of Dental Research, 92*(10), 899–904.

Kim, Y. M., Hwang, S., Kim, Y. M., Pyun, B. J., Kim, T. Y., Lee, S. T., et al. (2002). Endostatin blocks vascular endothelial growth factor-mediated signaling via direct interaction with KDR/Flk-1. *The Journal of Biological Chemistry, 277*(31), 27872–27879.

Kirfel, G., Rigort, A., Borm, B., Schulte, C., & Herzog, V. (2003). Structural and compositional analysis of the keratinocyte migration track. *Cell Motility and the Cytoskeleton, 55*(1), 1–13.

Kiritsi, D., Has, C., & Bruckner-Tuderman, L. (2013). Laminin 332 in junctional epidermolysis bullosa. *Cell Adhesion and Migration, 7*(1), 135–141.

Kiritsi, D., Kern, J. S., Schumann, H., Kohlhase, J., Has, C., & Bruckner-Tuderman, L. (2011). Molecular mechanisms of phenotypic variability in junctional epidermolysis bullosa. *Journal of Medical Genetics, 48*(7), 450–457.

Kiyozumi, D., Takeichi, M., Nakano, I., Sato, Y., Fukuda, T., & Sekiguchi, K. (2012). Basement membrane assembly of the integrin alpha8beta1 ligand nephronectin requires Fraser syndrome-associated proteins. *The Journal of Cell Biology, 197*(5), 677–689.

Koch, M., Olson, P. F., Albus, A., Jin, W., Hunter, D. D., Brunken, W. J., et al. (1999). Characterization and expression of the laminin gamma3 chain: a novel, non-basement membrane-associated, laminin chain. *The Journal of Cell Biology, 145*(3), 605–618.

Kopecki, Z., Arkell, R., Powell, B. C., & Cowin, A. J. (2009). Flightless I regulates hemidesmosome formation and integrin-mediated cellular adhesion and migration during wound repair. *Journal of Investigative Dermatology, 129*(8), 2031–2045.

Koshikawa, N., Minegishi, T., Sharabi, A., Quaranta, V., & Seiki, M. (2005). Membrane-type matrix metalloproteinase-1 (MT1-MMP) is a processing enzyme for human laminin gamma 2 chain. *The Journal of Biological Chemistry, 280*(1), 88–93.

Koster, J., van Wilpe, S., Kuikman, I., Litjens, S. H., & Sonnenberg, A. (2004). Role of binding of plectin to the integrin beta4 subunit in the assembly of hemidesmosomes. *Molecular Biology of the Cell, 15*(3), 1211–1223.
Kostka, G., Giltay, R., Bloch, W., Addicks, K., Timpl, R., Fassler, R., et al. (2001). Perinatal lethality and endothelial cell abnormalities in several vessel compartments of fibulin-1-deficient mice. *Molecular and Cellular Biology, 21*(20), 7025–7034.
Kreidberg, J. A., Donovan, M. J., Goldstein, S. L., Rennke, H., Shepherd, K., Jones, R. C., et al. (1996). Alpha 3 beta 1 integrin has a crucial role in kidney and lung organogenesis. *Development, 122*(11), 3537–3547.
Kuster, J. E., Guarnieri, M. H., Ault, J. G., Flaherty, L., & Swiatek, P. J. (1997). IAP insertion in the murine LamB3 gene results in junctional epidermolysis bullosa. *Mammalian Genome, 8*(9), 673–681.
Kuttner, V., Mack, C., Gretzmeier, C., Bruckner-Tuderman, L., & Dengjel, J. (2014). Loss of collagen VII is associated with reduced transglutaminase 2 abundance and activity. *Journal of Investigative Dermatology, 134*(9), 2381–2389.
Kuttner, V., Mack, C., Rigbolt, K. T., Kern, J. S., Schilling, O., Busch, H., et al. (2013). Global remodelling of cellular microenvironment due to loss of collagen VII. *Molecular Systems Biology, 9*, 657.
Lapiere, J. C., Chen, J. D., Iwasaki, T., Hu, L., Uitto, J., & Woodley, D. T. (1994). Type VII collagen specifically binds fibronectin via a unique subdomain within the collagenous triple helix. *Journal of Investigative Dermatology, 103*(5), 637–641.
Larjava, H., Salo, T., Haapasalmi, K., Kramer, R. H., & Heino, J. (1993). Expression of integrins and basement membrane components by wound keratinocytes. *Journal of Clinical Investigation, 92*(3), 1425–1435.
Li, J., Tzu, J., Chen, Y., Zhang, Y. P., Nguyen, N. T., Gao, J., et al. (2003). Laminin-10 is crucial for hair morphogenesis. *The EMBO Journal, 22*(10), 2400–2410.
Li, Q., Yang, X. H., Xu, F., Sharma, C., Wang, H. X., Knoblich, K., et al. (2013). Tetraspanin CD151 plays a key role in skin squamous cell carcinoma. *Oncogene, 32*(14), 1772–1783.
Libby, R. T., Champliaud, M. F., Claudepierre, T., Xu, Y., Gibbons, E. P., Koch, M., et al. (2000). Laminin expression in adult and developing retinae: evidence of two novel CNS laminins. *The Journal of Neuroscience, 20*(17), 6517–6528.
Lissitzky, J. C., Luis, J., Munzer, J. S., Benjannet, S., Parat, F., Chretien, M., et al. (2000). Endoproteolytic processing of integrin pro-alpha subunits involves the redundant function of furin and proprotein convertase (PC) 5A, but not paired basic amino acid converting enzyme (PACE) 4, PC5B or PC7. *Biochemical Journal, 346*(Pt 1), 133–138.
Litjens, S. H., de Pereda, J. M., & Sonnenberg, A. (2006). Current insights into the formation and breakdown of hemidesmosomes. *Trends in Cell Biology, 16*(7), 376–383.
Liu, S., Xu, S. W., Blumbach, K., Eastwood, M., Denton, C. P., Eckes, B., et al. (2010). Expression of integrin beta1 by fibroblasts is required for tissue repair in vivo. *Journal of Cell Science, 123*(Pt 21), 3674–3682.
Longmate, W. M., Monichan, R., Chu, M. L., Tsuda, T., Mahoney, M. G., & DiPersio, C. M. (2014). Reduced fibulin-2 contributes to loss of basement membrane integrity and skin blistering in mice lacking integrin alpha3beta1 in the epidermis. *Journal of Investigative Dermatology, 134*(6), 1609–1617.
Lu, W., Miyazaki, K., Mizushima, H., & Nemoto, N. (2001). Immunohistochemical distribution of laminin-5 gamma2 chain and its developmental change in human embryonic and foetal tissues. *Histochemical Journal, 33*(11–12), 629–637.
Ludwig, R. J., & Zillikens, D. (2011). Pathogenesis of epidermolysis bullosa acquisita. *Dermatologic Clinics, 29*(3), 493–501.
Margadant, C., Raymond, K., Kreft, M., Sachs, N., Janssen, H., & Sonnenberg, A. (2009). Integrin alpha3beta1 inhibits directional migration and wound re-epithelialization in the skin. *Journal of Cell Science, 122*(Pt 2), 278–288.

McLean, W. H., Irvine, A. D., Hamill, K. J., Whittock, N. V., Coleman-Campbell, C. M., Mellerio, J. E., et al. (2003). An unusual N-terminal deletion of the laminin alpha3a isoform leads to the chronic granulation tissue disorder laryngo-onycho-cutaneous syndrome. *Human Molecular Genetics, 12*(18), 2395–2409.
McMillan, J. R., Akiyama, M., & Shimizu, H. (2003). Ultrastructural orientation of laminin 5 in the epidermal basement membrane: an updated model for basement membrane organization. *Journal of Histochemistry and Cytochemistry, 51*(10), 1299–1306.
Meng, X., Klement, J. F., Leperi, D. A., Birk, D. E., Sasaki, T., Timpl, R., et al. (2003). Targeted inactivation of murine laminin gamma2-chain gene recapitulates human junctional epidermolysis bullosa. *Journal of Investigative Dermatology, 121*(4), 720–731.
Miner, J. H., Cunningham, J., & Sanes, J. R. (1998). Roles for laminin in embryogenesis: exencephaly, syndactyly, and placentopathy in mice lacking the laminin alpha5 chain. *The Journal of Cell Biology, 143*(6), 1713–1723.
Miner, J. H., Li, C., Mudd, J. L., Go, G., & Sutherland, A. E. (2004). Compositional and structural requirements for laminin and basement membranes during mouse embryo implantation and gastrulation. *Development, 131*(10), 2247–2256.
Miosge, N., Gotz, W., Sasaki, T., Chu, M. L., Timpl, R., & Herken, R. (1996). The extracellular matrix proteins fibulin-1 and fibulin-2 in the early human embryo. *Histochemical Journal, 28*(2), 109–116.
Mitchell, K., Szekeres, C., Milano, V., Svenson, K. B., Nilsen-Hamilton, M., Kreidberg, J. A., et al. (2009). Alpha3beta1 integrin in epidermis promotes wound angiogenesis and keratinocyte-to-endothelial-cell crosstalk through the induction of MRP3. *Journal of Cell Science, 122*(Pt 11), 1778–1787.
Mokkapati, S., Baranowsky, A., Mirancea, N., Smyth, N., Breitkreutz, D., & Nischt, R. (2008). Basement membranes in skin are differently affected by lack of nidogen 1 and 2. *Journal of Investigative Dermatology, 128*(9), 2259–2267.
Montanez, E., Ussar, S., Schifferer, M., Bosl, M., Zent, R., Moser, M., et al. (2008). Kindlin-2 controls bidirectional signaling of integrins. *Genes and Development, 22*(10), 1325–1330.
Muhle, C., Neuner, A., Park, J., Pacho, F., Jiang, Q., Waddington, S. N., et al. (2006). Evaluation of prenatal intra-amniotic LAMB3 gene delivery in a mouse model of Herlitz disease. *Gene Therapy, 13*(23), 1665–1676.
Murshed, M., Smyth, N., Miosge, N., Karolat, J., Krieg, T., Paulsson, M., et al. (2000). The absence of nidogen 1 does not affect murine basement membrane formation. *Molecular and Cellular Biology, 20*(18), 7007–7012.
Nanba, D., Hieda, Y., & Nakanishi, Y. (2000). Remodeling of desmosomal and hemidesmosomal adhesion systems during early morphogenesis of mouse pelage hair follicles. *Journal of Investigative Dermatology, 114*(1), 171–177.
van der Neut, R., Krimpenfort, P., Calafat, J., Niessen, C. M., & Sonnenberg, A. (1996). Epithelial detachment due to absence of hemidesmosomes in integrin beta 4 null mice. *Nature Genetics, 13*(3), 366–369.
Ng, Y. Z., Pourreyron, C., Salas-Alanis, J. C., Dayal, J. H., Cepeda-Valdes, R., Yan, W., et al. (2012 Jul 15). Fibroblast-derived dermal matrix drives development of aggressive cutaneous squamous cell carcinoma in patients with recessive dystrophic epidermolysis bullosa. *Cancer Research, 72*(14), 3522–3534.
Nicolaou, N., Margadant, C., Kevelam, S. H., Lilien, M. R., Oosterveld, M. J., Kreft, M., et al. (2012). Gain of glycosylation in integrin alpha3 causes lung disease and nephrotic syndrome. *Journal of Clinical Investigation, 122*(12), 4375–4387.
Niculescu, C., Ganguli-Indra, G., Pfister, V., Dupe, V., Messaddeq, N., De Arcangelis, A., et al. (2011). Conditional ablation of integrin alpha-6 in mouse epidermis leads to skin fragility and inflammation. *European Journal of Cell Biology, 90*(2–3), 270–277.
Nikolopoulos, S. N., Blaikie, P., Yoshioka, T., Guo, W., Puri, C., Tacchetti, C., et al. (2005). Targeted deletion of the integrin beta4 signaling domain suppresses laminin-

5-dependent nuclear entry of mitogen-activated protein kinases and NF-kappaB, causing defects in epidermal growth and migration. *Molecular and Cellular Biology, 25*(14), 6090–6102.
Nischt, R., Schmidt, C., Mirancea, N., Baranowsky, A., Mokkapati, S., Smyth, N., et al. (2007). Lack of nidogen-1 and -2 prevents basement membrane assembly in skin-organotypic coculture. *Journal of Investigative Dermatology, 127*(3), 545–554.
Nishie, W. (2013). Mouse models of autoimmune blistering diseases induced by the passive transfer of antibodies. *Methods in Molecular Biology, 961*, 363–369.
Nishie, W. (2014). Update on the pathogenesis of bullous pemphigoid: an autoantibody-mediated blistering disease targeting collagen XVII. *Journal of Dermatological Science, 73*(3), 179–186.
Nishie, W., Jackow, J., Hofmann, S. C., Franzke, C. W., & Bruckner-Tuderman, L. (2012). Coiled coils ensure the physiological ectodomain shedding of collagen XVII. *The Journal of Biological Chemistry, 287*(35), 29940–29948.
Nishie, W., Kiritsi, D., Nystrom, A., Hofmann, S. C., & Bruckner-Tuderman, L. (2011). Dynamic interactions of epidermal collagen XVII with the extracellular matrix: laminin 332 as a major binding partner. *American Journal of Pathology, 179*(2), 829–837.
Nishie, W., Lamer, S., Schlosser, A., Licarete, E., Franzke, C. W., Hofmann, S. C., et al. (2010). Ectodomain shedding generates Neoepitopes on collagen XVII, the major autoantigen for bullous pemphigoid. *The Journal of Immunology, 185*(8), 4938–4947.
Nishie, W., Sawamura, D., Goto, M., Ito, K., Shibaki, A., McMillan, J. R., et al. (2007). Humanization of autoantigen. *Natural Medicines, 13*(3), 378–383.
Nishiuchi, R., Murayama, O., Fujiwara, H., Gu, J., Kawakami, T., Aimoto, S., et al. (2003). Characterization of the ligand-binding specificities of integrin alpha3beta1 and alpha6-beta1 using a panel of purified laminin isoforms containing distinct alpha chains. *The Journal of Biochemistry, 134*(4), 497–504.
Nishiuchi, R., Takagi, J., Hayashi, M., Ido, H., Yagi, Y., Sanzen, N., et al. (2006). Ligand-binding specificities of laminin-binding integrins: a comprehensive survey of laminin-integrin interactions using recombinant alpha3beta1, alpha6beta1, alpha7beta1 and alpha6beta4 integrins. *Matrix Biology, 25*(3), 189–197.
Nystrom, A., Bruckner-Tuderman, L., & Kern, J. S. (2013). Cell- and protein-based therapy approaches for epidermolysis bullosa. *Methods in Molecular Biology, 961*, 425–440.
Nystrom, A., Buttgereit, J., Bader, M., Shmidt, T., Ozcelik, C., Hausser, I., et al. (2013). Rat model for dominant dystrophic epidermolysis bullosa: glycine substitution reduces collagen VII stability and shows gene-dosage effect. *PLoS One, 23;8*(5), e64243.
Nystrom, A., Velati, D., Mittapalli, V. R., Fritsch, A., Kern, J. S., & Bruckner-Tuderman, L. (2013). Collagen VII plays a dual role in wound healing. *Journal of Clinical Investigation, 123*(8), 3498–3509.
O'Reilly, M. S., Boehm, T., Shing, Y., Fukai, N., Vasios, G., Lane, W. S., et al. (1997). Endostatin: an endogenous inhibitor of angiogenesis and tumor growth. *Cell, 88*(2), 277–285.
O'Toole, E. A., Marinkovich, M. P., Hoeffler, W. K., Furthmayr, H., & Woodley, D. T. (1997). Laminin-5 inhibits human keratinocyte migration. *Experimental Cell Research, 233*(2), 330–339.
Occleston, N. L., Laverty, H. G., O'Kane, S., & Ferguson, M. W. (2008). Prevention and reduction of scarring in the skin by Transforming Growth Factor beta 3 (TGFbeta3): from laboratory discovery to clinical pharmaceutical. *Journal of Biomaterials Science, Polymer Edition, 19*(8), 1047–1063.
Odenthal, U., Haehn, S., Tunggal, P., Merkl, B., Schomburg, D., Frie, C., et al. (2004). Molecular analysis of laminin N-terminal domains mediating self-interactions. *The Journal of Biological Chemistry, 279*(43), 44504–44512.
Odorisio, T., Di Salvio, M., Orecchia, A., Di Zenzo, G., Piccinni, E., Cianfarani, F., et al. (2014 Aug 1). Monozygotic twins discordant for recessive dystrophic epidermolysis

bullosa phenotype highlight the role of TGF-β signalling in modifying disease severity. *Human Molecular Genetics, 23*(15), 3907–3922.
Ogawa, T., Tsubota, Y., Maeda, M., Kariya, Y., & Miyazaki, K. (2004). Regulation of biological activity of laminin-5 by proteolytic processing of gamma2 chain. *Journal of Cellular Biochemistry, 92*(4), 701–714.
Oksala, O., Salo, T., Tammi, R., Hakkinen, L., Jalkanen, M., Inki, P., et al. (1995). Expression of proteoglycans and hyaluronan during wound healing. *Journal of Histochemistry and Cytochemistry, 43*(2), 125–135.
Oommen, S., Francois, M., Kawasaki, M., Murrell, M., Kawasaki, K., Porntaveetus, T., et al. (2012). Cytoplasmic plaque formation in hemidesmosome development is dependent on SoxF transcription factor function. *PLoS One, 7*(9), e43857.
Otten, J. V., Hashimoto, T., Hertl, M., Payne, A. S., & Sitaru, C. (2014). Molecular diagnosis in autoimmune skin blistering conditions. *Current Molecular Medicine, 14*(1), 69–95.
Patel, H., Zich, J., Serrels, B., Rickman, C., Hardwick, K. G., Frame, M. C., et al. (2013). Kindlin-1 regulates mitotic spindle formation by interacting with integrins and Plk-1. *Nature Communications, 4*, 2056.
de Pereda, J. M., Lillo, M. P., & Sonnenberg, A. (2009). Structural basis of the interaction between integrin alpha6beta4 and plectin at the hemidesmosomes. *The EMBO Journal, 28*(8), 1180–1190.
de Pereda, J. M., Ortega, E., Alonso-Garcia, N., Gomez-Hernandez, M., & Sonnenberg, A. (2009). Advances and perspectives of the architecture of hemidesmosomes: lessons from structural biology. *Cell Adhesion and Migration, 3*(4), 361–364.
Piccinni, E., Di Zenzo, G., Maurelli, R., Dellambra, E., Teson, M., Has, C., et al. (2013). Induction of senescence pathways in Kindler syndrome primary keratinocytes. *British Journal of Dermatology, 168*(5), 1019–1026.
Poschl, E., Schlotzer-Schrehardt, U., Brachvogel, B., Saito, K., Ninomiya, Y., & Mayer, U. (2004). Collagen IV is essential for basement membrane stability but dispensable for initiation of its assembly during early development. *Development, 131*(7), 1619–1628.
Pouliot, N., Saunders, N. A., & Kaur, P. (2002). Laminin 10/11: an alternative adhesive ligand for epidermal keratinocytes with a functional role in promoting proliferation and migration. *Experimental Dermatology, 11*(5), 387–397.
Putnins, E. E., Firth, J. D., Lohachitranont, A., Uitto, V. J., & Larjava, H. (1999). Keratinocyte growth factor (KGF) promotes keratinocyte cell attachment and migration on collagen and fibronectin. *Cell Adhesion and Communication, 7*(3), 211–221.
Rabinovitz, I., Tsomo, L., & Mercurio, A. M. (2004). Protein kinase C-alpha phosphorylation of specific serines in the connecting segment of the beta 4 integrin regulates the dynamics of type II hemidesmosomes. *Molecular and Cellular Biology, 24*(10), 4351–4360.
Raghunath, M., Hopfner, B., Aeschlimann, D., Luthi, U., Meuli, M., Altermatt, S., et al. (1996). Cross-linking of the dermo-epidermal junction of skin regenerating from keratinocyte autografts. Anchoring fibrils are a target for tissue transglutaminase. *Journal of Clinical Investigation, 98*(5), 1174–1184.
Rapraeger, A., Jalkanen, M., & Bernfield, M. (1986). Cell surface proteoglycan associates with the cytoskeleton at the basolateral cell surface of mouse mammary epithelial cells. *The Journal of Cell Biology, 103*(6 Pt 2), 2683–2696.
Rattenholl, A., Pappano, W. N., Koch, M., Keene, D. R., Kadler, K. E., Sasaki, T., et al. (2002). Proteinases of the bone morphogenetic protein-1 family convert procollagen VII to mature anchoring fibril collagen. *The Journal of Biological Chemistry, 277*(29), 26372–26378.
Raymond, K., Kreft, M., Janssen, H., Calafat, J., & Sonnenberg, A. (2005). Keratinocytes display normal proliferation, survival and differentiation in conditional beta4-integrin knockout mice. *Journal of Cell Science, 118*(Pt 5), 1045–1060.

Remy, L., Trespeuch, C., Bachy, S., Scoazec, J. Y., & Rousselle, P. (2006). Matrilysin 1 influences colon carcinoma cell migration by cleavage of the laminin-5 beta3 chain. *Cancer Research, 66*(23), 11228–11237.

Rezniczek, G. A., de Pereda, J. M., Reipert, S., & Wiche, G. (1998). Linking integrin alpha6beta4-based cell adhesion to the intermediate filament cytoskeleton: direct interaction between the beta4 subunit and plectin at multiple molecular sites. *The Journal of Cell Biology, 141*(1), 209–225.

Richardson, R. J., Gebauer, J. M., Zhang, J. L., Kobbe, B., Keene, D. R., Karlsen, K. R., et al. (2014). AMACO is a component of the basement membrane-associated Fraser complex. *Journal of Investigative Dermatology, 134*(5), 1313–1322.

Rodgers, K. D., Sasaki, T., Aszodi, A., & Jacenko, O. (2007). Reduced perlecan in mice results in chondrodysplasia resembling Schwartz-Jampel syndrome. *Human Molecular Genetics, 16*(5), 515–528.

Rodius, S., Indra, G., Thibault, C., Pfister, V., & Georges-Labouesse, E. (2007). Loss of alpha6 integrins in keratinocytes leads to an increase in TGFbeta and AP1 signaling and in expression of differentiation genes. *Journal of Cellular Physiology, 212*(2), 439–449.

Roediger, M., Miosge, N., & Gersdorff, N. (2010). Tissue distribution of the laminin beta1 and beta2 chain during embryonic and fetal human development. *Journal of Molecular Histology, 41*(2–3), 177–184.

Rognoni, E., Widmaier, M., Jakobson, M., Ruppert, R., Ussar, S., Katsougkri, D., et al. (2014). Kindlin-1 controls Wnt and TGF-beta availability to regulate cutaneous stem cell proliferation. *Nature Medicine, 20*(4), 350–359.

Rossi, M., Morita, H., Sormunen, R., Airenne, S., Kreivi, M., Wang, L., et al. (2003). Heparan sulfate chains of perlecan are indispensable in the lens capsule but not in the kidney. *The EMBO Journal, 22*(2), 236–245.

Rousselle, P., & Beck, K. (2013). Laminin 332 processing impacts cellular behavior. *Cell Adhesion and Migration,* 7(1), 122–134.

Ryan, M. C., Christiano, A. M., Engvall, E., Wewer, U. M., Miner, J. H., Sanes, J. R., et al. (1996). The functions of laminins: lessons from in vivo studies. *Matrix Biology, 15*(6), 369–381.

Ryan, M. C., Lee, K., Miyashita, Y., & Carter, W. G. (1999). Targeted disruption of the LAMA3 gene in mice reveals abnormalities in survival and late stage differentiation of epithelial cells. *The Journal of Cell Biology, 145*(6), 1309–1323.

Ryan, M. C., Tizard, R., VanDevanter, D. R., & Carter, W. G. (1994). Cloning of the LamA3 gene encoding the alpha 3 chain of the adhesive ligand epiligrin. Expression in wound repair. *The Journal of Biological Chemistry, 269*(36), 22779–22787.

Saarela, J., Rehn, M., Oikarinen, A., Autio-Harmainen, H., & Pihlajaniemi, T. (1998). The short and long forms of type XVIII collagen show clear tissue specificities in their expression and location in basement membrane zones in humans. *American Journal of Pathology, 153*(2), 611–626.

Sachs, N., Kreft, M., van den Bergh Weerman, M. A., Beynon, A. J., Peters, T. A., Weening, J. J., et al. (2006). Kidney failure in mice lacking the tetraspanin CD151. *The Journal of Cell Biology, 175*(1), 33–39.

Saito, K., Chen, M., Bard, F., Chen, S., Zhou, H., Woodley, D., et al. (2009). TANGO1 facilitates cargo loading at endoplasmic reticulum exit sites. *Cell, 136*(5), 891–902.

Santoro, M. M., Gaudino, G., & Marchisio, P. C. (2003). The MSP receptor regulates alpha6beta4 and alpha3beta1 integrins via 14-3-3 proteins in keratinocyte migration. *Developmental Cell, 5*(2), 257–271.

Sasaki, T., Gohring, W., Mann, K., Brakebusch, C., Yamada, Y., Fassler, R., et al. (2001). Short arm region of laminin-5 gamma2 chain: structure, mechanism of processing and binding to heparin and proteins. *Journal of Molecular Biology, 314*(4), 751–763.

Schmidt, E., & Zillikens, D. (2013). Pemphigoid diseases. *Lancet, 381*(9863), 320–332.

Schon, M., Klein, C. E., Hogenkamp, V., Kaufmann, R., Wienrich, B. G., & Schon, M. P. (2000). Basal-cell adhesion molecule (B-CAM) is induced in epithelial skin tumors and inflammatory epidermis, and is expressed at cell-cell and cell-substrate contact sites. *Journal of Investigative Dermatology, 115*(6), 1047–1053.

Schumann, H., Kiritsi, D., Pigors, M., Hausser, I., Kohlhase, J., Peters, J., et al. (2013). Phenotypic spectrum of epidermolysis bullosa associated with alpha6beta4 integrin mutations. *British Journal of Dermatology, 169*(1), 115–124.

Schymeinsky, J., Nedbal, S., Miosge, N., Poschl, E., Rao, C., Beier, D. R., et al. (2002). Gene structure and functional analysis of the mouse nidogen-2 gene: nidogen-2 is not essential for basement membrane formation in mice. *Molecular and Cellular Biology, 22*(19), 6820–6830.

Seppinen, L., Sormunen, R., Soini, Y., Elamaa, H., Heljasvaara, R., & Pihlajaniemi, T. (2008). Lack of collagen XVIII accelerates cutaneous wound healing, while overexpression of its endostatin domain leads to delayed healing. *Matrix Biology, 27*(6), 535–546.

Sesarman, A., & Sitaru, C. (2013). Induction of experimental epidermolysis bullosa acquisita by immunization with murine collagen VII. *Methods in Molecular Biology, 961*, 371–387.

Sewry, C. A., Philpot, J., Sorokin, L. M., Wilson, L. A., Naom, I., Goodwin, F., et al. (1996). Diagnosis of merosin (laminin-2) deficient congenital muscular dystrophy by skin biopsy. *Lancet, 347*(9001), 582–584.

Sher, I., Zisman-Rozen, S., Eliahu, L., Whitelock, J. M., Maas-Szabowski, N., Yamada, Y., et al. (2006). Targeting perlecan in human keratinocytes reveals novel roles for perlecan in epidermal formation. *The Journal of Biological Chemistry, 281*(8), 5178–5187.

Shin, J. E., Oh, J. H., Kim, Y. K., Jung, J. Y., & Chung, J. H. (2011). Transcriptional regulation of proteoglycans and glycosaminoglycan chain-synthesizing glycosyltransferases by UV irradiation in cultured human dermal fibroblasts. *Journal of Korean Medical Science, 26*(3), 417–424.

Short, K., Wiradjaja, F., & Smyth, I. (2007). Let's stick together: the role of the Fras1 and Frem proteins in epidermal adhesion. *IUBMB Life, 59*(7), 427–435.

Sigle, R. O., Gil, S. G., Bhattacharya, M., Ryan, M. C., Yang, T. M., Brown, T. A., et al. (2004). Globular domains 4/5 of the laminin alpha3 chain mediate deposition of precursor laminin 5. *Journal of Cell Science, 117*(Pt 19), 4481–4494.

Sitaru, C. (2007). Experimental models of epidermolysis bullosa acquisita. *Experimental Dermatology, 16*(6), 520–531.

Smyth, N., Vatansever, H. S., Meyer, M., Frie, C., Paulsson, M., & Edgar, D. (1998). The targeted deletion of the LAMC1 gene. *Annals of the New York Academy of Sciences, 857*, 283–286.

Sollberg, S., Peltonen, J., & Uitto, J. (1992). Differential expression of laminin isoforms and beta 4 integrin epitopes in the basement membrane zone of normal human skin and basal cell carcinomas. *Journal of Investigative Dermatology, 98*(6), 864–870.

Sprenger, A., Weber, S., Zarai, M., Engelke, R., Nascimento, J. M., Gretzmeier, C., et al. (2013). Consistency of the proteome in primary human keratinocytes with respect to gender, age, and skin localization. *Molecular and Cellular Proteomics, 12*(9), 2509–2521.

Sproule, T. J., Bubier, J. A., Grandi, F. C., Sun, V. Z., Philip, V. M., McPhee, C. G., et al. (2014). Molecular identification of collagen 17a1 as a major genetic modifier of laminin gamma 2 mutation-induced junctional epidermolysis bullosa in mice. *PLoS Genetics, 10*(2), e1004068.

Stephens, L. E., Sutherland, A. E., Klimanskaya, I. V., Andrieux, A., Meneses, J., Pedersen, R. A., et al. (1995). Deletion of beta 1 integrins in mice results in inner cell mass failure and peri-implantation lethality. *Genes and Development, 9*(15), 1883–1895.

Sterk, L. M., Geuijen, C. A., van den Berg, J. G., Claessen, N., Weening, J. J., & Sonnenberg, A. (2002). Association of the tetraspanin CD151 with the laminin-binding integrins alpha3beta1, alpha6beta1, alpha6beta4 and alpha7beta1 in cells in culture and in vivo. *Journal of Cell Science, 115*(Pt 6), 1161–1173.
Sterk, L. M., Geuijen, C. A., Oomen, L. C., Calafat, J., Janssen, H., & Sonnenberg, A. (2000). The tetraspan molecule CD151, a novel constituent of hemidesmosomes, associates with the integrin alpha6beta4 and may regulate the spatial organization of hemidesmosomes. *The Journal of Cell Biology, 149*(4), 969–982.
Stum, M., Davoine, C. S., Vicart, S., Guillot-Noel, L., Topaloglu, H., Carod-Artal, F. J., et al. (2006). Spectrum of HSPG2 (Perlecan) mutations in patients with Schwartz-Jampel syndrome. *Human Mutation, 27*(11), 1082–1091.
Sudhakar, A., Sugimoto, H., Yang, C., Lively, J., Zeisberg, M., & Kalluri, R. (2003). Human tumstatin and human endostatin exhibit distinct antiangiogenic activities mediated by alpha v beta 3 and alpha 5 beta 1 integrins. *Proceedings of the National Academy of Sciences of the United States of America, 100*(8), 4766–4771.
Sugawara, K., Tsuruta, D., Ishii, M., Jones, J. C., & Kobayashi, H. (2008). Laminin-332 and -511 in skin. *Experimental Dermatology, 17*(6), 473–480.
Sundaram, G. M., Common, J. E., Gopal, F. E., Srikanta, S., Lakshman, K., Lunny, D. P., et al. (2013). 'See-saw' expression of microRNA-198 and FSTL1 from a single transcript in wound healing. *Nature, 495*(7439), 103–106.
Tanimura, S., Tadokoro, Y., Inomata, K., Binh, N. T., Nishie, W., Yamazaki, S., et al. (2011). Hair follicle stem cells provide a functional niche for melanocyte stem cells. *Cell Stem Cell, 8*(2), 177–187.
Tasanen, K., Eble, J. A., Aumailley, M., Schumann, H., Baetge, J., Tu, H., et al. (2000). Collagen XVII is destabilized by a glycine substitution mutation in the cell adhesion domain Col15. *The Journal of Biological Chemistry, 275*(5), 3093–3099.
Tiedemann, K., Sasaki, T., Gustafsson, E., Gohring, W., Batge, B., Notbohm, H., et al. (2005). Microfibrils at basement membrane zones interact with perlecan via fibrillin-1. *The Journal of Biological Chemistry, 280*(12), 11404–11412.
Timpl, R., Sasaki, T., Kostka, G., & Chu, M. L. (2003). Fibulins: a versatile family of extracellular matrix proteins. *Nature Reviews Molecular Cell Biology, 4*(6), 479–489.
Tran, M., Rousselle, P., Nokelainen, P., Tallapragada, S., Nguyen, N. T., Fincher, E. F., et al. (2008). Targeting a tumor-specific laminin domain critical for human carcinogenesis. *Cancer Research, 68*(8), 2885–2894.
Tsubota, Y., Yasuda, C., Kariya, Y., Ogawa, T., Hirosaki, T., Mizushima, H., et al. (2005). Regulation of biological activity and matrix assembly of laminin-5 by COOH-terminal, LG4-5 domain of alpha3 chain. *The Journal of Biological Chemistry, 280*(15), 14370–14377.
Urbano, J. M., Torgler, C. N., Molnar, C., Tepass, U., Lopez-Varea, A., Brown, N. H., et al. (2009). Drosophila laminins act as key regulators of basement membrane assembly and morphogenesis. *Development, 136*(24), 4165–4176.
Ussar, S., Moser, M., Widmaier, M., Rognoni, E., Harrer, C., Genzel-Boroviczeny, O., et al. (2008). Loss of Kindlin-1 causes skin atrophy and lethal neonatal intestinal epithelial dysfunction. *PLoS Genetics, 4*(12), e1000289.
Usui, M. L., Mansbridge, J. N., Carter, W. G., Fujita, M., & Olerud, J. E. (2008). Keratinocyte migration, proliferation, and differentiation in chronic ulcers from patients with diabetes and normal wounds. *Journal of Histochemistry and Cytochemistry, 56*(7), 687–696.
Utriainen, A., Sormunen, R., Kettunen, M., Carvalhaes, L. S., Sajanti, E., Eklund, L., et al. (2004). Structurally altered basement membranes and hydrocephalus in a type XVIII collagen deficient mouse line. *Human Molecular Genetics, 13*(18), 2089–2099.
de Vega, S., Iwamoto, T., & Yamada, Y. (2009). Fibulins: multiple roles in matrix structures and tissue functions. *Cellular and Molecular Life Sciences, 66*(11–12), 1890–1902.

Vogel, M. J., van Zon, P., Brueton, L., Gijzen, M., van Tuil, M. C., Cox, P., et al. (2012). Mutations in GRIP1 cause Fraser syndrome. *Journal of Medical Genetics, 49*(5), 303–306.

Walko, G., Vukasinovic, N., Gross, K., Fischer, I., Sibitz, S., Fuchs, P., et al. (2011). Targeted proteolysis of plectin isoform 1a accounts for hemidesmosome dysfunction in mice mimicking the dominant skin blistering disease EBS-Ogna. *PLoS Genetics,* 7(12), e1002396.

Wang, H., Leavitt, L., Ramaswamy, R., & Rapraeger, A. C. (2010). Interaction of syndecan and alpha6beta4 integrin cytoplasmic domains: regulation of ErbB2-mediated integrin activation. *The Journal of Biological Chemistry, 285*(18), 13569–13579.

Wang, X., Ghasri, P., Amir, M., Hwang, B., Hou, Y., Khalili, M., et al. (2013). Topical application of recombinant type VII collagen incorporates into the dermal-epidermal junction and promotes wound closure. *Molecular Therapy, 21*(7), 1335–1344.

Wegener, H., Leineweber, S., & Seeger, K. (2013). The vWFA2 domain of type VII collagen is responsible for collagen binding. *Biochemical and Biophysical Research, 430*(2), 449–453.

Wen, T., Zhang, Z., Yu, Y., Qu, H., Koch, M., & Aumailley, M. (2010). Integrin alpha3 subunit regulates events linked to epithelial repair, including keratinocyte migration and protein expression. *Wound Repair and Regeneration, 18*(3), 325–334.

Wilhelmsen, K., Litjens, S. H., Kuikman, I., Margadant, C., van Rheenen, J., & Sonnenberg, A. (2007). Serine phosphorylation of the integrin beta4 subunit is necessary for epidermal growth factor receptor induced hemidesmosome disruption. *Molecular Biology of the Cell, 18*(9), 3512–3522.

Winograd-Katz, S. E., Fassler, R., Geiger, B., & Legate, K. R. (2014). The integrin adhesome: from genes and proteins to human disease. *Nature Reviews Molecular Cell Biology, 15*(4), 273–288.

Woodall, B. P., Nystrom, A., Iozzo, R. A., Eble, J. A., Niland, S., Krieg, T., et al. (2008). Integrin alpha2beta1 is the required receptor for endorepellin angiostatic activity. *The Journal of Biological Chemistry, 283*(4), 2335–2343.

Wright, M. D., Geary, S. M., Fitter, S., Moseley, G. W., Lau, L. M., Sheng, K. C., et al. (2004). Characterization of mice lacking the tetraspanin superfamily member CD151. *Molecular and Cellular Biology, 24*(13), 5978–5988.

Xu, Z., Ichikawa, N., Kosaki, K., Yamada, Y., Sasaki, T., Sakai, L. Y., et al. (2010). Perlecan deficiency causes muscle hypertrophy, a decrease in myostatin expression, and changes in muscle fiber composition. *Matrix Biology, 29*(6), 461–470.

Yamada, M., & Sekiguchi, K. (2013). Disease-associated single amino acid mutation in the calf-1 domain of integrin alpha3 leads to defects in its processing and cell surface expression. *Biochemical and Biophysical Research, 441*(4), 988–993.

Yamaguchi, Y., Takihara, T., Chambers, R. A., Veraldi, K. L., Larregina, A. T., & Feghali-Bostwick, C. A. (2012). A peptide derived from endostatin ameliorates organ fibrosis. *Science Translational Medicine, 4*(136), 136ra171.

Yauch, R. L., Berditchevski, F., Harler, M. B., Reichner, J., & Hemler, M. E. (1998). Highly stoichiometric, stable, and specific association of integrin alpha3beta1 with CD151 provides a major link to phosphatidylinositol 4-kinase, and may regulate cell migration. *Molecular Biology of the Cell, 9*(10), 2751–2765.

Yuen, W. Y., Huizinga, J., & Jonkman, M. F. (2013). Punch grafting of chronic ulcers in patients with laminin-332-deficient, non-Herlitz junctional epidermolysis bullosa. *Journal of the American Academy of Dermatology, 68*(1), 93–97, 97 e91–92.

Yuen, W. Y., Sinke, R. J., & Jonkman, M. F. (2013). ITGB4-associated non-Herlitz junctional epidermolysis bullosa: report of two new cases carrying two novel ITGB4 mutations. *British Journal of Dermatology, 168*(2), 432–434.

Yurchenco, P. D. (2011). Basement membranes: cell scaffoldings and signaling platforms. *Cold Spring Harbor Perspectives in Biology, 3*(2).

Zhou, Z., Wang, J., Cao, R., Morita, H., Soininen, R., Chan, K. M., et al. (2004). Impaired angiogenesis, delayed wound healing and retarded tumor growth in perlecan heparan sulfate-deficient mice. *Cancer Research, 64*(14), 4699–4702.

Zimina, E. P., Bruckner-Tuderman, L., & Franzke, C. W. (2005). Shedding of collagen XVII ectodomain depends on plasma membrane microenvironment. *The Journal of Biological Chemistry, 280*(40), 34019–34024.

Zimina, E. P., Fritsch, A., Schermer, B., Bakulina, A. Y., Bashkurov, M., Benzing, T., et al. (2007). Extracellular phosphorylation of collagen XVII by ecto-casein kinase 2 inhibits ectodomain shedding. *The Journal of Biological Chemistry, 282*(31), 22737–22746.

Zoller, M. (2009). Tetraspanins: push and pull in suppressing and promoting metastasis. *Nature Reviews Cancer, 9*(1), 40–55.

Zweers, M. C., Davidson, J. M., Pozzi, A., Hallinger, R., Janz, K., Quondamatteo, F., et al. (2007). Integrin alpha2beta1 is required for regulation of murine wound angiogenesis but is dispensable for reepithelialization. *Journal of Investigative Dermatology, 127*(2), 467–478.

CHAPTER FIVE

Applying Proteomics to Investigate Extracellular Matrix in Health and Disease

Michael Randles[1,2] and Rachel Lennon[1,2,3,*]
[1]Wellcome Trust Centre for Cell-Matrix Research, Faculty of Life Sciences, University of Manchester, Manchester, UK
[2]Institute of Human Development, Faculty of Human Sciences, University of Manchester, Manchester, UK
[3]Department of Paediatric Nephrology, Central Manchester University Hospitals NHS Foundation Trust (CMFT), Manchester Academic Health Science Centre (MAHSC), Manchester, UK
*Corresponding author: E-mail: Rachel.Lennon@manchester.ac.uk

Contents

Abstract

The molecular composition of basement membranes (BMs) has traditionally been investigated by candidate-based approaches leading to the identification of key structural components as described in previous chapters. Laminins, collagen IV, nidogens, perlecan, and type XV/XVIII collagen are integral to BMs with isoforms showing tissue

Current Topics in Membranes, Volume 76
ISSN 1063-5823
http://dx.doi.org/10.1016/bs.ctm.2015.06.001

specificity. More recently the application of mass spectrometry (MS)-based proteomics has led to the discovery of many more structural and regulatory components of BMs and more broadly, extracellular matrix (ECM). These investigations have revealed tissue-specific signatures of between 100 and 150 ECM components, demonstrating the complexity of the extracellular niche. In addition to providing a structural scaffold for cells, ECM is a dynamic extracellular environment capable of regulating the physical properties of tissues. Global investigations of ECM with proteomics in turn enable systems level analyses and when applied to health and disease states these investigations provide insights into pathways regulating matrix dysregulation. This chapter focuses on the methods used to extract ECM and on the analysis of its composition using MS-based proteomics, and it provides examples of how these approaches have been used to investigate health and disease states.

1. INTRODUCTION

Multicellular organisms and the cells and tissues of which they are composed represent complex biological systems. Within these systems, fine regulation of complex signaling pathways is required for life. The extracellular matrix (ECM) is an integral part of all tissue systems, conferring physical properties in addition to controlling signaling cascades through sequestration of growth factors and engagement of cell surface receptors. As such, there is an increasing need to understand ECM biology at the systems level (Byron, Humphries, & Humphries, 2013). Basement membranes (BMs) are condensed networks of ECM, and their composition is likely to be more complex and dynamic than we currently appreciate. They are fundamental and ancient components of multicellular organisms. To date, candidate-based approaches have determined that type IV collagen, laminins, nidogens, perlecan, and type XV/XVIII collagen are conserved components of BMs (Hynes, 2009). Understanding the function of these molecules has been advanced through studies of genetic mutations in humans and animals in addition to targeted studies of these proteins at the level of tissue expression, single-molecule structure, and analysis of interactions with other biological molecules. These are irreplaceable scientific approaches. However, systems biology attempts to systematically study all concurrent processes in a tissue by the global measurement of differentially perturbed states, and this approach has enormous potential to complement targeted studies.

Immunoaffinity-based approaches such as immunofluorescence microscopy, flow cytometry, enzyme-linked immunosorbent assay (ELISAs), and immunoblotting are central to the study of proteins in biological systems

(Figure 1(A)). One caveat of these approaches is that they rely on antibody affinity, antigen accessibility, and the availability of reagents. Furthermore, these approaches do not enable the identification and quantification of hundreds of proteins in a single experiment or the discovery of novel proteins,

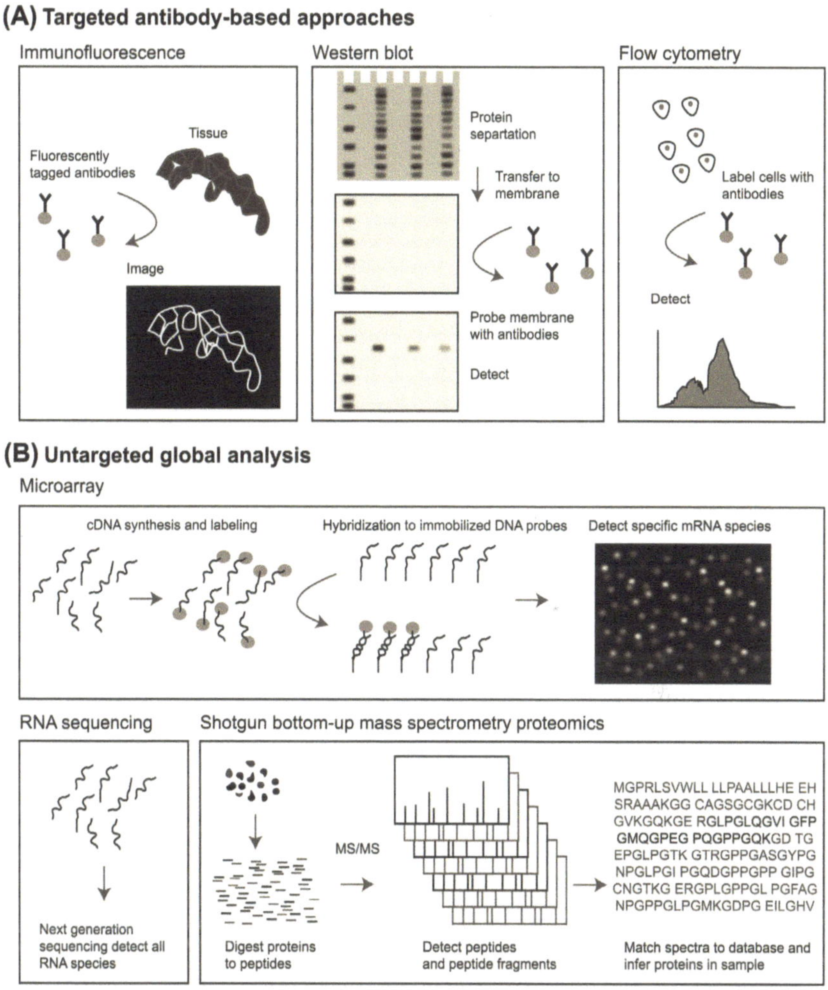

Figure 1 (A) Targeted antibody-based approaches. These include the labeling of tissues or cells with fluorescent antibodies to allow the localization of proteins with immunofluorescence microscopy, immunodetection of proteins in a sample with Western blotting, and the labeling of proteins for flow cytometry. (B) Untargeted global analysis. These include the analysis of transcription with microarray, involving the extraction of RNA, generation of cDNA, hybridization to selected probes, and the subsequent identification of RNAs. The direct sequencing of RNA and mass spectrometry-based proteomics based on the bottom-up analysis of peptides to determine the identifications of proteins in a sample.

as the target protein is predetermined. As a result, alternative techniques are required to enable the field of systems biology.

Analysis of the transcriptome with microarrays and RNA-sequencing experiments enables global quantification of gene expression, and these approaches are employed for systems level analyses (Figure 1(B)). These analyses are frequently utilized as surrogates for protein abundance. However, although powerful, these techniques do not represent the true presence and abundance of proteins within a biological sample, because protein expression is regulated through posttranscriptional mechanisms, such as synthesis and degradation (Walther & Mann, 2010). Furthermore, microarray and RNA sequencing are also incapable of probing the plethora of potential posttranslational modifications (PTMs) that alter protein behavior, including acetylation, phosphorylation, and glycosylation. Glycosylation is particularly important for BMs; the major BM components are highly glycosylated proteins and proteoglycans, and these modifications are integral to their physical, structural, and adhesive properties. Phosphorylation, a key PTM for intracellular signaling, could also be a novel regulatory mechanism of BM maintenance, as elegant recent work has revealed that kinases operate in the ECM (Bordoli et al., 2014).

Proteomics allows the unbiased study of protein composition and can therefore be used to examine multiple concurrent processes occurring within a biological system. Furthermore, proteomic data allows the interrogation of PTMs. This chapter will provide an overview of proteomic methodology applied to investigate ECM and describe a snapshot of the many global proteomic studies of the ECM, with a particular focus on those relevant to BMs.

2. MASS SPECTROMETRY-BASED PROTEOMICS

2.1 Data acquisition

Ideally, researchers would like to identify and quantify all proteins in a biological sample under different experimental or disease conditions. However, the desire to achieve maximal coverage including low-abundance proteins and to probe the huge array of protein modifications necessitates highly sensitive analytical techniques. Mass spectrometry (MS) enables the analysis of proteins in a global manner (Aebersold & Mann, 2003). The analytical instrument, a mass spectrometer, contains: an ion source, which generates charged gaseous analytes; a mass analyzer, which measures the mass-to-charge ratio (m/z) of analytes; and a detector that records the number of ions at each m/z value (Figure 1(B)).

For analysis of proteins, the process of generating ions in the gas phase is most commonly achieved by either electrospray ionization (ESI) (Fenn, Mann, Meng, Wong, & Whitehouse, 1989) or matrix-assisted laser desorption/ionization (MALDI) (Karas & Hillenkamp, 1988). These processes overcome the propensity of large molecules (such as proteins/peptides) to fragment when ionized and are referred to as "soft ionization" techniques. ESI ionizes the analytes out of a solution and is therefore readily coupled to prior liquid-based separation tools, for example, chromatography. MALDI sublimates and ionizes proteins or peptides out of a dry, crystalline matrix via laser pulses. MALDI-MS is normally used to analyze relatively simple protein/peptide mixtures, whereas integrated liquid-chromatography ESI-MS systems (LC-MS) are preferred for the analysis of complex samples. An additional benefit of ESI is the production of multiply charged ions extending the m/z range of the mass analyser. The development of ESI in 1989 led to John B. Fenn sharing the 2002 Nobel Prize in chemistry.

Following the generation of ions in the gaseous phase, ions are captured in mass analyzers. These devices operate by measuring the trajectories of ions in an electric field or by trapping ions for further manipulation. Examples of mass analyzers include quadrupole mass filters, time-of-flight instruments, fourier transform ion cyclotron analyzers, and the orbitrap. These analyzers can be stand alone or, more commonly, assembled in tandem to take advantage of their relative strengths.

Biological samples are complex and frequently contain thousands of proteins. In order to detect, and ideally quantify, all of the proteins present in a biological sample, proteins are digested to peptides (frequently using enzymatic trypsin digestion), and these peptides are analyzed. By breaking down a complex protein into its building blocks there is more chance of assigning a unique identification (ID). An analogy here is the accurate weighing of all staff at a University. It would not be possible to give a unique ID based on their total body weight alone due to the limitation of accuracy and precision of the device used to weigh them and the number of individuals with the same weight. However, if each individual was asked to remove their coat, shoes, and bag and each of these items were weighed separately, then the probability of assigning a unique ID to each member of staff would increase. To assign a unique ID to peptides, tandem MS (MS/MS) employs two mass analyzers in series. In the first analyzer, precursor peptides are selected and analyzed. The exact mass of this precursor peptide is determined, and it is then fragmented along its backbone, usually by collision-induced dissociation (CID) with an inert gas. These fragment ions are then detected in the second analyzer, giving rise to characteristic spectra. The fragment and precursor m/z measurements can be compared to protein database search engines. These databases have libraries of peptides predicted in silico from theoretical digests of the complete proteome of the organism of interest, and therefore the identity of a peptide is thus inferred. The next challenge is to determine the original proteins from which the identified peptides have arisen. This is trivial in the case of peptides that are unique to a given protein, but pose a complex problem for peptides with sequences that are shared between proteins. As a result a range of software is now available for processing MS proteomic data; the most popular include Mascot (Perkins, Pappin, Creasy, & Cottrell, 1999), SEQUEST (Eng, McCormack, & Yates, 1994), and X!Tandem (Craig & Beavis, 2004).

2.2 Quantification

Quantification of peptides and proteins using MS enables comparison between different biological scenarios. In addition, copurified contaminant proteins can be filtered from data sets using "subtractive proteomics" where proteins detected in similar quantities in control samples and experimental samples can be filtered from the data. It is not feasible to determine absolute quantification of peptides, since this would require the use of peptide standards for each of the peptides identified, and therefore relative quantification is performed.

One common method for relative protein quantification uses the incorporation of "heavy or light" isotope-labeled lysines and arginines into different biological samples. This approach, known as stable isotopic labeling using amino acids in cell culture (SILAC), enables identical peptides from different samples to be distinguished within a single MS analysis (Ong et al., 2002). This labeling technique has also been used for tissue analysis via the generation of the SILAC mouse (Zanivan, Krueger, & Mann, 2012). Another technique, termed iTRAQ, utilizes isobaric tags that have the same intact mass but have different masses when fragmented and detected at MS/MS (Ross et al., 2004). The fragment spectra from the isobaric tags are then used as a correlate of relative peptide abundance.

Label-free quantification strategies allow the comparison of two or more MS analyses. Methods of this relative quantification approach include spectral counting, which correlates the number of observed mass spectra matched to peptides from a protein to the abundance of that protein (Liu,

Sadygov, & Yates, 2004). Peptide counting (Ishihama et al., 2005) and peptide ion signal measurements (Silva et al., 2005) are further examples, and a combination of these parameters has also been used (Griffin et al., 2010). The continual development of instrumentation and software for MS will enable further refinement and precision in both labeled and label-free quantification, and this is requisite for systems-level analyses.

Despite advances in proteomic pipelines it is still a challenge to detect, resolve, and quantify peptides in very complex mixtures (Figure 2). One approach therefore is to simplify the peptide sample before it enters the mass spectrometer, and high-performance liquid chromatography is generally used to achieve this goal. Further upstream in the sample preparation, peptides or proteins can be separated by SDS-PAGE or off-gel electrophoresis and run as separate fractions. Even further upstream, if analysis of a specific subset of proteins is desirable, such as ECM, then isolation or enrichment of these protein compartments is a suitable sample preparation step to increase the resolution of the proteome of interest.

The majority of proteomic experiments analyze whole tissue lysates, which are saturated with cellular proteins that unavoidably mask the identification of proteins of the extracellular environment. This is further confounded by the fact that ECM proteins pose analytical difficulties to proteomic strategies due to their biochemical qualities. ECM proteins interact with many other components and are frequently highly covalently crosslinked. As a result, ECM proteins have low solubility. Extraction and analysis of ECM proteins is therefore a major challenge, which several research groups have started to address. Their studies have overcome the challenges associated with the proteomic analysis of ECM resulting in major discoveries in the context of health and disease.

2.3 Bioinformatic analyses

MS generates large data sets, and peptide identifications are first searched against protein databases. Importantly, variable protein modifications can be selected by the search engine enquiry, and this increases the coverage of ECM and BM proteins. Examples of variable modifications include oxidized methionine, deamidation of asparagine, pyroglutamic acid modification at N-terminal glutamine, and hydroxylation of proline and lysine. The latter are especially important for the identification of collagen peptides. Once proteins have been confidently identified from peptides, downstream bioinformatic analyses allow patterns to be visualized in the data. Gene ontology classifications can be used to identify and select protein groups

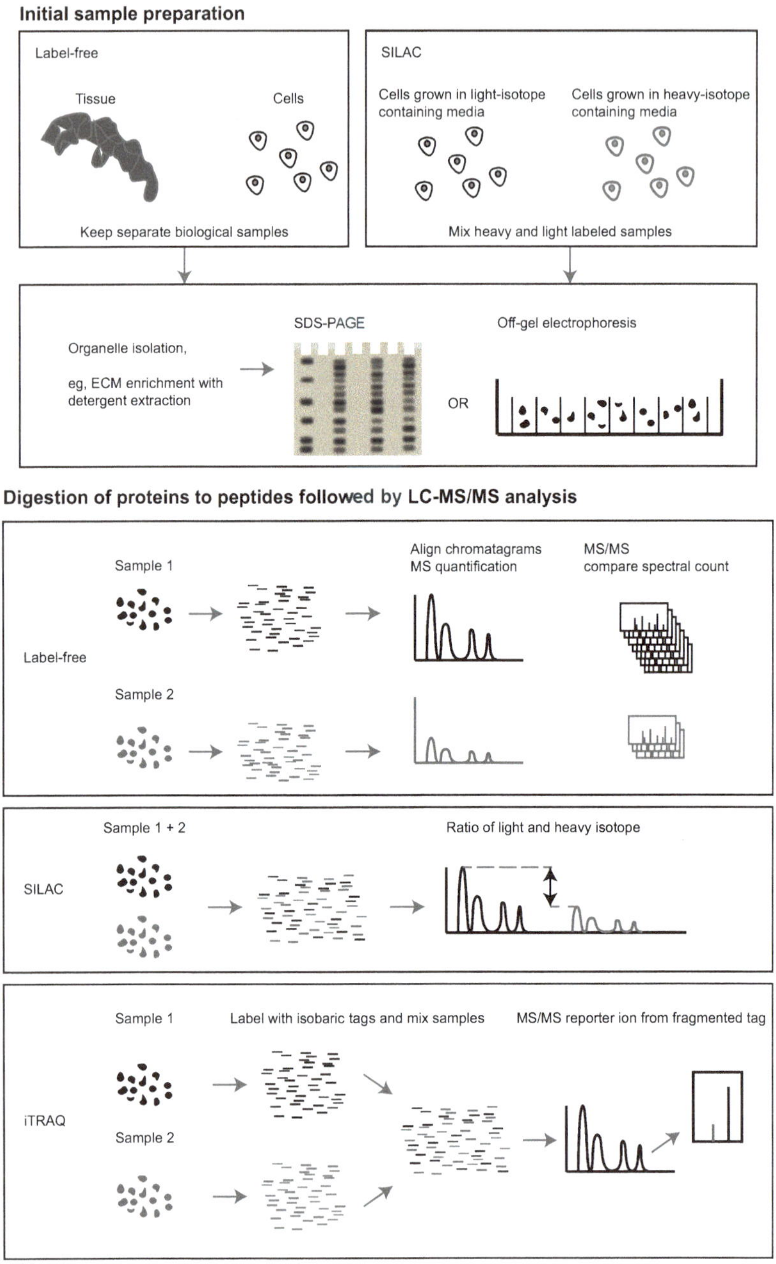

Figure 2 Mass spectrometry pipelines. Sample preparation for labeled and label-free quantification. Samples are further simplified or fractionated by 1-D SDS gel electrophoresis or off-gel electrophoresis. Once proteins are digested to peptides, typically with enzymatic trypsin, analysis for label-free quantification compares peptide or spectral counts. For SILAC, the ratio of light and heavy isotopes is determined and for ITRAQ, there is MS/MS reporting of the fragmented tags.

of interest, for example, those in the extracellular region. Clusters of proteins that change in a similar manner between biological samples can be appreciated via hierarchical clustering. Interactions between identified proteins can also be appreciated in the wider context of published data by mapping the experimental data onto networks of known protein–protein interactions, enabling the identification of key subnetworks (Sevimoglu & Arga, 2014). Finally, MS data can be used as the input into pathway analysis tools such as Ingenuity Pathway Analysis (Ingenuity Systems Inc., Redwood City, CA) and Pathway Studio (Ariadne Genomics, Rockville, MD). These tools enable data sets to be related to pathways and processes described in the published literature and therefore allow the generation of hypotheses about regulation within the biological system under analysis.

3. ENRICHMENT OF ECM FOR PROTEOMIC ANALYSIS

To achieve the highest detection of ECM components by MS, biological samples need to undergo enrichment. The basic principle of ECM extraction requires tissue dissociation through physical disruption followed by separation of the ECM from the cellular compartments by the use of sequential chemical extractions. A number of methods have now been described in the studies reported in Section 4, and there are variations of a common theme with the following sequence. Tissue samples are first diced into small pieces to increase the surface-area-to-volume ratio and therefore increase the penetration of buffers used for decellularization. The first extraction step typically uses a high concentration of NaCl to release proteins that are loosely bound to the ECM. The second step uses sodium dodecyl sulfate (SDS) and mild mechanical disruption for several hours to achieve removal of cellular components. In the final step the remaining proteins are solubilized and denatured.

Some protocols include guanidine at pH 5.8 in the final step, since this is more effective at solubilizing vascular proteoglycans, whereas a higher pH is required for the solubilization of the proteoglycans from other ECM-rich tissues, such as cartilage (Eisenstein, Larsson, Kuettner, Sorgente, & Hascal, 1975). The presence of guanidine, however, hampers downstream biochemical analysis, and therefore proteins from guanidine extracts must be precipitated using ethanol. Further sample preparation can include deglycosylation (removal of glycosaminoglycan side chains), and this can be achieved with a number of enzymes. Chondroitinase ABC from *Proteus vulgaris* catalyzes the removal of polysaccharides containing 134-D-hexosaminyl and 133-D-glucuronosyl or 133-L-iduronosyl linkages to disaccharides containing 4-deoxy-D-gluc-4-enuronosyl groups. The enzyme acts on chondroitin 4-sulfate, chondroitin 6-sulfate, and dermatan sulfate glycosaminoglycan side chains. Keratanase from *Bacteroides fragilis* cleaves internal 134-galactose linkages in unbranched, repeating poly-N-acetyl-lactosamine and acts on keratan sulfate, and heparinase II from *Flavobacterium heparinum* cleaves heparan sulfate.

The extracted ECM protein samples are then typically further reduced in complexity by one-dimensional SDS-PAGE. The gel containing the proteins can then be then sliced into separate fractions and the proteins are digested from the gel with trypsin as an "in-gel" digestion before analysis with LC-MS/MS. This process is known as GeLC−MS/MS. Theoretically, simplification of the tissue by removal of the cellular components and separation of protein fractions by molecular weight with 1-D SDS-PAGE should increase the coverage of the ECM proteome. However, this approach assumes complete extraction of peptides from the gels prior to analysis. Alternatively samples can be prepared using an "in-solution" digestion with trypsin. This approach does not allow the fractionation of the sample, and often the yield of proteins is less and this method is not typically used for the preparation of ECM samples. However, the in-solution method is necessary and superior for further enrichment of PTMs, and especially for the identification of phosphorylated peptides (Robertson et al., 2015).

4. DEFINING THE MATRISOME

Having acquired an ECM data set (the numerator), it is important to consider how effective the enrichment and identification processes are, and for this we require a denominator. A significant body of work involved an in silico prediction of the in vivo "matrisome" (Naba et al., 2012). The matrisome was defined as "the ensemble of ECM proteins and associated factors." The first stage of this huge task was to define the domains commonly found in known ECM proteins. These domains were associated with the following protein subgroups: glycoproteins and proteoglycans (55 domains); ECM-affiliated proteins (6 domains); ECM regulators, including ECM-remodeling enzymes (25 domains); and ECM regulators and secreted factors, including growth factors and cytokines (39 domains). These domain lists were compiled independently and based on previous knowledge, data from the literature, and iterative query of the UniProt protein database. The second phase was to define lists of "excluding domains" where the presence of an excluding domain would remove a protein from the previously defined groups: ECM glycoproteins and proteoglycans (20 domains), ECM regulators (12 domains), or secreted factors (17 domains). UniProt protein entries to InterPro domain information were searched independently for the presence of each set of "inclusion domains." This procedure was performed in parallel for both human and murine protein databases, and Naba et al. therefore defined a matrisome of 1065 proteins (Figure 3). Of these proteins 178 are categorized as BM proteins.

Following this in silico analysis the next step was to annotate the matrisome with experimental data from tissue using global proteomics. Naba et al. analyzed the proteome of enriched ECM extracted from murine lung and colon. The analyses identified 143 ECM proteins in lung and 106 ECM

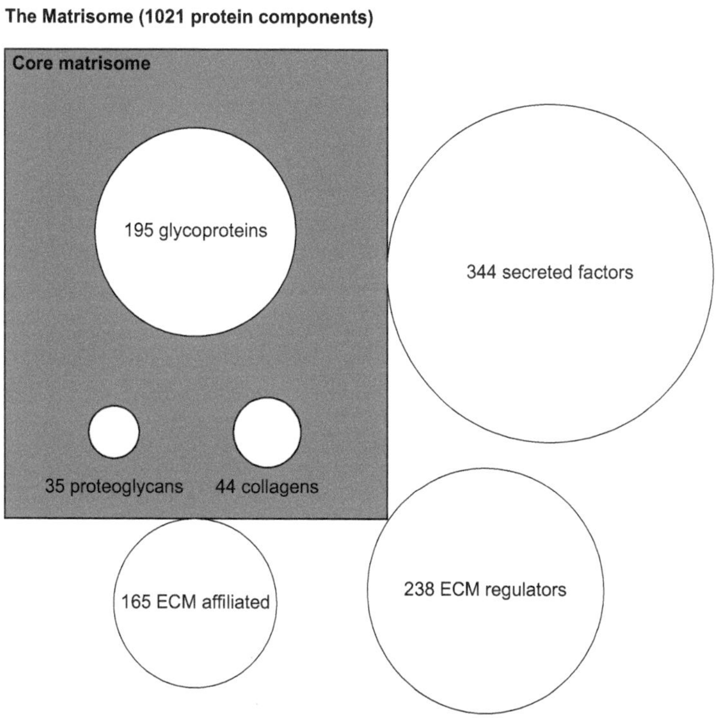

Figure 3 The predicted matrisome. The in silico prediction of the matrisome is comprised of core components (glycoproteins, collagens, proteoglycans) in addition to extracellular matrix (ECM)-affiliated proteins, ECM regulators, and ECM-secreted factors.

proteins in colon. Interestingly, there was a tissue-specific signature for each tissue type. These analyses detected many expected proteins, for example, thrombospondin 1, which was expressed in the lung but not detected in the colon, and mucin 2, a protein that is known to be secreted by the goblet cells of the colon, was specifically found in the ECM of colon and not in the ECM of lung. Eighty-four matrisome proteins were consistently found in replicate ECM samples from both lung and colon. The vast majority of collagens, proteoglycans, and expected BM components were found in both tissues. Differentially expressed proteins between the two tissue types belong mainly to the ECM glycoprotein, ECM-affiliated, ECM regulators, and secreted factor categories. This study not only demonstrated tissue specificity of the ECM niche but furthermore provided a predicted denominator available as an open access resource for investigators to use when analyzing ECM from different tissue types.

5. PROTEOMICS OF ECM IN HEALTH AND DISEASE

5.1 The tumor niche

The tumor microenvironment has potential for therapeutic targeting, and therefore further characterization of this niche is required. Histologically, the morphology of ECM associated with a tumor is known to correlate with patient outcome, and global analyses of ECM from tumors with differing metastatic potential may be beneficial for understanding the mechanisms by which the tumor microenvironment plays a role in disease progression. An example is the study of melanoma. A proteomic analysis revealed that the same matrisome proteins were expressed by melanoma tumors with both high and low metastatic potential. Most of the BM components (type IV and XV collagens, laminins, perlecan, nidogen 2) were expressed in the same proportions by the different types of tumor, although collagen XVIII and collagen IV $\alpha 3$ were expressed by poorly metastatic melanoma, whereas laminin $\beta 2$ and nidogen 1 were only detected in the ECM of metastatic melanoma.

In addition to the types and abundance of ECM proteins expressed by tumors, it is also interesting to understand the origin of the tumor ECM; is the tumor ECM produced and secreted by the tumor cells themselves, by the stromal cells, or by both compartments? The murine sequence of a given protein is, in most cases, sufficiently different from its human ortholog to be distinguished by proteomic analyses. Therefore, grafting tumors of human origin subcutaneously into mice enables the dissection of the origins of the tumor ECM proteins. Proteomic studies of the tumor microenvironment using melanoma xenografts in this context identified fibrillin-1, fibronectin, and several BM components: collagen IV α 1, 2, 3, and 5 chains, collagen XVIII, perlecan, and nidogen 1 as secreted by both the tumor cells and the host stromal cells. In contrast, hyaluronan and proteoglycan link protein-1 were found to be expressed exclusively in the metastatic tumors and specifically secreted by the tumor cells. Further studies of human mammary carcinoma xenografts showed that primary tumors of differing metastatic potential were distinct in ECM composition (Naba, Clauser, Lamar, Carr, & Hynes, 2014), hence defining ECM signatures of poorly and highly metastatic mammary carcinomas. These signatures revealed upregulation of signaling pathways including TGF-β and vascular endothelial growth factor (VEGF) by use of Ingenuity Pathway Analysis. Several proteins identified in this way are characteristic of highly metastatic tumors (LTBP3, SNED1, EGLN1, and S100A2). Moreover, they play causal roles in metastasis, and high expression of LTBP3 and SNED1 correlates with poor outcome for subgroups of patients with breast cancer.

The same study found a distinction between BM proteins expressed by tumors of different metastatic potential. Some components were only expressed by mammary carcinoma tumor cells (laminin chains α3, β3, γ2), others expressed by both compartments (laminin chains α5, γ1, collagen VI α3, and perlecan), and some only expressed by the stroma (nidogens 1 and 2). BM components (laminin chains β1 and β2, and the collagen chains α1(IV), α5(IV), α1(VI), α1(VI), and α1(XVIII)) were secreted in poorly metastatic tumors mostly by the stroma, whereas in highly metastatic tumors, they were secreted by both the tumor cells and the stromal cells. Tumor cells of differing metastatic potential not only synthesize distinct subsets of ECM proteins but also influence which ECM proteins are produced by the stroma. In addition, probing the tumor ECM with Ingenuity Pathway Analysis revealed the activation of major signaling cascades. The investigators found that 17 of the 43 ECM proteins (nearly 40%) found to be upregulated in metastatic tumors are downstream of the TGF-β signaling pathway. These and other analyses have the potential to identify therapeutic targets within the tumor niche.

5.2 Cardiovascular ECM

Cardiovascular disease and the associated changes in the ECM have also been investigated with proteomics. In an analysis of morphologically normal samples of human aorta, 272 and 321 proteins were detected by MS in the NaCl and guanidine extracts, respectively (Didangelos et al., 2010). According to Gene Ontology annotation 61 proteins in the NaCl extracts and 84 in the guanidine extracts were defined as "extracellular." Due to the overlap between the proteins identified in the two extracts, in total these analyses identified 103 extracellular proteins, and the investigators used spectral counting for the estimation of relative protein abundance. ECM proteins that had not previously been associated with the ECM of blood vessels (BVs) were identified in these aortic samples, and they included podocan, sclerostin, and agrin. Podocan is an atypical proteoglycan, which has so far been found in the glomerular BM as a product of kidney podocytes (Ross et al., 2003).

Having characterized the ECM of normal human aorta, the same group then applied this MS analysis pipeline to the study of abdominal aortic aneurysms (AAAs) from human patients. AAAs are characterized by pathological remodeling of the aortic ECM including well-characterized elastolysis and collagenolysis. Using spectral counting they identified six glycoproteins that were increased in AAA samples: collagen XII, tenascin, thrombospondin 2, fibronectin, ACLP, and periostin. In addition, MMP-12 was identified as the only metalloproteinase that was significantly increased in AAA in their proteomics analysis. Further analysis revealed that MMP-12 induced extensive fragmentation of fibronectin and cleaved collagen XII, tenascin, and periostin.

In a subsequent study, this group analyzed ECM in cardiovascular tissues of a porcine model of ischemia/reperfusion injury. ECM proteins such as cartilage intermediate layer protein 1, matrilin-4, extracellular adipocyte enhancer binding protein 1, collagen-1(XIV), and several members of the small leucine-rich proteoglycan family, including asporin and prolargin, were shown to contribute to cardiac remodeling in this context. Two distinct cardiac regions (the focal injury in the left ventricle and the border region close to the occluded coronary artery) revealed that regulation of mRNA levels did not correlate with regulation of ECM proteins. Gene expression of ECM proteins was similar in both regions, but the corresponding protein levels were much higher in the focal lesion. Further analysis based on 100 ECM proteins revealed characteristic early and late-stage cardiac remodeling with TGF-β signaling identified at the center of the interaction network generated using Pathway Studio. Overall these analyses further our understanding of the composition of the vascular extracellular environment and shed light on ECM remodeling and degradation during pathological processes. In addition these in-depth analyses have the potential to identify disease biomarkers, which may be less abundant in serum or urine and therefore elusive to detection in global analyses of these more accessible samples. However, once biomarkers are identified in detailed studies of disease tissue, appropriate assays could be generated for detection in more accessible samples.

5.3 ECM in the eye

Proteomic analysis of the ECM of the eye is simplified by the ability to separate and isolate the retinal BM (or inner limiting membrane, ILM) from the vitreous body. In addition, BMs of interest are thick. For example, in humans the thickness of ILMs ranges from 1 to 2 μm, whereas the lens capsule (LC) ranges from 10 to 15 μm. MS-based proteomic analysis of embryonic chick ILM identified 27 BM ECM proteins according to manual and Gene Ontology annotation (Balasubramani et al., 2010). Label-free quantification identified nidogens-1 and -2; laminin subunits α1, α5, β2, and γ1; agrin; collagen XVIII; perlecan; FRAS1; and FREM2 as the most abundant BM protein components of ILMs. In contrast, laminin subunits α3, β1, γ2, γ3 and collagen IV subunits α5 and α6 were minor constituents.

In further studies of BMs from adult human eyes, the ILM, the retinal vascular BMs, and the LC were isolated for analysis by 1D-SDS-PAGE and LC–MS/MS (GeLC–MS/MS). Peptide and protein identifications were performed using MaxQuant, and 1129 proteins were identified (Uechi, Sun, Schreiber, Halfter, & Balasubramani, 2014). According to MatrixDB, the in silico matrisome and the Uniprot database 202 proteins were identified as ECM. The dominant collagen IV trimer in LC and BV BMs had a chain composition of α1α1α2(IV), whereas the dominant collagen IV in the ILM had the α3α4α5(IV) chain composition. The data also demonstrated that the ratio of laminin and collagen IV varied among different BM types: the ratio of collagen IV to the other BM proteins was highest in LC, followed by BV, and lowest for the ILM.

5.4 Glomerular ECM

The glomerulus is the filtration unit of the kidney, and it comprises a bundle of specialized capillaries enclosed by Bowman's capsule. Filtration occurs across the capillary walls, and these are lined by endothelial cells, covered by podocytes and the two cell types are separated by the glomerular basement membrane (GBM). The ECM compartments in the glomerulus are the GBM of the capillary walls in addition to the mesangial matrix and the matrix associated with Bowman's capsule (Figure 4). In our own investigation of human

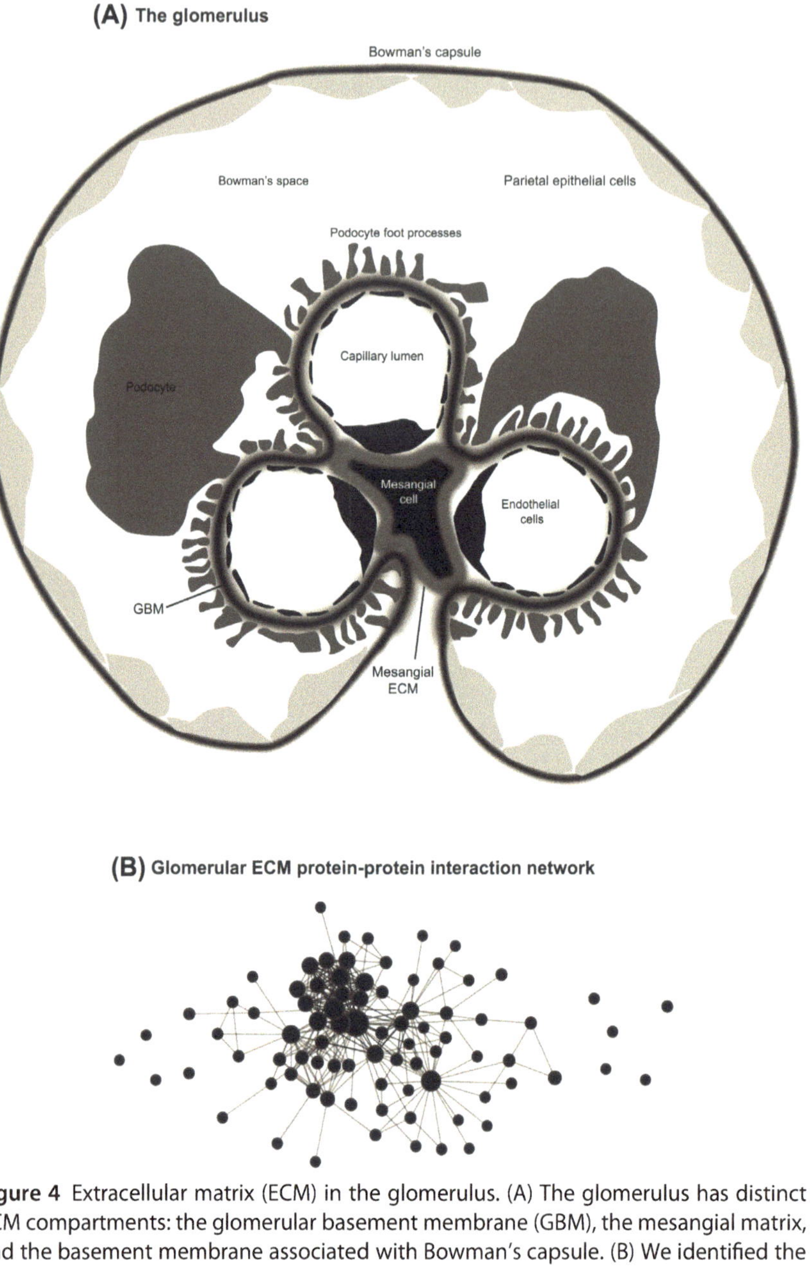

Figure 4 Extracellular matrix (ECM) in the glomerulus. (A) The glomerulus has distinct ECM compartments: the glomerular basement membrane (GBM), the mesangial matrix, and the basement membrane associated with Bowman's capsule. (B) We identified the glomerular ECM proteome, a complex network of 144 structural and regulatory ECM components.

glomerular ECM using proteomics we identified 144 structural and regulatory ECM proteins, and achieved a 38% enrichment of extracellular proteins (Lennon et al., 2014). This enrichment compared favorably with comparable ECM proteomic studies reporting an enrichment of between 12 and 30% (Byron et al., 2013). The ECM signature we described includes all previously identified glomerular ECM components, in addition to many new and abundant components. Using relative protein quantification with spectral counting and peptide abundance, we demonstrated a dominance of collagen IV, collagen I, and laminin isoforms in the glomerular ECM together with abundant collagen VI and TINAGL1. Protein network analysis enabled the creation of a glomerular ECM interactome, which revealed a core of highly connected structural components. Topological network analysis confirmed that BM and other structural ECM proteins formed a highly connected "core" subnetwork, whereas ECM-associated proteins were less clustered in the network.

To enable the characterization of specific cellular contributions to the glomerular basement GBM, we analyzed ECM isolated from podocytes and glomerular endothelial cells in vitro using proteomics (Byron et al., 2014). We identified 127 extracellular proteins in endothelial-derived matrices and 142 extracellular proteins in podocyte-derived matrices. These analyses identified cell type-specific differences in ECM composition, indicating distinct contributions to glomerular ECM assembly. Furthermore, coculture of podocytes and endothelial cells resulted in an altered composition and organization of ECM compared to monoculture ECMs from single cell types, suggesting a role for cell–cell cross talk in the production of glomerular ECM. This was supported by the identification of BM–like ECM deposition between cocultured cells. Importantly, compared to monoculture, the coculture ECM proteome better resembled a tissue-derived glomerular ECM data set, indicating its relevance to the GBM in vivo. Protein network analyses revealed a common core of 35 highly connected structural ECM proteins within both tissue-derived and cell-derived ECMs, and we speculate that these proteins are required for initial BM assembly.

Interestingly in cell-derived ECMs, we identified collagen IV α1 and α2 chains by MS, but there was minimal detection of α3, α4, or α5 chains. These findings suggest that ECM derived from glomerular cells in culture resembles a developmental ECM (Miner, 2005), and this concept was further supported by the low relative abundance of laminin β2, a predominant isoform in the mature glomerulus. While RT-PCR analysis revealed that both glomerular cell types expressed mRNA for *COL4A1*, *3*, *4*, and *5*, q-PCR demonstrated that the relative abundance of mRNA was significantly less than the amount isolated from human renal cortex. It is conceivable that glomerular cells have the capacity to synthesize both the collagen IV $\alpha1\alpha1\alpha2$ trimers and the mature $\alpha3\alpha4\alpha5$ trimers, but require the appropriate environmental context.

5.5 The stem cell niche

Cell fate is influenced by ECM composition. To investigate the stem cell niche, proteomics was performed on matrices derived from human embryonic stem cells (hESCs) and their feeder cells (Soteriou et al., 2013). This investigation compared ECM produced by mouse embryonic fibroblasts and hESCs to that produced by human placental stromal fibroblasts and hESCs. It found that supportive and unsupportive hESC substrates comprised distinct ECM compositions. Furthermore, this study examined candidates detected by the proteomic analysis for their ability to maintain hESC self-renewal in the absence of feeder cells. Fibrillin-1 was capable of supporting self-renewal alone, and when combined with fibronectin, perlecan, and fibulin-2 were also capable of supporting pluripotency in cultures without feeder cells.

5.6 ECM in the skin

Collagen VII is an important structural component of skin, forming anchoring fibrils, and it is lost in a hereditary condition known as recessive dystrophic epidermolysis bullosa (RDEB). Major clinical signs of severe RDEB include weak dermal-epidermal adhesion and dermal fibrosis. In a global (SILAC)-based quantitative MS of ECM produced by of primary human fibroblasts from normal and pathologically altered skin, proteomic analysis revealed that loss of collagen VII causes a global impact on ECM produced by fibroblasts (Kuttner et al., 2013). Changes included a reduction in BM components, and increase in the dermal matrix proteins, TGF-β and metalloproteases. This study clearly demonstrates that a lack of collagen VII does not simply represent a structural loss, but causes a number of parallel changes at the protein level. Reduced abundance of the BM components collagen IV, nidogen 1, or the laminin β1 and γ1 chains, which assemble to create laminin 511, weakens the overall structure of the BM and thereby amplifies the skin fragility caused by lack of functional anchoring fibrils. Moreover, the inherently high production and secretion of ECM proteins in RDEB fibroblasts, including tenascin C and interstitial collagens I, III, V, and VI, possibly contribute to excessive scarring and dermal fibrosis seen in RDEB.

6. EMERGING TECHNIQUES

6.1 Glycoproteomics

Over the past 10 years, MS-based proteomics has started to reveal the true extent of protein PTM. Thousands of residues can now be confidently identified and localized in the sequence of the protein. The quantification of PTM levels between different cellular states is likewise established, with label-free methods showing particular promise (Olsen & Mann, 2013). It is also becoming possible to determine the absolute occupancy or stoichiometry of PTM sites on a large scale. Powerful software for the bioinformatic analysis of thousands of PTM sites has been developed. However, a complete inventory of sites has not been established for any PTM. The central challenge for the field is to develop streamlined methods for determining biological functions for the myriad of modifications now known to exist. Glycosylation is important for ECM and BM function. However, glycosylated tryptic peptides can be too hydrophilic to be captured by conventional reversed-phase liquid chromatography (Figure 5). One option to avoid this problem is to use proteases with different cleavage specificities such as chymotrypsin, Lys-N, or endoproteinase Glu-C in addition to trypsin. N-linked glycosylation modifications can be extremely heterogeneous, and as a result determination of the structure of the glycosylation site is far more challenging than defining the modification site.

N-linked glycosylation modification sites have been mapped using a "filter-aided sample preparation" (FASP)-based method, in which glycopeptides are enriched by binding to lectins on the top of a filter. Using this methodology Mann and colleagues mapped 6367 N-glycosylation sites on 2352 proteins in four mouse tissues and blood plasma using MS. Interestingly, these sites almost always contained the sequence N-!P-[S|T]-!P (where !P is not proline) and rarely the N-X-C motif or nonconsensus sequences. Combining the FASP approach with analysis of subcellular glycosite localization demonstrated that sites are always oriented toward the extracellular space (Zielinska, Gnad, Wisniewski, & Mann, 2010).

The role of N-linked glycoproteins in ischemia and reperfusion injury was analyzed using glycopeptide enrichments with hydrazide capture, titanium dioxide, and hydrophilic interaction liquid chromatography with

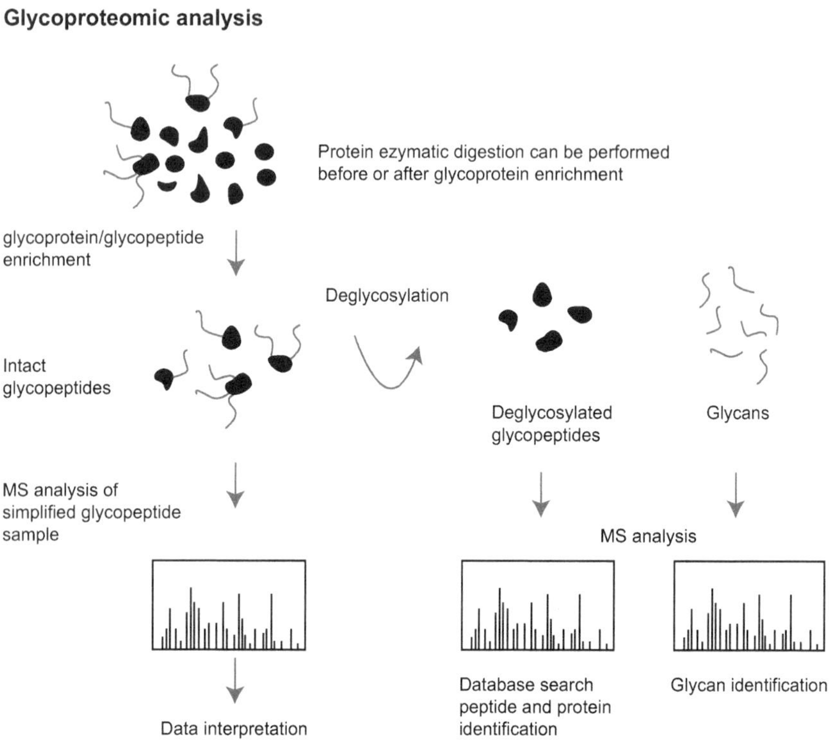

Figure 5 The glycoproteomics pipeline. Proteins are digested before or after glycoprotein enrichment. Samples are deglycosylated, and the global analysis of glycopeptides and glycans occurs in parallel.

and without an ion-pairing agent. Glycosylated peptides were treated with PNGase F and analyzed by liquid chromatography-MS/MS. This allowed the identification of 1556 nonredundant N-linked glycosylation sites, representing 972 protein groups from rat left ventricular myocardium ex vivo. Quantitation was via isobaric tags for relative and absolute quantitation (iTRAQ) following prolonged ischemia and reperfusion (40 min ischemia and 20 min reperfusion) indicative of myocardial infarction. The iTRAQ approach revealed 80 of 437 glycopeptides with altered abundance (Parker et al., 2011).

6.2 Degradomics

Within BMs proteolytic activity is crucial for the turnover of components, and these processes are also required for health and likely to be dysregulated in disease, for example, the rupture of atherosclerotic plaques. Proteomics has been therefore applied to understanding proteolytic enzymes that degrade ECM. Global analysis of protease cleavage sites, also known as degradomics, involves the analysis of N-terminal peptides, revealing information about both the protein targets and sites of cleavage of proteases. Positive-selection techniques to identify the N-termini generated by digestion with a protease of interest modify the free α-amines of proteins and the neo-N-termini of protease cleavage products for enrichment and exploit natural N-terminal acetylation to prevent copurification of these blocked peptides (Gevaert et al., 2003; Ji, Guo, & Li, 2005; Mahrus et al., 2008; McDonald, Robertson, Hurst, & Beynon, 2005; Timmer et al., 2007). Two such approaches are described below in further detail.

6.2.1 TAILS

Terminal amine isotopic labeling of substrates (TAILS) is a three-step quantitative proteomics approach for labeling and isolating N-terminal peptides before and after exposure to a protease of interest (Kleifeld et al., 2010). This approach allows the analysis of naturally acetylated, cyclized, or labeled N-termini from proteins and their protease cleavage products by tandem MS, and uses peptide isotope quantification to discriminate between the substrates of the protease of interest and the products of background proteolysis.

After proteome-wide proteolysis, the protease is inactivated and the sample is denatured and reduced. To quantify cleavage events specific to the protease of interest and to distinguish these from proteolysis products present in an untreated sample, stable isotopes can be introduced to determine the relative abundance of peptides in the protease-treated and control samples. Reductive dimethylation of primary amines using (d(2)C13)-formaldehyde (protease-treated, heavy) or (d(0)C12)-formaldehyde (control, light), catalyzed by sodium cyanoborohydride13, simultaneously labels and blocks all lysine primary amines, as well as the free (α)-amino groups of N-termini on all proteins and their protease cleavage products. The labeled proteomes are combined and trypsinized. As lysine dimethylation prevents trypsin cleavage, by cleaving with arginyl peptidase (ArgC)-like specificity after arginines only, the resultant generation of longer peptides by trypsin aids the identification of the protease-shortened neo-N-terminal peptides generated by cleavage. TAILS uses a new class of highly water-soluble polymers that selectively enrich the blocked N-terminal peptides by negative selection. Hyperbranched polyglycerols-aldehyde (HPG-ALD) polymer readily reacts with all unblocked internal and C-terminal tryptic peptides

Didangelos, A., Yin, X., Mandal, K., Baumert, M., Jahangiri, M., & Mayr, M. (2010). Proteomics characterization of extracellular space components in the human aorta. *Molecular and Cellular Proteomics, 9*(9), 2048–2062. http://dx.doi.org/10.1074/mcp.M110.001693.

Eisenstein, R., Larsson, S. E., Kuettner, K. E., Sorgente, N., & Hascal, V. C. (1975). The ground substance of the arterial wall. Part 1. Extractability of glycosaminoglycans and the isolation of a proteoglycan from bovine aorta. *Atherosclerosis, 22*(1), 1–17.

Eng, J. K., McCormack, A. L., & Yates, J. R. (1994). An approach to correlate tandem mass spectral data of peptides with amino acid sequences in a protein database. *Journal of the American Society for Mass Spectrometry, 5*(11), 976–989. http://dx.doi.org/10.1016/1044-0305(94)80016-2.

Fenn, J. B., Mann, M., Meng, C. K., Wong, S. F., & Whitehouse, C. M. (1989). Electrospray ionization for mass-spectrometry of large biomolecules. *Science, 246*(4926), 64–71. http://dx.doi.org/10.1126/science.2675315.

Gevaert, K., Goethals, M., Martens, L., Van Damme, J., Staes, A., Thomas, G. R., et al. (2003). Exploring proteomes and analyzing protein processing by mass spectrometric identification of sorted N-terminal peptides. *Nature Biotechnology, 21*(5), 566–569. http://dx.doi.org/10.1038/nbt810.

Griffin, N. M., Yu, J., Long, F., Oh, P., Shore, S., Li, Y., et al. (2010). Label-free, normalized quantification of complex mass spectrometry data for proteomic analysis. *Nature Biotechnology, 28*(1), 83–89. http://dx.doi.org/10.1038/nbt.1592.

Hynes, R. O. (2009). The extracellular matrix: not just pretty fibrils. *Science, 326*(5957), 1216–1219. http://dx.doi.org/10.1126/science.1176009.

Ishihama, Y., Oda, Y., Tabata, T., Sato, T., Nagasu, T., Rappsilber, J., et al. (2005). Exponentially modified protein abundance index (emPAI) for estimation of absolute protein amount in proteomics by the number of sequenced peptides per protein. *Molecular and Cellular Proteomics, 4*(9), 1265–1272. http://dx.doi.org/10.1074/mcp.M500061-MCP200.

Ji, C., Guo, N., & Li, L. (2005). Differential dimethyl labeling of N-termini of peptides after guanidination for proteome analysis. *Journal of Proteome Research, 4*(6), 2099–2108. http://dx.doi.org/10.1021/pr050215d.

Karas, M., & Hillenkamp, F. (1988). Laser desorption ionization of proteins with molecular masses exceeding 10,000 daltons. *Analytical Chemistry, 60*(20), 2299–2301. http://dx.doi.org/10.1021/Ac00171a028.

Kleifeld, O., Doucet, A., auf dem Keller, U., Prudova, A., Schilling, O., Kainthan, R. K., et al. (2010). Isotopic labeling of terminal amines in complex samples identifies protein N-termini and protease cleavage products. *Nature Biotechnology, 28*(3), 281–288. http://dx.doi.org/10.1038/nbt.1611.

Kuttner, V., Mack, C., Rigbolt, K. T., Kern, J. S., Schilling, O., Busch, H., et al. (2013). Global remodelling of cellular microenvironment due to loss of collagen VII. *Molecular Systems Biology, 9*, 657. http://dx.doi.org/10.1038/msb.2013.17.

Lennon, R., Byron, A., Humphries, J. D., Randles, M. J., Carisey, A., Murphy, S., et al. (2014). Global analysis reveals the complexity of the human glomerular extracellular matrix. *Journal of the American Society of Nephrology*. http://dx.doi.org/10.1681/ASN.2013030233.

Liu, H. B., Sadygov, R. G., & Yates, J. R. (2004). A model for random sampling and estimation of relative protein abundance in shotgun proteomics. *Analytical Chemistry, 76*(14), 4193–4201. http://dx.doi.org/10.1021/Ac0498563.

Mahrus, S., Trinidad, J. C., Barkan, D. T., Sali, A., Burlingame, A. L., & Wells, J. A. (2008). Global sequencing of proteolytic cleavage sites in apoptosis by specific labeling of protein N termini. *Cell, 134*(5), 866–876. http://dx.doi.org/10.1016/j.cell.2008.08.012.

McDonald, L., Robertson, D. H., Hurst, J. L., & Beynon, R. J. (2005). Positional proteomics: selective recovery and analysis of N-terminal proteolytic peptides. *Nature Methods, 2*(12), 955–957. http://dx.doi.org/10.1038/nmeth811.

Miner, J. H. (2005). Building the glomerulus: a matricentric view. *Journal of American Society of Nephrology, 16*(4), 857–861. http://dx.doi.org/10.1681/ASN.2004121139.
Naba, A., Clauser, K. R., Hoersch, S., Liu, H., Carr, S. A., & Hynes, R. O. (2012). The matrisome: in silico definition and in vivo characterization by proteomics of normal and tumor extracellular matrices. *Molecular and Cellular Proteomics, 11*(4). http://dx.doi.org/10.1074/mcp.M111.014647. M111 014647.
Naba, A., Clauser, K. R., Lamar, J. M., Carr, S. A., & Hynes, R. O. (2014). Extracellular matrix signatures of human mammary carcinoma identify novel metastasis promoters. *eLife, 3*. http://dx.doi.org/10.7554/eLife.01308. doi: ARTN e01308.
Olsen, J. V., & Mann, M. (2013). Status of large-scale analysis of Post-translational modifications by mass spectrometry. *Molecular and Cellular Proteomics, 12*(12), 3444–3452.
Ong, S. E., Blagoev, B., Kratchmarova, I., Kristensen, D. B., Steen, H., Pandey, A., et al. (2002). Stable isotope labeling by amino acids in cell culture, SILAC, as a simple and accurate approach to expression proteomics. *Molecular and Cellular Proteomics, 1*(5), 376–386. http://dx.doi.org/10.1074/mcp.M200025-MCP200.
Parker, B. L., Palmisano, G., Edwards, A. V. G., White, M. Y., Engholm-Keller, K., Lee, A., et al. (2011). Quantitative N-linked glycoproteomics of myocardial ischemia and reperfusion injury reveals early remodeling in the extracellular environment. *Molecular and Cellular Proteomics, 10*(8). http://dx.doi.org/10.1074/mcp.M110.006833.
Perkins, D. N., Pappin, D. J., Creasy, D. M., & Cottrell, J. S. (1999). Probability-based protein identification by searching sequence databases using mass spectrometry data. *Electrophoresis, 20*(18), 3551–3567. http://dx.doi.org/10.1002/(SICI)1522-2683(19991201)20:18<3551::AID-ELPS3551>3.0.CO;2–2.
Robertson, J., Jacquemet, G., Byron, A., Jones, M. C., Warwood, S., Selley, J. N., et al. (2015). Defining the phospho-adhesome through the phosphoproteomic analysis of integrin signalling. *Nature Communications, 13*(6), 6265. http://dx.doi.org/10.1038/ncomms7265.
Ross, M. D., Bruggeman, L. A., Hanss, B., Sunamoto, M., Marras, D., Klotman, M. E., et al. (2003). Podocan, a novel small leucine-rich repeat protein expressed in the sclerotic glomerular lesion of experimental HIV-associated nephropathy. *Journal of Biological Chemistry, 278*(35), 33248–33255. http://dx.doi.org/10.1074/jbc.M301299200.
Ross, P. L., Huang, Y. L. N., Marchese, J. N., Williamson, B., Parker, K., Hattan, S., et al. (2004). Multiplexed protein quantitation in *Saccharomyces cerevisiae* using amine-reactive isobaric tagging reagents. *Molecular and Cellular Proteomics, 3*(12), 1154–1169. http://dx.doi.org/10.1074/mcp.M400129-MCP200.
Schilling, O., Huesgen, P. F., Barre, O., Auf dem Keller, U., & Overall, C. M. (2011). Characterization of the prime and non-prime active site specificities of proteases by proteome-derived peptide libraries and tandem mass spectrometry. *Nature Protocols, 6*(1), 111–120. http://dx.doi.org/10.1038/nprot.2010.178.
Sevimoglu, T., & Arga, K. Y. (2014). The role of protein interaction networks in systems biomedicine. *Computational and Structural Biotechnology Journal, 11*(18), 22–27. http://dx.doi.org/10.1016/j.csbj.2014.08.008.
Silva, J. C., Denny, R., Dorschel, C. A., Gorenstein, M., Kass, I. J., Li, G. Z., et al. (2005). Quantitative proteomic analysis by accurate mass retention time pairs. *Analytical Chemistry, 77*(7), 2187–2200. http://dx.doi.org/10.1021/ac048455k.
Soteriou, D., Iskender, B., Byron, A., Humphries, J. D., Borg-Bartolo, S., Haddock, M. C., et al. (2013). Comparative proteomic analysis of supportive and unsupportive extracellular matrix substrates for human embryonic stem cell maintenance. *The Journal of Biological Chemistry, 288*(26), 18716–18731. http://dx.doi.org/10.1074/jbc.M113.463372.
Tholen, S., Biniossek, M. L., Gansz, M., Ahrens, T. D., Schlimpert, M., Kizhakkedathu, J. N., et al. (2014). Double deficiency of cathepsins B and L results in

massive secretome alterations and suggests a degradative cathepsin-MMP axis. *Cellular and Molecular Life Sciences, 71*(5), 899–916. http://dx.doi.org/10.1007/s00018-013-1406-1.

Timmer, J. C., Enoksson, M., Wildfang, E., Zhu, W., Igarashi, Y., Denault, J. B., et al. (2007). Profiling constitutive proteolytic events in vivo. *Biochemical Journal, 407*(1), 41–48. http://dx.doi.org/10.1042/BJ20070775.

Tucher, J., Linke, D., Koudelka, T., Cassidy, L., Tredup, C., Wichert, R., et al. (2014). LC-MS based cleavage site profiling of the proteases ADAM10 and ADAM17 using proteome-derived peptide libraries. *Journal of Proteome Research, 13*(4), 2205–2214. http://dx.doi.org/10.1021/pr401135u.

Uechi, G., Sun, Z., Schreiber, E. M., Halfter, W., & Balasubramani, M. (2014). Proteomic view of basement membranes from human retinal blood vessels, inner limiting membranes, and lens capsules. *Journal of Proteome Research*. http://dx.doi.org/10.1021/pr5002065.

Walther, T. C., & Mann, M. (2010). Mass spectrometry-based proteomics in cell biology. *The Journal of Cell Biology, 190*(4), 491–500. http://dx.doi.org/10.1083/jcb.201004052.

Zanivan, S., Krueger, M., & Mann, M. (2012). In vivo quantitative proteomics: the SILAC mouse. *Methods in Molecular Biology, 757*, 435–450. http://dx.doi.org/10.1007/978-1-61779-166-6_25.

Zielinska, D. F., Gnad, F., Wisniewski, J. R., & Mann, M. (2010). Precision mapping of an in vivo N-glycoproteome reveals rigid topological and sequence constraints. *Cell, 141*(5), 897–907. http://dx.doi.org/10.1016/j.cell.2010.04.012.

CHAPTER SIX

Molecular Basis of Laminin–Integrin Interactions

Masashi Yamada and Kiyotoshi Sekiguchi*
Laboratory of Extracellular Matrix Biochemistry, Institute for Protein Research, Osaka University, Suita, Osaka, Japan
*Corresponding author: E-mail: sekiguch@protein.osaka-u.ac.jp

Contents

Current Topics in Membranes, Volume 76
ISSN 1063-5823
http://dx.doi.org/10.1016/bs.ctm.2015.07.002

Abstract

Laminins are composed of three polypeptide chains, designated as α, β, and γ. The C-terminal region of laminin heterotrimers, containing coiled-coil regions, short tails, and laminin globular (LG) domains, is necessary and sufficient for binding to integrins, which are the major laminin receptor class. Laminin recognition by integrins critically requires the α chain LG domains and a glutamic acid residue of the γ chain at the third position from the C-terminus. Furthermore, the C-terminal region of the β chain contains a short amino acid sequence that modulates laminin affinity for integrins. Thus, all three of the laminin chains act cooperatively to facilitate integrin binding. Mammals possess 5 α (α1−5), 3 β (β1−3), and 3 γ (γ1−3) chains, combinations of which give rise to 16 distinct laminin isoforms. Each isoform is expressed in a tissue-specific and developmental stage-specific manner, exerting its functions through binding of integrins. In this review, we detail the current knowledge surrounding the molecular basis and physiological relevance of specific interactions between laminins and integrins, and describe the mechanisms underlying laminin action through integrins.

List of Abbreviations

2-BP 2-Bromopalmitate
AP Adaptor protein
ECM Extracellular matrix
ESC Embryonic stem cell
FAK Focal adhesion kinase
Ig Immunoglobulin
L4 Laminin 4
LCC Laminin coiled-coil
LE Laminin epidermal growth factor-like
LF Laminin four
LFx Modified LF
LG Laminin globular
LN Laminin N-terminal
Lu/BCAM Lutheran/basal cell adhesion molecule
Lβ Laminin β knob
MCAM Melanoma cell adhesion molecule
MIDAS Metal ion-dependent adhesion site
PI4K Phosphatidylinositol 4-kinase
PKC Protein kinase C
iPS Induced pluripotent stem cell
RGD Arg−Gly−Asp
SAM Substrate-attached material

1. INTRODUCTION

Laminins are major components of basement membranes, comprising an α, a β, and a γ chain that possess molecular masses of approximately 400, 200, and 200 kDa, respectively. There are 5 α (α1−5), 3 β (β1−3), and 3 γ (γ1–3) chains recognized in mammals (Aumailley, 2013; Domogatskaya, Rodin, & Tryggvason, 2012), and combinations of these give rise to 16 distinct isoforms which are expressed in tissue-specific and developmentally regulated manners (Aumailley et al., 2005). Laminin chains typically consist of a laminin N-terminal domain, laminin epidermal growth factor-like (LE) domains, laminin 4 (L4) domains, a laminin four (LF) domain (present in β chains), a modified LF domain (present in α3 and α5 chains), a laminin coiled-coil (LCC) domain, a laminin β knob domain (present in LCC domains of β chains), and a tandem repeat of five C-terminal laminin globular (LG) domains (present in α chains and designated as LG1–5; Figure 1). Laminin heterotrimers are stabilized by disulfide bonds between chains

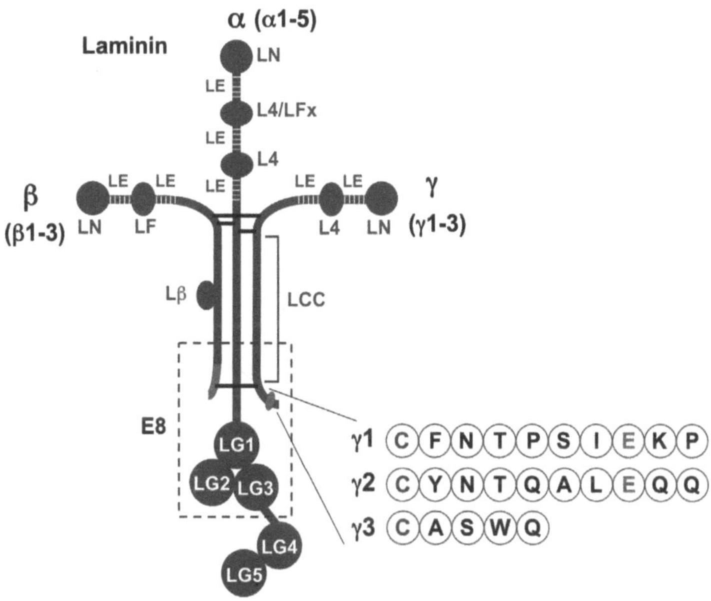

Figure 1 Structure of laminins and their integrin-binding sites. Laminins consist of α, β, and γ chains. These chains possess laminin N-terminal (LN) domains, repeating laminin epidermal growth factor-like (LE) domains, laminin 4 (L4) domains, a laminin four (LF) domain, a modified LF (LFx) domain, a laminin coiled-coil (LCC) domain, a laminin β knob (Lβ) domain, and C-terminal laminin globular (LG) domains. Interchain disulfide bonds exist at the boundary of LE and LCC domains and near the C-terminus (black lines). The E8 fragment is the smallest unit known to exhibit full integrin-binding activity. The integrin-binding activity of laminins requires both the LG1–3 domains of the α chain and the glutamic acid residue in the C-terminal tail of the γ chain (shown in red (light gray in print versions)). This glutamic acid residue is present in γ1 and γ2 chains but not in the γ3 chain. A short amino acid stretch within the β chain (shown in red (light gray in print versions)) can also modulate integrin affinities for laminins.

near the boundary of the LE and LCC domains, and between the β and γ chains near their C-termini (Aumailley, 2013; Durbeej, 2010).

Individual laminin chains exhibit tissue- and cell type-specific expression (Miner et al., 1997; Miner & Yurchenco, 2004). Information that details these patterns is available from the high-resolution image database "Mouse Basement Membrane Bodymap." This database contains immunohistochemical data concerning expression of 41 extracellular matrix (ECM) proteins, including laminins, that are present in the basement membranes of E16.5 mouse embryos (http://www.matrixome.com/bm/). For example, laminin α chains are differentially expressed throughout heart ventricles. The α2 chain is present in basement membranes that surround cardiac muscle cells, the α4 chain is found around both cardiac muscle cells and blood vessels, and the α5 chain has been identified around blood vessels and in the epicardium. At the same time, expression of the α1 and α3 chains is not observed within the ventricles (Figure 2). Because laminins function as both scaffold proteins and signal inducers through binding to their cell surface receptor components, differential expression of individual laminin isoforms may confer distinctive functions specific to cell types.

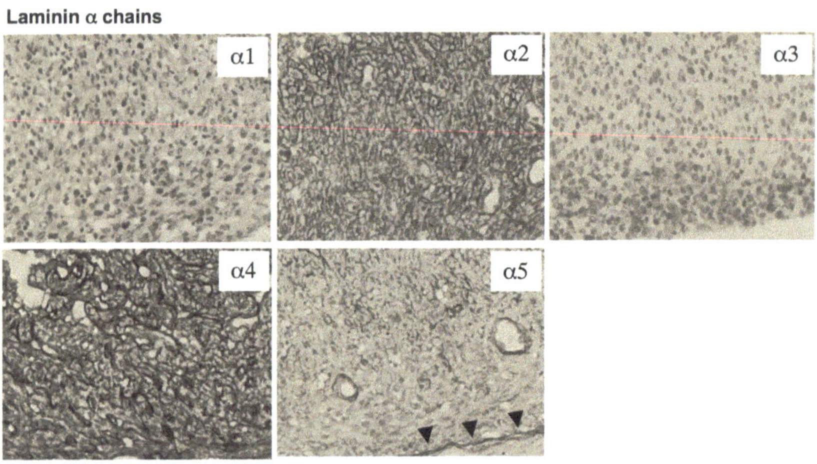

Figure 2 Distinct localization of laminin α chains in the heart. (*Data are taken from the Mouse Basement Membrane Database (http://www.matrixome.com/bm/).*) Frozen sections of heart ventricles of E16.5 mouse embryos immunostained with antibodies against laminin α1, α2, α3, α4, and α5 chains. Sections were counterstained with hematoxylin. Positive signals for α2 were observed surrounding myocardial cells, those for α4 around both myocardial cells and blood vessels, and those for α5 around blood vessels and within the epicardium (arrowheads). (See color plate)

2. LAMININ RECEPTORS

2.1 Membrane proteins possessing negatively charged carbohydrate chains

Dystroglycan is a cell surface laminin receptor. This receptor possesses a highly glycosylated extracellular α-dystroglycan subunit noncovalently linked to a β-dystroglycan subunit. The β subunit contains short extracellular, transmembrane, and intracellular domains. The intracellular domain ultimately links with the actin cytoskeleton via interactions with dystrophin and utrophin (Yurchenco, 2011). Through its O-linked carbohydrate chains, α-dystroglycan binds in a Ca^{2+}-dependent manner mainly to the LG4 domain present on the laminin α chain of multiple laminin isoforms, including laminin-111, laminin-211, and laminin-511 (Ido et al., 2004; Wizemann et al., 2003; Yu & Talts, 2003). α-Dystroglycan provides the primary anchorage to accumulate laminins, and thereby participates in basement membrane assembly in Reichert's membrane, breast epithelium, and skeletal muscle.

Besides dystroglycan, laminin LG4 domains also bind sulfatides and heparin through clusters of basic amino acid residues present in loops connecting β strands (Tisi, Talts, Timpl, & Hohenester, 2000). In Schwann cells, galactosyl sulfatides anchor laminins and facilitate assembly of basement membranes (Li et al., 2005). Syndecans are cell surface heparan sulfate proteoglycans that also function as laminin receptors through binding of their heparan sulfate chains to LG4, which promotes cell adhesion (Okamoto et al., 2003; Utani et al., 2001). Syndecan-1 also interacts with the N-terminal LE domains of the laminin γ2 chain to regulate cell adhesion and motility (Ogawa, Tsubota, Hashimoto, Kariya, & Miyazaki, 2007).

2.2 Integrins

Integrins are heterodimeric membrane proteins composed of noncovalently associated α and β subunits (Campbell & Humphries, 2011). To date, at least 24 separate integrins consisting of distinct combinations of α and β subunits have been identified in mammals. Among these, integrins α1β1, α2β1, α3β1, α6β1, α6β4, and α7β1 function as laminin receptors (Belkin & Stepp, 2000). Integrins α1β1 and α2β1 recognize N-terminal domains of laminin α1 and α2 chains (Calderwood, Tuckwell, Eble, Kuhn, & Humphries, 1997; Colognato, MacCarrick, O'Rear, & Yurchenco, 1997; Desban & Duband, 1997), while integrins α3β1, α6β1, α6β4, and α7β1 associate with C-terminal portions of laminins (Belkin & Stepp, 2000). Studies using knockout mice show that the absence of integrin α1 or α2 does not result in severe defects (Chen, Diacovo, Grenache, Santoro, & Zutter, 2002; Gardner, Kreidberg, Koteliansky, & Jaenisch, 1996; Pozzi, Wary, Giancotti, & Gardner, 1998). However, knock out of laminin chains does cause serious defects, which resemble those observed in mice deficient in integrins α3, α6, α7, or β6 (Belkin & Stepp, 2000; Miner & Yurchenco, 2004). Integrin αvβ3 also binds the L4 domain of laminin α5 chain in an Arg−Gly−Asp (RGD)-dependent manner (Sasaki & Timpl, 2001). Furthermore, an interaction between laminin-511/521 and integrin αvβ3 promotes growth factor-induced proliferation, but plays only a minor role in cell adhesion (Genersch, Ferletta, Virtanen, Haller, & Ekblom, 2003). Therefore, this review will focus on α3β1, α6β1, α6β4, and α7β1 as the major laminin-binding integrins.

2.3 Lutheran/basal cell adhesion molecule and melanoma cell adhesion molecule

Lutheran/basal cell adhesion molecule (Lu/BCAM) and melanoma cell adhesion molecule (MCAM) are also laminin receptors (El Nemer et al., 1998; Flanagan et al., 2012; Ishikawa et al., 2014; Kikkawa, Moulson, Virtanen, & Miner, 2002; Udani et al., 1998). Lu/BCAM and MCAM contain five extracellular immunoglobulin (Ig)-like (V−V−C2−C2−C2), single transmembrane and short intracellular domains (Campbell et al., 1994; Lehmann, Riethmuller, & Johnson, 1989; Parsons et al., 1995). Together with activated leukocyte cell adhesion molecule (ALCAM, CD166), Lu/BCAM and MCAM belong to a small subfamily of the Ig superfamily. Lu and BCAM arise from a single gene by alternative pre-mRNA splicing and possess identical extracellular and transmembrane domains, while differing in their cytoplasmic domains, with Lu being 40 amino acid residues longer than BCAM (Campbell et al., 1994; Rahuel, Le Van Kim, Mattei, Cartron, & Colin, 1996). Lu has been studied primarily as the Lutheran blood group system antigen and in the context of sickle cell disease (Hebbel, 1997; Wick & Eckman, 1996). Meanwhile, BCAM is upregulated in ovarian carcinoma, suggesting an involvement in tumor progression (Campbell et al., 1994). Lu/BCAM is a specific receptor for the laminin α5 chain-containing laminins, which include laminin-511 and laminin-521 (El Nemer et al., 1998; Kikkawa et al., 2002; Udani et al., 1998). Lu/BCAM and laminin-binding integrins share a binding site on laminins, which includes the LG1−3 domains. However, our unpublished data indicate that Lu/BCAM does not require the C-terminal Glu residue of laminin γ chains needed by laminin-binding integrins, the importance of which is described later in this review. Recently, Lu/BCAM has been reported to accelerate tumor cell migration by weakening cell adhesion through binding to laminin-511, competitively displacing laminin-binding integrins (Kikkawa, Ogawa, et al., 2013).

MCAM (CD146) is expressed by vascular endothelial cells, smooth muscle cells, Schwann cells, glomerular mesangial cells, intermediate trophoblasts, mesenchymal stem cells, and leukocytes (Daniel et al., 2005; Shih, Nesbit, Herlyn, & Kurman, 1998; Wang & Yan, 2013). MCAM was originally identified as a marker of melanoma, and is also expressed in tumors of the prostate and bone where it is closely associated with tumor invasion and metastasis (Lehmann et al., 1987; Wang & Yan, 2013). Recently, Flanagan et al. (2012) reported that MCAM interacts with laminin-411 to enable entry of circulating lymphocytes into the brain. Additionally, MCAM preferentially binds to laminin-421 to promote integrin α6β1-mediated melanoma cell migration (Ishikawa et al., 2014).

3. LIGAND SPECIFICITIES OF LAMININ-BINDING INTEGRINS

3.1 Integrin α3 and α6

Binding specificities of laminin-binding integrins have been examined primarily by cell adhesion assays using antibodies that functionally block integrins. Recently, recombinant soluble integrins lacking the transmembrane and cytoplasmic regions have also been used to determine ligand-binding specificities. These newer assays allow direct binding between integrins and laminins to be observed, facilitating precise evaluation of integrin–laminin interactions (Eble et al., 1998; Nishiuchi et al., 2006; von der Mark et al., 2002).

Laminin–integrin interaction specificity is mainly dependent on laminin α chains. Accumulating evidence indicates that integrin α3β1 displays strong binding activities toward α5-laminins (laminin-511/521) and α3-laminin (laminin-332) (Figure 3; Eble et al., 1998; Nishiuchi et al., 2006). Integrin α6β4 exhibits binding preferences similar to those of α3β1, but α6β1 shows broad binding specificities. The affinity of α6-containing integrins for α5-laminins is also distinct, with integrin α6β4 having decreased binding affinity compared with α6β1. These observations indicate that the β subunits play an auxiliary role in defining the ligand specificity and affinity of laminin-binding α6 integrins, although it is widely accepted that integrin ligand specificity is determined primarily by the α subunit.

Knockout mouse studies have shown that a lack of laminin α3, β3, or γ2 chains results in a severe skin blistering disease (Kuster, Guarnieri, Ault,

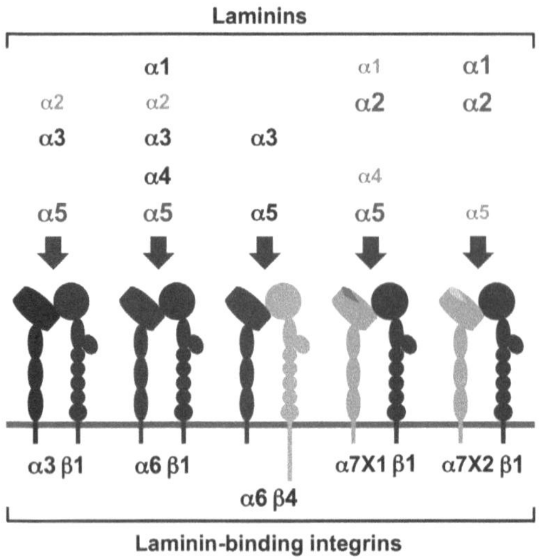

Figure 3 Binding specificities of laminin-binding integrins. Affinities of laminins containing distinct α chains to laminin-binding integrins are indicated by different colors: Red, high; black, medium; gray, low. Font size is proportional to the affinity strength. (See color plate)

Flaherty, & Swiatek, 1997; Meng et al., 2003; Ryan, Lee, Miyashita, & Carter, 1999). This condition is also observed in integrin α6- or β4-deficient mice and to lesser extent in integrin α3-deficient mice (Dowling, Yu, & Fuchs, 1996; Georges-Labouesse et al., 1996; van der Neut, Krimpenfort, Calafat, Niessen, & Sonnenberg, 1996), underscoring the physiological importance of laminin-332 interaction with these integrins in the epidermis. Integrin β4 is linked to intermediate filaments through a long cytoplasmic region which facilitates formation of hemidesmosomes for stable adhesion to basement membranes (Borradori & Sonnenberg, 1999; Hirako & Owaribe, 1998). Indeed, integrin α6β4—but not α3β1—plays a central role in the adhesion of epidermal cells to basement membranes through hemidesmosomes. Following knock out of the laminin α5 chain, mice display abnormalities in multiple organs, including kidneys, lungs, limbs, placenta, intestines, and hair follicles (Fukumoto et al., 2006; Li et al., 2003; Miner, Cunningham, & Sanes, 1998; Miner & Li, 2000; Nguyen, Miner, Pierce, & Senior, 2002; Rebustini et al., 2007). Similar defects in lung and kidney epithelia are also observed in mice lacking integrin α3 (Kreidberg et al., 1996), underscoring the importance of interactions between α5-containing laminins and integrin α3β1 in kidney and lung development.

3.2 Integrin α7

Two splicing variants of integrin α7β1 have been described: α7X1β1 and α7X2β1 (Ziober et al., 1993). These differ in the ~40 amino acid variable region at the boundary between blades 3 and 4 of the β-propeller domain of the α7 subunit and exhibit distinct ligand-binding specificities. Integrin α7X1β1 binds to almost all laminin isoforms except for α3-laminin (laminin-332), while integrin α7X2β1 preferentially binds to α1-laminin (laminin-111) and α2-laminins—and α5-laminins to a lesser extent—but not α3- or α4-laminins (Figure 3; von der Mark et al., 2007; Nishiuchi et al., 2006; Taniguchi et al., 2009). α7X1β1 and α7X2β1 differ from α3β1, α6β1, and α6β4 in that they do not bind to α3-laminin. Additionally, α7X1β1 and α7X2β1 are the only integrins that exhibit significant binding to α2-laminins. Given that integrin α7β1 is expressed predominantly by muscle cells, and laminin-211/221 are the major laminin isoforms present in muscle cell basement membranes (Burkin & Kaufman, 1999; Colognato & Yurchenco, 2000; Song, Wang, Foster, Bielser, & Kaufman, 1992), binding of integrin α7β1 to laminin-211/221 is likely important for attachment of muscle cells to their surrounding basement membrane. Consistently, mice lacking either integrin α7 or laminin α2 develop muscular dystrophy, although the phenotype of laminin α2-knockout mice is more severe than that of integrin α7-knockout mice (Kuang, Xu, Vachon, & Engvall, 1998; Mayer et al., 1997; Miyagoe et al., 1997).

The laminin-binding integrins contain one of three closely related α subunits—α3, α6, or α7—which are believed to have a common ancestral gene. Despite their close similarities, these α subunits exhibit distinct—albeit partly overlapping—specificities toward laminins upon dimerization with a common β1 subunit. Additionally, differences between the integrin β subunits β1 and β4 and between the α7X1 and α7X2 splice variants influence ligand preference. These specific interactions between laminins and integrins play important roles in the development and maintenance of various tissues and organs.

4. MECHANISMS UNDERLYING THE RECOGNITION OF LAMININS BY INTEGRINS

4.1 Integrin-binding sites on laminins

Multiple efforts to elucidate the region(s) of laminins responsible for integrin binding have been made. Collectively, these studies indicate that the C-terminal regions of laminins, especially α-chain LG1–3, are involved in laminin recognition by integrins (Hirosaki, Mizushima, Tsubota, Moriyama, & Miyazaki, 2000; Ido et al., 2004; Ido, Harada, Yagi, & Sekiguchi, 2006; Kikkawa et al., 2007). However, the specific recognition sites have not been narrowed down further, and the molecular mechanisms underlying the interaction between laminins and integrins remain unclear. In contrast, interactions between integrins and collagens or RGD-containing adhesion proteins have been clearly defined by X-ray crystallography. The RGD motif binds at an interface between the integrin α and β subunits, with the basic Arg residue fitting into a cleft within a β-propeller module in the α subunit, while the acidic Asp residue coordinates with a divalent metal ion present at the metal ion-dependent adhesion site (MIDAS) of the β-I domain (Nagae et al., 2012; Springer, Zhu, & Xiao, 2008; Xiong et al., 2002). Studies have tried to uncover integrin-binding sites on laminins using a synthetic peptide-based approach similar to that used to identify the RGD motif in fibronectin (Pierschbacher & Ruoslahti, 1984); however, no individual sequence has been found to interact with integrins to a degree comparable to that of intact laminins (Kikkawa, Hozumi, et al., 2013; Nomizu et al., 1995; Okazaki et al., 2002; Shang, Koshikawa, Schenk, & Quaranta, 2001; Yu & Talts, 2003).

The E8 fragment produced by brief elastase digestion of mouse laminin-111 retains almost the full integrin-dependent cell adhesion activity of the parental molecule (Deutzmann et al., 1990). This fragment comprises short coiled-coil stretches of α1, β1, and γ1 chains, in which the α1 chain also contains LG1—3—but not LG4—5—domains, and the β1 and γ1 chains are linked by a disulfide bond near their respective C-termini (Figure 1). The cell adhesive activity of the E8 fragment is lost when its α1 chain fragment is separated from the disulfide-linked β1—γ1 fragment (Deutzmann et al., 1990; Sung, O'Rear, & Yurchenco, 1993). However, adhesion activity is restored to the α1 chain fragment upon in vitro recombination and refolding with the β1—γ1 dimer, highlighting the importance of heterotrimerization of α chains with the β and γ chains for the integrin-binding activity of laminins. The importance of heterotrimerization has also been demonstrated with the recombinant E8 fragment of laminin-332, whose integrin-binding activity is restored only when the α3 chain fragment is assembled with the disulfide-bonded β3—γ2 dimer segment (Kunneken et al., 2004).

4.2 Involvement of laminin γ chains in integrin binding

The C-terminal region of laminin γ chains, particularly the glutamic acid residue at the third position from the C-terminus, is critically involved in laminin recognition by integrins (Figure 1; Ido et al., 2007). This conclusion is based on several observations. First, deletion of the C-terminal three—but not two—amino acids from the γ1 chain completely abrogates the integrin-binding activity of laminin-511, as does a Glu1607Gln substitution at the third position from the γ1 C-terminus. Moreover, identical mutations to the γ1 chain also inactivate the integrin-binding activity of laminin-311. These findings demonstrate the importance of the C-terminal Glu1607 residue for integrin binding by γ1-containing laminins. Second, the γ2 chain also contains a Glu residue (Glu1191) at the third position from the C-terminus, which is functionally equivalent to Glu1607 of the γ1 chain. Deletion of the C-terminal three amino acids or a Glu1191Gln substitution completely abrogates laminin-332 integrin-binding activity. Therefore, the nearly C-terminal Glu residues of both the γ1 and γ2 chains play crucial roles in the recognition of laminins by integrins.

The C-terminal tail of the γ3 laminin chain distal to the interchain disulfide bond with the β chain is shorter than that of either the γ1 or γ2 chain, being only four residues in length compared with nine for both γ1 and γ2 (Figure 1; Cserhalmi-Friedman et al., 2001; Iivanainen, Morita, & Tryggvason, 1999; Koch et al., 1999). Importantly, the γ3 chain does not possess the nearly C-terminal Glu residue corresponding to that conserved in the γ1 and γ2 chains. Consistently, γ3 chain-containing laminins (typically laminin-213) lack the ability to bind integrins (Ido et al., 2008). Integrin-binding activity can be conferred upon laminin-213 by substitution of the four C-terminal amino acids of the γ3 chain with the corresponding nine C-terminal amino acids of the γ1 chain. Furthermore, Glu–Gln substitution within this γ1-derived amino acid sequence eliminates the integrin-binding activity of recombinant laminin-213, further underscoring the importance of the C-terminal Glu residue in mediating the integrin-binding activity of γ1- and γ2-laminins.

The biological significance of the lack of integrin binding by γ3-laminins remains unknown. However, γ3-laminins may possess a dominant-negative function that negatively regulates cell–basement membrane interactions during embryonic development and tissue organization. Given that the LG4–5 domains of the laminin α chains are capable of binding to cell surface receptors other than integrins, such as α-dystroglycan, heparan sulfate proteoglycans, and sulfated glycolipids (Ido et al., 2004; Li et al., 2005; Okamoto et al., 2003; Tisi et al., 2000; Utani et al., 2001; Wizemann et al., 2003; Yu & Talts, 2003), it is likely that the γ3-laminins are specialized ECM proteins that exert their functions through nonintegrin laminin receptors. Laminin γ3-knockout mice show ectopic granule cells in the cerebellum and an apparent increase in capillary branching in the outer retina, although they are viable and otherwise apparently healthy (Denes et al., 2007; Li et al., 2012). These phenotypes may result from the loss of γ3 chain expression in a subset of vascular basement membranes within the brain, retina, and the pial surface of brain (Denes et al., 2007; Egles et al., 2007; Li et al., 2012; Libby et al., 2000). Additionally, mutations within the laminin γ3 gene (*LAMC3*) lead to human disease including autism and congenital malformation of the occipital lobe along with diminished mental capacity (Barak et al., 2011; O'Roak et al., 2011). Together, these findings demonstrate integrin-independent functions for γ3-laminins in the eye and brain.

Despite the demonstrated importance of γ chain C-terminal Glu residues for binding to integrins, precisely how the Glu residue is involved in the recognition of laminins by integrins remains unsettled. One possibility is that the Glu residue is directly recognized by integrins through coordination with the metal ion within the MIDAS of laminin-binding integrins. It has been well known that integrin ligands contain an acidic amino acid residue (i.e., Asp or Glu) in their integrin-binding site, as exemplified by the Asp residue of the RGD motif, the carboxylate group of which coordinates with a metal ion at MIDAS of the β-I domain (Nagae et al., 2012; Springer et al., 2008; Xiong et al., 2002). Likewise, collagen binds to integrin α2β1 through the GFOGER motif (in which O equals hydroxyproline), the Glu residue of which coordinates with the metal ion within the MIDAS of the α2-I domain, as revealed by the crystal structure of the α2-I domain complexed with the triple-helical GFOGER peptide (Emsley, Knight, Farndale, Barnes, & Liddington, 2000; Knight et al., 2000). Alternatively, the α chain LG1–3 domains—but not the γ chains—may provide the critical acidic residues that coordinate with the metal ion in the MIDAS. However, our preliminary results show that substitution of any of the acidic amino acid residues conserved within the LG1–3 domains with Ala does not eliminate the integrin-binding activity of the laminin-511 E8 fragment to the same degree as does the Glu–Gln mutation within the C-terminus of the γ1 or γ2 chain. These findings indicate that the Glu residue in γ1 and γ2 chains—but not any individual acidic amino acid residue in the LG1–3 domains—coordinates with the metal ion in the integrin MIDAS motif. Navdaev et al. (2008) used electron microscopy to demonstrate that C-terminal truncation of the γ2 chain induced opening of the compact assembly of the LG1–3 domains of the E8 fragment of laminin-332, suggesting that the C-terminal tail of the γ2 chain is involved in stabilization of the integrin-binding competent conformation of the LG1–3 domains. Therefore, the γ chain Glu residue may be required for stabilization of the functionally active conformation of the LG1–3 domains, possibly through direct interaction with the LG1–3 domains.

4.3 Involvement of laminin β chains in integrin binding

Laminin β chains are also involved in interactions between laminins and integrins. β1- and β2-laminins exhibit different binding affinities toward a subset of laminin-binding integrins that contain the X2 region within their α subunits (Taniguchi et al., 2009). The integrin α7 splice variants α7X1 and α7X2 demonstrate distinct dependency on laminin β1 and β2 chains in their affinity for laminins. While β2-laminins (laminin-121 and -221) are bound by integrin α7X2β1 with higher affinities than are β1-laminins (laminin-111 and -211), integrin α7X1β1 does not exhibit any preference for β2-laminins over β1-laminins (Taniguchi et al., 2009). Additionally, integrin α3β1, which contains an X2 exon-derived sequence (Ziober et al., 1993), binds β2-laminin (laminin-521) more strongly than β1-laminin (laminin-511), as is the case for integrin α7X2β1 (Taniguchi et al., 2009). In contrast, the X1-type integrins α6β1 and α6β4 do not exhibit any preference for β2-laminins. The integrin α6 gene also contains the X2-coding exon, but the integrin α6 mRNA species containing this exon is rarely generated (Ziober et al., 1993). An artificial integrin α6X2β1 has been produced by replacing the X1 sequence with X2, enabling the integrin to recognize differences in laminin β chains and conferring a higher affinity for β2-laminins over β1-laminins (Taniguchi et al., 2009). These observations indicate that X2-type (α3 and α7X2)—but not X1-type (α6 and α7X1)—integrins can discriminate between β1- and β2-laminins, with a higher affinity for β2-laminins.

The laminin β chain mutational study revealed that the region controlling integrin-binding affinity is within the C-terminal 20 amino acid residues of the β chains (Taniguchi et al., 2009). This suggests that the β chain C-terminal regions also participate in laminin recognition by integrins, along with the LG1–3 domains in α chains and the Glu residue in γ chains (Figure 1). However, the mechanism by which the laminin β2 chain regulates the affinity of laminin binding by X2-type integrins remains to be elucidated.

Electron microscopy of negatively stained laminin α1 chains and crystal structures of the LG4–5 domains of the laminin α2 chain suggests that the LG1–3 domains assume a cloverleaf configuration (Timpl et al., 2000). The integrity of the LG1–3 domains is critical for integrin binding by laminins, as swapping of individual LG domains between different laminin α chains disrupts integrin-binding activity (Ido et al., 2006; Kikkawa et al., 2007). The C-terminal 20 amino acid residues of laminin β chains, which are involved in laminin–integrin interactions, are believed to be in close proximity to and potentially in direct contact with the LG1–3 domains. Therefore, these residues may fine-tune the active conformation of the LG1–3 domains.

Alternatively, the modulation of integrin-binding affinity by laminin β chains may be explained by a direct interaction between the C-terminal region of the β2 chain and X2-type integrins. X2 regions contain conserved tyrosine residues that are equivalent to Tyr208 in the β-propeller domain of the integrin α5 subunit. This Tyr residue of α5 has been shown to interact with the so-called "synergy site" of fibronectin (von der Mark et al., 2007; Mould et al., 2003). Therefore, the X2 regions may well directly recognize the C-terminal regions of laminin β chains in a manner analogous to that of the high-affinity binding of integrin α5β1 to fibronectin through the synergy site. However, the synergy site reportedly influences the conformation of the cell-binding domain containing the RGD motif and thereby promotes binding to integrin α5β1 (Altroff, Choulier, & Mardon, 2003; Takagi, Strokovich, Springer, & Walz, 2003). If this is true, the influence of laminin β chain C-terminal region on LG domain integrity may indeed be the molecular mechanism underlying the different modes of integrin recognition by laminin β chains.

The differences in integrin binding between β1- and β2-laminins raise questions as to the physiological relevance of these features. The β2 chain is predominantly expressed in the basement membranes of kidney glomeruli and neuromuscular junctions (Miner et al., 1997; Patton, Miner, Chiu, & Sanes, 1997; Sasaki, Mann, Miner, Miosge, & Timpl, 2002). Moreover, laminin β2-knockout mice have aberrant neuromuscular junctions and dysfunctional glomerular basement membranes, even though upregulation of the β1 chain is observed (Noakes, Gautam, Mudd, Sanes, & Merlie, 1995; Noakes, Miner, et al., 1995; Patton et al., 1997). These observations indicate that the functions of β2-laminins cannot be fully compensated for by β1-laminins. Consistent with the expression patterns of β2-laminins, the X2-type integrin α3β1 is expressed by kidney glomerular podocytes adherent to glomerular basement membranes (Kikkawa, Virtanen, & Miner, 2003; Sterk et al., 1998). Additionally, α7β1 is the major integrin found in neuromuscular junctions (Martin, Kaufman, Kramer, & Sanes, 1996). While it remains unclear which type of α7 subunit—α7X1 or α7X2—is expressed in the junctions, α7X2 is the more likely of the two as it is the major integrin α subunit expressed in adult skeletal muscle (Ziober et al., 1993). Therefore, high-affinity interactions between X2-type integrins and β2-laminins appear to contribute to the maintenance of structural and/or functional integrity in both kidney glomeruli and neuromuscular junctions.

Taken together, the above findings detail how the C-terminal regions of the α, β, and γ chains of laminins act cooperatively to provide binding sites for integrins. However, precisely how the LG1–3 domains of α chains, C-terminal short stretches of β chains, and nearly C-terminal Glu residue of γ chains are recognized by integrins remains unresolved. Future studies should determine three-dimensional structures of laminin–integrin complexes to better understand the mechanistic basis of interactions between laminins and integrins.

5. MOLECULAR MECHANISMS OF LAMININ ACTION THROUGH INTEGRINS

5.1 Signaling by integrins

The binding of integrins to ECM proteins stimulates clustering of integrins and the formation of integrin-based multiprotein complexes called focal adhesions (Hynes, 2002). Focal adhesions are platforms that elicit both mechanical and chemical signals, which regulate cytoskeletal rearrangements and various cellular responses (Geiger & Yamada, 2011; Harburger & Calderwood, 2009). Mechanical signals are mediated by focal adhesion components including talin, kindlin, vinculin, and α-actinin, which functionally connect the actin cytoskeleton to the ECM. Chemical signals are mediated by other focal adhesion components such as focal adhesion kinase (FAK), Src, paxillin, and integrin-linked kinase, which regulate the activities of various enzymes including protein and lipid kinases and GTPases. Aberrations in these signaling events can precipitate various pathological conditions including cancer, immune disorders, thrombosis, and skeletal muscle dystrophy (Evans et al., 2009; Guo & Giancotti, 2004; Wickstrom, Radovanac, & Fassler, 2011).

5.2 Cell adhesion and migration on laminins

Laminins exert their functions in manners different from those of other ECM proteins. Laminins stimulate migration of lung adenocarcinoma cells, squamous carcinoma cells, and keratinocytes more efficiently than do other ECM proteins, including fibronectin and collagen (Frank & Carter, 2004; Gu, Sumida, Sanzen, & Sekiguchi, 2001; Winterwood, Varzavand, Meland, Ashman, & Stipp, 2006; Zhou & Kramer, 2005). Cells form thick actin stress fibers and prominent focal adhesions on fibronectin or collagen, but form thin stress fibers and fewer and smaller focal adhesions on laminins (Frank & Carter, 2004; Gu et al., 2001; Zhou & Kramer, 2005). Sensory and olfactory neurons also show preferential migration on laminins, compared with fibronectin and collagen (Calof & Lander, 1991). Moreover, as for carcinoma cells and keratinocytes, neuronal cells formed fewer and smaller focal contacts on laminins than fibronectin (Gomez, Roche, & Letourneau, 1996). These observations indicate that laminin-binding integrins mediate distinct cell adhesion behaviors characterized by smaller and fewer focal adhesions, and thinner and fewer actin stress fibers, thereby enhancing cell migration.

5.3 Signaling events on laminins

Several lines of evidence indicate that laminin-binding integrins transduce intracellular signals that are distinct from those of other integrins, thereby evoking distinct cell behaviors. Lung adenocarcinoma cell adhesion to fibronectin stimulates strong activation of RhoA small GTPase, but weak activation of Rac1 small GTPase (Gu et al., 2001). In contrast, adhesion to laminin-511/521 through integrin $\alpha3\beta1$ induces intense activation of Rac1, but weak RhoA activation. Additionally, FAK is significantly activated on fibronectin compared with laminin-511/521. Complex formation by p130CAS, CrkII, and DOCK180 is more potently induced by cell adhesion to laminins compared with fibronectin, resulting in prominent activation of Rac1 (Gu et al., 2001). Similarly, the adhesion of squamous carcinoma cells to laminin-332 through integrin $\alpha3\beta1$ leads to weak activation of RhoA and intensive activation of the Cdc42 small GTPase and its effector PAK1 compared with adhesion to collagen (Zhou & Kramer, 2005). Taken together, these observations indicate that laminins regulate the activities of Rho family small GTPases in manners different from those of other ECM proteins, thereby eliciting distinct cellular responses through the reorganization of the actin cytoskeleton.

The strong activation of Rac1 and Cdc42 by laminins augments formation of lamellipodia and filopodia, respectively. Meanwhile, activation of RhoA is less pronounced upon adhesion to laminins, resulting in thinner actin stress fibers and smaller focal adhesions than occur on fibronectin.

Reciprocal impacts on the Rho family small GTPases may account for enhanced cell migration on laminins. Laminin-511/521 have also been shown to promote survival of human lung adenocarcinoma cells more potently than fibronectin (Gu, Fujibayashi, Yamada, & Sekiguchi, 2002). Consistently, laminins are more potent activators of Akt signaling than fibronectin, but less potent activators of the mitogen-activated protein kinase (MAPK) pathway (Gu et al., 2002). Despite these advances in knowledge, the molecular mechanisms underlying the preferential activation of Rac1, Cdc42, and Akt by laminins remain undetermined. The identification of specific molecules associating directly with and/or working immediately downstream of laminin-binding integrins should resolve this question.

5.4 Actions and signals elicited by laminins in ESCs and iPSCs

In early mouse embryos, laminin-511 is predominantly localized in the basement membranes underlying epiblast cells (Miner, Li, Mudd, Go, & Sutherland, 2004). Consistently, mouse and human embryonic stem cells (ESCs) and induced pluripotent stem cells (iPSCs)—both of which have properties very similar to epiblast cells (Boroviak, Loos, Bertone, Smith, & Nichols, 2014)—are maintained on laminin-511 more robustly than on other cell adhesion proteins, such as fibronectin and vitronectin (Domogatskaya, Rodin, Boutaud, & Tryggvason, 2008; Miyazaki et al., 2008, 2012; Rodin et al., 2010). ESCs and iPSCs are highly anchorage-dependent and readily undergo apoptosis when dissociated into single cells unless they quickly secure anchorage onto substrates (Watanabe et al., 2007). Integrin $\alpha6\beta1$ is the major integrin type expressed by ESCs and iPSCs, and binds to laminin-511/521 with the highest affinity among all laminin isoforms with an apparent dissociation constant of $\sim$0.8 nM (Nishiuchi et al., 2006). This high-affinity interaction of integrin $\alpha6\beta1$ with laminin-511/521 underlies the robust proliferation of ESCs/iPSCs on laminin-511. Furthermore, laminin-511 activates Akt better than do fibronectin or vitronectin, though laminin-511 also stimulates phosphorylation of MAPKs (Miyazaki et al., 2012). It appears likely that the strong activation of these intracellular signaling pathways by laminin-511 promotes survival and robust growth of ESCs and iPSCs. Recently, recombinant laminin-511 and -521 and the laminin-511 E8 fragment have become commercially available as substrates for the culture of ESCs and iPSCs. These products will facilitate future clinical applications of pluripotent stem cells. Moreover, elucidation of the molecular mechanisms underlying the action of laminins on these stem cells will aid further development of regenerative medicine.

6. THE ROLES OF TETRASPANINS IN THE REGULATION OF LAMININ-BINDING INTEGRINS

6.1 Tetraspanin-enriched microdomains are associated with laminin-binding integrins

The laminin-binding integrins α3β1, α6β1, α6β4, and α7β1 form stable complexes with CD151, a tetraspanin family protein (Figure 4; Charrin et al., 2009; Hemler, 2005; Stipp, 2010; Yanez-Mo, Barreiro, Gordon-Alonso, Sala-Valdes, & Sanchez-Madrid, 2009). The interaction of integrins with CD151 enables an indirect association with other tetraspanins such as CD9 and CD81 through the interaction of CD151 with these tetraspanins (Takeda et al., 2007; Winterwood et al., 2006; Yamada et al., 2008; Yang et al., 2008). The association between the tetraspanins—but not that

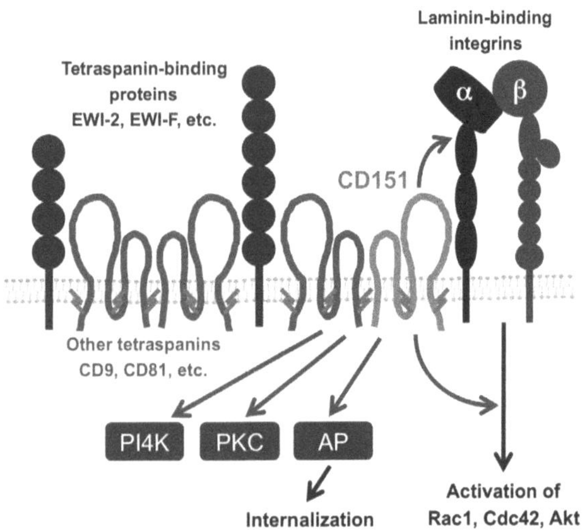

Figure 4 Tetraspanin-enriched microdomains are associated with laminin-binding integrins. Laminin-binding integrins are associated with the tetraspanin CD151 through their α subunits. CD151 also interacts with other tetraspanins such as CD9 and CD81, which in turn bind to several transmembrane proteins including EWI-2, EWI-F, and CD19 to form tetraspanin-enriched microdomains. Interactions between tetraspanins require palmitoyl groups (indicated by red (gray in print versions) polygonal lines) linked to Cys residues of the intracellular regions of tetraspanins. Laminin-binding integrins are incorporated into these microdomains via association with CD151. Tetraspanins interact with several intracellular proteins including PI4K, classical protein kinase C (PKC), and adaptor protein (AP) complexes. Laminin-binding integrins preferentially activate Rac1, Cdc42, and Akt, thereby promoting cell migration and survival. CD151 regulates ligand-binding activity and intracellular signaling by laminin-binding integrins.

between laminin-binding integrins and CD151—is dependent on the palmitoylation of tetraspanin cytoplasmic domains (Berditchevski, Odintsova, Sawada, & Gilbert, 2002; Charrin et al., 2002; Yang et al., 2002; Yang et al., 2004). Furthermore, CD9 and CD81 also interact with other membrane proteins including Ig superfamily proteins, EWI-2, EWI-F, and CD19 (Charrin et al., 2001; Horvath et al., 1998; Stipp, Kolesnikova, & Hemler, 2001). Accordingly, laminin-binding integrins are incorporated into multimolecular membrane microdomains, commonly referred to as tetraspanin-enriched microdomains, via their binding to CD151 (Charrin et al., 2009; Hemler, 2005).

6.2 Role of CD151 in the functions of laminin-binding integrins

The physiological importance of CD151 has been highlighted by several in vivo studies. A nonsense mutation in CD151 causes end-stage hereditary nephropathy, pretibial epidermolysis bullosa, and sensorineural deafness in humans (Karamatic Crew et al., 2004). Furthermore, CD151-null mice exhibit defective platelet aggregation, keratinocyte migration, and T cell proliferation, as well as decreased pathological angiogenesis, although these mice are viable and fertile (Cowin et al., 2006; Lau et al., 2004; Takeda et al., 2007; Wright et al., 2004). Sachs et al. (2012, 2006) reported that CD151-deficient mice show severe renal defects caused by abnormalities of the glomerular basement membrane, loss of podocyte foot processes, glomerulosclerosis, and cystic tubular dilation, though the initiation and progression of these defects are dependent on blood pressure. Conditional knock out of integrin α3 in podocytes results in renal defects similar to those in CD151 knockout mice (Sachs et al., 2006), suggesting a close functional association between CD151 and integrin α3β1.

Studies performed in vitro using cultured cells support a role for CD151 in integrin α3β1-mediated events, including cell adhesion, migration, cellular cable formation, and cell–cell adhesion (Chattopadhyay, Wang, Ashman, Brady-Kalnay, & Kreidberg, 2003; Hasegawa et al., 2007; Kazarov, Yang, Stipp, Sehgal, & Hemler, 2002; Nishiuchi et al., 2005; Shigeta et al., 2003; Winterwood et al., 2006; Yamada et al., 2008). RNAi-based knockdown studies have revealed that CD151 positively regulates the ligand-binding strength of integrin α3β1, thereby influencing the morphology of cells on laminins (Yamada et al., 2008). Indeed, CD151 potentiates integrin α3β1 activity by stabilizing its active conformation, and promotes cell adhesion to laminins (Nishiuchi et al., 2005). Lammerding, Kazarov, Huang, Lee, and Hemler (2003) demonstrated that CD151 augments integrin α6β1-mediated adhesion strength through enhancement of cytoskeletal linkages and/or integrin clustering, based upon measurement of adhesion forces of NIH3T3 cells expressing a cytoplasmic C-terminal mutant of CD151 to laminin-coated magnetic beads. CD151 may, therefore, stimulate both the ligand-binding activity of integrins and the clustering and cytoskeletal engagement of integrins, thereby promoting integrin-dependent cell adhesion.

6.3 Regulation of laminin-evoked signaling by tetraspanins

RNAi-silencing and gene-targeting experiments in mice have demonstrated a role for CD151 in the regulation of signals transduced by laminin-binding integrins. CD151 promotes activation of FAK, Src, Akt, MAPK, Rac1, and Cdc42 but suppresses activation of nitric oxide synthase 3 (Sadej et al., 2009; Takeda et al., 2007; Yamada et al., 2008; Yang et al., 2008). However, precisely how CD151 regulates integrin-dependent intracellular signaling events remains unclear. There are some reports that classical protein kinase C (PKC) and phosphatidylinositol 4-kinases are associated with tetraspanins (Figure 4; Berditchevski, Tolias, Wong, Carpenter, & Hemler, 1997; Yauch, Berditchevski, Harler, Reichner, & Hemler, 1998; Yauch & Hemler, 2000; Zhang, Bontrager, & Hemler, 2001). Zhang, Bontrager, and Hemler, (2001) reported that tetraspanins—including CD151, CD9, and CD81—associate with phorbol ester-activated PKC. Phosphorylation of integrin α3 is induced indirectly by PKC through an unknown protein kinase(s) in a manner dependent on the stimulation of phorbol ester (Zhang, Bontrager, Stipp, et al., 2001). Moreover, Chinese hamster ovary cells expressing an integrin α3 harboring a mutation at the phosphorylation site exhibit decreased levels of tyrosine-phosphorylated FAK, paxillin, and p130Cas upon integrin α3β1 engagement. These observations indicate that tetraspanins promote tyrosine phosphorylation-based signaling through the recruitment of activated PKC to integrin α3, and the subsequent induction of PKC-dependent phosphorylation of integrin α3.

6.4 Regulation of integrin trafficking by tetraspanins

Accumulating evidence suggests that tetraspanins regulate the trafficking of their binding proteins. For example, plating of CD151-depleted epidermoid carcinoma cells on laminin-332 led to impaired internalization of integrin α3 compared with control cells (Winterwood et al., 2006). Several tetraspanins including CD151, CD82, and CD63 possess potential tyrosine-based sorting motifs (YXXϕ; Y and ϕ represent tyrosine and bulky hydrophobic amino acid residues, respectively) in their C-termini (Berditchevski & Odintsova, 2007; Stipp, Kolesnikova, & Hemler, 2003). Through these motifs, tetraspanins appear to associate with adaptor protein (AP) complexes, which link tetraspanins to clathrins, thereby facilitating endocytosis via clathrin-coated vesicles (Figure 4). The YXXϕ motif in CD63 is recognized by the AP-3 complex, and is involved in the determination of its subcellular localization (Rous et al., 2002). Mutation of the YXXϕ motif in CD151 abrogates CD151 internalization from the cell surface, suggesting that CD151 regulates integrin trafficking through this motif, which may also interact with AP complexes similar to CD63 (Liu et al., 2007).

The TspanC8 subgroup of tetraspanins participates in the maturation and cell surface expression of ADAM10 through the regulation of its exit from the endoplasmic reticulum (Dornier et al., 2012). However, this is not the case for CD151, as studies using CD151 knockout mice and CD151-depleted cells have shown that CD151 is not required for cell surface expression of integrins (Winterwood et al., 2006; Wright et al., 2004; Yamada et al., 2008).

6.5 Role of tetraspanins in rear-end retraction of migrating cells

High concentrations of tetraspanins are present within substrate-attached material (SAM) (Yamada et al., 2013), which are the cellular feet remaining on substrates following detachment of cells with EGTA (Terry & Culp, 1974). SAM is similar in nature to the retraction fibers and footprints that appear during rear-end retraction in migrating cells, as SAM formation is suppressed following inhibition of ROCK, myosin II, or dynamin, which are involved with retraction of the trailing edges of migrating cells (Yamada et al., 2013). Tetraspanins such as CD9 and CD81 are present at high levels in the footprints of keratinocytes and in both the footprints and retraction fibers of human prostate cancer Du145 cells (Penas, Garcia-Diez, Sanchez-Madrid, & Yanez-Mo, 2000; Zhang & Huang, 2012). These observations suggest that tetraspanins regulate trailing edge adhesion. Moreover, cells adhering to laminin-511 form a larger number of finer and longer SAMs compared with those grown on fibronectin (Figure 5). Consistently, formation of retraction fibers and front-rear polarity is more prominent in migrating cells when grown on laminin-511 compared with those on fibronectin (Figure 5). Preliminary results from our laboratory indicate that SAM formation on laminin-511 is delayed by CD151 knockdown, or inhibition of palmitoylation by 2-bromopalmitate (2-BP), which can disrupt interactions

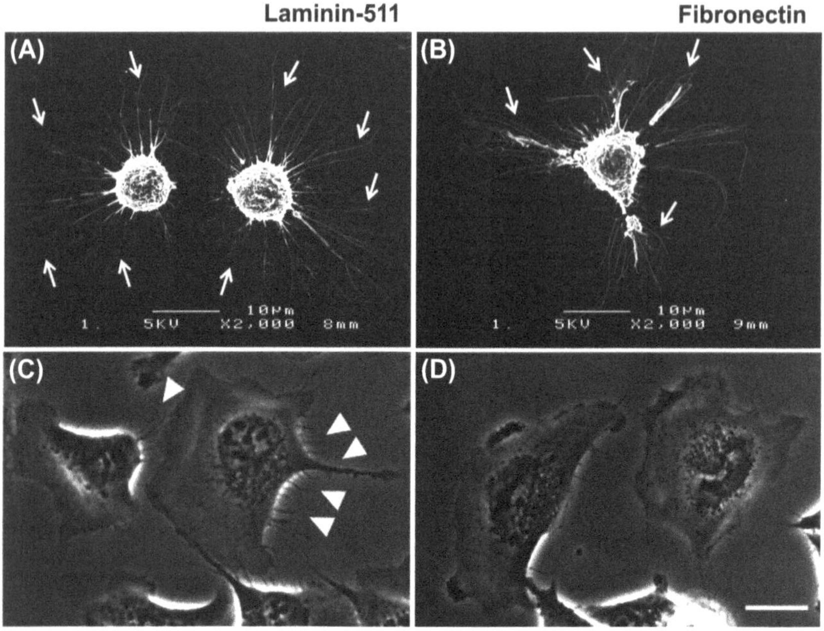

Figure 5 Morphology of substrate-attached material (SAM) and migrating cells on laminin and fibronectin. (A and B) Human lung adenocarcinoma A549 cells adherent to dishes coated with laminin-511 (A) or fibronectin (B) were treated with EGTA, and the resulting SAM was subjected to scanning electron microscopy. Arrows indicate SAM. SAM deposits on laminin-511 were fine, long, and large in number compared with those on fibronectin. (C and D) A549 cells on laminin-511 (C), but not fibronectin (D), show clear front-rear polarity and retraction fibers (arrowheads). Bar represents 20 μm.

among tetraspanins (Kovalenko, Yang, Kolesnikova, & Hemler, 2004). Accordingly, aberrant trailing edge detachment following knockdown of CD151 or treatment with 2-BP was observed on laminin-511. Furthermore, an anti-CD151 mAb prevents rear-end detachment during the migration of human epidermoid carcinoma cells, possibly through induction of CD151 dysfunction during rear-end retraction (Zijlstra, Lewis, Degryse, Stuhlmann, & Quigley, 2008). Collectively, these results reveal a role for tetraspanins in the regulation of laminin-binding integrins, which influences trailing edge adhesion in migrating cells grown on laminins. As CD151 is involved in laminin-binding integrin internalization (Liu et al., 2007; Winterwood et al., 2006), tetraspanins may regulate the turnover of laminin-binding integrins in trailing edges, thereby facilitating disassembly of the integrin-based adhesion complex and rear-end retraction in migrating cells. Further study of tetraspanins and tetraspanin-enriched microdomains associated with laminin-binding integrins should elucidate the molecular mechanisms underlying various cellular functions on laminins.

7. CONCLUSIONS

Almost all types of parenchymal cells adhere to basement membranes to facilitate survival, growth regulation, differentiation, and polarity. Adhesive interactions between cells and basement membranes are primarily mediated by laminins and laminin-binding integrins on the cell surface. These interactions are diverse in specificity and affinity, and depend upon multiple combinations of laminin and integrin isoforms. Expression of these isoforms is intricately regulated in embryonic development and fine-tuned for individual cell types. Knowledge of where and when each laminin–integrin interaction operates—and their characteristic functions—is therefore important for a more complete understanding of cell adhesion in development and maintenance of tissue and organ homeostasis, and should contribute to a better understanding of fundamental mechanisms underlying physiological and pathological processes in multicellular organisms.

ACKNOWLEDGMENTS

We thank Drs Ryoko Sato-Nishiuchi and Yukimasa Taniguchi for valuable comments. Our research was supported by Grants-in-Aid for Scientific Research from the Japan Society for Promotion of Science (to M.Y. and K.S.) and for Scientific Research on Priority Areas from the Ministry of Education, Culture, Sports, Science and Technology of Japan (to K.S.).

REFERENCES

Altroff, H., Choulier, L., & Mardon, H. J. (2003). Synergistic activity of the ninth and tenth FIII domains of human fibronectin depends upon structural stability. *Journal of Biological Chemistry, 278*(1), 491–497.

Aumailley, M. (2013). The laminin family. *Cell Adhesion and Migration, 7*(1), 48–55.

Aumailley, M., Bruckner-Tuderman, L., Carter, W. G., Deutzmann, R., Edgar, D., Ekblom, P., et al. (2005). A simplified laminin nomenclature. *Matrix Biology, 24*(5), 326–332.

Barak, T., Kwan, K. Y., Louvi, A., Demirbilek, V., Saygi, S., Tuysuz, B., et al. (2011). Recessive LAMC3 mutations cause malformations of occipital cortical development. *Nature Genetics, 43*(6), 590–594.

Belkin, A. M., & Stepp, M. A. (2000). Integrins as receptors for laminins. *Microscopy Research and Technique, 51*(3), 280–301.

Berditchevski, F., & Odintsova, E. (2007). Tetraspanins as regulators of protein trafficking. *Traffic, 8*(2), 89–96.

Berditchevski, F., Odintsova, E., Sawada, S., & Gilbert, E. (2002). Expression of the palmitoylation-deficient CD151 weakens the association of $\alpha 3\beta 1$ integrin with the tetraspanin-enriched microdomains and affects integrin-dependent signaling. *Journal of Biological Chemistry, 277*(40), 36991–37000.

Berditchevski, F., Tolias, K. F., Wong, K., Carpenter, C. L., & Hemler, M. E. (1997). A novel link between integrins, transmembrane-4 superfamily proteins (CD63 and CD81), and phosphatidylinositol 4-kinase. *Journal of Biological Chemistry, 272*(5), 2595–2598.

Boroviak, T., Loos, R., Bertone, P., Smith, A., & Nichols, J. (2014). The ability of inner-cell-mass cells to self-renew as embryonic stem cells is acquired following epiblast specification. *Nature Cell Biology, 16*(6), 516–528.

Borradori, L., & Sonnenberg, A. (1999). Structure and function of hemidesmosomes: more than simple adhesion complexes. *Journal of Investigative Dermatology, 112*(4), 411–418.

Burkin, D. J., & Kaufman, S. J. (1999). The α7β1 integrin in muscle development and disease. *Cell and Tissue Research, 296*(1), 183–190.

Calderwood, D. A., Tuckwell, D. S., Eble, J., Kuhn, K., & Humphries, M. J. (1997). The integrin α1 A-domain is a ligand binding site for collagens and laminin. *Journal of Biological Chemistry, 272*(19), 12311–12317.

Calof, A. L., & Lander, A. D. (1991). Relationship between neuronal migration and cell-substratum adhesion: laminin and merosin promote olfactory neuronal migration but are anti-adhesive. *Journal of Cell Biology, 115*(3), 779–794.

Campbell, I. G., Foulkes, W. D., Senger, G., Trowsdale, J., Garin-Chesa, P., & Rettig, W. J. (1994). Molecular cloning of the B-CAM cell surface glycoprotein of epithelial cancers: a novel member of the immunoglobulin superfamily. *Cancer Research, 54*(22), 5761–5765.

Campbell, I. D., & Humphries, M. J. (2011). Integrin structure, activation, and interactions. *Cold Spring Harbor Perspectives in Biology, 3*(3).

Charrin, S., Le Naour, F., Oualid, M., Billard, M., Faure, G., Hanash, S. M., et al. (2001). The major CD9 and CD81 molecular partner. Identification and characterization of the complexes. *Journal of Biological Chemistry, 276*(17), 14329–14337.

Charrin, S., Le Naour, F., Silvie, O., Milhiet, P. E., Boucheix, C., & Rubinstein, E. (2009). Lateral organization of membrane proteins: tetraspanins spin their web. *Biochemical Journal, 420*(2), 133–154.

Charrin, S., Manie, S., Oualid, M., Billard, M., Boucheix, C., & Rubinstein, E. (2002). Differential stability of tetraspanin/tetraspanin interactions: role of palmitoylation. *FEBS Letters, 516*(1–3), 139–144.

Chattopadhyay, N., Wang, Z., Ashman, L. K., Brady-Kalnay, S. M., & Kreidberg, J. A. (2003). α3β1 integrin-CD151, a component of the cadherin-catenin complex, regulates PTPmu expression and cell-cell adhesion. *Journal of Cell Biology, 163*(6), 1351–1362.

Chen, J., Diacovo, T. G., Grenache, D. G., Santoro, S. A., & Zutter, M. M. (2002). The α2 integrin subunit-deficient mouse: a multifaceted phenotype including defects of branching morphogenesis and hemostasis. *American Journal of Pathology, 161*(1), 337–344.

Colognato, H., MacCarrick, M., O'Rear, J. J., & Yurchenco, P. D. (1997). The laminin α2-chain short arm mediates cell adhesion through both the α1β1 and α2β1 integrins. *Journal of Biological Chemistry, 272*(46), 29330–29336.

Colognato, H., & Yurchenco, P. D. (2000). Form and function: the laminin family of heterotrimers. *Developmental Dynamics, 218*(2), 213–234.

Cowin, A. J., Adams, D., Geary, S. M., Wright, M. D., Jones, J. C., & Ashman, L. K. (2006). Wound healing is defective in mice lacking tetraspanin CD151. *Journal of Investigative Dermatology, 126*(3), 680–689.

Cserhalmi-Friedman, P. B., Olson, P. F., Koch, M., Champliaud, M. F., Brunken, W. J., Burgeson, R. E., et al. (2001). Structural analysis and mutation detection strategy for the human LAMC3 gene. *Biochemical and Biophysical Research, 280*(1), 39–44.

Daniel, L., Bardin, N., Moal, V., Dignat-George, F., Berland, Y., & Figarella-Branger, D. (2005). Tubular CD146 expression in nephropathies is related to chronic renal failure. *Nephron Experimental Nephrology, 99*(4), e105–111.

Denes, V., Witkovsky, P., Koch, M., Hunter, D. D., Pinzon-Duarte, G., & Brunken, W. J. (2007). Laminin deficits induce alterations in the development of dopaminergic neurons in the mouse retina. *Visual Neuroscience, 24*(4), 549–562.
Desban, N., & Duband, J. L. (1997). Avian neural crest cell migration on laminin: interaction of the α1β1 integrin with distinct laminin-1 domains mediates different adhesive responses. *Journal of Cell Science, 110*(Pt 21), 2729–2744.
Deutzmann, R., Aumailley, M., Wiedemann, H., Pysny, W., Timpl, R., & Edgar, D. (1990). Cell adhesion, spreading and neurite stimulation by laminin fragment E8 depends on maintenance of secondary and tertiary structure in its rod and globular domain. *European Journal of Biochemistry, 191*(2), 513–522.
Domogatskaya, A., Rodin, S., Boutaud, A., & Tryggvason, K. (2008). Laminin-511 but not -332, -111, or -411 enables mouse embryonic stem cell self-renewal in vitro. *Stem Cells, 26*(11), 2800–2809.
Domogatskaya, A., Rodin, S., & Tryggvason, K. (2012). Functional diversity of laminins. *Annual Review of Cell and Developmental Biology, 28*, 523–553.
Dornier, E., Coumailleau, F., Ottavi, J. F., Moretti, J., Boucheix, C., Mauduit, P., et al. (2012). TspanC8 tetraspanins regulate ADAM10/Kuzbanian trafficking and promote notch activation in flies and mammals. *Journal of Cell Biology, 199*(3), 481–496.
Dowling, J., Yu, Q. C., & Fuchs, E. (1996). β4 integrin is required for hemidesmosome formation, cell adhesion and cell survival. *Journal of Cell Biology, 134*(2), 559–572.
Durbeej, M. (2010). Laminins. *Cell and Tissue Research, 339*(1), 259–268.
Eble, J. A., Wucherpfennig, K. W., Gauthier, L., Dersch, P., Krukonis, E., Isberg, R. R., et al. (1998). Recombinant soluble human α3β1 integrin: purification, processing, regulation, and specific binding to laminin-5 and invasin in a mutually exclusive manner. *Biochemistry, 37*(31), 10945–10955.
Egles, C., Claudepierre, T., Manglapus, M. K., Champliaud, M. F., Brunken, W. J., & Hunter, D. D. (2007). Laminins containing the β2 chain modulate the precise organization of CNS synapses. *Molecular and Cellular Neuroscience, 34*(3), 288–298.
El Nemer, W., Gane, P., Colin, Y., Bony, V., Rahuel, C., Galacteros, F., et al. (1998). The Lutheran blood group glycoproteins, the erythroid receptors for laminin, are adhesion molecules. *Journal of Biological Chemistry, 273*(27), 16686–16693.
Emsley, J., Knight, C. G., Farndale, R. W., Barnes, M. J., & Liddington, R. C. (2000). Structural basis of collagen recognition by integrin α2β1. *Cell, 101*(1), 47–56.
Evans, R., Patzak, I., Svensson, L., De Filippo, K., Jones, K., McDowall, A., et al. (2009). Integrins in immunity. *Journal of Cell Science, 122*(Pt 2), 215–225.
Flanagan, K., Fitzgerald, K., Baker, J., Regnstrom, K., Gardai, S., Bard, F., et al. (2012). Laminin-411 is a vascular ligand for MCAM and facilitates TH17 cell entry into the CNS. *PLoS One, 7*(7), e40443.
Frank, D. E., & Carter, W. G. (2004). Laminin 5 deposition regulates keratinocyte polarization and persistent migration. *Journal of Cell Science, 117*(Pt 8), 1351–1363.
Fukumoto, S., Miner, J. H., Ida, H., Fukumoto, E., Yuasa, K., Miyazaki, H., et al. (2006). Laminin α5 is required for dental epithelium growth and polarity and the development of tooth bud and shape. *Journal of Biological Chemistry, 281*(8), 5008–5016.
Gardner, H., Kreidberg, J., Koteliansky, V., & Jaenisch, R. (1996). Deletion of integrin α1 by homologous recombination permits normal murine development but gives rise to a specific deficit in cell adhesion. *Developmental Biology, 175*(2), 301–313.
Geiger, B., & Yamada, K. M. (2011). Molecular architecture and function of matrix adhesions. *Cold Spring Harbor Perspectives in Biology, 3*(5).
Genersch, E., Ferletta, M., Virtanen, I., Haller, H., & Ekblom, P. (2003). Integrin αvβ3 binding to human α5-laminins facilitates FGF-2- and VEGF-induced proliferation of human ECV304 carcinoma cells. *European Journal of Cell Biology, 82*(3), 105–117.

Georges-Labouesse, E., Messaddeq, N., Yehia, G., Cadalbert, L., Dierich, A., & Le Meur, M. (1996). Absence of integrin α6 leads to epidermolysis bullosa and neonatal death in mice. *Nature Genetics, 13*(3), 370–373.

Gomez, T. M., Roche, F. K., & Letourneau, P. C. (1996). Chick sensory neuronal growth cones distinguish fibronectin from laminin by making substratum contacts that resemble focal contacts. *Journal of Neurobiology, 29*(1), 18–34.

Gu, J., Fujibayashi, A., Yamada, K. M., & Sekiguchi, K. (2002). Laminin-10/11 and fibronectin differentially prevent apoptosis induced by serum removal via phosphatidylinositol 3-kinase/Akt- and MEK1/ERK-dependent pathways. *Journal of Biological Chemistry, 277*(22), 19922–19928.

Gu, J., Sumida, Y., Sanzen, N., & Sekiguchi, K. (2001). Laminin-10/11 and fibronectin differentially regulate integrin-dependent Rho and Rac activation via p130Cas-CrkII-DOCK180 pathway. *Journal of Biological Chemistry, 276*(29), 27090–27097.

Guo, W., & Giancotti, F. G. (2004). Integrin signalling during tumour progression. *Nature Reviews Molecular Cell Biology, 5*(10), 816–826.

Harburger, D. S., & Calderwood, D. A. (2009). Integrin signalling at a glance. *Journal of Cell Science, 122*(Pt 2), 159–163.

Hasegawa, M., Furuya, M., Kasuya, Y., Nishiyama, M., Sugiura, T., Nikaido, T., et al. (2007). CD151 dynamics in carcinoma-stroma interaction: integrin expression, adhesion strength and proteolytic activity. *Laboratory Investigation, 87*(9), 882–892.

Hebbel, R. P. (1997). Perspectives series: cell adhesion in vascular biology. Adhesive interactions of sickle erythrocytes with endothelium. *Journal of Clinical Investigation, 99*(11), 2561–2564.

Hemler, M. E. (2005). Tetraspanin functions and associated microdomains. *Nature Reviews Molecular Cell Biology, 6*(10), 801–811.

Hirako, Y., & Owaribe, K. (1998). Hemidesmosomes and their unique transmembrane protein BP180. *Microscopy Research and Technique, 43*(3), 207–217.

Hirosaki, T., Mizushima, H., Tsubota, Y., Moriyama, K., & Miyazaki, K. (2000). Structural requirement of carboxyl-terminal globular domains of laminin α3 chain for promotion of rapid cell adhesion and migration by laminin-5. *Journal of Biological Chemistry, 275*(29), 22495–22502.

Horvath, G., Serru, V., Clay, D., Billard, M., Boucheix, C., & Rubinstein, E. (1998). CD19 is linked to the integrin-associated tetraspans CD9, CD81, and CD82. *Journal of Biological Chemistry, 273*(46), 30537–30543.

Hynes, R. O. (2002). Integrins: bidirectional, allosteric signaling machines. *Cell, 110*(6), 673–687.

Ido, H., Harada, K., Futaki, S., Hayashi, Y., Nishiuchi, R., Natsuka, Y., et al. (2004). Molecular dissection of the α-dystroglycan- and integrin-binding sites within the globular domain of human laminin-10. *Journal of Biological Chemistry, 279*(12), 10946–10954.

Ido, H., Harada, K., Yagi, Y., & Sekiguchi, K. (2006). Probing the integrin-binding site within the globular domain of laminin-511 with the function-blocking monoclonal antibody 4C7. *Matrix Biology, 25*(2), 112–117.

Ido, H., Ito, S., Taniguchi, Y., Hayashi, M., Sato-Nishiuchi, R., Sanzen, N., et al. (2008). Laminin isoforms containing the γ3 chain are unable to bind to integrins due to the absence of the glutamic acid residue conserved in the C-terminal regions of the γ1 and γ2 chains. *Journal of Biological Chemistry, 283*(42), 28149–28157.

Ido, H., Nakamura, A., Kobayashi, R., Ito, S., Li, S., Futaki, S., et al. (2007). The requirement of the glutamic acid residue at the third position from the carboxyl termini of the laminin γ chains in integrin binding by laminins. *Journal of Biological Chemistry, 282*(15), 11144–11154.

Iivanainen, A., Morita, T., & Tryggvason, K. (1999). Molecular cloning and tissue-specific expression of a novel murine laminin γ3 chain. *Journal of Biological Chemistry, 274*(20), 14107–14111.

Ishikawa, T., Wondimu, Z., Oikawa, Y., Gentilcore, G., Kiessling, R., Egyhazi Brage, S., et al. (2014). Laminins 411 and 421 differentially promote tumor cell migration via α6β1 integrin and MCAM (CD146). *Matrix Biology, 38*, 69–83.

Karamatic Crew, V., Burton, N., Kagan, A., Green, C. A., Levene, C., Flinter, F., et al. (2004). CD151, the first member of the tetraspanin (TM4) superfamily detected on erythrocytes, is essential for the correct assembly of human basement membranes in kidney and skin. *Blood, 104*(8), 2217–2223.

Kazarov, A. R., Yang, X., Stipp, C. S., Sehgal, B., & Hemler, M. E. (2002). An extracellular site on tetraspanin CD151 determines α3 and α6 integrin-dependent cellular morphology. *Journal of Cell Biology, 158*(7), 1299–1309.

Kikkawa, Y., Hozumi, K., Katagiri, F., Nomizu, M., Kleinman, H. K., & Koblinski, J. E. (2013). Laminin-111-derived peptides and cancer. *Cell Adhesion and Migration,* 7(1), 150–256.

Kikkawa, Y., Moulson, C. L., Virtanen, I., & Miner, J. H. (2002). Identification of the binding site for the Lutheran blood group glycoprotein on laminin α5 through expression of chimeric laminin chains in vivo. *Journal of Biological Chemistry, 277*(47), 44864–44869.

Kikkawa, Y., Ogawa, T., Sudo, R., Yamada, Y., Katagiri, F., Hozumi, K., et al. (2013). The lutheran/basal cell adhesion molecule promotes tumor cell migration by modulating integrin-mediated cell attachment to laminin-511 protein. *Journal of Biological Chemistry, 288*(43), 30990–31001.

Kikkawa, Y., Sasaki, T., Nguyen, M. T., Nomizu, M., Mitaka, T., & Miner, J. H. (2007). The LG1-3 tandem of laminin α5 harbors the binding sites of Lutheran/basal cell adhesion molecule and α3β1/α6β1 integrins. *Journal of Biological Chemistry, 282*(20), 14853–14860.

Kikkawa, Y., Virtanen, I., & Miner, J. H. (2003). Mesangial cells organize the glomerular capillaries by adhering to the G domain of laminin α5 in the glomerular basement membrane. *Journal of Cell Biology, 161*(1), 187–196.

Knight, C. G., Morton, L. F., Peachey, A. R., Tuckwell, D. S., Farndale, R. W., & Barnes, M. J. (2000). The collagen-binding A-domains of integrins α1β1 and α2β1 recognize the same specific amino acid sequence, GFOGER, in native (triple-helical) collagens. *Journal of Biological Chemistry, 275*(1), 35–40.

Koch, M., Olson, P. F., Albus, A., Jin, W., Hunter, D. D., Brunken, W. J., et al. (1999). Characterization and expression of the laminin γ3 chain: a novel, non-basement membrane-associated, laminin chain. *Journal of Cell Biology, 145*(3), 605–618.

Kovalenko, O. V., Yang, X., Kolesnikova, T. V., & Hemler, M. E. (2004). Evidence for specific tetraspanin homodimers: inhibition of palmitoylation makes cysteine residues available for cross-linking. *Biochemical Journal, 377*(Pt 2), 407–417.

Kreidberg, J. A., Donovan, M. J., Goldstein, S. L., Rennke, H., Shepherd, K., Jones, R. C., et al. (1996). α3β1 integrin has a crucial role in kidney and lung organogenesis. *Development, 122*(11), 3537–3547.

Kuang, W., Xu, H., Vachon, P. H., & Engvall, E. (1998). Disruption of the lama2 gene in embryonic stem cells: laminin α2 is necessary for sustenance of mature muscle cells. *Experimental Cell Research, 241*(1), 117–125.

Kunneken, K., Pohlentz, G., Schmidt-Hederich, A., Odenthal, U., Smyth, N., Peter-Katalinic, J., et al. (2004). Recombinant human laminin-5 domains. Effects of heterotrimerization, proteolytic processing, and N-glycosylation on α3β1 integrin binding. *Journal of Biological Chemistry, 279*(7), 5184–5193.

Kuster, J. E., Guarnieri, M. H., Ault, J. G., Flaherty, L., & Swiatek, P. J. (1997). IAP insertion in the murine LamB3 gene results in junctional epidermolysis bullosa. *Mammalian Genome, 8*(9), 673–681.

Lammerding, J., Kazarov, A. R., Huang, H., Lee, R. T., & Hemler, M. E. (2003). Tetraspanin CD151 regulates α6β1 integrin adhesion strengthening. *Proceedings of the National Academy of Sciences of the United States of America, 100*(13), 7616–7621.

Lau, L. M., Wee, J. L., Wright, M. D., Moseley, G. W., Hogarth, P. M., Ashman, L. K., et al. (2004). The tetraspanin superfamily member CD151 regulates outside-in integrin αIIbβ3 signaling and platelet function. *Blood, 104*(8), 2368–2375.

Lehmann, J. M., Holzmann, B., Breitbart, E. W., Schmiegelow, P., Riethmuller, G., & Johnson, J. P. (1987). Discrimination between benign and malignant cells of melanocytic lineage by two novel antigens, a glycoprotein with a molecular weight of 113,000 and a protein with a molecular weight of 76,000. *Cancer Research, 47*(3), 841–845.

Lehmann, J. M., Riethmuller, G., & Johnson, J. P. (1989). MUC18, a marker of tumor progression in human melanoma, shows sequence similarity to the neural cell adhesion molecules of the immunoglobulin superfamily. *Proceedings of the National Academy of Sciences of the United States of America, 86*(24), 9891–9895.

Li, S., Liquari, P., McKee, K. K., Harrison, D., Patel, R., Lee, S., et al. (2005). Laminin-sulfatide binding initiates basement membrane assembly and enables receptor signaling in Schwann cells and fibroblasts. *Journal of Cell Biology, 169*(1), 179–189.

Li, Y. N., Radner, S., French, M. M., Pinzon-Duarte, G., Daly, G. H., Burgeson, R. E., et al. (2012). The γ3 chain of laminin is widely but differentially expressed in murine basement membranes: expression and functional studies. *Matrix Biology, 31*(2), 120–134.

Li, J., Tzu, J., Chen, Y., Zhang, Y. P., Nguyen, N. T., Gao, J., et al. (2003). Laminin-10 is crucial for hair morphogenesis. *EMBO Journal, 22*(10), 2400–2410.

Libby, R. T., Champliaud, M. F., Claudepierre, T., Xu, Y., Gibbons, E. P., Koch, M., et al. (2000). Laminin expression in adult and developing retinae: evidence of two novel CNS laminins. *Journal of Neuroscience, 20*(17), 6517–6528.

Liu, L., He, B., Liu, W. M., Zhou, D., Cox, J. V., & Zhang, X. A. (2007). Tetraspanin CD151 promotes cell migration by regulating integrin trafficking. *Journal of Biological Chemistry, 282*(43), 31631–31642.

von der Mark, H., Poschl, E., Lanig, H., Sasaki, T., Deutzman, R., & von der Mark, K. (2007). Distinct acidic clusters and hydrophobic residues in the alternative splice domains X1 and X2 of α7 integrins define specificity for laminin isoforms. *Journal of Molecular Biology, 371*(5), 1188–1203.

von der Mark, H., Williams, I., Wendler, O., Sorokin, L., von der Mark, K., & Poschl, E. (2002). Alternative splice variants of α7β1 integrin selectively recognize different laminin isoforms. *Journal of Biological Chemistry, 277*(8), 6012–6016.

Martin, P. T., Kaufman, S. J., Kramer, R. H., & Sanes, J. R. (1996). Synaptic integrins in developing, adult, and mutant muscle: selective association of α1, α7A, and α7B integrins with the neuromuscular junction. *Developmental Biology, 174*(1), 125–139.

Mayer, U., Saher, G., Fassler, R., Bornemann, A., Echtermeyer, F., von der Mark, H., et al. (1997). Absence of integrin α7 causes a novel form of muscular dystrophy. *Nature Genetics, 17*(3), 318–323.

Meng, X., Klement, J. F., Leperi, D. A., Birk, D. E., Sasaki, T., Timpl, R., et al. (2003). Targeted inactivation of murine laminin γ2-chain gene recapitulates human junctional epidermolysis bullosa. *Journal of Investigative Dermatology, 121*(4), 720–731.

Miner, J. H., Cunningham, J., & Sanes, J. R. (1998). Roles for laminin in embryogenesis: exencephaly, syndactyly, and placentopathy in mice lacking the laminin α5 chain. *Journal of Cell Biology, 143*(6), 1713–1723.

Miner, J. H., & Li, C. (2000). Defective glomerulogenesis in the absence of laminin α5 demonstrates a developmental role for the kidney glomerular basement membrane. *Developmental Biology, 217*(2), 278–289.

Miner, J. H., Li, C., Mudd, J. L., Go, G., & Sutherland, A. E. (2004). Compositional and structural requirements for laminin and basement membranes during mouse embryo implantation and gastrulation. *Development, 131*(10), 2247–2256.

Miner, J. H., Patton, B. L., Lentz, S. I., Gilbert, D. J., Snider, W. D., Jenkins, N. A., et al. (1997). The laminin α chains: expression, developmental transitions, and chromosomal

locations of α1-5, identification of heterotrimeric laminins 8-11, and cloning of a novel α3 isoform. *Journal of Cell Biology, 137*(3), 685–701.
Miner, J. H., & Yurchenco, P. D. (2004). Laminin functions in tissue morphogenesis. *Annual Review of Cell and Developmental Biology, 20*, 255–284.
Miyagoe, Y., Hanaoka, K., Nonaka, I., Hayasaka, M., Nabeshima, Y., Arahata, K., et al. (1997). Laminin α2 chain-null mutant mice by targeted disruption of the Lama2 gene: a new model of merosin (laminin 2)-deficient congenital muscular dystrophy. *FEBS Letters, 415*(1), 33–39.
Miyazaki, T., Futaki, S., Hasegawa, K., Kawasaki, M., Sanzen, N., Hayashi, M., et al. (2008). Recombinant human laminin isoforms can support the undifferentiated growth of human embryonic stem cells. *Biochemical and Biophysical Research, 375*(1), 27–32.
Miyazaki, T., Futaki, S., Suemori, H., Taniguchi, Y., Yamada, M., Kawasaki, M., et al. (2012). Laminin E8 fragments support efficient adhesion and expansion of dissociated human pluripotent stem cells. *Nature Communications, 3*, 1236.
Mould, A. P., Symonds, E. J., Buckley, P. A., Grossmann, J. G., McEwan, P. A., Barton, S. J., et al. (2003). Structure of an integrin-ligand complex deduced from solution x-ray scattering and site-directed mutagenesis. *Journal of Biological Chemistry, 278*(41), 39993–39999.
Nagae, M., Re, S., Mihara, E., Nogi, T., Sugita, Y., & Takagi, J. (2012). Crystal structure of α5β1 integrin ectodomain: atomic details of the fibronectin receptor. *Journal of Cell Biology, 197*(1), 131–140.
Navdaev, A., Heitmann, V., Desantana Evangelista, K., Morgelin, M., Wegener, J., & Eble, J. A. (2008). The C-terminus of the γ2 chain but not of the β3 chain of laminin-332 is indirectly but indispensably necessary for integrin-mediated cell reactions. *Experimental Cell Research, 314*(3), 489–497.
van der Neut, R., Krimpenfort, P., Calafat, J., Niessen, C. M., & Sonnenberg, A. (1996). Epithelial detachment due to absence of hemidesmosomes in integrin β4 null mice. *Nature Genetics, 13*(3), 366–369.
Nguyen, N. M., Miner, J. H., Pierce, R. A., & Senior, R. M. (2002). Laminin α5 is required for lobar septation and visceral pleural basement membrane formation in the developing mouse lung. *Developmental Biology, 246*(2), 231–244.
Nishiuchi, R., Sanzen, N., Nada, S., Sumida, Y., Wada, Y., Okada, M., et al. (2005). Potentiation of the ligand-binding activity of integrin α3β1 via association with tetraspanin CD151. *Proceedings of the National Academy of Sciences of the United States of America, 102*(6), 1939–1944.
Nishiuchi, R., Takagi, J., Hayashi, M., Ido, H., Yagi, Y., Sanzen, N., et al. (2006). Ligand-binding specificities of laminin-binding integrins: a comprehensive survey of laminin-integrin interactions using recombinant α3β1, α6β1, α7β1 and α6β4 integrins. *Matrix Biology, 25*(3), 189–197.
Noakes, P. G., Gautam, M., Mudd, J., Sanes, J. R., & Merlie, J. P. (1995). Aberrant differentiation of neuromuscular junctions in mice lacking s-laminin/laminin β2. *Nature, 374*(6519), 258–262.
Noakes, P. G., Miner, J. H., Gautam, M., Cunningham, J. M., Sanes, J. R., & Merlie, J. P. (1995). The renal glomerulus of mice lacking s-laminin/laminin β2: nephrosis despite molecular compensation by laminin β1. *Nature Genetics, 10*(4), 400–406.
Nomizu, M., Kim, W. H., Yamamura, K., Utani, A., Song, S. Y., Otaka, A., et al. (1995). Identification of cell binding sites in the laminin α1 chain carboxyl-terminal globular domain by systematic screening of synthetic peptides. *Journal of Biological Chemistry, 270*(35), 20583–20590.
Ogawa, T., Tsubota, Y., Hashimoto, J., Kariya, Y., & Miyazaki, K. (2007). The short arm of laminin γ2 chain of laminin-5 (laminin-332) binds syndecan-1 and regulates cellular adhesion and migration by suppressing phosphorylation of integrin β4 chain. *Molecular Biology of the Cell, 18*(5), 1621–1633.

Okamoto, O., Bachy, S., Odenthal, U., Bernaud, J., Rigal, D., Lortat-Jacob, H., et al. (2003). Normal human keratinocytes bind to the α3LG4/5 domain of unprocessed laminin-5 through the receptor syndecan-1. *Journal of Biological Chemistry, 278*(45), 44168–44177.
Okazaki, I., Suzuki, N., Nishi, N., Utani, A., Matsuura, H., Shinkai, H., et al. (2002). Identification of biologically active sequences in the laminin α4 chain G domain. *Journal of Biological Chemistry, 277*(40), 37070–37078.
O'Roak, B. J., Deriziotis, P., Lee, C., Vives, L., Schwartz, J. J., Girirajan, S., et al. (2011). Exome sequencing in sporadic autism spectrum disorders identifies severe de novo mutations. *Nature Genetics, 43*(6), 585–589.
Parsons, S. F., Mallinson, G., Holmes, C. H., Houlihan, J. M., Simpson, K. L., Mawby, W. J., et al. (1995). The Lutheran blood group glycoprotein, another member of the immunoglobulin superfamily, is widely expressed in human tissues and is developmentally regulated in human liver. *Proceedings of the National Academy of Sciences of the United States of America, 92*(12), 5496–5500.
Patton, B. L., Miner, J. H., Chiu, A. Y., & Sanes, J. R. (1997). Distribution and function of laminins in the neuromuscular system of developing, adult, and mutant mice. *Journal of Cell Biology, 139*(6), 1507–1521.
Penas, P. F., Garcia-Diez, A., Sanchez-Madrid, F., & Yanez-Mo, M. (2000). Tetraspanins are localized at motility-related structures and involved in normal human keratinocyte wound healing migration. *Journal of Investigative Dermatology, 114*(6), 1126–1135.
Pierschbacher, M. D., & Ruoslahti, E. (1984). Cell attachment activity of fibronectin can be duplicated by small synthetic fragments of the molecule. *Nature, 309*(5963), 30–33.
Pozzi, A., Wary, K. K., Giancotti, F. G., & Gardner, H. A. (1998). Integrin α1β1 mediates a unique collagen-dependent proliferation pathway in vivo. *Journal of Cell Biology, 142*(2), 587–594.
Rahuel, C., Le Van Kim, C., Mattei, M. G., Cartron, J. P., & Colin, Y. (1996). A unique gene encodes spliceoforms of the B-cell adhesion molecule cell surface glycoprotein of epithelial cancer and of the Lutheran blood group glycoprotein. *Blood, 88*(5), 1865–1872.
Rebustini, I. T., Patel, V. N., Stewart, J. S., Layvey, A., Georges-Labouesse, E., Miner, J. H., et al. (2007). Laminin α5 is necessary for submandibular gland epithelial morphogenesis and influences FGFR expression through β1 integrin signaling. *Developmental Biology, 308*(1), 15–29.
Rodin, S., Domogatskaya, A., Strom, S., Hansson, E. M., Chien, K. R., Inzunza, J., et al. (2010). Long-term self-renewal of human pluripotent stem cells on human recombinant laminin-511. *Nature Biotechnology, 28*(6), 611–615.
Rous, B. A., Reaves, B. J., Ihrke, G., Briggs, J. A., Gray, S. R., Stephens, D. J., et al. (2002). Role of adaptor complex AP-3 in targeting wild-type and mutated CD63 to lysosomes. *Molecular Biology of the Cell, 13*(3), 1071–1082.
Ryan, M. C., Lee, K., Miyashita, Y., & Carter, W. G. (1999). Targeted disruption of the LAMA3 gene in mice reveals abnormalities in survival and late stage differentiation of epithelial cells. *Journal of Cell Biology, 145*(6), 1309–1323.
Sachs, N., Claessen, N., Aten, J., Kreft, M., Teske, G. J., Koeman, A., et al. (2012). Blood pressure influences end-stage renal disease of Cd151 knockout mice. *Journal of Clinical Investigation, 122*(1), 348–358.
Sachs, N., Kreft, M., van den Bergh Weerman, M. A., Beynon, A. J., Peters, T. A., Weening, J. J., et al. (2006). Kidney failure in mice lacking the tetraspanin CD151. *Journal of Cell Biology, 175*(1), 33–39.
Sadej, R., Romanska, H., Baldwin, G., Gkirtzimanaki, K., Novitskaya, V., Filer, A. D., et al. (2009). CD151 regulates tumorigenesis by modulating the communication between tumor cells and endothelium. *Molecular Cancer Research, 7*(6), 787–798.
Sasaki, T., Mann, K., Miner, J. H., Miosge, N., & Timpl, R. (2002). Domain IV of mouse laminin β1 and β2 chains. *European Journal of Biochemistry, 269*(2), 431–442.

Sasaki, T., & Timpl, R. (2001). Domain IVa of laminin α5 chain is cell-adhesive and binds β1 and αVβ3 integrins through Arg-Gly-Asp. *FEBS Letters, 509*(2), 181–185.
Shang, M., Koshikawa, N., Schenk, S., & Quaranta, V. (2001). The LG3 module of laminin-5 harbors a binding site for integrin α3β1 that promotes cell adhesion, spreading, and migration. *Journal of Biological Chemistry, 276*(35), 33045–33053.
Shigeta, M., Sanzen, N., Ozawa, M., Gu, J., Hasegawa, H., & Sekiguchi, K. (2003). CD151 regulates epithelial cell-cell adhesion through PKC- and Cdc42-dependent actin cytoskeletal reorganization. *Journal of Cell Biology, 163*(1), 165–176.
Shih, I. M., Nesbit, M., Herlyn, M., & Kurman, R. J. (1998). A new Mel-CAM (CD146)-specific monoclonal antibody, MN-4, on paraffin-embedded tissue. *Modern Pathology, 11*(11), 1098–1106.
Song, W. K., Wang, W., Foster, R. F., Bielser, D. A., & Kaufman, S. J. (1992). H36-α7 is a novel integrin α chain that is developmentally regulated during skeletal myogenesis. *Journal of Cell Biology, 117*(3), 643–657.
Springer, T. A., Zhu, J., & Xiao, T. (2008). Structural basis for distinctive recognition of fibrinogen γC peptide by the platelet integrin αIIbβ3. *Journal of Cell Biology, 182*(4), 791–800. http://dx.doi.org/10.1083/jcb.200801146.
Sterk, L. M., de Melker, A. A., Kramer, D., Kuikman, I., Chand, A., Claessen, N., et al. (1998). Glomerular extracellular matrix components and integrins. *Cell Adhesion and Communication, 5*(3), 177–192.
Stipp, C. S. (2010). Laminin-binding integrins and their tetraspanin partners as potential antimetastatic targets. *Expert Reviews in Molecular Medicine, 12*, e3.
Stipp, C. S., Kolesnikova, T. V., & Hemler, M. E. (2001). EWI-2 is a major CD9 and CD81 partner and member of a novel Ig protein subfamily. *Journal of Biological Chemistry, 276*(44), 40545–40554.
Stipp, C. S., Kolesnikova, T. V., & Hemler, M. E. (2003). Functional domains in tetraspanin proteins. *Trends in Biochemical Sciences, 28*(2), 106–112.
Sung, U., O'Rear, J. J., & Yurchenco, P. D. (1993). Cell and heparin binding in the distal long arm of laminin: identification of active and cryptic sites with recombinant and hybrid glycoprotein. *Journal of Cell Biology, 123*(5), 1255–1268.
Takagi, J., Strokovich, K., Springer, T. A., & Walz, T. (2003). Structure of integrin α5β1 in complex with fibronectin. *EMBO Journal, 22*(18), 4607–4615.
Takeda, Y., Kazarov, A. R., Butterfield, C. E., Hopkins, B. D., Benjamin, L. E., Kaipainen, A., et al. (2007). Deletion of tetraspanin Cd151 results in decreased pathologic angiogenesis in vivo and in vitro. *Blood, 109*(4), 1524–1532.
Taniguchi, Y., Ido, H., Sanzen, N., Hayashi, M., Sato-Nishiuchi, R., Futaki, S., et al. (2009). The C-terminal region of laminin β chains modulates the integrin binding affinities of laminins. *Journal of Biological Chemistry, 284*(12), 7820–7831.
Terry, A. H., & Culp, L. A. (1974). Substrate-attached glycoproteins from normal and virus-transformed cells. *Biochemistry, 13*(3), 414–425.
Timpl, R., Tisi, D., Talts, J. F., Andac, Z., Sasaki, T., & Hohenester, E. (2000). Structure and function of laminin LG modules. *Matrix Biology, 19*(4), 309–317.
Tisi, D., Talts, J. F., Timpl, R., & Hohenester, E. (2000). Structure of the C-terminal laminin G-like domain pair of the laminin α2 chain harbouring binding sites for α-dystroglycan and heparin. *EMBO Journal, 19*(7), 1432–1440.
Udani, M., Zen, Q., Cottman, M., Leonard, N., Jefferson, S., Daymont, C., et al. (1998). Basal cell adhesion molecule/lutheran protein. The receptor critical for sickle cell adhesion to laminin. *Journal of Clinical Investigation, 101*(11), 2550–2558.
Utani, A., Nomizu, M., Matsuura, H., Kato, K., Kobayashi, T., Takeda, U., et al. (2001). A unique sequence of the laminin α3 G domain binds to heparin and promotes cell adhesion through syndecan-2 and -4. *Journal of Biological Chemistry, 276*(31), 28779–28788.

Wang, Z., & Yan, X. (2013). CD146, a multi-functional molecule beyond adhesion. *Cancer Letters, 330*(2), 150–162.

Watanabe, K., Ueno, M., Kamiya, D., Nishiyama, A., Matsumura, M., Wataya, T., et al. (2007). A ROCK inhibitor permits survival of dissociated human embryonic stem cells. *Nature Biotechnology, 25*(6), 681–686.

Wick, T. M., & Eckman, J. R. (1996). Molecular basis of sickle cell-endothelial cell interactions. *Current Opinion in Hematology, 3*(2), 118–124.

Wickstrom, S. A., Radovanac, K., & Fassler, R. (2011). Genetic analyses of integrin signaling. *Cold Spring Harbor Perspectives in Biology, 3*(2).

Winterwood, N. E., Varzavand, A., Meland, M. N., Ashman, L. K., & Stipp, C. S. (2006). A critical role for tetraspanin CD151 in α3β1 and α6β4 integrin-dependent tumor cell functions on laminin-5. *Molecular Biology of the Cell, 17*(6), 2707–2721.

Wizemann, H., Garbe, J. H., Friedrich, M. V., Timpl, R., Sasaki, T., & Hohenester, E. (2003). Distinct requirements for heparin and α-dystroglycan binding revealed by structure-based mutagenesis of the laminin α2 LG4-LG5 domain pair. *Journal of Molecular Biology, 332*(3), 635–642.

Wright, M. D., Geary, S. M., Fitter, S., Moseley, G. W., Lau, L. M., Sheng, K. C., et al. (2004). Characterization of mice lacking the tetraspanin superfamily member CD151. *Molecular and Cellular Biology, 24*(13), 5978–5988.

Xiong, J. P., Stehle, T., Zhang, R., Joachimiak, A., Frech, M., Goodman, S. L., et al. (2002). Crystal structure of the extracellular segment of integrin αVβ3 in complex with an Arg-Gly-Asp ligand. *Science, 296*(5565), 151–155.

Yamada, M., Mugnai, G., Serada, S., Yagi, Y., Naka, T., & Sekiguchi, K. (2013). Substrate-attached materials are enriched with tetraspanins and are analogous to the structures associated with rear-end retraction in migrating cells. *Cell Adhesion and Migration, 7*(3), 304–314.

Yamada, M., Sumida, Y., Fujibayashi, A., Fukaguchi, K., Sanzen, N., Nishiuchi, R., et al. (2008). The tetraspanin CD151 regulates cell morphology and intracellular signaling on laminin-511. *FEBS Journal, 275*(13), 3335–3351.

Yanez-Mo, M., Barreiro, O., Gordon-Alonso, M., Sala-Valdes, M., & Sanchez-Madrid, F. (2009). Tetraspanin-enriched microdomains: a functional unit in cell plasma membranes. *Trends in Cell Biology, 19*(9), 434–446.

Yang, X., Claas, C., Kraeft, S. K., Chen, L. B., Wang, Z., Kreidberg, J. A., et al. (2002). Palmitoylation of tetraspanin proteins: modulation of CD151 lateral interactions, subcellular distribution, and integrin-dependent cell morphology. *Molecular Biology of the Cell, 13*(3), 767–781.

Yang, X., Kovalenko, O. V., Tang, W., Claas, C., Stipp, C. S., & Hemler, M. E. (2004). Palmitoylation supports assembly and function of integrin-tetraspanin complexes. *Journal of Cell Biology, 167*(6), 1231–1240.

Yang, X. H., Richardson, A. L., Torres-Arzayus, M. I., Zhou, P., Sharma, C., Kazarov, A. R., et al. (2008). CD151 accelerates breast cancer by regulating α6 integrin function, signaling, and molecular organization. *Cancer Research, 68*(9), 3204–3213.

Yauch, R. L., Berditchevski, F., Harler, M. B., Reichner, J., & Hemler, M. E. (1998). Highly stoichiometric, stable, and specific association of integrin α3β1 with CD151 provides a major link to phosphatidylinositol 4-kinase, and may regulate cell migration. *Molecular Biology of the Cell, 9*(10), 2751–2765.

Yauch, R. L., & Hemler, M. E. (2000). Specific interactions among transmembrane 4 superfamily (TM4SF) proteins and phosphoinositide 4-kinase. *Biochemical Journal, 351*(Pt 3), 629–637.

Yu, H., & Talts, J. F. (2003). B1 integrin and α-dystroglycan binding sites are localized to different laminin-G-domain-like (LG) modules within the laminin α5 chain G domain. *Biochemical Journal, 371*(Pt 2), 289–299.

Yurchenco, P. D. (2011). Basement membranes: cell scaffoldings and signaling platforms. *Cold Spring Harbor Perspectives in Biology, 3*(2).
Zhang, X. A., Bontrager, A. L., & Hemler, M. E. (2001). Transmembrane-4 superfamily proteins associate with activated protein kinase C (PKC) and link PKC to specific β1 integrins. *Journal of Biological Chemistry, 276*(27), 25005–25013.
Zhang, X. A., Bontrager, A. L., Stipp, C. S., Kraeft, S. K., Bazzoni, G., Chen, L. B., et al. (2001). Phosphorylation of a conserved integrin α3 QPSXXE motif regulates signaling, motility, and cytoskeletal engagement. *Molecular Biology of the Cell, 12*(2), 351–365.
Zhang, X. A., & Huang, C. (2012). Tetraspanins and cell membrane tubular structures. *Cellular and Molecular Life Sciences, 69*(17), 2843–2852.
Zhou, H., & Kramer, R. H. (2005). Integrin engagement differentially modulates epithelial cell motility by RhoA/ROCK and PAK1. *Journal of Biological Chemistry, 280*(11), 10624–10635.
Zijlstra, A., Lewis, J., Degryse, B., Stuhlmann, H., & Quigley, J. P. (2008). The inhibition of tumor cell intravasation and subsequent metastasis via regulation of in vivo tumor cell motility by the tetraspanin CD151. *Cancer Cell, 13*(3), 221–234.
Ziober, B. L., Vu, M. P., Waleh, N., Crawford, J., Lin, C. S., & Kramer, R. H. (1993). Alternative extracellular and cytoplasmic domains of the integrin α7 subunit are differentially expressed during development. *Journal of Biological Chemistry, 268*(35), 26773–26783.

CHAPTER SEVEN

Cell Receptor–Basement Membrane Interactions in Health and Disease: A Kidney-Centric View

Corina M. Borza[1,5,*], Xiwu Chen[1], Roy Zent[1,2,4,5,6] and Ambra Pozzi[1,2,3,5,6,*]

[1]Division of Nephrology, Department of Medicine, Vanderbilt University Medical Center, Nashville, TN, USA
[2]Department of Cancer Biology, Vanderbilt University Medical Center, Nashville, TN, USA
[3]Department of Molecular Physiology and Biophysics, Vanderbilt University Medical Center, Nashville, TN, USA
[4]Department of Cell and Developmental Biology, Vanderbilt University Medical Center, Nashville, TN, USA
[5]Vanderbilt Center for Kidney Disease, Vanderbilt University Medical Center, Nashville, TN, USA
[6]Department of Medicine, Veterans Administration Hospital, Nashville, TN, USA
*Corresponding authors: E-mail: ambra.pozzi@vanderbilt.edu and corina.borza@vanderbilt.edu

Contents

Abstract

Cell–extracellular matrix (ECM) interactions are essential for tissue development, homeostasis, and response to injury. Basement membranes (BMs) are specialized ECMs that separate epithelial or endothelial cells from stromal components and interact

Current Topics in Membranes, Volume 76
ISSN 1063-5823
http://dx.doi.org/10.1016/bs.ctm.2015.07.003

with cells via cellular receptors, including integrins and discoidin domain receptors. Disruption of cell–BM interactions due to either injury or genetic defects in either the ECM components or cellular receptors often lead to irreversible tissue injury and loss of organ function. Animal models that lack specific BM components or receptors either globally or in selective tissues have been used to help with our understanding of the molecular mechanisms whereby cell–BM interactions regulate organ function in physiological and pathological conditions. We review recently published works on animal models that explore how cell–BM interactions regulate kidney homeostasis in both health and disease.

Basement membranes (BMs) are specialized sheetlike extracellular matrix (ECM) structures, which lie beneath epithelial or endothelial cells. In addition to providing mechanical stability, BMs regulate essential cell functions, including cell polarity, proliferation, apoptosis, and matrix synthesis/remodeling. These effects are mediated by the integrin, discoidin (DS), and dystroglycan transmembrane family receptors.

The kidney is formed by functional units called nephrons, which consist of the glomerular filtering unit and specialized tubules that reabsorb and secrete the filtrate. In the glomerulus, there is a specialized BM called the glomerular BM (GBM) that separates endothelial cells from podocytes and is an essential component of the glomerular filtration barrier. In the tubules, there is a BM that separates a monolayer of tubular epithelial cells from the stroma. Defects of these BM components as well as the cellular receptors required for cells to interact with these BMs have been associated with kidney diseases. This review highlights recent findings on animal models with perturbations in BM components or cellular receptors that have significantly contributed to our understanding of kidney disease.

1. BASEMENT MEMBRANE COMPONENTS IN HEALTHY AND DISEASED KIDNEY

The main BM components are collagen IV, laminins (LMs), nidogen, and heparan sulfate proteoglycans (see below for details on their structure). BMs in the glomerulus provide support for mesangial cells, and the GBM is a physical separation between endothelial cells and podocytes. The GBM contains specific isoforms of BM components, such as the α3α4α5 collagen IV network, LM-521, and agrin (reviewed in Borza & Pozzi, 2012; Miner, 2012; Pozzi, Voziyan, Hudson, & Zent, 2009; Suh & Miner, 2013). Mutations in genes encoding some of the key GBM components cause severe kidney abnormalities, which underscore their importance for tissue development, homeostasis, and response to injury. In this regard, mutations in collagen IV or LM cause Alport and Pierson syndromes in humans, respectively. The availability of mice either lacking or carrying mutated BM components has allowed investigation of the molecular mechanisms whereby these matrices regulate glomerular and tubular kidney function. We will review only the most recent findings related to these mouse models, as the renal phenotype of some of these mice has already been extensively reviewed (Abrahamson, 2012; Cosgrove, Kalluri, Miner, Segal, & Borza, 2007; Gross & Kashtan, 2009; Kashtan & Segal, 2011; Suh & Miner, 2013).

2. THE ALPORT MOUSE MODELS OF KIDNEY DISEASE

Collagen IV, the major component of BMs, is a triple helical protein which contains a short 7S domain at the N-terminus, a long collagenous domain that occupies the midsection of the molecule, and a noncollagenous domain (NC1) positioned at the C-terminus (Hudson, Tryggvason, Sundaramoorthy, & Neilson, 2003). There are six genetically distinct α chains (α1-α6), which assemble into three specific heterotrimeric molecules; the $\alpha1\alpha1\alpha2$, $\alpha3\alpha4\alpha5$, and $\alpha5\alpha5\alpha6$ protomers. These protomers form three distinct networks by dimerization via NC1-to-NC1 interactions and by tetramerization via 7S-to-7S domain interactions (Hudson et al., 2003). In the adult kidney, the $\alpha1\alpha1\alpha2$ network is found primarily in the mesangium of the glomerulus and in the tubular BM; the $\alpha3\alpha4\alpha5$ network is mainly present in the GBM; and the $\alpha5\alpha5\alpha6$ network in the Bowman's capsule BM (Hudson et al., 2003).

Mutations in either COL4A3, COL4A4, or COL4A5 chains that result in the absence of $\alpha3\alpha4\alpha5$(IV) network and persistence of $\alpha1\alpha1\alpha2$(IV) network in the GBM cause Alport syndrome (Figure 1). The $\alpha1\alpha1\alpha2$(IV) network, which is not as highly cross-linked or resistant to proteases as the $\alpha3\alpha4\alpha5$(IV) network, provides less mechanical stability and is insufficient to maintain normal kidney function. Patients present with either macroscopic or microscopic hematuria and thickening and splitting of the GBM, and many will ultimately develop end-stage glomerulosclerosis. Mice deficient in COL4A3 (Cosgrove et al., 1996; Miner & Sanes, 1996), COL4A4 (Arnold et al., 2011), COL4A5 (Rheault et al., 2004), and COL4A3/COL4A4 (Lu et al., 1999) recapitulate human pathology, but the disease penetrance is highly strain-dependent (Cosgrove et al., 2007). For instance, COL4A3-null mice reach end-stage renal failure (ESRF)

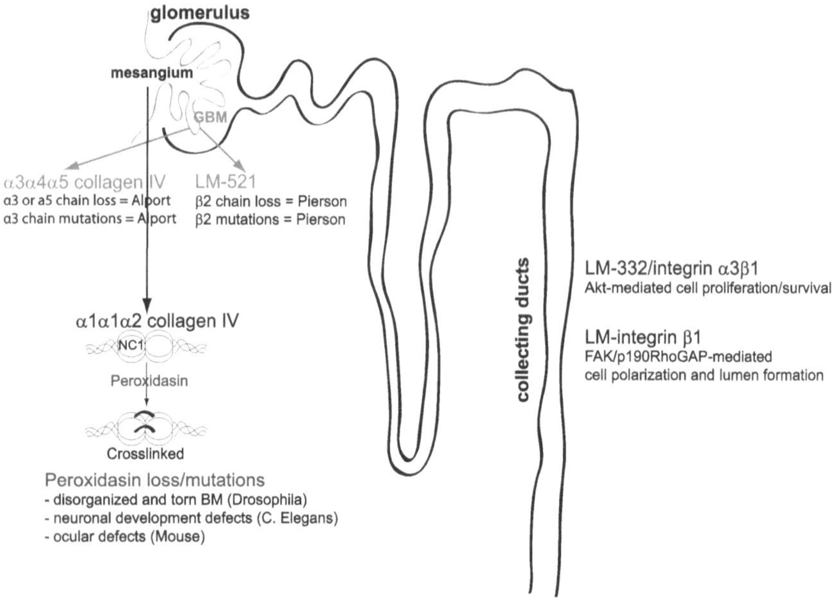

Figure 1 Schematic representation of the nephron highlighting the contribution of some basement membrane components in glomerular and tubular homeostasis. GBM, glomerular basement membrane; NC1, noncollagenous domain 1; LM, laminin; BM, basement membrane.

around 66 days of age on the 129X1/SvJ background, while on the C57BL/6J background, the mean age at ESRF was 194 days of age, which suggests the existence of modifier genes that influence disease progression (Andrews, Mudd, Li, & Miner, 2002). Strain-specific ectopic expression of the α5α5α6(IV) network in the GBM of C57/BL6 but not 129X1/SvJ COL4A3-null mice may also contribute to the milder Alport renal phenotype on the C57Bl/6 genetic background (Kang et al., 2006).

Although the defects in Alport syndrome are attributed largely to a defective GBM, it has been proposed that the unfolded protein response of endoplasmic reticulum (ER) stress in podocytes induced by defective collagen IV chains might also contribute to the pathogenesis. In support of this, a COL4A3 chain carrying the G1332E mutation overexpressed in podocytes in vitro or expressed in vivo in mice caused chain retention in the ER. This resulted in the activation of unfolded protein response-related markers of ER stress and the development of an Alport syndrome phenotype (Pieri et al., 2014). Interestingly, heterozygous COL4A3-G1334E mutations are observed in patients with thin GBM disease (Pieri et al., 2014).

Recently, a mouse model for Alport syndrome was reported in which the collagen α3α4α5(IV) network is synthesized and incorporated into the GBM but is abnormal and produced at low levels resulting in Alport-like lesions (Korstanje et al., 2014). These mice, identified in a colony of NONcNZO recombinant inbred mice, arose through a spontaneous mutation, localized to chromosome 1, which results in skipping of the *Col4a4* exon 30 but maintains the mRNA reading frame and generates a shorter collagen α4(IV) chain. As abnormal collagen α3α4α5(IV) is also found in a subset of Alport patients, this mouse represents an excellent model to analyze how abnormal collagen IV structure and assembly leads to Alport syndrome.

Generation of COL4A3-null mice has allowed not only investigation of the mechanisms responsible for Alport syndrome, but also testing of potential therapies. In this context, COL4A3-null mice engineered to express an inducible α3(IV) chain in the podocytes formed an α3α4α5(IV) network in the GBM with intact podocyte foot process architecture, reduced glomerulosclerosis and albuminuria, and a longer life span (Lin, Suh, Go, & Miner, 2014). Thus, the α3(IV) chain and the resulting α3α4α5(IV) protomers produced by podocytes are sufficient to promote proper collagen network formation in the GBM. In other studies, intracardial injection of amniotic fluid stem cells in COL4A5-null mice before the onset of proteinuria delayed the progression of glomerular sclerosis and prolonged animal survival (Garcia et al., 2013). This protective effect was due to recruitment and activation of antifibrotic and pro-tissue remodeling M2 macrophages, rather than stem cells differentiating into podocyte-like cells or collagen α5(IV) chain production. Finally, miR-21 silencing in COL4A3-null mice has been shown to improve survival and to reduce glomerulosclerosis, interstitial fibrosis, tubular injury and inflammation (Gomez et al., 2015). This protective effect was due to reduced transforming growth factor (TGF)-β-induced fibrogenesis and inflammation in glomerular and interstitial cells as well as improved mitochondrial function in both glomerular and tubular cells. This study suggests that inhibition of miR-21 represents a potential therapeutic strategy for chronic kidney diseases including Alport nephropathy.

3. LAMININS AND MOUSE MODELS OF GLOMERULAR KIDNEY DISEASE

LMs are large heterotrimeric glycoproteins that are essential for BM assembly. Each trimer is composed of one α, one β, and one γ chain

(Colognato & Yurchenco, 2000) and there are currently five α, four β, and three γ chain genes described in vertebrates which can assemble into 15 different heterotrimers (reviewed in Miner, 2008). Many of the LM chains are expressed during kidney development under strict temporal control. For instance, LM-111 is expressed in the presumptive GBM, LM-511 and LM-521 in the semimature GBM, and LM-521 is the sole trimer present in mature GBM (reviewed in Miner, 2008).

The importance of LM-521 in the GBM is demonstrated by the fact that patients carrying null mutations in the Lamβ2 gene develop Pierson syndrome characterized by mesangial sclerosis and diffuse alterations of the GBM (Zenker et al., 2004). Consistent with this finding, mice lacking the Lamβ2 chain develop massive proteinuria and glomerulosclerosis (Noakes et al., 1995) (Figure 1).

In addition to Lamβ2 null mutations, certain Lamβ2 missense mutations, including C321R, cause congenital nephrotic syndrome. To determine how this mutation leads to glomerular disease, a Lamβ2-null mice expressing the rat C321R-Lamβ2 was generated (Chen et al., 2013). During the first postnatal month, C321R-Lamβ2 attenuated proteinuria in Lamβ2-null mice in a dose-dependent fashion, however, as the mice aged they developed proteinuria and renal failure. This phenotype occurs because the C321R mutation leads to improper secretion of LM-521, podocyte ER stress, and apoptosis. The finding that in vitro treatment with chemicals that facilitate protein folding and trafficking increased the secretion of the mutant Lamβ2 (Chen et al., 2013) suggests that therapies that improve protein folding might be beneficial for the treatment of mild forms of Pierson syndrome. These data are very similar to those described for mice harboring the COL4A3-G1334E mutation and support the hypothesis that ER stress is a general mechanism of podocyte injury in mice harboring point mutations in BM proteins that cannot be properly secreted.

One feature of patients with Pierson syndrome and the Lamβ2-null mice is ectopic expression of the Lamβ1 chain in the GBM. However, the expression of this chain is only marginally increased and fails to compensate for the loss of Lamβ2. Interestingly, Lamβ2-null mice engineered to express high levels of Lamβ1 selectively in podocytes are spared from the development of nephrotic syndrome and show a greatly extended life span (Suh, Jarad, VanDeVoorde, & Miner, 2011). The finding that the levels of Lamβ1 inversely correlate to albuminuria and defects in the GBM suggests that maneuvers to increase Lamβ1 expression in patients with Lamβ2 null mutations could ameliorate the severity of nephrotic syndrome.

4. LAMININS AND RENAL EPITHELIAL CELL HOMEOSTASIS

In addition to functioning in BM assembly, LMs interact with cellular receptors known as integrins to provide polarity cues and to control cell function. We recently reported that kidney collecting duct cells interact with LM-332, a major component of kidney tubular BMs, via integrin α3β1. This interaction is key in promoting integrin α3β1-dependent Akt activation and tubular cell function. Interestingly, K63-linked polyubiquitination, but not the classical PI3K, is necessary for promoting LM-322/integrin α3β1-dependent cell signaling required for the proper development of the collecting system (Yazlovitskaya et al., 2015). Using MDCK cells grown as hollow cysts in Matrigel, Bryant and colleagues recently identified a molecular switch mechanism controlling polarity orientation whereby ECM signals through an integrin β1/FAK/p190RhoGAP complex to promote trafficking of podocalyxin from a basal to an apical membrane position thus allowing lumen formation (Bryant et al., 2014). Thus, interactions between BM components and tubular cells play a key role in governing the proper development, polarization, and lumen formation of kidney tubules (Figure 1).

5. NIDOGENS AND HEPARAN SULFATE PROTEOGLYCANS IN KIDNEY HOMEOSTASIS

Nidogen-1 and nidogen-2 are widely expressed in BMs, interact with both LMs and collagen IV, and are hypothesized to function as a bridge between the two networks. However, nidogen-1- or nidogen-2-deficient mice are normal with no obvious kidney defects. Interestingly, deletion of both nidogen-1 and nidogen-2 results in mice that die shortly after birth, although their BMs show only mild abnormalities (Bader et al., 2005).

Proteoglycans like agrin and perlecan contain heparan sulfate polysaccharide side chains covalently attached to a core protein. Because heparan sulfate side chains are negatively charged, proteoglycans serve in conferring the GBM with a net negative charge. Surprisingly, podocyte-specific deletion of agrin alone or in combination with loss of perlecan, the two predominant proteoglycans in GBM, does not affect the GBM's structure (Goldberg, Harvey, Cunningham, Tryggvason, & Miner, 2009). These studies indicate that unlike collagen IV and LMs, these proteins do not play a role in the maintenance of the kidney BMs at both physiological and pathological levels.

6. BASEMENT MEMBRANE MODIFYING ENZYMES IN HEALTH AND DISEASE

In addition to mutation and/or loss of key BM components, loss of enzymes required for posttranslational modification of BM components affects tissue mechanical stability and are implicated in tissue biogenesis and maintenance. Collagen IV forms a network that is stabilized by a sulfilimine bond between the alpha chains (Vanacore et al., 2009). This bond is catalyzed by peroxidasin, a BM-bound extracellular heme peroxidase that requires bromine as a key cofactor for its activity (Bhave et al., 2012; Fidler et al., 2014; McCall et al., 2014). Loss or mutation in peroxidasin has been associated with disorganized collagen IV networks and torn BMs in *Drosophila* (Bhave et al., 2012), neuronal developmental defects in *Caenorhabditis elegans* (Lee et al., 2015), and severe ocular defects in mice (Yan et al., 2014) (Figure 1). The finding that peroxidasin expression is increased in a murine model of kidney fibrosis and is organized into a fibril-like network suggests that it promotes matrix formation in response to injury (Peterfi et al., 2009). Whether increased peroxidasin expression contributes to physiological or pathological fibrogenic responses is unclear. Thus, the development of a mouse lacking peroxidasin in selective part of the kidney is needed to determine the role of this cross-linking enzyme in kidney repair following injury.

7. INTEGRINS IN HEALTHY AND DISEASED KIDNEY

Integrins are transmembrane receptors for ECM components that consist of noncovalently associated α and β subunits. There are 18α and 8β subunits in mammals, which form 24 unique heterodimers (Fu, Wang, & Luo, 2012) with distinct specificities for the ECM. In this chapter, we will focus on integrins that function as receptors for major BMs components, namely, collagen IV and LM.

In addition to their functioning in anchoring cells to ECM, integrins are signaling molecules that regulate cell migration, differentiation, proliferation, and survival under both physiological and pathological conditions (Askari, Buckley, Mould, & Humphries, 2009). Integrins modulate these diverse cellular functions by interacting with the cytoskeleton of the cell and by mediating bidirectional cell signaling from the outside of the cell inwards and from the inside of the cell outwards (Fu et al., 2012). Thus, integrins act as a bridge for cells to bind to and transduce signals from the ECM into the cell as well as for cells to modify the extracellular environment. The observation that the expression of some of the integrin family members is altered in the course of kidney diseases has initiated studies to analyze the contribution of these matrix receptors to kidney function under both physiological and pathological conditions. Generation of mice lacking integrin subunits in selective kidney cells has contributed to our understanding of the roles these matrix receptors play in kidney homeostasis in health and disease. Collagen receptors, α1β1 and α2β1, and LM receptors, α3β1 and α6β1, are highly expressed in kidney (Mathew, Chen, Pozzi, & Zent, 2012). As global deletion of the integrin β1 subunit leads to embryonic lethality at peri-implantation stage (Fassler & Meyer, 1995), the role of this subunit in kidney homeostasis has been made possible only by the recent generation of conditionally mutant integrin $\beta1^{fl/fl}$ mice. These, together with various Cre transgenic mice, have enabled specific deletion of the β1 integrin in various kidney cells at early and late stages of development.

7.1 Integrin β1 in glomerular homeostasis

Selective deletion of the β1 subunit in the podocytes by crossing the $\beta1^{fl/fl}$ mice with the podocin-Cre mice has resulted in mice with podocyte abnormalities and proteinuria at birth, despite a grossly normal GBM (Pozzi et al., 2008) (Figure 2). Following the advent of glomerular filtration, these mice show progressive podocyte loss as well as capillary loop and mesangium degeneration with little evidence of glomerulosclerosis (Pozzi et al., 2008). By 3 weeks of age, the mice develop severe ESRF characterized by both tubulointerstitial and glomerular pathology (Pozzi et al., 2008). In contrast to this data, Kanasaki and colleagues showed that deleting the β1 subunit in the podocytes by crossing the $\beta1^{fl/fl}$ mice with the nephrin-Cre mice results in detectable proteinuria on day 1 and death within a week of birth (Kanasaki et al., 2008). The kidneys of these mice exhibit normal glomerular endothelium, but show severe GBM defects with multilaminations and splitting as well as podocyte foot process effacement (Pozzi et al., 2008). The difference in phenotypes in these two studies may be because: (1) the nephrin-Cre promoter is stronger than the podocin-Cre; (2) the use of different mouse backgrounds; (3) partial deletion of the β1 subunit due to incomplete efficiency of the Cre; (4) long half-life of the β1 subunit protein, despite complete efficiency of the Cre; and (5) compensation by non β1 containing integrins. Despite these differences, these two studies demonstrate that podocyte β1 integrin is critical for postnatal development and maintaining the structural integrity of the glomerulus, especially the filtration barrier.

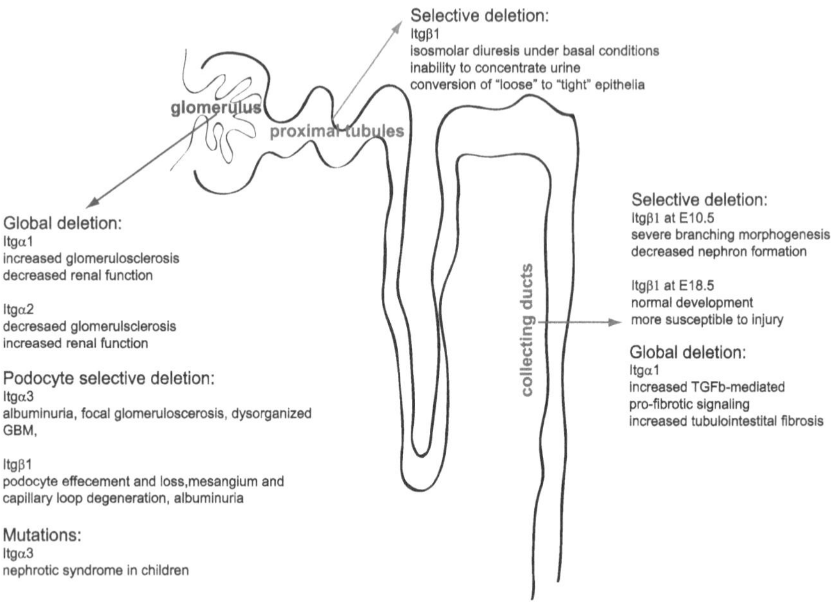

Figure 2 Schematic representation of the nephron highlighting the contribution of some basement membrane-binding integrins in glomerular and tubular homeostasis. GBM, glomerular basement membrane.

7.2 Integrin β1 in tubular cell homeostasis

Integrin β1 is also required for the development of the ureteric bud (UB). Deletion of the integrin β1 subunit using a Hoxb7-cre mouse, expressed in the UB from its initial outgrowth at ~E10.5, results in a severe branching morphogenesis defect with decreased nephron formation and death at 4–6 weeks of age (Zhang et al., 2009). Interestingly, deleting the β1 subunit in the collecting ducts at E18.5 using an aquaporin-2-cre mouse did not result in developmental defects, although the mice were more susceptible to injury (Zhang et al., 2009). These results show that integrin β1 is required for governing cell growth and branching during the early stages of UB development and structural integrity of the collecting duct at later stages of UB development (Zhang et al., 2009) (Figure 2). Recently, we showed that the integrin β1 controls the fate of kidney proximal and distal epithelial cells by regulating the composition and function of tight and adherent junctions. Deletion of the integrin β1 subunit in proximal tubules using the γGT-cre mouse (expressed in proximal tubules at P10) has minimal impact in kidney morphology, but results in iso-osmolar diuresis under basal conditions and an inability to concentrate urine following water deprivation

(Elias et al., 2014). This defect is due to the fact that deleting the integrin β1 subunit in proximal tubular cells converts them from a "loose" to a "tight" epithelium with features similar to those seen in distal tubular cells (Figure 2). Thus, this study suggests that cell—matrix interactions might regulate terminal differentiation and function of polarized epithelial cells.

7.3 Integrins α1β1 and α2β1 in glomerular homeostasis

While deletion of the integrin β1 subunit eliminates multiple integrin heterodimers, the availability of mice lacking specific integrin α subunits has allowed investigation of the contribution of specific alpha subunits to kidney function in health and disease. Integrins α1β1 and α2β1 are the two major collagen binding receptors in the kidney. Both integrins can bind collagen IV; however, deleting these two major receptors does not affect normal glomerular development (Chen, Diacovo, Grenache, Santoro, & Zutter, 2002; Gardner, Kreidberg, Koteliansky, & Jaenisch, 1996; Holtkotter et al., 2002). By contrast, both integrins play an important role in regulating ECM production and degradation in the course of kidney fibrosis (Borza & Pozzi, 2012; Pozzi et al., 2009) (Figure 2).

Integrin α1β1, which binds with high affinity to collagen IV, is expressed by podocytes, endothelial cells, and mesangial cells of the glomerulus (Korhonen, Ylanne, Laitinen, & Virtanen, 1990; Voigt et al., 1995). This receptor is associated with renal disease and is overexpressed in the proliferating mesangium in glomerulonephritis (Kuhara, Kagami, & Kuroda, 1997; Shikata et al., 1995). In addition, integrin α1 antibodies reduce scarring in rat models of glomerular injury by inhibiting integrin α1β1-dependent (VLA-1) leukocyte function (Cook et al., 2002). This is likely due to the inability of leukocytes to traffic to the sites of injury. Despite this beneficial effect of inhibiting VLA-1 in a model of glomerulonephritis, mice lacking the integrin α1 subunit appear to develop more glomerulosclerosis than wild-type mice in multiple models of glomerular injury (Chen et al., 2004; Yu et al., 2012; Zent et al., 2006). These results can be explained by the fact that integrin α1β1 is a negative regulator of collagen synthesis, as this receptor is required to sense extracellular collagen levels and downregulate both endogenous collagen I and collagen IV synthesis (Gardner, Broberg, Pozzi, Laato, & Heino, 1999). Consistent with these findings, integrin α1-null mice develop worse glomerulosclerosis than wild-type mice. Interestingly, integrin α1-null mesangial cells produce more profibrotic reactive oxygen species (ROS) than wild-type cells, which leads to decreased cell proliferation and increased glomerular collagen IV accumulation (Chen et al., 2007).

Thus, integrin $\alpha1\beta1$ negatively modulates glomerulosclerosis by either directly altering collagen production or by negatively regulating the production of ROS, which in turn control collagen turnover and ultimately fibrosis. In vitro studies have demonstrated that integrin $\alpha1\beta1$ negatively controls ROS production by downregulating the activation state of the profibrotic epidermal growth factor (EGF) receptor (Chen et al., 2007), and it does so by controlling the levels and phosphorylation state of caveolin-1, a scaffolding protein involved in receptor signaling and localization (Borza et al., 2010; Chen et al., 2010). Thus, integrin $\alpha1\beta1$ is a negative regulator of collagen production and its engagement is beneficial in the setting of fibrosis.

Integrin $\alpha2\beta1$ is another collagen IV receptor, although it binds this ligand with lower affinity than integrin $\alpha1\beta1$. Like integrin $\alpha1\beta1$, integrin $\alpha2\beta1$ is expressed by mesangial cells and podocytes (Borza et al., 2008; Chen et al., 2004). Expression of integrin $\alpha2\beta1$ increases in the kidneys of patients with diabetic nephropathy (Jin et al., 1996) and rapidly progressive glomerulonephritis (Baraldi et al., 1995). However, whether increased expression of this collagen receptor contributes to or counteracts the development of glomerulosclerosis is unclear. Integrin $\alpha2$-null mice develop mild proteinuria at 6 months of age and mild glomerular damage due to increased expression of the profibrotic TGF-β and connective tissue growth factor (Girgert et al., 2010). Although this result suggests that integrin $\alpha2\beta1$ is a negative regulator of glomerulosclerosis, in vitro studies with nonrenal cells suggest that integrin $\alpha2\beta1$ is a positive regulator of collagen I and ROS synthesis (Honore, Kovacic, Pichard, Briand, & Rognoni, 2003; Ivaska et al., 1999). Furthermore, crossing the COL4A3-null mice, a mouse model of Alport disease, with the integrin $\alpha2$-null mouse results in increased survival, improved renal function, and decreased glomerular matrix deposition and scarring (Rubel et al., 2014). We investigated the role of integrin $\alpha2\beta1$ in glomerulosclerosis and found that integrin $\alpha2$-null mice developed significantly less proteinuria and glomerulosclerosis than wild-type mice following adriamycin-mediated injury (Borza et al., 2012). In agreement with the observation that loss of integrin $\alpha2\beta1$ plays a protective role in glomerular injury, treatment of wild-type mice with a selective integrin $\alpha2\beta1$ inhibitor (Miller et al., 2009) decreases albuminuria and glomerular injury following adriamycin injection (Borza et al., 2012). The profibrotic role of integrin $\alpha2\beta1$ can be explained by the fact that binding of this receptor to collagen I induces activation of STAT3, a latent transcription factor involved fibrotic diseases (Chuang & He, 2010; Pechkovsky et al., 2012). Consistent with this finding, genetic deletion or inhibition of integrin α2β1 blocks integrin–collagen interactions, thus resulting in decreased STAT3 activation with consequent decreased collagen IV deposition and amelioration of glomerular damage (Borza et al., 2012). Thus, integrin α2β1 positively regulates collagen IV synthesis, which contributes to glomerular injury, and its inhibition is beneficial in the setting of fibrosis.

7.4 Integrin α1β1 in tubular homeostasis

As mentioned above, selective deletion of the integrin β1 subunit in proximal or distal tubules of the nephron identified a role for this subunit in cell fate, proliferation, branching, and response to injury (Zhang et al., 2009). In addition, the LM receptor integrin α3β1 has been shown to play a role in collecting system development of the kidney (Yazlovitskaya et al., 2015). The role of the BM collagen-binding integrins in tubular homeostasis is unknown. We recently found that integrin α1-null mice develop more tubulointerstitial fibrosis than wild type mice after unilateral ureter obstruction-mediated injury (Chen et al., 2014) (Figure 2). This effect was due to increased activation of profibrotic signaling downstream of TGF-β receptor II in integrin α1-null mice or collecting duct cells. Interestingly, we found that integrin α1β1 counteracts activation of TGF-β receptor II-mediated profibrotic signaling by negatively regulating the tyrosine phosphorylation levels of TGF-β receptor II (Chen et al., 2014). This protective effect is mediated by selective integrin α1β1-dependent recruitment and activation of the tyrosine phosphatase TCPTP. Although this study was focused on kidney injury, given the wide expression of integrin α1β1, TCPTP, and TGF-β receptor II, this finding suggests that integrin α1β1/TCPTP-mediated prevention of tyrosine phosphorylation of TGF-β receptor II might be viewed as a valid tool to control unwanted activation of TGF-β signaling. This approach could be therapeutic in situations such as inflammation, cancer, and fibrotic diseases.

7.5 Integrin α3β1 and glomerular homeostasis

Integrin α3β1 is highly expressed by podocytes and facilitates tight binding to the GBM in order to maintain a functional filtration barrier (Sachs et al., 2012, 2006). Global deletion of the integrin α3 subunit in mice results in abnormalities in glomerular development and alterations in the GBM, and integrin α3-null mice die soon after birth (Kreidberg et al., 1996). In contrast, deletion of the same subunit specifically in podocytes leads to massive proteinuria caused by focal glomerulosclerosis and disorganization of the GBM (Sachs et al., 2006) (Figure 2). Together with the kidney phenotype of the mice lacking the integrin β1 subunit in podocytes (see above), these studies indicate that integrin α3β1 is the major receptor required to maintain the glomerular filtration barrier. More importantly, the phenotypes of α3-deficient mice have recently been validated in humans where severe renal abnormalities and premature death are associated with absence of or mutations in the integrin α3 subunit (Has et al., 2012; Nicolaou et al., 2012; Shukrun et al., 2014; Yalcin et al., 2015). The importance of integrin α3β1 in the maintenance of podocyte stability is also demonstrated by the finding that podocyte expression of the integrin α3 subunit in patients with primary focal segmental glomerulosclerosis is significantly lower than in normal controls, and the expression of this subunit negatively correlates with the degree of glomerular sclerosis score (Chen et al., 2006). Thus, analysis of patients with mutations in integrin α3 which result in lethality and α3-deficient mice indicates that integrin α3β1 is crucial for BM organization and kidney function.

A possible mechanism whereby integrin α3β1 controls podocyte stability is via its interaction with the tetraspanin protein CD151 (Sachs et al., 2012), which mediates integrin α3β1-dependent adhesion. Interestingly, CD151-null mice show severe alterations of the GBM consisting of massive thickening and splitting and consequent kidney failure (Baleato, Guthrie, Gubler, Ashman, & Roselli, 2008; Sachs et al., 2006), thus mimicking the phenotype of mice lacking the integrin α3 subunit in podocytes (Sachs et al., 2006). Selective deletion of CD151 in cultured mouse podocytes results in redistribution of integrin α3β1 from diffuse/strong focal adhesions to large/weak focal adhesions, thus decreasing binding to LM substrata (Sachs et al., 2012). Consistent with this in vitro finding, in vivo podocyte-specific deletion of CD151 results in proteinuria, podocyte loss, and glomerulopathy (Sachs et al., 2012). Thus, CD151 is a crucial modifier of integrin-mediated adhesion of podocytes to the GBM and plays a critical role in ensuring tight adhesion of podocytes to the GBM for maintaining glomerular integrity (Pozzi & Zent, 2012). Overall, diseases associated with mutations in integrin α3β1, CD151, and LM β2 suggest a key role for LMs and their principal receptors in normal glomerular function.

7.6 Nonintegrin receptors in kidney homeostasis

In addition to integrins, cells interact with BMs through nonintegrin receptors such as dystroglycan, syndecans, and discoidin domain receptors (DDR).

Dystroglycan is a transmembrane receptor that consists of two subunits: the extracellular α subunit that binds BM components such as LM and agrin, and the transmembrane β subunit that binds to cytoskeletal proteins. This receptor is highly expressed in muscle, and skeletal muscle-targeted deletion of dystroglycan or fukutin, one of several enzymes required for dystroglycan processing, results in muscular dystrophy in mice (Beedle et al., 2012; Cohn et al., 2002). Interestingly, dystroglycan is also expressed at high levels in podocytes; however, podocyte-specific deletion of this receptor does not result in significant renal abnormalities either at baseline or following injury (Jarad, Pippin, Shankland, Kreidberg, & Miner, 2011). This study clearly indicates that, unlike muscle, dystroglycan is not a primary receptor in kidney glomerular cells, and other BM receptors contribute to glomerular homeostasis.

In contrast to the studies described above, a role for the transmembrane heparan sulfate proteoglycan syndecan-4 in promoting injury of proximal and collecting tubular cells was recently described. In tubular cells, syndecan-4 interacts with the ECM cross-linking enzyme transglutaminase type 2 (TG2). This interaction is necessary for cell surface trafficking, localization, and activity of TG2 (Scarpellini et al., 2009). Interestingly, loss of syndecan-4 protects mice from injury-induced tubular interstitial fibrosis due to reduced TG2 activation and excessive cross-linking and accumulation of ECM (Scarpellini et al., 2014) (Figure 3). Although this study suggests that preventing syndecan-4/TG2 interaction or inhibitingTG2 action might be beneficial for the treatment of kidney fibrosis, the picture is complicated by the multiple functional properties of TG2. In addition to its role in promoting matrix cross-linking, TG2 promotes clearance of necrotic cells and plays a protective role in promoting hepatic repair following injury (Nardacci et al., 2003). Thus, the prospects for successfully targeting TG2 to reduce fibrosis and/or injury seems to be highly tissue-specific.

DDR1 and DDR2 are receptor tyrosine kinases that bind to and are activated by collagen. While both DDRs bind to fibrillar collagens, only DDR1 binds collagen IV and so will be discussed further. DDR1 is composed of an extracellular DS homology domain which contains the collagen binding site, a DS-like domain, an extracellular juxtamembrane region that contains N- and O-glycosylation sites and matrix metalloproteinase cleavage sites, a transmembrane domain that mediates collagen-independent receptor dimerization, a large intracellular juxtamembrane region that contains tyrosines that may serve as docking sites upon phosphorylation, and an

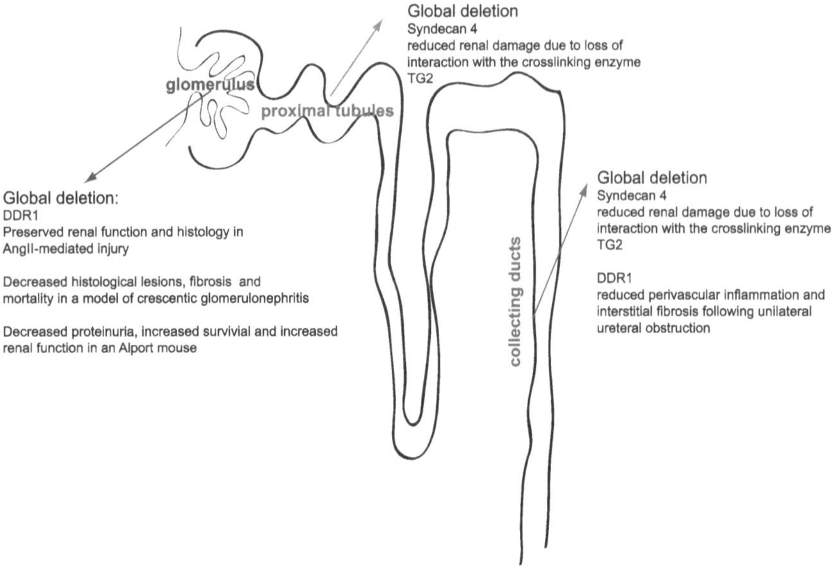

Figure 3 Schematic representation of the nephron highlighting the contribution of nonintegrin receptors in glomerular and tubular homeostasis. DDR, discoidin domain receptor.

intracellular tyrosine kinase domain (reviewed in Borza & Pozzi, 2014). DDR1 is expressed at low levels in healthy adult kidney, but DDR1 expression increases in patients with lupus nephritis and Goodpasture's syndrome as well as in a mouse model of crescentic glomerulonephritis (Kerroch et al., 2012). Similarly, DDR1 expression increases in the glomeruli of rats that have undergone partial renal ablation (Lee, Eidman, Kren, Hostetter, & Segal, 2004) and in tubules of mice that have undergone unilateral ureteral obstruction (Guerrot et al., 2011), suggesting that DDR1 plays a role in renal injury. Older DDR1-null mice show focal thickening of the GBM and mild proteinuria (Gross et al., 2004), suggesting that DDR1 might play a protective role in the maintenance of kidney homeostasis. However, extensive analysis of DDR1-null mice in several mouse models of kidney injury indicated that compared to wild-type mice, DDR1-null mice have increased survival, improved renal function, as well as reduced fibrosis and inflammation (Flamant et al., 2006; Gross et al., 2010; Guerrot et al., 2011; Kerroch et al., 2012). Moreover, DDR1-null mice show reduced macrophage infiltration following kidney injury, suggesting that DDR1 contributes to kidney damage and fibrosis by promoting inflammatory responses (Guerrot et al., 2011; Kerroch et al., 2012) (Figure 3).

8. CONCLUSIONS

In this review chapter, we highlighted recent findings supporting the role of ECM components and integrins in regulating kidney function. The availability of transgenic mice has enabled recapitulation of the features of some human kidney diseases and, in some cases, diseases were identified because phenotypes were initially identified in animals.

Selective molecular targeting of matrix components and their receptors has proven to be problematic in kidney disease. In this regard, although forced expression of Lamβ1 in podocytes ameliorates feature of Pierson syndrome in Lamβ2-null mice (Suh et al., 2011) and injection of amniotic stem cells or silencing microRNA ameliorates renal damage in a mouse model of Alport syndrome (Garcia et al., 2013), whether forced expression of Lamβ1 or stem cell therapy can be achieved in humans to treat these devastating diseases is unclear. Finally, although we provide evidence that activation of the integrin α1β1/TCPTP plays a key role in protecting from glomerular and tubular fibrosis, the generation and tissue-targeted delivery of integrin α1β1 and TCPTP activators might not be easy to achieve. Despite these difficulties, the current mouse models available have clearly strengthened our understanding of how integrins and BM components not only control kidney function, but also can be targeted to selectively reduce and ideally prevent kidney diseases.

ACKNOWLEDGMENTS

This work was in part supported by the Veterans Affairs Merit Reviews 1I01BX002025-01 (A.P.) and 1I01BX002196-01 (R.Z.); the National Institutes of Health grants DK095761 (A.P.), DK075594 (R.Z.), DK069221 (R.Z.), DK083187 (R.Z.).

REFERENCES

Abrahamson, D. R. (2012). Role of the podocyte (and glomerular endothelium) in building the GBM. *Seminars in Nephrology, 32*, 342–349.

Andrews, K. L., Mudd, J. L., Li, C., & Miner, J. H. (2002). Quantitative trait loci influence renal disease progression in a mouse model of Alport syndrome. *The American Journal of Pathology, 160*, 721–730.

Arnold, C. N., Xia, Y., Lin, P., Ross, C., Schwander, M., Smart, N. G., et al. (2011). Rapid identification of a disease allele in mouse through whole genome sequencing and bulk segregation analysis. *Genetics, 187*, 633–641.

Askari, J. A., Buckley, P. A., Mould, A. P., & Humphries, M. J. (2009). Linking integrin conformation to function. *Journal of Cell Science, 122*, 165–170.

Bader, B. L., Smyth, N., Nedbal, S., Miosge, N., Baranowsky, A., Mokkapati, S., et al. (2005). Compound genetic ablation of nidogen 1 and 2 causes basement membrane defects and perinatal lethality in mice. *Molecular and Cellular Biology, 25*, 6846–6856.

Baleato, R. M., Guthrie, P. L., Gubler, M. C., Ashman, L. K., & Roselli, S. (2008). Deletion of CD151 results in a strain-dependent glomerular disease due to severe alterations of the glomerular basement membrane. *The American Journal of Pathology, 173*, 927–937.
Baraldi, A., Zambruno, G., Furci, L., Ballestri, M., Tombesi, A., Ottani, D., et al. (1995). Beta 1 and beta 3 integrin upregulation in rapidly progressive glomerulonephritis. *Nephrology, Dialysis, Transplantation, 10*, 1155–1161.
Beedle, A. M., Turner, A. J., Saito, Y., Lueck, J. D., Foltz, S. J., Fortunato, M. J., et al. (2012). Mouse fukutin deletion impairs dystroglycan processing and recapitulates muscular dystrophy. *The Journal of Clinical Investigation, 122*, 3330–3342.
Bhave, G., Cummings, C. F., Vanacore, R. M., Kumagai-Cresse, C., Ero-Tolliver, I. A., Rafi, M., et al. (2012). Peroxidasin forms sulfilimine chemical bonds using hypohalous acids in tissue genesis. *Nature Chemical Biology, 8*, 784–790.
Borza, C. M., Borza, D. B., Pedchenko, V., Saleem, M. A., Mathieson, P. W., Sado, Y., et al. (2008). Human podocytes adhere to the KRGDS motif of the alpha3alpha4alpha5 collagen IV network. *Journal of the American Society of Nephrology, 19*, 677–684.
Borza, C. M., Chen, X., Mathew, S., Mont, S., Sanders, C. R., Zent, R., et al. (2010). Integrin {alpha}1{beta}1 promotes caveolin-1 dephosphorylation by activating T cell protein-tyrosine phosphatase. *The Journal of Biological Chemistry, 285*, 40114–40124.
Borza, C. M., & Pozzi, A. (2012). The role of cell-extracellular matrix interactions in glomerular injury. *Experimental Cell Research, 318*, 1001–1010.
Borza, C. M., & Pozzi, A. (2014). Discoidin domain receptors in disease. *Matrix Biology, 34*, 185–192.
Borza, C. M., Su, Y., Chen, X., Yu, L., Mont, S., Chetyrkin, S., et al. (2012). Inhibition of integrin alpha2beta1 ameliorates glomerular injury. *Journal of the American Society of Nephrology, 23*, 1027–1038.
Bryant, D. M., Roignot, J., Datta, A., Overeem, A. W., Kim, M., Yu, W., et al. (2014). A molecular switch for the orientation of epithelial cell polarization. *Developmental Cell, 31*, 171–187.
Chen, C. A., Hwang, J. C., Guh, J. Y., Chang, J. M., Lai, Y. H., & Chen, H. C. (2006). Reduced podocyte expression of alpha3beta1 integrins and podocyte depletion in patients with primary focal segmental glomerulosclerosis and chronic PAN-treated rats. *The Journal of Laboratory and Clinical Medicine, 147*, 74–82.
Chen, J., Diacovo, T. G., Grenache, D. G., Santoro, S. A., & Zutter, M. M. (2002). The alpha(2) integrin subunit-deficient mouse: a multifaceted phenotype including defects of branching morphogenesis and hemostasis. *The American Journal of Pathology, 161*, 337–344.
Chen, X., Abair, T. D., Ibanez, M. R., Su, Y., Frey, M. R., Dise, R. S., et al. (2007). Integrin {alpha}1{beta}1 controls reactive oxygen species synthesis by negatively regulating epidermal growth factor receptor-mediated Rac activation. *Molecular and Cellular Biology, 27*, 3313–3326.
Chen, X., Moeckel, G., Morrow, J. D., Cosgrove, D., Harris, R. C., Fogo, A. B., et al. (2004). Lack of integrin alpha1beta1 leads to severe glomerulosclerosis after glomerular injury. *The American Journal of Pathology, 165*, 617–630.
Chen, X., Wang, H., Liao, H. J., Hu, W., Gewin, L., Mernaugh, G., et al. (2014). Integrin-mediated type II TGF-beta receptor tyrosine dephosphorylation controls SMAD-dependent profibrotic signaling. *The Journal of Clinical Investigation, 124*, 3295–3310.
Chen, X., Whiting, C., Borza, C., Hu, W., Mont, S., Bulus, N., et al. (2010). Integrin alpha1beta1 regulates epidermal growth factor receptor activation by controlling peroxisome proliferator-activated receptor gamma-dependent caveolin-1 expression. *Molecular and Cellular Biology, 30*, 3048–3058.
Chen, Y. M., Zhou, Y., Go, G., Marmerstein, J. T., Kikkawa, Y., & Miner, J. H. (2013). Laminin beta2 gene missense mutation produces endoplasmic reticulum stress in podocytes. *Journal of the American Society of Nephrology, 24*, 1223–1233.

Chuang, P. Y., & He, J. C. (2010). JAK/STAT signaling in renal diseases. *Kidney International, 78*, 231–234.
Cohn, R. D., Henry, M. D., Michele, D. E., Barresi, R., Saito, F., Moore, S. A., et al. (2002). Disruption of DAG1 in differentiated skeletal muscle reveals a role for dystroglycan in muscle regeneration. *Cell, 110*, 639–648.
Colognato, H., & Yurchenco, P. D. (2000). Form and function: the laminin family of heterotrimers. *Developmental Dynamics, 218*, 213–234.
Cook, H. T., Khan, S. B., Allen, A., Bhangal, G., Smith, J., Lobb, R. R., et al. (2002). Treatment with an antibody to VLA-1 integrin reduces glomerular and tubulointerstitial scarring in a rat model of crescentic glomerulonephritis. *The American Journal of Pathology, 161*, 1265–1272.
Cosgrove, D., Kalluri, R., Miner, J. H., Segal, Y., & Borza, D. B. (2007). Choosing a mouse model to study the molecular pathobiology of Alport glomerulonephritis. *Kidney International, 71*, 615–618.
Cosgrove, D., Meehan, D. T., Grunkemeyer, J. A., Kornak, J. M., Sayers, R., Hunter, W. J., et al. (1996). Collagen COL4A3 knockout: a mouse model for autosomal Alport syndrome. *Genes and Development, 10*, 2981–2992.
Elias, B. C., Mathew, S., Srichai, M. B., Palamuttam, R., Bulus, N., Mernaugh, G., et al. (2014). The integrin beta1 subunit regulates paracellular permeability of kidney proximal tubule cells. *The Journal of Biological Chemistry, 289*, 8532–8544.
Fassler, R., & Meyer, M. (1995). Consequences of lack of β1 integrin gene expression in mice. *Genes and Development, 9*, 1896–1908.
Fidler, A. L., Vanacore, R. M., Chetyrkin, S. V., Pedchenko, V. K., Bhave, G., Yin, V. P., et al. (2014). A unique covalent bond in basement membrane is a primordial innovation for tissue evolution. *Proceedings of the National Academy of Sciences of the United States of America, 111*, 331–336.
Flamant, M., Placier, S., Rodenas, A., Curat, C. A., Vogel, W. F., Chatziantoniou, C., et al. (2006). Discoidin domain receptor 1 null mice are protected against hypertension-induced renal disease. *Journal of the American Society of Nephrology, 17*, 3374–3381.
Fu, G., Wang, W., & Luo, B. H. (2012). Overview: structural biology of integrins. *Methods in Molecular Cell Biology, 757*, 81–99.
Garcia, O., Carraro, G., Turcatel, G., Hall, M., Sedrakyan, S., Roche, T., et al. (2013). Amniotic fluid stem cells inhibit the progression of bleomycin-induced pulmonary fibrosis via CCL2 modulation in bronchoalveolar lavage. *PLoS One, 8*, e71679.
Gardner, H., Broberg, A., Pozzi, A., Laato, M., & Heino, J. (1999). Absence of integrin alpha1beta1 in the mouse causes loss of feedback regulation of collagen synthesis in normal and wounded dermis. *Journal of Cell Science, 112*, 263–272.
Gardner, H., Kreidberg, J., Koteliansky, V., & Jaenisch, R. (1996). Deletion of integrin alpha 1 by homologous recombination permits normal murine development but gives rise to a specific deficit in cell adhesion. *Developmental Biology, 175*, 301–313.
Girgert, R., Martin, M., Kruegel, J., Miosge, N., Temme, J., Eckes, B., et al. (2010). Integrin alpha2-deficient mice provide insights into specific functions of collagen receptors in the kidney. *Fibrogenesis and Tissue Repair, 3*, 19.
Goldberg, S., Harvey, S. J., Cunningham, J., Tryggvason, K., & Miner, J. H. (2009). Glomerular filtration is normal in the absence of both agrin and perlecan-heparan sulfate from the glomerular basement membrane. *Nephrology, Dialysis, Transplantation, 24*, 2044–2051.
Gomez, I. G., MacKenna, D. A., Johnson, B. G., Kaimal, V., Roach, A. M., Ren, S., et al. (2015). Anti-microRNA-21 oligonucleotides prevent Alport nephropathy progression by stimulating metabolic pathways. *The Journal of Clinical Investigation, 125*, 141–156.
Gross, O., Beirowski, B., Harvey, S. J., McFadden, C., Chen, D., Tam, S., et al. (2004). DDR1-deficient mice show localized subepithelial GBM thickening with focal loss of slit diaphragms and proteinuria. *Kidney International, 66*, 102–111.

Gross, O., Girgert, R., Beirowski, B., Kretzler, M., Kang, H. G., Kruegel, J., et al. (2010). Loss of collagen-receptor DDR1 delays renal fibrosis in hereditary type IV collagen disease. *Matrix Biology, 29*, 346–356.

Gross, O., & Kashtan, C. E. (2009). Treatment of Alport syndrome: beyond animal models. *Kidney International, 76*, 599–603.

Guerrot, D., Kerroch, M., Placier, S., Vandermeersch, S., Trivin, C., Mael-Ainin, M., et al. (2011). Discoidin domain receptor 1 is a major mediator of inflammation and fibrosis in obstructive nephropathy. *The American Journal of Pathology, 179*, 83–91.

Has, C., Sparta, G., Kiritsi, D., Weibel, L., Moeller, A., Vega-Warner, V., et al. (2012). Integrin alpha3 mutations with kidney, lung, and skin disease. *The New England Journal of Medicine, 366*, 1508–1514.

Holtkotter, O., Nieswandt, B., Smyth, N., Muller, W., Hafner, M., Schulte, V., et al. (2002). Integrin alpha 2-deficient mice develop normally, are fertile, but display partially defective platelet interaction with collagen. *The Journal of Biological Chemistry, 277*, 10789–10794.

Honore, S., Kovacic, H., Pichard, V., Briand, C., & Rognoni, J. B. (2003). Alpha2beta1-integrin signaling by itself controls G1/S transition in a human adenocarcinoma cell line (Caco-2): implication of NADPH oxidase-dependent production of ROS. *Experimental Cell Research, 285*, 59–71.

Hudson, B. G., Tryggvason, K., Sundaramoorthy, M., & Neilson, E. G. (2003). Alport's syndrome, Goodpasture's syndrome, and type IV collagen. *The New England Journal of Medicine, 348*, 2543–2556.

Ivaska, J., Reunanen, H., Westermarck, J., Koivisto, L., Kahari, V. M., & Heino, J. (1999). Integrin alpha2beta1 mediates isoform-specific activation of p38 and upregulation of collagen gene transcription by a mechanism involving the alpha2 cytoplasmic tail. *The Journal of Cell Biology, 147*, 401–416.

Jarad, G., Pippin, J. W., Shankland, S. J., Kreidberg, J. A., & Miner, J. H. (2011). Dystroglycan does not contribute significantly to kidney development or function, in health or after injury. *American Journal of Physiology. Renal Physiology, 300*, F811–F820.

Jin, D. K., Fish, A. J., Wayner, E. A., Mauer, M., Setty, S., Tsilibary, E., et al. (1996). Distribution of integrin subunits in human diabetic kidneys. *Journal of the American Society of Nephrology, 7*, 2636–2645.

Kanasaki, K., Kanda, Y., Palmsten, K., Tanjore, H., Lee, S. B., Lebleu, V. S., et al. (2008). Integrin beta1-mediated matrix assembly and signaling are critical for the normal development and function of the kidney glomerulus. *Developmental Biology, 313*, 584–593.

Kang, J. S., Wang, X. P., Miner, J. H., Morello, R., Sado, Y., Abrahamson, D. R., et al. (2006). Loss of alpha3/alpha4(IV) collagen from the glomerular basement membrane induces a strain-dependent isoform switch to alpha5alpha6(IV) collagen associated with longer renal survival in Col4a3-/- Alport mice. *Journal of the American Society of Nephrology, 17*, 1962–1969.

Kashtan, C. E., & Segal, Y. (2011). Glomerular basement membrane disorders in experimental models for renal diseases: impact on understanding pathogenesis and improving diagnosis. *Contributions to Nephrology, 169*, 175–182.

Kerroch, M., Guerrot, D., Vandermeersch, S., Placier, S., Mesnard, L., Jouanneau, C., et al. (2012). Genetic inhibition of discoidin domain receptor 1 protects mice against crescentic glomerulonephritis. *FASEB Journal, 26*, 4079–4091.

Korhonen, M., Ylanne, J., Laitinen, L., & Virtanen, I. (1990). The alpha 1-alpha 6 subunits of integrins are characteristically expressed in distinct segments of developing and adult human nephron. *The Journal of Cell Biology, 111*, 1245–1254.

Korstanje, R., Caputo, C. R., Doty, R. A., Cook, S. A., Bronson, R. T., Davisson, M. T., et al. (2014). A mouse Col4a4 mutation causing Alport glomerulosclerosis with abnormal collagen alpha3alpha4alpha5(IV) trimers. *Kidney International, 85*, 1461–1468.

Kreidberg, J., Donovan, M., Goldstein, S., Rennke, H., Shepherd, K., Jones, R., et al. (1996). Alpha 3 beta 1 integrin has a crucial role in kidney and lung organogenesis. *Development, 122*, 3537–3547.

Kuhara, T., Kagami, S., & Kuroda, Y. (1997). Expression of beta 1-integrins on activated mesangial cells in human glomerulonephritis. *Journal of the American Society of Nephrology, 8*, 1679–1687.

Lee, J., Bandyopadhyay, J., Lee, J. I., Cho, I., Park, D., & Cho, J. H. (2015). A role for peroxidasin PXN-1 in aspects of *C. elegans* development. *Molecules and Cells, 38*, 51–57.

Lee, R., Eidman, K. E., Kren, S. M., Hostetter, T. H., & Segal, Y. (2004). Localization of discoidin domain receptors in rat kidney. *Nephron. Experimental Nephrology, 97*, e62–70.

Lin, X., Suh, J. H., Go, G., & Miner, J. H. (2014). Feasibility of repairing glomerular basement membrane defects in Alport syndrome. *Journal of the American Society of Nephrology, 25*, 687–692.

Lu, W., Phillips, C. L., Killen, P. D., Hlaing, T., Harrison, W. R., Elder, F. F., et al. (1999). Insertional mutation of the collagen genes Col4a3 and Col4a4 in a mouse model of Alport syndrome. *Genomics, 61*, 113–124.

Mathew, S., Chen, X., Pozzi, A., & Zent, R. (2012). Integrins in renal development. *Pediatric Nephrology, 27*, 891–900.

McCall, A. S., Cummings, C. F., Bhave, G., Vanacore, R., Page-McCaw, A., & Hudson, B. G. (2014). Bromine is an essential trace element for assembly of collagen IV scaffolds in tissue development and architecture. *Cell, 157*, 1380–1392.

Miller, M. W., Basra, S., Kulp, D. W., Billings, P. C., Choi, S., Beavers, M. P., et al. (2009). Small-molecule inhibitors of integrin alpha2beta1 that prevent pathological thrombus formation via an allosteric mechanism. *Proceedings of the National Academy of Sciences of the United States of America, 106*, 719–724.

Miner, J. H. (2008). Laminins and their roles in mammals. *Microscopy Research and Technique, 71*, 349–356.

Miner, J. H. (2012). The glomerular basement membrane. *Experimental Cell Research, 318*, 973–978.

Miner, J. H., & Sanes, J. R. (1996). Molecular and functional defects in kidneys of mice lacking collagen alpha 3(IV): implications for Alport syndrome. *The Journal of Cell Biology, 135*, 1403–1413.

Nardacci, R., Lo Iacono, O., Ciccosanti, F., Falasca, L., Addesso, M., Amendola, A., et al. (2003). Transglutaminase type II plays a protective role in hepatic injury. *The American Journal of Pathology, 162*, 1293–1303.

Nicolaou, N., Margadant, C., Kevelam, S. H., Lilien, M. R., Oosterveld, M. J., Kreft, M., et al. (2012). Gain of glycosylation in integrin alpha3 causes lung disease and nephrotic syndrome. *The Journal of Clinical Investigation, 122*, 4375–4387.

Noakes, P. G., Miner, J. H., Gautam, M., Cunningham, J. M., Sanes, J. R., & Merlie, J. P. (1995). The renal glomerulus of mice lacking s-laminin/laminin beta 2: nephrosis despite molecular compensation by laminin beta 1. *Nature Genetics, 10*, 400–406.

Pechkovsky, D. V., Prele, C. M., Wong, J., Hogaboam, C. M., McAnulty, R. J., Laurent, G. J., et al. (2012). STAT3-mediated signaling dysregulates lung fibroblast-myofibroblast activation and differentiation in UIP/IPF. *The American Journal of Pathology, 180*, 1398–1412.

Peterfi, Z., Donko, A., Orient, A., Sum, A., Prokai, A., Molnar, B., et al. (2009). Peroxidasin is secreted and incorporated into the extracellular matrix of myofibroblasts and fibrotic kidney. *The American Journal of Pathology, 175*, 725–735.

Pieri, M., Stefanou, C., Zaravinos, A., Erguler, K., Stylianou, K., Lapathitis, G., et al. (2014). Evidence for activation of the unfolded protein response in collagen IV nephropathies. *Journal of the American Society of Nephrology, 25*, 260–275.

Pozzi, A., Jarad, G., Moeckel, G. W., Coffa, S., Zhang, X., Gewin, L., et al. (2008). Beta1 integrin expression by podocytes is required to maintain glomerular structural integrity. *Developmental Biology, 316*, 288–301.
Pozzi, A., Voziyan, P. A., Hudson, B. G., & Zent, R. (2009). Regulation of matrix synthesis, remodeling and accumulation in glomerulosclerosis. *Current Pharmaceutical Design, 15*, 1318–1333.
Pozzi, A., & Zent, R. (2012). Hold tight or you'll fall off: CD151 helps podocytes stick in high-pressure situations. *The Journal of Clinical Investigation, 122*, 13–16.
Rheault, M. N., Kren, S. M., Thielen, B. K., Mesa, H. A., Crosson, J. T., Thomas, W., et al. (2004). Mouse model of X-linked Alport syndrome. *Journal of the American Society of Nephrology, 15*, 1466–1474.
Rubel, D., Frese, J., Martin, M., Leibnitz, A., Girgert, R., Miosge, N., et al. (2014). Collagen receptors integrin alpha2beta1 and discoidin domain receptor 1 regulate maturation of the glomerular basement membrane and loss of integrin alpha2beta1 delays kidney fibrosis in COL4A3 knockout mice. *Matrix Biology, 34*, 13–21.
Sachs, N., Claessen, N., Aten, J., Kreft, M., Teske, G. J., Koeman, A., et al. (2012). Blood pressure influences end-stage renal disease of Cd151 knockout mice. *The Journal of Clinical Investigation, 122*, 348–358.
Sachs, N., Kreft, M., van den Bergh Weerman, M. A., Beynon, A. J., Peters, T. A., Weening, J. J., et al. (2006). Kidney failure in mice lacking the tetraspanin CD151. *The Journal of Cell Biology, 175*, 33–39.
Scarpellini, A., Germack, R., Lortat-Jacob, H., Muramatsu, T., Billett, E., Johnson, T., et al. (2009). Heparan sulfate proteoglycans are receptors for the cell-surface trafficking and biological activity of transglutaminase-2. *The Journal of Biological Chemistry, 284*, 18411–18423.
Scarpellini, A., Huang, L., Burhan, I., Schroeder, N., Funck, M., Johnson, T. S., et al. (2014). Syndecan-4 knockout leads to reduced extracellular transglutaminase-2 and protects against tubulointerstitial fibrosis. *Journal of the American Society of Nephrology, 25*, 1013–1027.
Shikata, K., Makino, H., Morioka, S., Kashitani, T., Hirata, K., Ota, Z., et al. (1995). Distribution of extracellular matrix receptors in various forms of glomerulonephritis. *American Journal of Kidney Diseases, 25*, 680–688.
Shukrun, R., Vivante, A., Pleniceanu, O., Vax, E., Anikster, Y., Dekel, B., et al. (2014). A human integrin-alpha3 mutation confers major renal developmental defects. *PLoS One, 9*, e90879.
Suh, J. H., Jarad, G., VanDeVoorde, R. G., & Miner, J. H. (2011). Forced expression of laminin beta1 in podocytes prevents nephrotic syndrome in mice lacking laminin beta2, a model for Pierson syndrome. *Proceedings of the National Academy of Sciences of the United States of America, 108*, 15348–15353.
Suh, J. H., & Miner, J. H. (2013). The glomerular basement membrane as a barrier to albumin. *Nature Reviews Nephrology, 9*, 470–477.
Vanacore, R., Ham, A. J., Voehler, M., Sanders, C. R., Conrads, T. P., Veenstra, T. D., et al. (2009). A sulfilimine bond identified in collagen IV. *Science, 325*, 1230–1234.
Voigt, S., Gossrau, R., Baum, O., Loster, K., Hofmann, W., & Reutter, W. (1995). Distribution and quantification of alpha 1-integrin subunit in rat organs. *The Histochemical Journal, 27*, 123–132.
Yalcin, E. G., He, Y., Orhan, D., Pazzagli, C., Emiralioglu, N., & Has, C. (2015). Crucial role of posttranslational modifications of integrin alpha3 in interstitial lung disease and nephrotic syndrome. *Human Molecular Genetics, 24*, 3679–3688.
Yan, X., Sabrautzki, S., Horsch, M., Fuchs, H., Gailus-Durner, V., Beckers, J., et al. (2014). Peroxidasin is essential for eye development in the mouse. *Human Molecular Genetics, 23*, 5597–5614.

Yazlovitskaya, E. M., Tseng, H. Y., Viquez, O., Tu, T., Mernaugh, G., McKee, K. K., et al. (2015). Integrin alpha3beta1 regulates kidney collecting duct development via TRAF6-dependent K63-linked polyubiquitination of Akt. *Molecular Biology of the Cell, 26*, 1857–1874.

Yu, L., Su, Y., Paueksakon, P., Cheng, H., Chen, X., Wang, H., et al. (2012). Integrin alpha1/Akita double-knockout mice on a Balb/c background develop advanced features of human diabetic nephropathy. *Kidney International, 81*, 1086–1097.

Zenker, M., Aigner, T., Wendler, O., Tralau, T., Muntefering, H., Fenski, R., et al. (2004). Human laminin beta2 deficiency causes congenital nephrosis with mesangial sclerosis and distinct eye abnormalities. *Human Molecular Genetics, 13*, 2625–2632.

Zent, R., Yan, X., Su, Y., Hudson, B. G., Borza, D. B., Moeckel, G. W., et al. (2006). Glomerular injury is exacerbated in diabetic integrin alpha1-null mice. *Kidney International, 70*, 460–470.

Zhang, X., Mernaugh, G., Yang, D. H., Gewin, L., Srichai, M. B., Harris, R. C., et al. (2009). beta1 integrin is necessary for ureteric bud branching morphogenesis and maintenance of collecting duct structural integrity. *Development, 136*, 3357–3366.

CHAPTER EIGHT

The Basement Membrane Proteoglycans Perlecan and Agrin: Something Old, Something New

Kevin J. McCarthy
Department of Pathology, Louisiana State University Health Sciences Center-Shreveport, Shreveport, LA, USA
E-mail: kmccar2@lsuhsc.edu

Contents

Abstract

Several members of the proteoglycan family are integral components of basement membranes; other proteoglycan family members interact with or bind to molecular residents of the basement membrane. Proteoglycans are polyfunctional molecules, for they derive their inherent bioactivity from the amino acid motifs embedded in the core protein structure as well as the glycosaminoglycan (GAG) chains that are covalently attached to the core protein. The presence of the covalently attached GAG chains

Current Topics in Membranes, Volume 76
ISSN 1063-5823
http://dx.doi.org/10.1016/bs.ctm.2015.09.001

significantly expands the "partnering" potential of proteoglycans, permitting them to interact with a broad spectrum of targets, including growth factors, cytokines, chemokines, and morphogens. Thus proteoglycans in the basement membrane are poised to exert diverse effects on the cells intimately associated with basement membranes.

1. INTRODUCTION

Members of the proteoglycan superfamily of connective tissue components achieve their functional diversity via (1) the large number of core proteins in the superfamily (>40) and (2) the posttranslational assembly of glycosaminoglycan chains (GAG) on the core protein, carbohydrate structures unique to the proteoglycan family. Research conducted over the past 35 years has made significant advances in uncovering the biological activities intrinsic to both the core proteins and GAG chains. When one examines the body of work that has evolved from proteoglycan research, those outside of the proteoglycan field can gain an appreciation how this single family of molecules can exert their many profound effects on biological systems. If one takes a reductionist perspective, both the GAG chains and core proteins have bioactivities inherent to their structures; however, the biological binding partners of GAG chains are now known to exceed >400 different molecular species, far superseding the currently identified binding partners for all of the core proteins associated with the superfamily combined. Because of their ability to engage multiple binding partners in many different fashions, proteoglycans can be viewed as the metaphorical "Swiss Army Knives" of biosystems. Albeit their biological activities are complex, the basis of understanding of these molecules from a holistic perspective lies in a very simple concept—i.e., a key, critical function of the amino acid motifs present in the proteoglycan core proteins facilitates the precise delivery/localization of GAG chains to the right place, at the right time.

The occurrence of proteoglycans as intrinsic components of basement membranes can be found early in evolution. Their presence in these highly specialized matrices can be traced back to the development of the mesoglea, a primordial basement membrane-like structure interposed between the endoderm and ectoderm in the cnidarian, *Hydra vulgaris* (Sarras, Madden, et al., 1991; Sarras, Meador, & Zhang, 1991). More recent studies have since isolated and characterized the GAGs from hydra, reporting that the major GAG species present in total hydra homogenates was chondroitin sulfate (CS), present as both nonsulfated and sulfated CS chains; heparan sulfate

(HS) was also identified but present to a lesser extent (Yamada, Morimoto, Fujisawa, & Sugahara, 2007). Finding both CS and HS in *H. vulgaris* also implies that the key enzymes necessary for the assembly for the respective carbohydrate structures also developed early in evolution. Although the specific proteoglycan core proteins in the hydra mesoglea have yet to be fully characterized at the molecular level, recent sequencing of the hydra genome revealed a partial mRNA (XM_004211434) that has been predicted to encode for a putative basement membrane-specific HS proteoglycan (HSPG). Interestingly, the predicted organization of the core protein reveals a modular organization similar to that described for the basement membrane proteoglycan perlecan: the hydra molecule has six IgG-like domains, followed by three laminin G-like domains, with each domain separated from one another by an EGF-like domain. From a functional aspect, basement membrane proteoglycans are important in homeostasis within *H. vulgaris*; inhibition of proteoglycan assembly in the hydra via the use of xylosides (see below) prevents regeneration of structure after injury of the animal (Sarras, Meador, et al., 1991).

2. A SUCCINCT HISTORICAL PERSPECTIVE

The basement membrane proteoglycan field has evolved significantly since its inception in the early 1970s. Searching the databases with a generic search string such as "basement membrane proteoglycan" will retrieve >2500 references pertaining to the topic; if one broadens the focus of the search to incorporate the studies of HS or CS, two of the GAGs that are covalently attached to proteoglycan core proteins, a significant, exponential increase in the list of retrieved references occurs; the list stretches back to the discovery of a thromboplastic substance first described in a paper in 1916 (McLean, 1916) that would eventually be known as heparin. Since the proteoglycan literature straddles both the fields of protein biology and glycobiology, citing all of the key studies that contributed to the genesis, growth, and maturation of the basement membrane proteoglycan field would in itself be too formidable task. The following paragraphs serve to highlight just a few of the many steps that have contributed to the genesis and growth of the basement membrane proteoglycan field.

2.1 HS and CS GAGs are found in basement membranes

Routine tissue staining methods used by the early histologists, such as Periodic Acid Schiff (Lev & Spicer, 1964; Rambourg & Leblond, 1967; Wislocki,

Fawcett, & Dempsey, 1951; Yamada, 1963) and Alcian Blue (Muller, 1959; Rizzoli & Gliozzi, 1955; Vialli, 1951), demonstrated the existence of acid mucopolysaccharides (GAGs) in tissue basement membranes. Through the use of electron dense stains/probes such as lysozyme (Caulfield & Farquhar, 1976), cationized ferritin (Rennke, Cotran, & Venkatachalam, 1975; Rennke & Venkatachalam, 1977), and ruthenium red (Bernfield & Banerjee, 1972; Copley & Scheinthal, 1970; Wight & Ross, 1975), ultrastructural studies further refined the earlier light microscopy observations by discerning both the presence and pattern of distribution of anionic sites within basement membranes. Using a combined approach of transmission electron microscopy of ruthenium red-labeled specimens and ^{3}H glucosamine radiolabeling, the association of CS was demonstrated by Cohn, Banerjee, and Bernfield (1977) in the basal lamina of developing salivary gland. Concurrently, studies by Kanwar and Farquhar (1979a, 1979b) attributed the presence of ruthenium red-labeled anionic sites in the glomerular basement membrane (GBM) to the presence of HS. Antibody-based detection of GAGs added a degree of refinement to the identification of GAG carbohydrates in basement membranes; the studies of Couchman, Caterson, Christner, and Baker (1984) first employed this approach to demonstrate the presence of CS GAGs in various basement membranes in normal adult tissues.

2.2 Discerning the role of proteoglycans in basement membranes in tissues

Because GAGs are a carbohydrate species unique to proteoglycans, one of the earliest approaches used as a means to discern the function of proteoglycans/GAGs in tissue focused upon interfering with the assembly of intact proteoglycans. Since proteoglycans have a duality of function that is mediated by both core protein and the attached GAGs, investigators exploring basement membrane proteoglycan biology adapted a strategy that had been identified and utilized by colleagues in the cartilage field (Gibson, Segen, & Audhya, 1977; Lohmander, Hascall, & Caplan, 1979; Schwartz, 1976, 1977; Stearns & Goetinck, 1979). The approach essentially "uncoupled" the assembly of intact proteoglycan by divorcing the process of GAG assembly from the proteoglycan core protein. The method took advantage of earlier observations that GAG assembly for HS and CS both began with the initial attachment of a xylose residue to a serine present within the sequence of a proteoglycan core protein (Gregory, Laurent, & Roden, 1964; Lindahl & Roden, 1964, 1965), followed by the subsequent addition of a galactose residue to xylose via the actions of a galactosyltransferase (Helting, 1972; Helting & Roden, 1969). Okayama, Kimata, and Suzuki (1973) subsequently identified ρ-nitrophenyl-β-D-xyloside as a substrate for the action of the galactosyltransferase enzyme. When introduced into a biological system, ρ-nitrophenyl-β-D-xyloside was shown to compete as a substrate for the same GAG assembly machinery that the native proteoglycan core proteins synthesized by cells, diverting GAG assembly onto the xyloside at the expense of proteoglycan assembly.

Studies by embryologists and teratologists (Gibson, Segen, & Doller, 1979; Kanke, Doller, Segen, & Gibson, 1982; Oohira, Nogami, & Nakanishi, 1981) began to utilize xylosides as tools in their exploration of the development of soft tissues of the embryo. Perhaps the first association of xyloside-induced effects on basement membrane proteoglycan assembly was made by Morris-Kay and Crutch (1982) in their studies of neurulation in rodent embryos. Data from that study showed a significant decrease in Alcian Blue staining of basement membranes in embryos treated with ρ-nitrophenyl-β-D-xyloside. In turn, defects in somitic differentiation, neurulation, and enlargement of the heart and large vessels were reported. Later studies used xyloside-based interference to explore the role of basement membranes in other organ systems including salivary gland morphogenesis (Spooner, Bassett, & Stokes, 1985; Spooner, Paulsen, & Sullins, 1989; Thompson & Spooner, 1982, 1983), kidney (Kanwar, Hascall, Jakubowski, & Gibbons, 1984; Kanwar, Rosenzweig, & Jakubowski, 1987; Klein, Brown, Moran, Oegema, & Platt, 1989; Lelongt, Makino, Dalecki, & Kanwar, 1988; Platt, Brown, Granlund, Oegema, & Klein, 1987), and lung (Smith, Hilfer, Searls, Nathanson, & Allodoli, 1990; Smith, Webster, Nathanson, Searls, & Hilfer, 1990). The results of those studies shared a common theme; the presence of xylosides significantly reduced key developmental processes through their effects on basement membrane proteoglycan assembly. Although the mechanism(s) by which proteoglycans influenced these processes was not specifically identified in these early investigations, the results laid the foundation for subsequent investigations that would ultimately lead those in the field to explore the role of proteoglycans in tethering/sequestering/presenting developmentally important morphogens and growth factors.

2.3 The search for basement membrane proteoglycan core proteins

Although the existence of both HS and CS had been known for several decades prior to 1960, the biochemistry studies by Lindahl and Roden

(1964, 1965, 1966) (Lindahl, Cifonelli, Lindahl, & Roden, 1965), Toole and Lowther (1965), and Jansson and Lindahl (1970) provided the initial characterization of the carbohydrate linkage of heparan and chondroitin to proteoglycan core proteins. However, the characterization of the protein component of basement membrane proteoglycans lagged behind the carbohydrate biochemistry studies and the morphology studies. The delay in core protein discovery was primarily due to (1) the lack of a source of large quantities of basement membrane material that could be used for on the isolation and characterization of biochemical quantities of proteoglycan core proteins and (2) the difficulties in obtaining protein sequence information. An initial, key step that moved the protein side of the field forward occurred with eventual reclassification/characterization of the EHS tumor (Englebreth-Holm-Swarm) and its associated matrix (Orkin et al., 1977) from a chondrosarcoma to a basement membrane-secreting tumor.

Extraction of the extracellular matrix from the EHS tumor led to the isolation and initial characterization of a high molecular weight HSPG referred to as BM-1 (Hassell et al., 1980). Antibodies raised against the BM-1 proteoglycan core protein were shown to immunostain basement membranes in vasculature, kidney, cornea, and skin (Hassell et al., 1980). Subsequently J.L. Stow, Sawada, and Farquhar (1985) isolated basement membrane proteoglycans from glomerular isolates, a nontumor tissue source. The polyclonal antiserum that was developed using their approach recognized a core protein of a smaller HSPG (~130 kD) that localized to the GBM. The Hassell laboratory (Hassell et al., 1985) subsequently reported data suggesting that the core proteins of a large (400 kD) and small (130 kD) HSPG isolated from the EHS tumor were related, based on the observation that antibodies raised against the large 400 kD protein recognized epitopes present within the smaller 130 kD protein. An interpretation of that data suggested that the two species of basement membrane proteoglycans might possibly be derived from the same gene. Studies from D.J. Klein et al. (1988) provided additional evidence suggesting that the different HSPGs identified in the GBM were the result of proteolytic processing of the larger core proteins to smaller HSPGs.

Two additional cell-based sources of basement membrane proteoglycans were identified subsequent to the discovery of the EHS tumor, the murine parietal yolk sac (PYS) cell lines (Couchman, Woods, Hook, & Christner, 1985; Oohira, Wight, McPherson, & Bornstein, 1982; Tyree, Horigan, Klippenstein, & Hassell, 1984) and the rat L-2 cell line (Fenger, Wewer, & Albrechtsen, 1984; Wewer, 1982). U. M. Wewer, Albrechtsen, and Hassell (1985) showed that both the L-2 cell line and the EHS tumor cells produced a repertoire of proteoglycans that were remarkably similar, both in the size of the core protein produced by the cells (400 kD) and in the nature of the cross-reactivity between antibodies raised against the proteoglycans from the two sources. Additional studies (Couchman, 1987) reported the development of a monospecific polyclonal antiserum against an HSPG isolated from the PYS2 cell line having a different molecular weight core protein. Antibodies raised against this core protein showed minimal cross-reactivity against those isolated from the EHS tumor. The pattern of staining of the anti-HSPG antibodies raised by the Couchman laboratory (Couchman, 1987) was similar to the pattern of staining reported earlier from the Farquhar laboratory (Stow et al., 1985). Taken together, the data from the Couchman laboratory (Couchman, 1987) added evidence to further substantiate the concept that a second distinct species of HSPG core protein existed in basement membranes.

Previous studies also had demonstrated that there was CS in basement membranes (Cohn et al., 1977; Couchman et al., 1984; Klein, Brown, & Oegema, 1986; Stow, Glasgow, Handley, & Hascall, 1983), raising the question as to whether or not CS had been substituted onto the core protein of one of the recently identified HSPGs or, alternatively, if a proteoglycan carrying CS (i.e., CS proteoglycan, CSPG) was actually a unique entity. Iozzo and Clark (1986, 1987) initially isolated and characterized a CSPG having an estimated size of 125–135 kD from the rat parietal endoderm basement membrane. Subsequently, the Couchman laboratory (McCarthy, Accavitti, & Couchman, 1989) isolated a similarly sized CSPG from the same tissue source and developed a panel of monoclonal antibodies directed against the proteoglycan core protein. The immunostaining pattern of several of those antibodies differed significantly in some tissues (McCarthy, Bynum, St John, Abrahamson, & Couchman, 1993; McCarthy, Bynum, St John, Abrahamson, & Couchman, 1994) from that reported for the basement membrane HSPGs, providing further evidence that the CSPG might represent a novel species of basement membrane proteoglycan.

2.4 Getting to the core of basement membrane proteoglycans

The development and use of cDNA libraries (based on the λGT11 backbone) was a second key step that propelled the search for core proteins significantly forward. With this new tool, the molecular characterization of several core proteins of other members of the proteoglycan superfamily had begun, and the molecular characterization of the first proteoglycan core protein, nee decorin, was completed in 1986 (Krusius & Ruoslahti, 1986). The Hassel laboratory (Noonan et al., 1988) reported the initial identification and characterization of several cDNA clones representing several regions of the mRNA encoding the large HSPG from the EHS tumor (BM-1). At the time, two cDNA clones had been isolated, one clone representing a portion of the laminin homology domain, and the other representing a portion of the IgG repeat like motifs (see below). The mRNA encoding the proteoglycan core protein was estimated by northern blot analysis to be approximately 12 kB in size. At the time, automated DNA sequencing technology was not yet in widespread use, so the task to complete the molecular characterization of the cDNA encoding for BM-1 (perlecan) was formidable. Sequencing of the entire cDNA for perlecan from the murine EHS tumor was completed (Noonan et al., 1991), followed shortly afterward by reports of the coding sequence for human perlecan (Murdoch, Dodge, Cohen, Tuan, & Iozzo, 1992; Kallunki & Tryggvason, 1992).

The discovery of the other three core proteins of the basement membrane proteoglycan family occurred in the reverse fashion from that for perlecan; i.e., the core proteins were initially characterized at the molecular level, and their identification as proteoglycans occurred at a later date. Agrin had been identified as a protein associated with the aggregation of acetylcholine receptors in the CNS and in muscle (Nitkin et al., 1987; Smith et al., 1987), and its initial molecular characterization had been reported in 1991 (Rupp, Payan, Magill-Solc, Cowan, & Scheller, 1991). The first report that agrin was a member of the basement membrane proteoglycan family was by Tsen, Halfter, Kroger, and Cole (1995). The data showed that agrin, isolated from chicken brain, existed as a HSPG having a molecular mass of approximately 400 kD; after heparitinase removal of the HS chains, the size of the core protein was resolved to approximately 250 kD. Type XV collagen was initially identified by Myers, Kivirikko, Gordon, Dion, and Pihlajaniemi (1992) and the characterization completed by Kivirikko et al. (1994). Six years later, Li, Clark, & Myers (2000) demonstrated that type XV collagen had CS chains covalently attached to the core protein. In a similar sequence of events, type XVIII collagen was initially identified by Abe, Muragaki, Yoshioka, Inoue, and Ninomiya (1993), subsequently characterized at the molecular level (Muragaki, Abe, Ninomiya, Olsen, & Ooshima, 1994; Oh et al., 1994) and eventually identified as a HSPG (Halfter, Dong, Schurer, & Cole, 1998).

3. BASEMENT MEMBRANE PROTEOGLYCAN CORE PROTEINS

Although the four members of the basement membrane proteoglycan play important roles in maintaining basement membrane organization, most of the body of work on types XV and XVIII collagens has a primary focus on their functions in the context of collagen biology; research into their functional roles as proteoglycans is not as extensive as it is for perlecan and agrin. For this reason, this review will focus upon the structure and function of perlecan and agrin.

3.1 Perlecan

It is hard to believe that the formal study of perlecan biology, starting from its initial molecular characterization to the present, now encompasses a quarter century of research. There exists a set of excellent literature reviews published during the past 10 years that serve to highlight key areas of perlecan research and the complex biological properties of this molecule (Farach-Carson & Carson, 2007; Farach-Carson, Warren, Harrington, & Carson, 2014; Iozzo, 2005; Iozzo, Zoeller, & Nystrom, 2009; Kirn-Safran, Farach-Carson, & Carson, 2009; Knox & Whitelock, 2006; Melrose, Hayes, Whitelock, & Little, 2008; Whitelock, Melrose, & Iozzo, 2008).

3.1.1 Perlecan gene

The location of the gene encoding human perlecan (*HSPG2*) resides on chromosome 1 (1p36.1-1p34) (Dodge et al., 1991; Kallunki et al., 1991); chromosome 4 is the location for the *Hspg2* gene in mouse (Chakravarti, Phillips, & Hassell, 1991). The organization of the human perlecan gene showed the mRNA to be represented at the genomic level by 94 exons spanning approximately 120 kB (Cohen, Grassel, Murdoch, & Iozzo, 1993). The exon organization of the perlecan gene corresponds with the domain organization of the intact core protein, the relationship of the exons encoding each domain indicated below. Besides containing elements within the promoter region that permit the expression of perlecan in a wide variety of cell lines, the promoter region of the perlecan gene has been reported to contain one or more TGF-β/SMAD3 response elements and NFκB response elements (Iozzo et al., 1997; Warren, Grindel, Francis, Carson, & Farach-Carson, 2014).

3.1.2 Perlecan is composed of modular elements

The murine and human perlecan cDNA sequences were published by several groups within a relatively short period of time (Kallunki & Tryggvason, 1992; Murdoch et al., 1992; Noonan et al., 1991). These revealed a high degree of homology between the proteoglycans from the two species. The molecular characterization of the perlecan core protein revealed a protein having a multidomain structure composed of modular elements, with the domains identified by the Roman numerals I–V (from amino to carboxyl end). Domain I (exons 1–6), representing the only region of the molecule that is unique to perlecan, contains three GAG acceptor sites and a SEA module (sperm protein/enterokinase/agrin (Bork & Patthy, 1995)). Domain II (exons 7–11) encompasses a region with four repeats, having significant homology to the LDL receptor, followed by a short segment (95 amino acids) having a structure comparable to the IgG repeats found in Domain IV. Domain III (exons 12–38) is the first of two regions of perlecan having homology to the short arm of the laminin α1 chain, and has three repetitive subdomain elements, with each element consisting of a cysteine-free globular region followed by a series of cysteine-rich repeats. Domain IV (exons 39–78) consists of a series of IgG-like repeats having structural homology to the IgG-like repeats found in neural cell adhesion molecule (N-CAM). The number of these IgG-like repeats differs between the mouse (14 consecutive repeats plus 1 repeat in Domain II) and human (21 consecutive repeats plus 1 repeat in Domain II), accounting for the reported differences in the estimated molecular mass of the core proteins, 396 kD and 467 kD, respectively. Domain V (exons 79–94) is a region homologous to the globular regions present in the LG modules of laminin α1; the region consists of three globular regions, the first two in the sequence followed by two EGF-like repeats, the third globular domain being the terminal portion of the molecule. A second GAG acceptor site, located in the second globular region of Domain V, was shown to permit the assembly of CS/dermatan sulfate (DS) GAG on the core protein (Tapanadechopone, Hassell, Rigatti, & Couchman, 1999).

3.1.3 Bioactivities of the perlecan domains

The perlecan core protein has its own intrinsic bioactivities, and because of its overall size, one would surmise that multiple functions would be encoded within the core protein. Expression of full-length perlecan in vitro is difficult due to the large size of the molecule, but expression of wild-type or mutated perlecan domains as fusion proteins has allowed several laboratories to discern key aspects of perlecan core protein function.

Domain I was shown to serve as the primary GAG acceptor site in perlecan (Costell, Mann, Yamada, & Timpl, 1997; Dolan, Horchar, Rigatti, & Hassell, 1997; Graham, Whitelock, & Underwood, 1999; Kokenyesi & Silbert, 1995; Sasaki, Costell, Mann, & Timpl, 1998). The presence of either HS or CS significantly expands its potential to engage a wide variety of ligands via their binding to the covalently attached GAG structures. Within Domain I are three Ser-Gly-Asp sites, each of which is capable of promoting the assembly of either HS or CS. Three elements within Domain I have been shown to facilitate the assembly of HS: (1) the close proximity of the three Ser-Gly-Asp sites to one another; (2) an acidic cluster of amino acids located N-terminal to the GAG acceptor sites; and (3) the presence of the SEA module on the carboxyl end of Domain I (Dolan et al., 1997). Mutating one or more of the serine residues was demonstrated to lead to a decrease or loss of GAG substitution on Domain I constructs (Dolan et al., 1997; Sasaki et al., 1998). In vitro, the loss of recombinant Domain I GAG substitution via site-directed mutagenesis was shown to increase the susceptibility of Domain I to proteolysis and also decrease the ability of Domain I to bind to extracellular ligands such as laminin (Sasaki et al., 1998). Outside of promoting GAG assembly, Domain I (and II) has also been shown to directly interact with fibrillin (Tiedemann et al., 2005) and, along with the interactions of the HS chains with tropoelastin, promote its colocalization with elastic fibers within vascular walls (Hayes et al., 2011).

Domain II has been shown to encompass four modules having homology to the LDL receptor, and the motif structure predicts that this region of perlecan should have the potential to bind to LDL. Although there is a body of literature showing that HS itself attached to Domain I binds to LDL either directly or via the engagement of "bridging" molecules that can engage both HS and LDL (e.g., lipoprotein lipase), thus permitting perlecan to sequester LDL, there are few studies that have explored the ability of Domain II to bind LDL directly. Costell et al. (1996) expressed a Domain II fusion protein in studies that focused on defining the structural organization of murine perlecan Domain II, identifying one putative N-glycosylation site in the LA3 module and seven O-glycosylation sites overall. A recent report (Xu et al., 2015) confirmed those observations, showing that O-linked glycosylation of the second (T47, T54, T63, T64, and T72) and fifth (T164, S201) repeat structures occurred in Domain II of human perlecan. The work of Xu et al. further extended the understanding of the biology of Domain II, showing that the glycosylation of these sites in Domain II with carbohydrate containing (α2-3)-linked sialic acids was important for the direct binding of ApoB-100. The study further showed an additive effect occurred with regard to ApoB-100 binding when a construct consisting of Domain 1 (HS-substitution)—Domain 2 (glycosylated) was expressed in CHO cells. Moreover, colocalization of (α2-3)-linked sialic acids and perlecan core protein occurred in atherosclerotic plaques in areas coincident with ApoB-100 accumulation (Xu et al., 2015). These findings add further support for the role of perlecan in accelerating pathologic processes in atherosclerosis.

Domain III shares structural similarities with the short arm of the laminin α1 chain (Murdoch et al., 1992); it consists of three globular regions (L4 modules—III1, III2, and III3) separated from one another by cysteine-rich laminin EGF-like motifs (LE modules) (Schulze, Sasaki, Costell, Mann, & Timpl, 1996). The homology between the human and murine sequence in this region is high (85%, Schulze et al., 1996), but one notable difference is that the murine sequence has an RGDS motif embedded within, but this motif is lacking in the human sequence. Although early studies showed that this RGDS motif could support cell adhesion (Chakravarti, Horchar, Jefferson, Laurie, & Hassell, 1995), the results from subsequent studies suggest that this cell-binding site may be held in a cryptic configuration, unavailable for the engagement of the RGD-binding integrins (Schulze et al., 1996).

Domain III has been shown to be capable of engaging multiple ligands, including the promotion of self-association via this site (Mongiat et al., 2000). The region of Domain III encompassing L4 module III1 and three successive LE modules were shown to bind to FGF-7 (Mongiat et al., 2000); FGF-BP bound nearby, to the second LE module of Domain III (Mongiat et al., 2001). The L4 module III2 was shown to bind PDGF-AA and PDGF-BB, with little or no affinity for III1 or IIIc (Gohring, Sasaki, Heldin, & Timpl, 1998). FGF-18 was also shown to bind the perlecan core protein in Domain III, and in this context the interaction of FGF-18 with perlecan significantly decreased its mitogenic effects on chondrocytes (Smith, West, & Hassell, 2007). The protein von Willibrand factor A domain-related protein (WARP), first isolated and characterized from cartilage extracts (Allen et al., 2006), interacts with the HS chains on Domain I of perlecan and Domain III-2 of perlecan. The distribution of WARP in tissues has since been extended from skeletal tissues and developing limb bud to encompass peripheral nerve and the endomysium of muscle (Allen, Brachvogel, Farlie, Fitzgerald, & Bateman, 2008).

Domain IV consists of 14 (mouse) or 21 (human) IgG-type modules (IG) similar in structure to the modules found in N-CAM (Kallunki & Tryggvason, 1992; Murdoch et al., 1992; Noonan et al., 1991). The mRNA encoding this region of perlecan has been suggested to undergo alternative splicing to yield several possible variant forms of perlecan in mammalian systems (Noonan & Hassell, 1993). The interactions of Domain IV points to it having a key role in the association of perlecan with other basement membrane components/matrix molecules; the IG modules in Domain IV were shown to bind to nidogens, fibulin-2, fibronectin, and heparin (Hopf, Gohring, Kohfeldt, Yamada, & Timpl, 1999). Using expressed proteins representing different regions of Domain IV, these intermolecular interactions were subsequently shown to occur at different locations along the length of the IG repeat structure of Domain IV. The IG3 module was found to serve as a high-affinity binding site for nidogen-1 and nidogen-2, allowing for formation of the ternary complex between perlecan, mediated by the G2 domain of nidogen, with the $\gamma 1$ chain of laminin-111 (Hopf, Gohring, Mann, & Timpl, 2001). Fibulin-2 binding was demonstrated for IG2-3 modules, with a second lower-affinity binding site to occur at IG4-5. Fibronectin binding was shown to occur between the IG4–IG5 modules, whereas heparin (and sulfatide)-binding sites were demonstrated at IG5 (Hopf et al., 2001). A second heparin-binding site, having a lower affinity for heparin, was reported for the region spanning modules IG10-IG15 (Hopf et al., 2001). A more recent study (Farach-Carson, Brown, Lynam, Safran, & Carson, 2008) has shown that a peptide (TWSKVGGHLRPGIVQSG) derived from a loop within IG14 was capable of promoting cell-matrix adhesion and focal adhesion kinase activation, with the interaction capable of being blocked by heparin but not CS.

Domain V consists of three LG modules separated by EGF-like repeats arranged in the pattern LG_1-EGF_1-EGF_2-LG_2-EGF_3-EGF_4-LG_3, the LG modules having homology to the LG modules present at the carboxyl end of the laminin $\alpha 1$ chain (Kallunki & Tryggvason, 1992; Murdoch et al., 1992; Noonan et al., 1991). Both HS and CS substitution has been reported to occur within Domain V (Brown, Sasaki, Gohring, Yamada, & Timpl, 1997; Friedrich et al., 1999; Tapanadechopone et al., 1999), but the type of GAG and substitution site within the region may be contextually variable and cell-type dependent. Within Domain V are three potential GAG acceptor sites: SGAG (LG2, 3250–3253 (Tapanadechopone et al., 1999)) and EGSGG and LGSGE (LG3, 3509–3512, 3593–3596, respectively

(Brown et al., 1997; Tapanadechopone et al., 1999)). Expression of full-- length Domain V in 293 cells (Brown et al., 1997) resulted in a product capable of bearing both HS and CS chains. In a subsequent report using rotary shadowing studies, it was shown that perlecan extracted from the EHS tumor and from fibroblast cultures carried three GAG chains at one of the molecule and a single GAG chain (27% of the time) at the Domain V (carboxyl terminus) end (Friedrich et al., 1999). Using the same 293 cell expression system but singly expressing a repertoire of smaller fragments (e.g., LG_1-EGF_1-EGF_2, EGF_1-EGF_2-LG_2, etc.), HS and CS glycanation were shown to occur in fragments containing LG_2-EGF_3-EGF_4-LG_3, but the EGSGG (3509–3512) sequence in the LG_3 module was the GAG acceptor site in this particular cell line (Friedrich et al., 1999). Tapanadechopone et al. (1999), using a similar fusion protein expression approach in COS-7 and CHO cells, demonstrated that the sequence LGSGE (LG_3, 3593–3596) could be utilized by those cells as an HS acceptor site; deletion of the SGE sequence eliminated the HS substitution in Domain V but promoted the assembly of CS/DS at the SGAG (LG_2, 3250–3253).

Besides serving as a secondary site for the addition of GAGs, Domain V was shown to be capable of binding heparin, sulfatides, nidogen, fibulin-2, α dystroglycan, β1 integrins, and progranulin (Brown et al., 1997; Friedrich et al., 1999; Gonzalez, Mongiat, Slater, Baffa, & Iozzo, 2003; Talts, Andac, Gohring, Brancaccio, & Timpl, 1999). Intact Domain V promoted β1 integrin-dependent, heparin-insensitive cell adhesion comparable to that seen with intact perlecan; the degree of adhesion to either perlecan or Domain V occurred in a cell-dependent manner (Brown et al., 1997). Domain V engaged α dystroglycan in a heparin-insensitive, EDTA (10 mM)-sensitive fashion (Talts et al., 1999), a behavior similar to that seen for the laminin α1 LG modules. A key requirement for this binding was presence of at least two Domain V LG modules (Friedrich et al., 1999) held in the proper conformation. Modification of the secondary structure of Domain V by reduction/alkylation or GAG substitution in Domain V reduced binding affinity to α dystroglycan by 20-fold (Friedrich et al., 1999). The binding of Domain V to heparin also was found to be conformation-sensitive and dependent upon the presence of EGF_1-EGF_2 within the fragment of Domain V being used in the assay (Friedrich et al., 1999). Sulfatides appear to engage Domain V differently from heparin, possibly engaging EGF_3-EGF_4 and one or the other flanking LG modules (Friedrich et al., 1999). High-affinity nidogen or laminin–nidogen complex binding to Domain V was shown to require the presence of the LG_2 module in the fragment; fibulin-2 binding appears to require either one or the other LG modules (LG_2 or LG_3) and flanking EGF_3-EGF_4 tandem (Friedrich et al., 1999). Progranulin, a mitogen and growth factor which has been associated with development, wound repair, neoplasia, and dementia (Bateman & Bennett, 2009), interacts with the LG_1-EGF_1-EGF_2 of Domain V, or alternatively with LG_2-EGF_3-EGF_4 at lower affinity (Gonzalez et al., 2003). Domain V also interacts with the C-terminal half of type XVIII collagen (Mongiat, Sweeney, San Antonio, Fu, & Iozzo, 2003), the region of type XVIII collagen that has been shown to be proteolytically cleaved to generate the antiangiogenic factor endostatin (Folkman, 2006; O'Reilly et al., 1997). This interaction was also shown to occur in vivo in several basement membranes, including those in the kidney (Miosge, Simniok, Sprysch, & Herken, 2003). The LG1-LG2 region of Domain V (and intact perlecan) was also shown to interact with VEGF receptors 1 and 2 with high affinity, but outside of the active site of the receptor (Willis, Poluzzi, Mongiat, & Iozzo, 2013). The actions of Domain V, also known as endorepellin, will be discussed in the following section.

3.1.4 Perlecan function in tissues

Interestingly, a question posed in the title of the review "Perlecan: how does one molecule do so many things?" (Knox & Whitelock, 2006) can be answered by the title of another review "Life without perlecan has its problems" (Olsen, 1999). Because of the overall abundance of perlecan in most organ/tissue systems and, from the former section, the high biological connectivity inherent in the intact perlecan molecule, it would be predicted that loss of perlecan would have catastrophic consequences on embryonic development. Complete knockout of perlecan from the murine genome has no perceptible effect on the earliest stages of embryonic development, but does result in an increased incidence of embryonic death between E10 and E12 (Arikawa-Hirasawa, Watanabe, Takami, Hassell, & Yamada, 1999; Costell et al., 1999). Those animals surviving until birth died shortly afterward due to respiratory distress (Arikawa-Hirasawa et al., 1999). The major anomalies seen in the prenatal animals dying between E10 and E12 were due to cardiac defects (Costell et al., 1999) or to defects in cephalic development (Arikawa-Hirasawa et al., 1999; Costell et al., 1999). The cardiomyocytes in the E10.5 animals were found to either entirely lack basement membranes or possess small focal aggregates of basement membrane material associated with their cell surfaces (Costell et al., 1999). The basement membranes of other tissues examined in the study were found intact (Costell et al., 1999), suggesting that basement membranes lacking perlecan were susceptible to damage due the onset of mechanical stress at the time of the initiation of active heart rhythms. In the CNS, compared to that seen in control animals, the surface ectoderm of the developing brain of the *Hspg2*$^{-/-}$ embryos had numerous clefts in the basement membrane underlying the neuroectoderm (Costell et al., 1999). The exencephaly commonly seen in the *Hspg2*$^{-/-}$ embryos was thought to be due in part to the inability of the basement membrane in those areas to resist the mechanical force in areas of neural tissue where enhanced intraventricular pressure occurs during brain vesicle expansion (Costell et al., 1999).

Perlecan had been first identified in the cartilage in the pericellular matrix (PCM) by SundarRaj, Fite, Ledbetter, Chakravarti, and Hassell (1995) and was shown to promote chondrocyte adhesion and differentiation in vitro (SundarRaj et al., 1995 and French et al., 1999). Based on these observations, it was not surprising that a key finding in the former studies (Arikawa-Hirasawa et al., 1999; Costell et al., 1999) was the development of a severe osteochondrodysplasia found in embryos starting at E14.5–15. The animals that survived until birth were dwarfed compared to controls, had craniofacial defects (e.g., cleft palate, short snouts, lack of calvarial bones with exencephaly or, when present, domed skulls), shortened limbs, and vertebral column defects (French et al., 1999). Although cartilage is not considered to be a basement membrane matrix per se, the PCM surrounding each chondrocyte has been shown to contain the full complement of basement membrane molecules (Kvist et al., 2008). Besides basement membrane components, the PCM also contains the molecule WARP (see above), which has the ability to interact with both perlecan and collagen VI (a microfibril forming collagen), potentially serving a bridging function between perlecan and type VI collagen (Hansen et al., 2012). At the same time, deletion of WARP does not recapitulate the osteochondrodystrophic phenotype seen in the *Hspg2*$^{-/-}$ mice, suggesting that the WARP–perlecan interaction is not critical for normal chondrocyte behavior within the PCM (Allen et al., 2009).

Although the PCM appears morphologically homogenous, recent immunofluorescence-guided atomic force microscopy mapping studies of this region (Wilusz, Defrate, & Guilak, 2012) have divided this matrix into three zones: (1) an outermost zone (adjacent to the cartilage matrix proper) having an increased abundance of type VI collagen and lacking perlecan; (2) an innermost zone (adjacent to the chondrocyte) having an increased abundance of perlecan and lacking type VI collagen; and (3) an intermediate zone (between zone 1 and 2) having both perlecan and type VI collagen within the matrix. The PCM elastic modulus (i.e., the relative resistance to stretch) has been shown to vary among the three zones; the intermediate zone has a lower elastic modulus (more compliant), being sandwiched between the inner and outer zones, which have comparably higher elastic moduli (Wilusz et al., 2012). In light of the fact that the HS chains on perlecan have been shown to sequester and present FGF-2 to cognate receptors on chondrocytes (Smith, West, Govindraj, et al., 2007), one hypothetical function for perlecan in the PCM is that it serves in a mechanotransductive capacity, binding and releasing FGFs via interactions with HS in response to the compression and unloading of cartilage (Vincent, McLean, Full, Peston, & Saklatvala, 2007). However, signaling to and from chondrocytes via the FGF-FGFR system may have several levels of contextual complexity. The signaling activity of perlecan-HS bound FGF-2 was shown to be dependent on the presence of CS attached to perlecan (i.e., a hybrid perlecan proteoglycan bearing both CS and HS); the presence of CS was shown to diminish the interaction between FGF-2 and its cognate receptors FGFR1 and FGFR3, thus decreasing the ability of the growth factors to enhance proliferation (Smith, West, Govindraj, et al., 2007). Removal of CS via treatment with the enzyme chondroitinase ABC enhanced the proliferative effects of perlecan in these studies (Vincent et al., 2007). Perlecan isolated from human fetal chondrocyte cultures has been shown to carry HS, CS, and KS (keratan sulfate) (Chuang et al., 2010); perlecan isolated from developing bovine growth plates have perlecan substituted in a 25/75 HS/CS ratio (Govindraj et al., 2002). Data from a more recent report (Chuang et al., 2010) suggest that the HS on chondrocyte-derived perlecan is capable of binding FGF-2 and forming ternary complexes with FGFR1 (or FGFR3), but the complexes are biologically inactive. The actions of FGF-18, which has been shown to colocalize with perlecan in the hypertrophic region in growth plates (Shu, Smith, Little, & Melrose, 2013), have been shown to be dependent on the nature of its interaction with perlecan. Binding of FGF-18 to Domain III of perlecan was shown to prevent its mitogenic activity in growth plate chondrocyte cultures (Smith et al., 2007), whereas binding of FGF-18 to chondrocyte-derived perlecan via HS was shown to enhance FGF-18 signaling via FGFR-3 and promote BaF32 cell proliferation (Chuang et al., 2010). To add yet another layer of complexity upon the chondrocyte-PCM system, the perlecan isolated from cartilage extracts has been found via western blotting to occur as several bands of differing molecular weights and GAG

I are not critical for the function of perlecan in cartilage, but the studies did not conclusively exclude the possibility that chondrocytes did not add HS to the other acceptor sites in Domain V. Given the multiple sources of HS within tissues, it was suggested that the pleiotropy reported in *Hspg*$^{\Delta 3/\Delta 3}$ mice was due to the retention of HS on other basement membrane HSPGs (e.g., agrin, type XVIII collagen) or cell surface HSPGs (e.g., syndecans and glypicans), which probably confers a degree of functional redundancy. Loss of the Domain I HS acceptor site in the *Hspg2*$^{\Delta 3/\Delta 3}$ mice did result in impaired angiogenesis and wound repair due to the loss of the ability to bind FGF-2 via the HS chains (Zhou et al., 2004). In the lens of *Hspg*$^{\Delta 3/\Delta 3}$ mutant mice (Rossi et al., 2003) ostensible changes, such as congenital cataracts and microphthalmia, were found. To explore the possibility of functional redundancy due to the presence of HS associated with other basement membrane proteoglycans, the overlay of a second genetic mutation, deletion of collagen XVIII, did accelerate the onset of the lens damage and did increase the extent and severity of the defects (Zhou et al., 2004).

Recent studies identified domain V, subsequently named endorepellin (Mongiat, Sweeney, San Antonio, Fu, & Iozzo, 2003), as a regulator of angiogenesis. Initial studies showed that the LG2 module of domain V interacted with the carboxyl terminal antiangiogenic fragment of type XVIII collagen, known as endostatin; the interaction of the two molecules significantly decreasing the antiangiogenic activity of endostatin (Mongiat et al., 2003). Endorepellin in nanomolar quantities inhibited VEGF-induced endothelial cell migration in vitro and in vivo in a VEGF-induced capillary ingrowth in a chick chorioallantoic membrane assay and matrigel dermal plug assays (Mongiat et al., 2003). The LG3 module of endorepellin, which can be proteolytically cleaved from intact endorepellin by BMP-1/tolloid family of metalloproteinases (Gonzalez et al., 2005), was shown to interact with α2β1 integrin at the cell surface, causing disassembly of focal adhesions and stress fibers in endothelial cells via rapid activation (within 5 min) of cAMP, PKA (protein kinase A), and FAK (focal adhesion kinase) (Bix et al., 2004). This interaction was shown to be mediated via the interaction of the LG3 module of endorepellin with the α2 integrin I domain (Woodall et al., 2008). The ability of endorepellin to affect α2β1 integrin-mediated endothelial cell migration was further demonstrated in vitro using siRNA knockdown of α2 integrin subunit in HUVEC cells or in microvascular endothelial cells derived from $\alpha2^{-/-}$ mice (Woodall et al., 2008). In the same report, endorepellin was further demonstrated to interact with α2β1 integrin on vascular endothelial cells in vivo; the absence of α2β1 integrin in a $\alpha2^{-/-}$ murine tumor xenograft model led to a significant decrease in the ability of systemically administered endorepellin to block angiogenesis and growth of tumors (Woodall et al., 2008).

Endorepellin binding to α2β1 integrin was shown to increase the activity of SHP-1 phosphatase both in vitro and in vivo, the activity of which caused a rapid decrease in phosphorylation of several RTK (receptor tyrosine kinases), leading to a decrease in signaling through at least one RTK pathway (VEGFR) (Nystrom et al., 2009). Subsequently, it was shown that endorepellin antagonism of VEGFR2 signaling was far more complex in nature. The full antiangiogenic effects of endorepellin required the coengagement of the VEGFR2 and α2β1 integrin simultaneously via intact endorepellin (Goyal et al., 2011); the LG3 module engaging the α2β1 integrin and the LG1-LG2 modules binding the Ig3−5 domains of VEGFR2 to inhibit its function in an allosteric fashion (Willis et al., 2013). This "dual receptor antagonism" (Goyal et al., 2011) affects VEGFR2 signaling via several modes of action: (1) The binding of endorepelling increases the rate of internalization and degradation of both receptors thus causing a net decrease of receptor abundance at the cell surface; (2) By bringing the α2β1 integrin/SHP signaling axis in close proximity to VEGFR2 leads to the dephosphorylation of VEGFR2 at key tyrosine residues. The latter action, in turn, has been shown to dow nregulate signaling through several pathways including pathways associated with PI3K signaling, PKC signaling, and calcineurin/NFAT signaling (Goyal et al., 2012); the net outcome of these events leads to the transcriptional repression of both HIF1α and VEGFA expression.

Interestingly, domain V has been shown to exert the opposite effects on the endothelial cells of the brain microvasculature, promoting angiogenesis after stroke in a rodent model thus providing significant neuroprotective effects (Lee et al., 2011). The proangiogenic effects for domain V were demonstrated to be mediated via the engagement of α5β1 integrin that results in enhanced signaling through ERK pathways (Clarke et al., 2012; Lee et al., 2011). Although domain V does not have a formal RGD-based recognition site for α5β1 integrin; however studies by Clarke et al. (2012) identified a DGR sequence in the EGF2 repeat that is responsible in part for the signaling via α5β1 integrin. The apparent dichotomous response of domain V/endorepellin effects between the brain microvasculature and other microvasculature has been proposed by Clark et al. (2012) to be due to (1) the lack of α2β1 integrin on the endothelial cells in the microvasculature of the CNS; (2) the expression of α2β1 integrin, the higher affinity domain V receptor, in the microvasculature in other organ systems. These latter findings serve to further highlight the fact that signaling from endorepellin/domain V is highly contextually dependent.

In the renal biology field, the HS associated with proteoglycans of the GBM has been thought to contribute in part to the charge selectivity associated with the process of ultrafiltration. When injected into the circulatory system of rat, antibodies directed against HS were found to cause a rapid and significant increase in urinary albumin excretion that was transient in nature, returning to normal levels within 24 h (Van Den Born et al., 1992). Conversely, injection of a polyclonal antiserum against the 130 kD core protein of the GBM HSPG led to the gradual thickening of the GBM and a dose-dependent proteinuria that persisted several months postinjection (Miettinen, Stow, Mentone, & Farquhar, 1986). Although these studies are part of the historical record in the renal proteoglycan biology field, their results serve to show that perturbing either the HS or HSPGs of the glomerular ultrafiltration barrier led to functional changes in the ultrafiltration process.

Ultrastructure studies suggested the possibility of a differential distribution of HSPGs within the trilaminar structure of the GBM. The immunogold labeling studies by Rada and Carlson (1991a) had localized the large 400 kD HSPG from the EHS tumor (perlecan) to the entire GBM. The smaller 130 kD GBM HSPG immunolocalized to the lamina rarae confirming a similar finding made earlier by the Farquhar laboratory (Miettinen et al., 1986). A second report by Rada and Carlson (1991b) found that over the timecourse between 5 and 15 days in a rodent model of puromycin aminonucleoside nephrosis (PAN), a significant reduction in the immunogold labeling for perlecan core protein did occur in the GBM, with only a modest reduction of immunolabeling for the 130 kD HSPG occurring at one time interval. However, the anionic charge density as indicated by polyethylenimine (PEI) staining in the PAN model did not change over the timecourse of the study.

In the adult mammal, the basement membrane proteoglycans that are known to persist in the GBM are agrin (see below, Groffen, Ruegg, et al., 1998), perlecan and type XVIII collagen (Kinnunen et al., 2011). Given the emphasis that the field of renal biology had placed upon the HSPGs of the GBM as active participants in glomerular ultrafiltration (for review see McCarthy and Wassenhove-McCarthy, 2012) it was surprising to find that the effects of the *Hspg*$^{\Delta 3/\Delta 3}$ deletion in the kidney were unremarkable (Rossi et al., 2003). The kidneys of *Hspg*$^{\Delta 3/\Delta 3}$ mutant mice had no significant changes in renal morphology, GBM anionic charge density, or in albuminuria/proteinuria. Albuminuria was, however, able to be induced in the *Hspg*$^{\Delta 3/\Delta 3}$ mutant mice compared to control mice when the mice were physiologically challenged using an albumin overload model (Morita et al., 2005). Anionic charge density of the GBM as indicated by PEI labeling was maintained in the *Hspg*$^{\Delta 3/\Delta 3}$ mutant mice. No compensatory increase in the expression of agrin within the GBM was seen, although there was evidence that CS substitution might be enhanced at alternative sites within Domain V (Morita et al., 2005). Although collagen type XVIII is also a basement membrane HSPG and is present in the GBM, given the redundancy of HS-bearing proteoglycans associated with the GBM the lack of damage in the double-mutant mice (*Hspg*$^{\Delta 3/\Delta 3}$/*Col18a1*$^{-/-}$) was also surprising (Rossi et al., 2003). Collagen type XVIII has been shown to be present in the GBM in developing kidneys, in the adult glomerulus the distribution of type XVIII was first reported be limited to the mesangial matrix and Bowman's capsule (Saarela, Rehn, Oikarinen, Autio-Harmainen, & Pihlajaniemi, 1998). In subsequent studies (Kinnunen et al., 2011) demonstrated the presence of type XVIII collagen in the GBM. Deletion of type XVIII collagen did result in the development of podocyte foot process effacement but the loss of type XVIII collagen did not significantly affect PEI labeling (i.e. anionic charge). Although these findings seemed to cast some doubt with regard to the notion that basement membrane HSPGs serve as the charge selective barrier for ultrafiltration, the studies as performed did not eliminate the contribution of agrin to the presence of anionic charge within the GBM.

3.2 Agrin

Agrin was not "officially" recognized as a basement membrane proteoglycan until reported as such by Tsen et al. (1995). However, an earlier body of work, evolving from studies exploring the role of a proteoglycan isolated from the electric organ of *Torpedo* originally referred to as TAP-1 (Carlson, Caroni, & Kelly, 1986; Carlson & Wight, 1987; Iwata & Carlson, 1991), provides a historical, but remote, association with the role of proteoglycans and the synaptic junction. Although the research into TAP-1 biology eventually moved in another direction, at the same time that TAP-1 was being characterized biochemically as a proteoglycan, Nitkin et al. identified several peptides isolated from the same tissue that possessed the ability to aggregate acetylcholine receptors in a chick myotube-based assay. The properties of these peptides prompted the authors to name the molecule "agrin," derived from the Greek word "ageirein" which means "to gather" (Nitkin et al.,

1987). Unlike the molecular characterization of perlecan, which was reported nearly 11 years after the initial discovery of the proteoglycan core protein, the initial molecular characterization of agrin by Rupp et al. (1991) took place within 4 years of the initial isolation of the core protein.

3.2.1 Agrin gene and mRNA

The gene encoding agrin (*AGRN*) in humans is on chromosome 1 (1p36.33), and the mouse ortholog is on chromosome 4 (Rupp et al., 1992). The human gene contains 40 exons spanning 42,691 bases; the murine ortholog (*Agrn*) contains 43 exons (Rupp et al., 1992). As with perlecan, the exon organization reflects the domain organization of the final protein product (Rupp et al., 1992, 1991). The mRNA encoding agrin has been shown to undergo extensive alternative splicing at both the 5′ and 3′ ends, resulting in the possibility of a minimum of six different isoforms (Burgess, Skarnes, & Sanes, 2000; Neumann et al., 2001; Ruegg et al., 1992). The 5′ end of the *Agrn* gene encodes for two possible transcriptional start sites, the use of which will give rise to an mRNA which encodes for the secreted form of agrin (SSNtA agrin) or a membrane-anchored agrin (TM agrin) (Neumann et al., 2001); alternative splicing yields mRNAs which ultimately encode for a protein having a unique incorporation of either 49 amino acids (SN or short amino form) or 150 amino acids (LN or long amino form) at the amino terminus of the protein (Burgess et al., 2000). In the LN form of agrin, the SSNtA domain is encoded by three exons, and in the SN form the TM domain is encoded by one exon; each splice form is capable of being added to a common linker exon (Burgess, Nguyen, Son, Lichtman, & Sanes, 1999; Neumann et al., 2001) that precedes the first exon reported by Rupp et al. (1992) in their original characterization of the agrin gene. Expression of alternative splice variants encoding for TM agrin (SN form) appears to be restricted to the CNS, whereas expression of SSNtA agrin (i.e., basal lamina associated or LN form) has been shown to occur in most tissues throughout the body (Burgess et al., 2000; Neumann et al., 2001). This mode of alternative splicing places agrin in a unique position in terms of function within the proteoglycan superfamily; it can serve as either a cell surface or basement membrane proteoglycan depending on which splice variant is produced by the cell.

Near the 3′ end of the gene, the exon 19 donor site can be spliced to one of two acceptor sites within exon 20; those variants are referred to as X+ or X−. When incorporated into the resultant mRNA, X+ results in the addition of nine amino acids, but the function of this splice variant in mammalian systems is uncertain (Rupp et al., 1992). Alternative splicing of exon 28 gives rise to Y+ and Y− isoforms; splicing events among exons 31, 32, 33, and 34 are capable of giving rise to four alternative splice variants within this region collectively referred to as Z+ or Z−. The protein variants encoded by alternative splicing within the Z region (discussed in detail below) result in a final protein product having 0, 8, 11, or 19 (8 + 11) additional amino acids incorporated into the agrin sequence (Campanelli, Gayer, & Scheller, 1996; Ferns, Campanelli, Hoch, Scheller, & Hall, 1993; Gesemann et al., 1996; O'Toole et al., 1996; Rupp et al., 1992, 1991).

3.2.2 The domain organization of agrin

The mRNA for agrin encodes for a core protein ~200−220 kD in size. As with perlecan, the structural organization of agrin is modular. After the common linkage region discussed above, the original sequence information reported by Rupp et al. (1992, 1991) indicates a series of follistatin-type repeats (nine total) that follow. The first seven of these repeats are represented in the gene by exons 1−7 (Rupp et al., 1992, 1991), and repeats 8 (exons 8 and 9) and 9 (exons 13 and 14) are separated by intervening sequences (Rupp et al., 1992, 1991). The intervening sequence immediately following the 8th follistatin repeat is a cysteine-rich region (exon 10) followed by two laminin-type EGF repeats (Groffen, Buskens, et al., 1998; Rupp et al., 1992, 1991) encoded by exons 11 and 12. Exon 15 encodes the first of two serine/threonine-rich regions followed by a second cysteine-rich region (exon 16) partially homologous (22%) to the previous cysteine-rich region (Rupp et al., 1992). This is followed by a SEA module (Bork & Patthy, 1995) and the second serine/threonine region; as referred to above, alternative splicing of this second serine/threonine-rich region (X+ at the 5′ end of exon 20) also has been reported to occur.

The most characterized and heavily investigated region of agrin lies between exons 20 and 35 (Rupp et al., 1992). This is the part of the gene encoding the critical region of the agrin core protein that regulates synaptic organization and aggregation of acetylcholine receptors at the neuromuscular junction (Burgess et al., 1999; Campanelli et al., 1996; Gesemann et al., 1996; O'Toole et al., 1996). This region of agrin contains four EGF-like repeats flanking three laminin G modules in the order EGF1-LG1-EGF2-EGF3-LG2-EGF4 (Burgess et al., 1999; Campanelli et al., 1996; Gesemann et al., 1996; O'Toole et al., 1996; Rupp et al., 1992). These elements are represented within the gene starting for EGF1 within exon 20, LG1 (exons 21–22), EGF2-EGF3 (exons 23–24), LG2 (exons 26–30), EGF4 (exon 30), and LG3 (exons 34–35) (Rupp et al., 1992). Splicing exon 28 into the agrin mRNA results in the addition of four amino acids having the sequence KSRK near the end of the LG2 module, which inserts an additional positively charged sequence that permits the engagement of HS (Burgess et al., 1999; Campanelli et al., 1996; Gesemann et al., 1996; O'Toole et al., 1996). Exons 32 and 33 encode for mRNAs representing either an 8 (HLSNEIPA) or 11 (PDALDYPAEPS) amino acid sequence insert, respectively. Alternative splicing can result in either no insertion of intervening sequences (generating an mRNA with exons 31 and 34), the insertion each exon singly (exons 31-32-34 or 31-33-34), or inclusion of both exons 32 and 33 (Burgess et al., 1999; Campanelli et al., 1996; Gesemann et al., 1996; O'Toole et al., 1996).

3.2.3 Bioactivities of agrin domains

As with perlecan, the agrin core protein is inherently bioactive. Some of its activities have been well defined, especially for the carboxyl half of agrin, which is known to be a key element in synapse formation. Unlike perlecan, there are no numerical designations assigned for each region of agrin. The discussion below is an outline of activity starting from the amino end of the molecule.

3.2.3.1 Activities surrounding the agrin amino end

The mRNA encoding the amino terminus of agrin has been shown to undergo alternative splicing (as described above), thus allowing the cell to produce a common protein (~1900 amino acids) but having two uniquely functional amino terminal ends (Burgess et al., 2000; Neumann et al., 2001). Besides containing the necessary signal peptide for extracellular secretion, the LN form (SSNtA/secreted) of agrin has been shown to interact with laminin-111(Denzer, Brandenberger, Gesemann, Chiquet, & Ruegg, 1997). Rotary shadowing studies placed the interaction at one single point on the laminin long arm within the laminin coiled coil domain (Denzer et al., 1998); subsequent protein expression studies were able to identify a 20 amino acid region present in the laminin $\gamma1$ chain that served as the agrin-binding site; similar binding to the $\gamma2$ chain was not seen (Kammerer et al., 1999).

SN-agrin (TM-agrin) is produced when alternative splicing adds the exon encoding for a 49 amino acid segment to the front of the common agrin protein component, the addition of which switches the functional orientation of agrin from a basement membrane proteoglycan to that of a cell surface proteoglycan having a type II transmembrane orientation (Burgess et al., 1999; Neumann et al., 2001). In neuronal cell cultures, antibody-induced clustering of TM-agrin leads to the development of filopodia on axons and dendrites, an activity shown to be calcium-dependent but not directly dependent upon the activity of PKC or phosphatase activity (Annies et al., 2006). Unlike other cell surface proteoglycans such as members of the syndecan family (Multhaupt et al., 2009), the cytoplasmic tail of SN-agrin does not directly participate/activate signaling events associated with the formation of filopodia (McCroskery, Chaudhry, Lin, & Daniels, 2006; Porten et al., 2010). The engagement of heparan side chains on TM-agrin was shown to facilitate filopodia formation via enhanced intracellular CDC42 and Rac activity (Lin et al., 2010). Besides the activity of the HS chains in facilitating filopodia formation, the follistatin region in agrin, in particular the 7th domain, has been shown to play a role in this process (Porten et al., 2010). Given the observation that the cytoplasmic tail of TM-agrin does not appear to have any inherent signaling activity, the mechanism by which it influences neuronal filopodia formation may be indirect, possibly enhancing the formation of lipid rafts and thus setting up discrete signaling loci in neurons (Ramseger, White, & Kroger, 2009).

Besides its recently reported role in filopodia formation, the follistatin region of agrin has been reported to bind/engage TGF-β family members, N-CAM, tenascin, laminin, nephronectin, and FGF-2 via HS (Banyai, Sonderegger, & Patthy, 2010; Cotman, Halfter, & Cole, 1999; Sato et al., 2013). As mentioned earlier, HS substitution on the agrin core protein was demonstrated by Tsen et al. (1995). There are 20 possible SG residues within the entire amino acid sequence of chick agrin (Winzen, Cole, & Halfter, 2003). Two of those possible sites occur such that the SG motifs are in the proper context for HS substitution (Wang, Julenius, Hryhorenko, & Hagen, 2007; Zhang, David, & Esko, 1995; Zhang & Esko, 1994); one occurs between the 7th and 8th follistatin domain, and the other within the serine/threonine-rich domain immediately before the SEA module. Two putative sites having the same location for HS substitution were reported for murine agrin (Harvey et al., 2007). For human agrin, three possible sites in the proper context for GAG substitution were identified, but the two sites having the same location as seen for murine and ovine agrin were considered to be optimal sites for GAG substitution (Groffen, Buskens, et al., 1998). Although in the ovine model agrin was reported to exist as a hybrid proteoglycan capable of bearing both HS and CS chains (Winzen et al., 2003), in mammals agrin is routinely found substituted with HS, especially in the basement membranes of the glomerular capillaries (Groffen, Buskens, et al., 1998; Groffen, Ruegg, et al., 1998; Harvey et al., 2007).

3.2.3.2 Activities surrounding the agrin carboxyl end

Investigations into the bioactivity of agrin domains within the 95 kD region comprising the carboxyl end of the molecule have had heavy focus in the body of the agrin literature, as this region was first identified as playing a key role in agrin's ability to regulate the development and organization of the neuromuscular junction. As mentioned above, this region consists of four EGF-like repeats flanking three laminin G modules in the order EGF1-LG1-EGF2-EGF3-LG2-EGF4-LG3. This segment can be differentially modified by alternative splicing at two sites that are present within the LG2 module (Y site) or directly adjacent to the start of the LG3 module (Z site) (Burgess et al., 1999; Campanelli et al., 1996; Gesemann et al., 1996; O'Toole et al., 1996; Rupp et al., 1992). As also mentioned above, alternative splicing of the Y site exon leads to the insertion of four amino acids, KSRK, the presence of which permit HS binding. Dystroglycan, one cell surface receptor capable of binding agrin (Yamada et al., 1996; for review see Moore & Winder, 2010) in muscle and other tissues such as kidney and lung (Gesemann, Brancaccio, Schumacher, & Ruegg, 1998), engages agrin via the region spanning the LG1-EGF2-EGF3-LG2 modules of agrin in a heparan-independent fashion (Gesemann et al., 1996). The interaction between dystroglycan and agrin in this region is inhibited when the Y+ is present and binding heparan (O'Toole et al., 1996). LRP4, a member of the LDL receptor family, has been shown to bind both agrin and the receptor tyrosine kinase MuSK and facilitate MuSK receptor phosphorylation (Zhang et al., 2008). LRP4 interacts at its amino terminus (Zhang, Coldefy, Hubbard, & Burden, 2011) with the Z+ agrin alternative splice variant (i.e., the Z8 or exon 32 splice variant) (Kim et al., 2008). Integrins, in particular $\alpha v\beta 1$, interact with the LG2 domain; a second site for $\beta 1$ integrin binding falls within the EGF4 domain (Burgess, Dickman, Nunez, Glass, & Sanes, 2002).

3.2.4 Agrin function in tissues

The majority of the research with regard to understanding the function of agrin in tissues/tissue structure has focused on identifying the mechanism of action of agrin in acetylcholine receptor clustering in neuromuscular junction formation (for reviews, see Bezakova & Ruegg, 2003; Daniels,

2012; Ngo, Noakes, & Phillips, 2007). A thorough discussion of this rather formidable body of work would merit a separate chapter on its own. Outside of that particular body of work, our understanding of agrin function in other basement membranes is an area of future growth.

Agrin has been shown to play a role in the maintenance of the hematopoietic stem cell niche (Mazzon et al., 2011) and in the maturation of monocytes (Mazzon et al., 2012). In lymphocytes, agrin was identified to play a role in receptor clustering at the "immunological synapse" by inducing clustering of lipid rafts (Khan, Bose, Yam, Soloski, & Rupp, 2001), its expression upregulated upon lymphocyte activation (Zhang et al., 2006). Although the agrin receptor α-dystroglycan is found on cells of monocytic lineage and signaling via the engagement of agrin was associated with monocyte maturation and survival (Mazzon et al., 2012), the role of α-dystroglycan in the formation of the immunological synapse (Zhang et al., 2006) needs further investigation (Kabouridis et al., 2012).

Interestingly, as with perlecan, agrin is also expressed by chondrocytes, agrin being found in the resting, prehypertrophic, and proliferating chondrocyte zones of growth plates (Hausser, Ruegg, Brenner, & Ksiazek, 2007). Since the phenotype of the *Agrn*$^{-/-}$ mouse is lethal, a transgenic animal model was developed in which chick agrin (y+z+) was expressed using a motor neuron-specific promoter (Hausser et al., 2007). The "rescued" mutant mice (Tg/agrn$^{-/-}$ mice) had growth retardation that manifested itself in shortening of the long bones and spinal column; the thickness of the hypertrophic zone in the growth plate was decreased due to decreased proliferation of chondrocytes in the mutant mice (Hausser et al., 2007). However, the Tg/agrn$^{-/-}$ mice did not recapitulate the severe osteochondrodysplasia seen in the *Hspg2*$^{-/-}$ mutant mouse. A recent report explored the role of agrin in chondrocytes found in articular cartilage in normal and osteoarthritic (OA) cartilage (Eldridge et al., 2015). Results of microarray analysis studies showed that several of the known molecular components involved in agrin signaling, such as LRP4, α dystroglycan, and MusK were present in cartilage. In human OA cartilage, agrin immunostaining was decreased within the PCM; similar results were seen in a murine model of induced OA (Eldridge et al., 2015). Overexpression of agrin in chondrocytes in vitro led to an increase in SOX9 expression and an increase in the expression of mRNAs encoding for type II collagen and aggrecan, which are downstream targets of SOX9 (Eldridge et al., 2015). The enhanced expression of SOX9 was able to be blocked using antibodies against the agrin receptor α dystroglycan; however the increase in SOX9 expression could be blocked using an siRNA against Lrp4. Although these observations are interesting, what remains unclear is the manner by which agrin signaling is accomplished in this system; since LRP4 binds z+ agrin, how does the interaction between LRP4 and agrin (y-z-) occur?

In the kidney, Groffen, Ruegg, et al. (1998) first showed that agrin was a major HSPG within the GBM. Immunogold labeling/electron microscopy studies showed agrin labeling to be uniform within the GBM, whereas labeling for perlecan was limited to the endothelial side of the GBM (Groffen, Ruegg, et al., 1998). Using a panel of antibodies directed against the N and C termini of agrin, the agrin present in GBM was characterized as occurring as the full-length molecule having both N and C termini; in other renal basement membranes antibodies directed against the C terminus of agrin were nonreactive, suggesting the possibility that posttranslational processing of agrin occurred in some renal basement membranes (Raats et al., 1998). A recent study which correlated STORM (Stochastic Optical Reconstruction Microscopy) with electron microscopy imaging, Suleiman et al (Suleiman et al., 2013) used antibodies directed against the agrin N and C terminus to further refine the distribution of agrin within the GBM. The findings showed that agrin localized in a bilaminar fashion, perpendicular or slightly oblique to the plane of the GBM, with the C terminus of agrin nearest the cells (both podocytes and endothelial cells) and the N terminus of agrin nearest the central region of the GBM. In proteinuric diseases, such as diabetic nephropathy (Yard et al., 2001) or in animal models of renal disease (Raats et al., 1999), the intensity of agrin immunostaining was decreased compared to controls. In a murine model of Alport's syndrome, Suleiman et al (ibid) showed that the localization of agrin within the GBM was disrupted in focal areas, showing a breakdown of the discrete bilayer organization of agrin into a diffuse labeling pattern across the GBM proper.

In light of the fact that the Hspg2Δ3/Δ3 mouse did not develop significant albuminuria/proteinuria, Harvey et al. (2007) revisited the concept of HSPGs serving as a charge-selective barrier for ultrafiltration by generating a murine model of podocyte-specific knockout for agrin. Using antibodies directed against glomerular HS, agrin mutant mice showed a significant decrease in HS staining of the GBM compared to that seen in control animals (Harvey et al., 2007). PEI staining, a marker of the presence of anionic sites within the GBM, was significantly decreased in the agrin mutant animals compared to control. No upregulation of perlecan nor changes in the expression of nidogen, dystroglycan, or α3 integrin were seen (Harvey et al., 2007). Despite the loss of anionic charge, the mutant mice did not develop proteinuria/albuminuria compared to controls. Therefore, to once again address the issue of redundancy/functional compensation by another basement membrane HS, a double-mutant mouse, having both the *Hspg*$^{\Delta 3/\Delta 3}$ modification and lacking agrin in the GBM, was developed (Goldberg, Harvey, Cunningham, Tryggvason, & Miner, 2009). Ironically, this double-mutant mouse also showed no substantial changes in morphology, GBM charge, or glomerular function beyond what was reported for the agrin mutant mouse model (Harvey et al., 2007).

Now we understand the handles……what about the bristles?…Final thoughts

Whenever explaining to those outside the proteoglycan field what makes a molecular species a member of the proteoglycan family, I quickly revert to using the well-worn analogy which employs the common laboratory bottle brush as an example, assigning the handle to represent the core protein and the bristles to represent the linear carbohydrate structures of the glycosaminoglycan (GAG) chains.

Although the majority of investigators who choose to explore the field of proteoglycan biology appear to gravitate first to examining the various facets of the structure and biology associated with the core proteins, many seem to overlook the importance of the highly bioactive GAG chains. There have been numerous reviews over the past several decades that have summarized key milestones in our understanding of the GAG biosynthetic pathway (Bishop, Schuksz, & Esko, 2007; Esko, Kimata, & Lindahl, 2009; Esko & Lindahl, 2001; Esko & Selleck, 2002; Kreuger, Spillmann, Li, & Lindahl, 2006; Lindahl, Kusche-Gullberg, & Kjellen, 1998; Sugahara et al., 2003). The majority of the key enzymes (25+) involved with GAG synthesis and modification have been now been characterized at the molecular level. The assembly system for GAGs is inherently complicated; unlike DNA/RNA or protein synthesis, GAG synthesis is not template driven. Rather GAG synthesis relies on a series of highly orchestrated steps between the actions of the biosynthetic and modifying enzymes. Several laboratories within the glycobiology field have continued their efforts to discern the rules that govern GAG assembly, but there still remain significant gaps in the knowledge with regard to the events surrounding GAG assembly and modification.

Current research in the area of GAG biology has highlighted the fact that specific modifications to the GAGs (e.g., sulfation) regulate the binding of growth factors, morphogens, and cytokines. It is also known that the cell is capable of changing the pattern of these modications in response to the underlying disease process, possibly changing the context for intermolecular binding and potentially changing the inherent bioactivity of the intact proteoglycan molecule. Given the fact that GAGs have the potential to interact with >400 different targets, it becomes important to discern the manner by which GAG structures are modified by the cell in response to disease. In the protein field, obtaining the basic amino acid sequence information of a protein accelerated tremendously when the technology moved from amino acid sequencing (via Edman degradation methods) to deduce the amino acid sequence from the mRNA sequence encoding for the protein. A similar leap forward in technology that would allow the field to readily and rapidly gain information about specific GAG structures made by cells in both homeostasis and disease is the next great step forward.

REFERENCES

Abe, N., Muragaki, Y., Yoshioka, H., Inoue, H., & Ninomiya, Y. (1993). Identification of a novel collagen chain represented by extensive interruptions in the triple-helical region. *Biochemical and Biophysical Research Communications, 196*(2), 576–582. http://dx.doi.org/10.1006/bbrc.1993.2288.

Allen, J. M., Bateman, J. F., Hansen, U., Wilson, R., Bruckner, P., Owens, R. T., et al. (2006). WARP is a novel multimeric component of the chondrocyte pericellular matrix that interacts with perlecan. *Journal of Biological Chemistry, 281*(11), 7341–7349. http://dx.doi.org/10.1074/jbc.M513746200.

Allen, J. M., Brachvogel, B., Farlie, P. G., Fitzgerald, J., & Bateman, J. F. (2008). The extracellular matrix protein WARP is a novel component of a distinct subset of basement membranes. *Matrix Biology, 27*(4), 295–305. http://dx.doi.org/10.1016/j.matbio.2008.01.005.

Allen, J. M., Zamurs, L., Brachvogel, B., Schlotzer-Schrehardt, U., Hansen, U., Lamande, S. R., et al. (2009). Mice lacking the extracellular matrix protein WARP develop normally but have compromised peripheral nerve structure and function. *Journal of Biological Chemistry, 284*(18), 12020–12030. http://dx.doi.org/10.1074/jbc.M806968200.

Annies, M., Bittcher, G., Ramseger, R., Loschinger, J., Woll, S., Porten, E., et al. (2006). Clustering transmembrane-agrin induces filopodia-like processes on axons and dendrites. *Molecular and Cellular Neuroscience, 31*(3), 515–524. http://dx.doi.org/10.1016/j.mcn.2005.11.005.

Arikawa-Hirasawa, E., Le, A. H., Nishino, I., Nonaka, I., Ho, N. C., Francomano, C. A., et al. (2002). Structural and functional mutations of the perlecan gene cause Schwartz-Jampel syndrome, with myotonic myopathy and chondrodysplasia. *American Journal of Human Genetics, 70*(5), 1368–1375. http://dx.doi.org/10.1086/340390.

Arikawa-Hirasawa, E., Rossi, S. G., Rotundo, R. L., & Yamada, Y. (2002). Absence of acetylcholinesterase at the neuromuscular junctions of perlecan-null mice. *Nature Neuroscience, 5*(2), 119–123. http://dx.doi.org/10.1038/nn801.

Arikawa-Hirasawa, E., Watanabe, H., Takami, H., Hassell, J. R., & Yamada, Y. (1999). Perlecan is essential for cartilage and cephalic development. *Nature Genetics, 23*(3), 354–358. http://dx.doi.org/10.1038/15537.

Arikawa-Hirasawa, E., Wilcox, W. R., Le, A. H., Silverman, N., Govindraj, P., Hassell, J. R., et al. (2001). Dyssegmental dysplasia, Silverman-Handmaker type, is caused by functional null mutations of the perlecan gene. *Nature Genetics, 27*(4), 431–434. http://dx.doi.org/10.1038/86941.

Bangratz, M., Sarrazin, N., Devaux, J., Zambroni, D., Echaniz-Laguna, A., Rene, F., et al. (2012). A mouse model of Schwartz-Jampel syndrome reveals myelinating Schwann cell dysfunction with persistent axonal depolarization in vitro and distal peripheral nerve hyperexcitability when perlecan is lacking. *American Journal of Pathology, 180*(5), 2040–2055. http://dx.doi.org/10.1016/j.ajpath.2012.01.035.

Banyai, L., Sonderegger, P., & Patthy, L. (2010). Agrin binds BMP2, BMP4 and TGFbeta1. *PLoS One, 5*(5), e10758. http://dx.doi.org/10.1371/journal.pone.0010758.

Bateman, A., & Bennett, H. P. (2009). The granulin gene family: from cancer to dementia. *Bioessays, 31*(11), 1245–1254. http://dx.doi.org/10.1002/bies.200900086.

Bauche, S., Boerio, D., Davoine, C. S., Bernard, V., Stum, M., Bureau, C., et al. (2013). Peripheral nerve hyperexcitability with preterminal nerve and neuromuscular junction remodeling is a hallmark of Schwartz-Jampel syndrome. *Neuromuscular Disorders, 23*(12), 998–1009. http://dx.doi.org/10.1016/j.nmd.2013.07.005.

Bernfield, M. R., & Banerjee, S. D. (1972). Acid mucopolysaccharide (glycosaminoglycan) at the epithelial-mesenchymal interface of mouse embryo salivary glands. *Journal of Cell Biology, 52*(3), 664–673. Retrieved from http://www.ncbi.nlm.nih.gov/entrez/query.fcgi?cmd=Retrieve&db=PubMed&dopt=Citation&list_uids=4109689.

Bishop, J. R., Schuksz, M., & Esko, J. D. (2007). Heparan sulphate proteoglycans fine-tune mammalian physiology. *Nature, 446*(7139), 1030–1037. Retrieved from http://www.ncbi.nlm.nih.gov/entrez/query.fcgi?cmd=Retrieve&db=PubMed&dopt=Citation&list_uids=17460664.

Bix, G., Fu, J., Gonzalez, E. M., Macro, L., Barker, A., Campbell, S., et al. (2004). Endorepellin causes endothelial cell disassembly of actin cytoskeleton and focal adhesions through alpha2beta1 integrin. *Journal of Cell Biology, 166*(1), 97–109. http://dx.doi.org/10.1083/jcb.200401150.

Bork, P., & Patthy, L. (1995). The SEA module: a new extracellular domain associated with O-glycosylation. *Protein Science, 4*(7), 1421–1425. Retrieved from http://www.ncbi.nlm.nih.gov/entrez/query.fcgi?cmd=Retrieve&db=PubMed&dopt=Citation&list_uids=7670383.

Brown, J. C., Sasaki, T., Gohring, W., Yamada, Y., & Timpl, R. (1997). The C-terminal domain V of perlecan promotes beta1 integrin-mediated cell adhesion, binds heparin, nidogen and fibulin-2 and can be modified by glycosaminoglycans. *European Journal of Biochemistry, 250*(1), 39–46. Retrieved from http://www.ncbi.nlm.nih.gov/pubmed/9431988.

Burgess, R. W., Dickman, D. K., Nunez, L., Glass, D. J., & Sanes, J. R. (2002). Mapping sites responsible for interactions of agrin with neurons. *J Neurochem, 83*(2), 271–284. Retrieved from http://www.ncbi.nlm.nih.gov/pubmed/12423238.

Burgess, R. W., Nguyen, Q. T., Son, Y. J., Lichtman, J. W., & Sanes, J. R. (1999). Alternatively spliced isoforms of nerve- and muscle-derived agrin: their roles at the neuromuscular junction. *Neuron, 23*(1), 33–44. Retrieved from http://www.ncbi.nlm.nih.gov/pubmed/10402191.

Burgess, R. W., Skarnes, W. C., & Sanes, J. R. (2000). Agrin isoforms with distinct amino termini: differential expression, localization, and function. *Journal of Cell Biology, 151*(1), 41–52. Retrieved from http://www.ncbi.nlm.nih.gov/entrez/query.fcgi?cmd=Retrieve&db=PubMed&dopt=Citation&list_uids=11018052.

Bezakova, G., & Ruegg, M. A. (2003). New insights into the roles of agrin. *Nature Reviews Molecular Cell Biology, 4*(4), 295–308. http://dx.doi.org/10.1038/nrm1074.

Campanelli, J. T., Gayer, G. G., & Scheller, R. H. (1996). Alternative RNA splicing that determines agrin activity regulates binding to heparin and alpha-dystroglycan. *Development, 122*(5), 1663–1672. Retrieved from http://www.ncbi.nlm.nih.gov/entrez/query.fcgi?cmd=Retrieve&db=PubMed&dopt=Citation&list_uids=8625852.

Carlson, S. S., Caroni, P., & Kelly, R. B. (1986). A nerve terminal anchorage protein from electric organ. *Journal of Cell Biology, 103*(2), 509–520. Retrieved from http://www.ncbi.nlm.nih.gov/pubmed/2426282.

Carlson, S. S., & Wight, T. N. (1987). Nerve terminal anchorage protein 1 (TAP-1) is a chondroitin sulfate proteoglycan: biochemical and electron microscopic characterization. *Journal of Cell Biology, 105*(6 Pt 2), 3075–3086. Retrieved from http://www.ncbi.nlm.nih.gov/pubmed/3693407.

Caulfield, J. P., & Farquhar, M. G. (1976). Distribution of annionic sites in glomerular basement membranes: their possible role in filtration and attachment. *Proceedings of the National Academy of Sciences of the United States of America, 73*(5), 1646–1650. Retrieved from http://www.ncbi.nlm.nih.gov/pubmed/1064037.

Chakravarti, S., Horchar, T., Jefferson, B., Laurie, G. W., & Hassell, J. R. (1995). Recombinant domain III of perlecan promotes cell attachment through its RGDS sequence. *Journal of Biological Chemistry, 270*(1), 404–409. Retrieved from http://www.ncbi.nlm.nih.gov/pubmed/7814401.

Chakravarti, S., Phillips, S. L., & Hassell, J. R. (1991). Assignment of the perlecan (heparan sulfate proteoglycan) gene to mouse chromosome 4. *Mammalian Genome, 1*(4), 270–272. Retrieved from http://www.ncbi.nlm.nih.gov/pubmed/1686572.

Chuang, C. Y., Lord, M. S., Melrose, J., Rees, M. D., Knox, S. M., Freeman, C., et al. (2010). Heparan sulfate-dependent signaling of fibroblast growth factor 18 by chondrocyte-derived perlecan. *Biochemistry, 49*(26), 5524–5532. http://dx.doi.org/10.1021/bi1005199.

Clarke, D. N., Al Ahmad, A., Lee, B., Parham, C., Auckland, L., Fertala, A., et al. (2012). Perlecan domain V induces VEGf secretion in brain endothelial cells through integrin alpha5beta1 and ERK-dependent signaling pathways. *PLoS One, 7*(9), e45257. http://dx.doi.org/10.1371/journal.pone.0045257.

Cohen, I. R., Grassel, S., Murdoch, A. D., & Iozzo, R. V. (1993). Structural characterization of the complete human perlecan gene and its promoter. *Proceedings of the National Academy of Sciences of the United States of America, 90*(21), 10404–10408. Retrieved from http://www.ncbi.nlm.nih.gov/pubmed/8234307.

Cohn, R. H., Banerjee, S. D., & Bernfield, M. R. (1977). Basal lamina of embryonic salivary epithelia. Nature of glycosaminoglycan and organization of extracellular materials. *Journal of Cell Biology, 73*(2), 464–478. Retrieved from http://www.ncbi.nlm.nih.gov/entrez/query.fcgi?cmd=Retrieve&db=PubMed&dopt=Citation&list_uids=870502.

Copley, A. L., & Scheinthal, B. M. (1970). Nature of the endoendothelial layer as demonstrated by ruthenium red. *Experimental Cell Research, 59*(3), 491–492. Retrieved from http://www.ncbi.nlm.nih.gov/pubmed/4190850.

Costell, M., Gustafsson, E., Aszodi, A., Morgelin, M., Bloch, W., Hunziker, E., et al. (1999). Perlecan maintains the integrity of cartilage and some basement membranes. *Journal of Cell Biology, 147*(5), 1109–1122. Retrieved from http://www.ncbi.nlm.nih.gov/pubmed/10579729.

Costell, M., Mann, K., Yamada, Y., & Timpl, R. (1997). Characterization of recombinant perlecan domain I and its substitution by glycosaminoglycans and oligosaccharides. *European Journal of Biochemistry, 243*(1–2), 115–121. Retrieved from http://www.ncbi.nlm.nih.gov/pubmed/9030729.

Costell, M., Sasaki, T., Mann, K., Yamada, Y., & Timpl, R. (1996). Structural characterization of recombinant domain II of the basement membrane proteoglycan perlecan. *FEBS*

Letters, 396(2–3), 127–131. Retrieved from http://www.ncbi.nlm.nih.gov/pubmed/8914972.

Cotman, S. L., Halfter, W., & Cole, G. J. (1999). Identification of extracellular matrix ligands for the heparan sulfate proteoglycan agrin. *Experimental Cell Research, 249*(1), 54–64. http://dx.doi.org/10.1006/excr.1999.4463.

Couchman, J. R. (1987). Heterogenous distribution of a basement membrane heparan sulfate proteoglycan in rat tissues. *Journal of Cell Biology, 105*, 1901–1916.

Couchman, J. R., Caterson, B., Christner, J. E., & Baker, J. R. (1984). Mapping by monoclonal antibody detection of glycosaminoglycans in connective tissues. *Nature, 307*, 650–652.

Couchman, J. R., Woods, A., Hook, M., & Christner, J. E. (1985). Characterization of a dermatan sulfate proteoglycan synthesized by murine parietal yolk sac (PYS-2) cells. *Journal of Biological Chemistry, 260*, 13755–13762.

Daniels, M. P. (2012). The role of agrin in synaptic development, plasticity and signaling in the central nervous system. *Neurochemistry International, 61*(6), 848–853. http://dx.doi.org/10.1016/j.neuint.2012.02.028.

Denzer, A. J., Brandenberger, R., Gesemann, M., Chiquet, M., & Ruegg, M. A. (1997). Agrin binds to the nerve-muscle basal lamina via laminin. *Journal of Cell Biology, 137*(3), 671–683. Retrieved from http://www.ncbi.nlm.nih.gov/pubmed/9151673.

Denzer, A. J., Schulthess, T., Fauser, C., Schumacher, B., Kammerer, R. A., Engel, J., et al. (1998). Electron microscopic structure of agrin and mapping of its binding site in laminin-1. *EMBO Journal, 17*(2), 335–343. http://dx.doi.org/10.1093/emboj/17.2.335.

Dodge, G. R., Kovalszky, I., Chu, M. L., Hassell, J. R., McBride, O. W., Yi, H. F., et al. (1991). Heparan sulfate proteoglycan of human colon: partial molecular cloning, cellular expression, and mapping of the gene (HSPG2) to the short arm of human chromosome 1. *Genomics, 10*(3), 673–680. Retrieved from http://www.ncbi.nlm.nih.gov/pubmed/1679749.

Dolan, M., Horchar, T., Rigatti, B., & Hassell, J. R. (1997). Identification of sites in domain I of perlecan that regulate heparan sulfate synthesis. *Journal of Biological Chemistry, 272*(7), 4316–4322. Retrieved from http://www.ncbi.nlm.nih.gov/pubmed/9020150.

Eldridge, S., Nalesso, G., Ismail, H., Vicente-Greco, K., Kabouridis, P., Ramachandran, M., & Dell'Accio, F. (2015). Agrin mediates chondrocyte homeostasis and requires both LRP4 and alpha-dystroglycan to enhance cartilage formation in vitro and in vivo. *Annals of the Rheumatic Diseases*. http://dx.doi.org/10.1136/annrheumdis-2015-207316.

Esko, J. D., Kimata, K., & Lindahl, U. (2009). *Proteoglycans and sulfated glycosaminoglycans* (2nd ed.). New York: Cold Spring Harbor Press.

Esko, J., & Lindahl, U. (2001). Molecular diversity of heparan sulfate. *Journal of Clinical Investigation, 108*, 169–173.

Esko, J. D., & Selleck, S. B. (2002). Order out of chaos: assembly of ligand binding sites in heparan sulfate. *Annual Review of Biochemistry, 71*, 435–471. Retrieved from http://www.ncbi.nlm.nih.gov/entrez/query.fcgi?cmd=Retrieve&db=PubMed&dopt=Citation&list_uids=12045103.

Farach-Carson, M. C., Brown, A. J., Lynam, M., Safran, J. B., & Carson, D. D. (2008). A novel peptide sequence in perlecan domain IV supports cell adhesion, spreading and FAK activation. *Matrix Biology, 27*(2), 150–160. http://dx.doi.org/10.1016/j.matbio.2007.09.007.

Farach-Carson, M. C., & Carson, D. D. (2007). Perlecan—a multifunctional extracellular proteoglycan scaffold. *Glycobiology, 17*(9), 897–905. http://dx.doi.org/10.1093/glycob/cwm043.

Farach-Carson, M. C., Warren, C. R., Harrington, D. A., & Carson, D. D. (2014). Border patrol: insights into the unique role of perlecan/heparan sulfate proteoglycan 2 at cell and tissue borders. *Matrix Biology, 34*, 64–79. http://dx.doi.org/10.1016/j.matbio.2013.08.004.

Fenger, M., Wewer, U., & Albrechtsen, R. (1984). Basement membrane heparan sulfate proteoglycan from the L2 rat yolk sac carcinoma. *FEBS Letters, 173*, 75–79.

Ferns, M. J., Campanelli, J. T., Hoch, W., Scheller, R. H., & Hall, Z. (1993). The ability of agrin to cluster AChRs depends on alternative splicing and on cell surface proteoglycans. *Neuron, 11*(3), 491–502. Retrieved from http://www.ncbi.nlm.nih.gov/pubmed/8398142.

Folkman, J. (2006). Antiangiogenesis in cancer therapy—endostatin and its mechanisms of action. *Experimental Cell Research, 312*(5), 594–607. http://dx.doi.org/10.1016/j.yexcr.2005.11.015.

French, M. M., Smith, S. E., Akanbi, K., Sanford, T., Hecht, J., Farach-Carson, M. C., et al. (1999). Expression of the heparan sulfate proteoglycan, perlecan, during mouse embryogenesis and perlecan chondrogenic activity in vitro. *Journal of Cell Biology, 145*(5), 1103–1115. Retrieved from http://www.ncbi.nlm.nih.gov/pubmed/10352025.

Friedrich, M. V., Gohring, W., Morgelin, M., Brancaccio, A., David, G., & Timpl, R. (1999). Structural basis of glycosaminoglycan modification and of heterotypic interactions of perlecan domain V. *Journal of Molecular Biology, 294*(1), 259–270. http://dx.doi.org/10.1006/jmbi.1999.3259.

Gesemann, M., Brancaccio, A., Schumacher, B., & Ruegg, M. A. (1998). Agrin is a high-affinity binding protein of dystroglycan in non-muscle tissue. *Journal of Biological Chemistry, 273*(1), 600–605. Retrieved from http://www.ncbi.nlm.nih.gov/pubmed/9417121.

Gesemann, M., Cavalli, V., Denzer, A. J., Brancaccio, A., Schumacher, B., & Ruegg, M. A. (1996). Alternative splicing of agrin alters its binding to heparin, dystroglycan, and the putative agrin receptor. *Neuron, 16*(4), 755–767. Retrieved from http://www.ncbi.nlm.nih.gov/entrez/query.fcgi?cmd=Retrieve&db=PubMed&dopt=Citation&list_uids=8607994.

Gibson, K. D., Segen, B. J., & Audhya, T. K. (1977). The effect of beta-D-xylosides on chondroitin sulphate biosynthesis in embryonic chicken cartilage in the absence of protein synthesis inhibitors. *Biochemical Journal, 162*(2), 217–233. Retrieved from http://www.ncbi.nlm.nih.gov/pubmed/849282.

Gibson, K. D., Segen, B. J., & Doller, H. J., Jr. (1979). Changes in chemical composition of chick embryos treated with a beta-xyloside and a lathyrogen. *Teratology, 19*(3), 345–356. http://dx.doi.org/10.1002/tera.1420190311.

Gohring, W., Sasaki, T., Heldin, C. H., & Timpl, R. (1998). Mapping of the binding of platelet-derived growth factor to distinct domains of the basement membrane proteins BM-40 and perlecan and distinction from the BM-40 collagen-binding epitope. *European Journal of Biochemistry, 255*(1), 60–66. Retrieved from http://www.ncbi.nlm.nih.gov/pubmed/9692901.

Goldberg, S., Harvey, S. J., Cunningham, J., Tryggvason, K., & Miner, J. H. (2009). Glomerular filtration is normal in the absence of both agrin and perlecan-heparan sulfate from the glomerular basement membrane. *Nephrology Dialysis Transplantation, 24*(7), 2044–2051. http://dx.doi.org/10.1093/ndt/gfn758.

Gonzalez, E. M., Mongiat, M., Slater, S. J., Baffa, R., & Iozzo, R. V. (2003). A novel interaction between perlecan protein core and progranulin: potential effects on tumor growth. *Journal of Biological Chemistry, 278*(40), 38113–38116. http://dx.doi.org/10.1074/jbc.C300310200.

Gonzalez, E. M., Reed, C. C., Bix, G., Fu, J., Zhang, Y., Gopalakrishnan, B., et al. (2005). BMP-1/tolloid-like metalloproteases process endorepellin, the angiostatic C-terminal fragment of perlecan. *Journal of Biological Chemistry, 280*(8), 7080–7087. http://dx.doi.org/10.1074/jbc.M409841200.

Govindraj, P., West, L., Koob, T. J., Neame, P., Doege, K., & Hassell, J. R. (2002). Isolation and identification of the major heparan sulfate proteoglycans in the developing bovine rib growth plate. *Journal of Biological Chemistry, 277*(22), 19461–19469. http://dx.doi.org/10.1074/jbc.M200786200.

Goyal, A., Pal, N., Concannon, M., Paul, M., Doran, M., Poluzzi, C., et al. (2011). Endorepellin, the angiostatic module of perlecan, interacts with both the alpha2beta1 integrin and vascular endothelial growth factor receptor 2 (VEGFR2): a dual receptor antagonism. *Journal of Biological Chemistry, 286*(29), 25947–25962. http://dx.doi.org/10.1074/jbc.M111.243626.

Goyal, A., Poluzzi, C., Willis, C. D., Smythies, J., Shellard, A., Neill, T., et al. (2012). Endorepellin affects angiogenesis by antagonizing diverse vascular endothelial growth factor receptor 2 (VEGFR2)-evoked signaling pathways: transcriptional repression of hypoxia-inducible factor 1alpha and VEGFA and concurrent inhibition of nuclear factor of activated T cell 1 (NFAT1) activation. *Journal of Biological Chemistry, 287*(52), 43543–43556. http://dx.doi.org/10.1074/jbc.M112.401786.

Graham, L. D., Whitelock, J. M., & Underwood, P. A. (1999). Expression of human perlecan domain I as a recombinant heparan sulfate proteoglycan with 20-kDa glycosaminoglycan chains. *Biochemical and Biophysical Research Communications, 256*(3), 542–548. http://dx.doi.org/10.1006/bbrc.1999.0377.

Gregory, J. D., Laurent, T. C., & Roden, L. (1964). Enzymatic degradation of chondromucoprotein. *Journal of Biological Chemistry, 239*, 3312–3320. Retrieved from http://www.ncbi.nlm.nih.gov/pubmed/14245379.

Groffen, A. J., Buskens, C. A., van Kuppevelt, T. H., Veerkamp, J. H., Monnens, L. A., & van den Heuvel, L. P. (1998). Primary structure and high expression of human agrin in basement membranes of adult lung and kidney. *European Journal of Biochemistry, 254*(1), 123–128. Retrieved from http://www.ncbi.nlm.nih.gov/entrez/query.fcgi?cmd=Retrieve&db=PubMed&dopt=Citation&list_uids=9652404.

Groffen, A. J., Ruegg, M. A., Dijkman, H., van de Velden, T. J., Buskens, C. A., van den Born, J., & van den Heuvel, L. P. (1998). Agrin is a major heparan sulfate proteoglycan in the human glomerular basement membrane. *Journal of Histochemistry and Cytochemistry, 46*(1), 19–27. Retrieved from http://www.ncbi.nlm.nih.gov/entrez/query.fcgi?cmd=Retrieve&db=PubMed&dopt=Citation&list_uids=9405491.

Halfter, W., Dong, S., Schurer, B., & Cole, G. J. (1998). Collagen XVIII is a basement membrane heparan sulfate proteoglycan. *Journal of Biological Chemistry, 273*(39), 25404–25412. Retrieved from http://www.ncbi.nlm.nih.gov/entrez/query.fcgi?cmd=Retrieve&db=PubMed&dopt=Citation&list_uids=9738008.

Hansen, U., Allen, J. M., White, R., Moscibrocki, C., Bruckner, P., Bateman, J. F., et al. (2012). WARP interacts with collagen VI-containing microfibrils in the pericellular matrix of human chondrocytes. *PLoS One, 7*(12), e52793. http://dx.doi.org/10.1371/journal.pone.0052793.

Harvey, S. V., Jarad, G., Cunningham, J., Rops, A. L., van der Vlag, J., Berden, J. H., et al. (2007). Disruption of glomerular basement membrane charge through podocyte-specific mutation of agrin does not alter glomerular permselectivity. *American Journal of Pathology, 171*(1), 139–152.

Hassell, J. R., Leyshon, W. C., Ledbetter, S. R., Tyree, B., Suzuki, S., Kato, M., et al. (1985). Isolation of two forms of basement membrane proteoglycans. *Journal of Biological Chemistry, 260*, 8098–8105.

Hassell, J. R., Robey, P. G., Barrach, H. J., Wilczek, J., Rennard, S. I., & Martin, G. R. (1980). Isolation of a heparan sulfate-containing proteoglycan from basement membrane. *Proceedings of the National Academy of Sciences of the United States of America, 77*, 4504–4508.

Hausser, H. J., Ruegg, M. A., Brenner, R. E., & Ksiazek, I. (2007). Agrin is highly expressed by chondrocytes and is required for normal growth. *Histochemistry and Cell Biology, 127*(4), 363–374. http://dx.doi.org/10.1007/s00418-006-0258-2.

Hayes, A. J., Lord, M. S., Smith, S. M., Smith, M. M., Whitelock, J. M., Weiss, A. S., et al. (2011). Colocalization in vivo and association in vitro of perlecan and elastin. *Histochemistry and Cell Biology, 136*(4), 437–454. http://dx.doi.org/10.1007/s00418-011-0854-7.

Helting, T. (1972). Biosynthesis of heparin. Solubilization and partial purification of uridine diphosphate glucuronic acid: acceptor glucuronosyltransferase from mouse mastocytoma. *Journal of Biological Chemistry, 247*(13), 4327–4332. Retrieved from http://www.ncbi.nlm.nih.gov/pubmed/4260846.

Helting, T., & Roden, L. (1969). Biosynthesis of chondroitin sulfate. II. Glucuronosyl transfer in the formation of the carbohydrate-protein linkage region. *Journal of Biological Chemistry, 244*(10), 2799–2805. Retrieved from http://www.ncbi.nlm.nih.gov/pubmed/5770003.

Hopf, M., Gohring, W., Kohfeldt, E., Yamada, Y., & Timpl, R. (1999). Recombinant domain IV of perlecan binds to nidogens, laminin-nidogen complex, fibronectin, fibulin-2 and heparin. *European Journal of Biochemistry, 259*(3), 917–925. Retrieved from http://www.ncbi.nlm.nih.gov/pubmed/10092882.

Hopf, M., Gohring, W., Mann, K., & Timpl, R. (2001). Mapping of binding sites for nidogens, fibulin-2, fibronectin and heparin to different IG modules of perlecan. *Journal of Molecular Biology, 311*(3), 529–541. http://dx.doi.org/10.1006/jmbi.2001.4878.

Iozzo, R. V. (2005). Basement membrane proteoglycans: from cellar to ceiling. *Nature Reviews Molecular Cell Biology, 6*(8), 646–656. http://dx.doi.org/10.1038/nrm1702.

Iozzo, R. V., & Clark, C. C. (1986). Biosynthesis of proteoglycans by rat embryo parietal yolk sacs in organ culture. *Journal of Biological Chemistry, 261*(15), 6658–6669. Retrieved from http://www.ncbi.nlm.nih.gov/pubmed/3084487.

Iozzo, R. V., & Clark, C. C. (1987). Chondroitin sulfate proteoglycan is a constituent of the basement membrane in the rat embryo parietal yolk sac. *Histochemistry, 88*(1), 23–29. Retrieved from http://www.ncbi.nlm.nih.gov/pubmed/3436841.

Iozzo, R. V., Pillarisetti, J., Sharma, B., Murdoch, A. D., Danielson, K. G., Uitto, J., et al. (1997). Structural and functional characterization of the human perlecan gene promoter. Transcriptional activation by transforming growth factor-beta via a nuclear factor 1-binding element. *Journal of Biological Chemistry, 272*(8), 5219–5228. Retrieved from http://www.ncbi.nlm.nih.gov/pubmed/9030592.

Iozzo, R. V., Zoeller, J. J., & Nystrom, A. (2009). Basement membrane proteoglycans: modulators par excellence of cancer growth and angiogenesis. *Molecules and Cells, 27*(5), 503–513. http://dx.doi.org/10.1007/s10059-009-0069-0.

Iwata, M., & Carlson, S. S. (1991). A large chondroitin sulfate basement membrane-associated proteoglycan exists as a disulfide-stabilized complex of several proteins. *Journal of Biological Chemistry, 266*(1), 323–333. Retrieved from http://www.ncbi.nlm.nih.gov/pubmed/1985903.

Iwata, S., Ito, M., Nakata, T., Noguchi, Y., Okuno, T., Ohkawara, B., et al. (2015). A missense mutation in domain III in HSPG2 in Schwartz-Jampel syndrome compromises secretion of perlecan into the extracellular space. *Neuromuscular Disorders, 25*(8), 667–671. http://dx.doi.org/10.1016/j.nmd.2015.05.002.

Jansson, L., & Lindahl, U. (1970). Evidence for the existence of a multichain proteoglycan of heparan sulphate. *Biochemical Journal, 117*(4), 699–702. Retrieved from http://www.ncbi.nlm.nih.gov/pubmed/4247363.

Kabouridis, P. S., Pimentel, T. A., Brancaleone, V., D'Acquisto, F., Oliani, S. M., Perretti, M., et al. (2012). Distinct localization of T cell agrin during antigen presentation—evidence for the expression of agrin receptor(s) in antigen-presenting cells. *FEBS Journal, 279*(13), 2368–2380. http://dx.doi.org/10.1111/j.1742-4658.2012.08615.x.

Kallunki, P., Eddy, R. L., Byers, M. G., Kestila, M., Shows, T. B., & Tryggvason, K. (1991). Cloning of human heparan sulfate proteoglycan core protein, assignment of the gene (HSPG2) to 1p36.1–p35 and identification of a BamHI restriction fragment length polymorphism. *Genomics, 11*(2), 389–396. Retrieved from http://www.ncbi.nlm.nih.gov/pubmed/1685141.

Kallunki, P., & Tryggvason, K. (1992). Human basement membrane heparan sulfate proteoglycan core protein: a 467-kD protein containing multiple domains resembling elements of the low density lipoprotein receptor, laminin, neural cell adhesion molecules, and epidermal growth factor. *Journal of Cell Biology, 116*, 559–571.

Kammerer, R. A., Schulthess, T., Landwehr, R., Schumacher, B., Lustig, A., Yurchenco, P. D., et al. (1999). Interaction of agrin with laminin requires a coiled-coil conformation of the agrin-binding site within the laminin gamma1 chain. *EMBO Journal, 18*(23), 6762–6770. Retrieved from http://www.ncbi.nlm.nih.gov/entrez/query.fcgi?cmd=Retrieve&db=PubMed&dopt=Citation&list_uids=10581249.

Kanke, Y., Doller, H. J., Jr., Segen, B. J., & Gibson, K. D. (1982). Development of the biochemical and morphological changes induced by administration of a beta-xyloside to chick embryos. *Teratology, 25*(1), 81–88. http://dx.doi.org/10.1002/tera.1420250111.

Kanwar, Y. S., & Farquhar, M. G. (1979a). Anionic sites in the glomerular basement membrane. In vivo and in vitro localization to the lamina rarae by cationic probes. *Journal of Cell Biology, 81*, 137–153.

Kanwar, Y. S., & Farquhar, M. G. (1979b). Isolation of glycosaminoglycans (heparan sulfate) from glomerular basement membranes. *Proceedings of the National Academy of Sciences of the United States of America, 76*, 4493–4497.

Kanwar, Y. S., Hascall, V. C., Jakubowski, M. L., & Gibbons, J. T. (1984). Effect of beta-D-xyloside on the glomerular proteoglycans. I. Biochemical studies. *Journal of Cell Biology, 99*(2), 715–722. Retrieved from http://www.ncbi.nlm.nih.gov/pubmed/6746744.

Kanwar, Y. S., Rosenzweig, L. J., & Jakubowski, M. L. (1987). Effect of beta-D-xyloside on the renal glomerular cells. II. Morphological studies. *Laboratory Investigation, 56*(2), 160–169. Retrieved from http://www.ncbi.nlm.nih.gov/pubmed/3807315.

Khan, A. A., Bose, C., Yam, L. S., Soloski, M. J., & Rupp, F. (2001). Physiological regulation of the immunological synapse by agrin. *Science, 292*(5522), 1681–1686. http://dx.doi.org/10.1126/science.1056594.

Kim, N., Stiegler, A. L., Cameron, T. O., Hallock, P. T., Gomez, A. M., Huang, J. H., et al. (2008). Lrp4 is a receptor for Agrin and forms a complex with MuSK. *Cell, 135*(2), 334–342. http://dx.doi.org/10.1016/j.cell.2008.10.002.

Kirn-Safran, C., Farach-Carson, M. C., & Carson, D. D. (2009). Multifunctionality of extracellular and cell surface heparan sulfate proteoglycans. *Cellular and Molecular Life Sciences, 66*(21), 3421–3434. http://dx.doi.org/10.1007/s00018-009-0096-1.

Kinnunen, A. I., Sormunen, R., Elamaa, H., Seppinen, L., Miller, R. T., Ninomiya, Y., & Pihlajaniemi, T. (2011). Lack of collagen XVIII long isoforms affects kidney podocytes, whereas the short form is needed in the proximal tubular basement membrane. *J Biol Chem, 286*(10), 7755–7764. http://dx.doi.org/10.1074/jbc.M110.166132.

Kivirikko, S., Heinamaki, P., Rehn, M., Honkanen, N., Myers, J. C., & Pihlajaniemi, T. (1994). Primary structure of the alpha 1 chain of human type XV collagen and exon-intron organization in the 3′ region of the corresponding gene. *Journal of Biological Chemistry, 269*(7), 4773–4779. Retrieved from http://www.ncbi.nlm.nih.gov/pubmed/8106446.

Klein, D. J., Brown, D. M., Moran, A., Oegema, T. R., Jr., & Platt, J. L. (1989). Chondroitin sulfate proteoglycan synthesis and reutilization of beta-D-xyloside-initiated chondroitin/

basement membrane (GBM) and induce subepithelial basement membrane thickening. *Journal of Experimental Medicine, 163*, 1064–1084.

Miosge, N., Simniok, T., Sprysch, P., & Herken, R. (2003). The collagen type XVIII endostatin domain is co-localized with perlecan in basement membranes in vivo. *Journal of Histochemistry and Cytochemistry, 51*(3), 285–296. Retrieved from http://www.ncbi.nlm.nih.gov/pubmed/12588956.

Mongiat, M., Otto, J., Oldershaw, R., Ferrer, F., Sato, J. D., & Iozzo, R. V. (2001). Fibroblast growth factor-binding protein is a novel partner for perlecan protein core. *Journal of Biological Chemistry, 276*(13), 10263–10271. http://dx.doi.org/10.1074/jbc.M011493200.

Mongiat, M., Sweeney, S. M., San Antonio, J. D., Fu, J., & Iozzo, R. V. (2003). Endorepellin, a novel inhibitor of angiogenesis derived from the C terminus of perlecan. *Journal of Biological Chemistry, 278*(6), 4238–4249. http://dx.doi.org/10.1074/jbc.M210445200.

Mongiat, M., Taylor, K., Otto, J., Aho, S., Uitto, J., Whitelock, J. M., et al. (2000). The protein core of the proteoglycan perlecan binds specifically to fibroblast growth factor-7. *Journal of Biological Chemistry, 275*(10), 7095–7100. Retrieved from http://www.ncbi.nlm.nih.gov/pubmed/10702276.

Moore, C. J., & Winder, S. J. (2010). Dystroglycan versatility in cell adhesion: a tale of multiple motifs. *Cell Communication Signal, 8*, 3. http://dx.doi.org/10.1186/1478-811X-8-3.

Morita, H., Yoshimura, A., Inui, K., Ideura, T., Watanabe, H., Wang, L., et al. (2005). Heparan sulfate of perlecan is involved in glomerular filtration. *Journal of the American Society of Nephrology, 16*, 1703–1710.

Morriss-Kay, G. M., & Crutch, B. (1982). Culture of rat embryos with beta-D-xyloside: evidence of a role for proteoglycans in neurulation. *Journal of Anatomy, 134*(Pt 3), 491–506. Retrieved from http://www.ncbi.nlm.nih.gov/pubmed/6809722.

Muller, G. (1959). [Demonstration of acid mucopolysaccharides by toluidin blue metachromasia, alcian blue staining, PAS reaction and the ferrihydroxidesol-PAS reaction]. *Acta Histochemica, 6*(5–8), 218–224. Retrieved from http://www.ncbi.nlm.nih.gov/pubmed/13649222.

Multhaupt, H. A., Yoneda, A., Whiteford, J. R., Oh, E. S., Lee, W., & Couchman, J. R. (2009). Syndecan signaling: when, where and why? *Journal of Physiology and Pharmacology, 60*(Suppl 4), 31–38. Retrieved from http://www.ncbi.nlm.nih.gov/pubmed/20083849.

Muragaki, Y., Abe, N., Ninomiya, Y., Olsen, B. R., & Ooshima, A. (1994). The human alpha 1(XV) collagen chain contains a large amino-terminal non-triple helical domain with a tandem repeat structure and homology to alpha 1(XVIII) collagen. *Journal of Biological Chemistry, 269*(6), 4042–4046. Retrieved from http://www.ncbi.nlm.nih.gov/pubmed/8307960.

Murdoch, A. D., Dodge, G. R., Cohen, I., Tuan, R. S., & Iozzo, R. V. (1992). Primary structure of the human heparan sulfate proteoglycan from basement membrane (HSPG2/perlecan). A chimeric molecule with multiple domains homologous to the low density lipoprotein receptor, laminin, neural cell adhesion molecules, and epidermal growth factor. *Journal of Biological Chemistry, 267*(12), 8544–8557. Retrieved from http://www.ncbi.nlm.nih.gov/pubmed/1569102.

Myers, J. C., Kivirikko, S., Gordon, M. K., Dion, A. S., & Pihlajaniemi, T. (1992). Identification of a previously unknown human collagen chain, alpha 1(XV), characterized by extensive interruptions in the triple-helical region. *Proceedings of the National Academy of Sciences of the United States of America, 89*(21), 10144–10148. Retrieved from http://www.ncbi.nlm.nih.gov/pubmed/1279671.

Neumann, F. R., Bittcher, G., Annies, M., Schumacher, B., Kroger, S., & Ruegg, M. A. (2001). An alternative amino-terminus expressed in the central nervous system converts agrin to a type II transmembrane protein. *Molecular and Cellular Neuroscience, 17*(1), 208–225. http://dx.doi.org/10.1006/mcne.2000.0932.

Nicole, S., Davoine, C. S., Topaloglu, H., Cattolico, L., Barral, D., Beighton, P., et al. (2000). Perlecan, the major proteoglycan of basement membranes, is altered in patients with Schwartz-Jampel syndrome (chondrodystrophic myotonia). *Nature Genetics, 26*(4), 480–483. http://dx.doi.org/10.1038/82638.

Nitkin, R. M., Smith, M. A., Magill, C., Fallon, J. R., Yao, Y. M., Wallace, B. G., et al. (1987). Identification of agrin, a synaptic organizing protein from Torpedo electric organ. *Journal of Cell Biology, 105*(6 Pt 1), 2471–2478. Retrieved from http://www.ncbi.nlm.nih.gov/pubmed/2826489.

Ngo, S. T., Noakes, P. G., & Phillips, W. D. (2007). Neural agrin: a synaptic stabiliser. *The International Journal of Biochemistry and Cell Biology, 39*(5), 863–867. http://dx.doi.org/10.1016/j.biocel.2006.10.012.

Nonaka, R., Iesaki, T., de Vega, S., Daida, H., Okada, T., Sasaki, T., et al. (2015). Perlecan deficiency causes endothelial dysfunction by reducing the expression of endothelial nitric oxide synthase. *Physiological Reports, 3*(1). http://dx.doi.org/10.14814/phy2.12272.

Noonan, D. M., Fulle, A., Valente, P., Cai, S., Horigan, E., Sasaki, M., et al. (1991). The complete sequence of Perlecan, a basement membrane heparan sulfate proteoglycan, reveals extensive similarity with laminin A chain, low density lipoprotein-receptor, and the neural cell adhesion molecule. *Journal of Biological Chemistry, 266*, 22939–22947.

Noonan, D. M., & Hassell, J. R. (1993). Perlecan, the large low-density proteoglycan of basement membranes: structure and variant forms. *Kidney International, 43*(1), 53–60. Retrieved from http://www.ncbi.nlm.nih.gov/pubmed/8433569.

Noonan, D. M., Horigan, E. A., Ledbetter, S. R., Vogeli, G., Sasaki, M., Yamada, Y., et al. (1988). Identification of cDNA clones encoding different domains of the basement membrane heparan sulfate proteoglycan. *Journal of Biological Chemistry, 263*(31), 16379–16387. Retrieved from http://www.ncbi.nlm.nih.gov/pubmed/2972708.

Nystrom, A., Shaik, Z. P., Gullberg, D., Krieg, T., Eckes, B., Zent, R., et al. (2009). Role of tyrosine phosphatase SHP-1 in the mechanism of endorepellin angiostatic activity. *Blood, 114*(23), 4897–4906. http://dx.doi.org/10.1182/blood-2009-02-207134.

O'Reilly, M. S., Boehm, T., Shing, Y., Fukai, N., Vasios, G., Lane, W. S., et al. (1997). Endostatin: an endogenous inhibitor of angiogenesis and tumor growth. *Cell, 88*(2), 277–285. Retrieved from http://www.ncbi.nlm.nih.gov/pubmed/9008168.

O'Toole, J. J., Deyst, K. A., Bowe, M. A., Nastuk, M. A., McKechnie, B. A., & Fallon, J. R. (1996). Alternative splicing of agrin regulates its binding to heparin alpha-dystroglycan, and the cell surface. *Proceedings of the National Academy of Sciences of the United States of America, 93*(14), 7369–7374. Retrieved from http://www.ncbi.nlm.nih.gov/pubmed/8693000.

Oh, S. P., Kamagata, Y., Muragaki, Y., Timmons, S., Ooshima, A., & Olsen, B. R. (1994). Isolation and sequencing of cDNAs for proteins with multiple domains of Gly-Xaa-Yaa repeats identify a distinct family of collagenous proteins. *Proceedings of the National Academy of Sciences of the United States of America, 91*(10), 4229–4233. Retrieved from http://www.ncbi.nlm.nih.gov/pubmed/8183893.

Okayama, M., Kimata, K., & Suzuki, S. (1973). The influence of p-nitrophenyl beta-d-xyloside on the synthesis of proteochondroitin sulfate by slices of embryonic chick cartilage. *Journal of Biochemistry, 74*(5), 1069–1073. Retrieved from http://www.ncbi.nlm.nih.gov/pubmed/4770369.

Olsen, B. R. (1999). Life without perlecan has its problems. *Journal of Cell Biology, 147*(5), 909–912. Retrieved from http://www.ncbi.nlm.nih.gov/pubmed/10579711.

Oohira, A., Nogami, H., & Nakanishi, Y. (1981). Abnormal overgrowth of chick embryos treated with p-nitrophenyl beta-D-xyloside at early stages of development. *Journal of Embryology & Experimental Morphology, 61*, 221–232. Retrieved from http://www.ncbi.nlm.nih.gov/pubmed/7264543.

Oohira, A., Wight, T. N., McPherson, J., & Bornstein, P. (1982). Biochemical and ultrastructural studies of proteoheparan sulfates synthesized by PYS-2, a basement membrane-producing cell line. *Journal of Cell Biology, 92*(2), 357–367. Retrieved from http://www.ncbi.nlm.nih.gov/pubmed/6174529.

Orkin, R. W., Gehron, P., McGoodwin, E. B., Martin, G. R., Valentine, T., & Swarm, R. (1977). A murine tumor producing a matrix of basement membrane. *Journal of Experimental Medicine, 145*(1), 204–220. Retrieved from http://www.ncbi.nlm.nih.gov/pubmed/830788.

Platt, J. L., Brown, D. M., Granlund, K., Oegema, T. R., & Klein, D. J. (1987). Proteoglycan metabolism associated with mouse metanephric development: morphologic and biochemical effects of beta-D-xyloside. *Developmental Biology, 123*(2), 293–306. Retrieved from http://www.ncbi.nlm.nih.gov/pubmed/3653508.

Porten, E., Seliger, B., Schneider, V. A., Woll, S., Stangel, D., Ramseger, R., et al. (2010). The process-inducing activity of transmembrane agrin requires follistatin-like domains. *Journal of Biological Chemistry, 285*(5), 3114–3125. http://dx.doi.org/10.1074/jbc.M109.039420.

Raats, C. J., Bakker, M. A., Hoch, W., Tamboer, W. P., Groffen, A. J., van den Heuvel, L. P., et al. (1998). Differential expression of agrin in renal basement membranes as revealed by domain-specific antibodies. *Journal of Biological Chemistry, 273*(28), 17832–17838. Retrieved from http://www.ncbi.nlm.nih.gov/entrez/query.fcgi?cmd=Retrieve&db=PubMed&dopt=Citation&list_uids=9651386.

Raats, C. J., Luca, M. E., Bakker, M. A., Van Der Wal, A., Heeringa, P., Van Goor, H., et al. (1999). Reduction in glomerular heparan sulfate correlates with complement deposition and albuminuria in active Heymann nephritis. *Journal of the American Society of Nephrology, 10*(8), 1689–1699. Retrieved from http://www.ncbi.nlm.nih.gov/entrez/query.fcgi?cmd=Retrieve&db=PubMed&dopt=Citation&list_uids=10446936.

Rada, J. A., & Carlson, E. C. (1991a). Electron microscopic histochemical and immunochemical analyses of heparan sulfate proteoglycan distribution in renal glomerular basement membranes. *Histology and Histopathology, 6*(2), 149–160. Retrieved from http://www.ncbi.nlm.nih.gov/pubmed/1724933.

Rada, J. A., & Carlson, E. C. (1991b). Anionic site and immunogold quantitation of heparan sulfate proteoglycans in glomerular basement membranes of puromycin aminonucleoside nephrotic rats. *Anatomical Record, 231*, 35–47.

Rambourg, A., & Leblond, C. P. (1967). Staining of basement membranes and associated structures by the periodic acid-Schiff and periodicacid-silver methenamine techniques. *Journal of Ultrastructure Research, 20*(3), 306–309. Retrieved from http://www.ncbi.nlm.nih.gov/pubmed/4169782.

Ramseger, R., White, R., & Kroger, S. (2009). Transmembrane form agrin-induced process formation requires lipid rafts and the activation of Fyn and MAPK. *Journal of Biological Chemistry, 284*(12), 7697–7705. http://dx.doi.org/10.1074/jbc.M806719200.

Rennke, H. G., Cotran, R. S., & Venkatachalam, M. A. (1975). Role of molecular charge in glomerular permeability. Tracer studies with cationized ferritins. *Journal of Cell Biology, 67*(3), 638–646. Retrieved from http://www.ncbi.nlm.nih.gov/entrez/query.fcgi?cmd=Retrieve&db=PubMed&dopt=Citation&list_uids=1202017.

Rennke, H. G., & Venkatachalam, M. A. (1977). Glomerular permeability: in vivo tracer studies with polyanionic and polycationic ferritins. *Kidney International, 11*(1), 44–53. Retrieved from http://www.ncbi.nlm.nih.gov/entrez/query.fcgi?cmd=Retrieve&db=PubMed&dopt=Citation&list_uids=839653.

Rizzoli, C., & Gliozzi, M. A. (1955). [Alcian blue 8 G.N. staining of mucopolysaccharide acids separated by paper electrophoresis]. *Bollettino Societa Italiana Biologia Sperimentale, 31*(5), 426–428. Retrieved from http://www.ncbi.nlm.nih.gov/pubmed/13269460.

Rossi, M., Morita, H., Sormunen, R., Airenne, S., Kreivi, M., Wang, L., et al. (2003). Heparan sulfate chains of perlecan are indispensable in the lens capsule but not in the kidney. *EMBO Journal, 22*(2), 236–245.

Ruegg, M. A., Tsim, K. W., Horton, S. E., Kroger, S., Escher, G., Gensch, E. M., et al. (1992). The agrin gene codes for a family of basal lamina proteins that differ in function and distribution. *Neuron, 8*(4), 691–699. Retrieved from http://www.ncbi.nlm.nih.gov/pubmed/1314621.

Rupp, F., Ozcelik, T., Linial, M., Peterson, K., Francke, U., & Scheller, R. (1992). Structure and chromosomal localization of the mammalian agrin gene. *Journal of Neuroscience, 12*(9), 3535–3544. Retrieved from http://www.ncbi.nlm.nih.gov/pubmed/1326608.

Rupp, F., Payan, D. G., Magill-Solc, C., Cowan, D. M., & Scheller, R. H. (1991). Structure and expression of a rat agrin. *Neuron, 6*(5), 811–823. Retrieved from http://www.ncbi.nlm.nih.gov/pubmed/1851019.

Saarela, J., Rehn, M., Oikarinen, A., Autio-Harmainen, H., & Pihlajaniemi, T. (1998). The short and long forms of type XVIII collagen show clear tissue specificities in their expression and location in basement membrane zones in humans. *American Journal of Pathology, 153*(2), 611–626. http://dx.doi.org/10.1016/S0002-9440(10)65603-9.

Sarras, M. P., Jr., Madden, M. E., Zhang, X. M., Gunwar, S., Huff, J. K., & Hudson, B. G. (1991). Extracellular matrix (mesoglea) of *Hydra vulgaris*. I. Isolation and characterization. *Developmental Biology, 148*(2), 481–494. Retrieved from http://www.ncbi.nlm.nih.gov/pubmed/1743396.

Sarras, M. P., Jr., Meador, D., & Zhang, X. M. (1991). Extracellular matrix (mesoglea) of *Hydra vulgaris*. II. Influence of collagen and proteoglycan components on head regeneration. *Developmental Biology, 148*(2), 495–500. Retrieved from http://www.ncbi.nlm.nih.gov/pubmed/1743397.

Sasaki, T., Costell, M., Mann, K., & Timpl, R. (1998). Inhibition of glycosaminoglycan modification of perlecan domain I by site-directed mutagenesis changes protease sensitivity and laminin-1 binding activity. *FEBS Letters, 435*(2–3), 169–172. Retrieved from http://www.ncbi.nlm.nih.gov/pubmed/9762901.

Sato, Y., Shimono, C., Li, S., Nakano, I., Norioka, N., Sugiura, N., et al. (2013). Nephronectin binds to heparan sulfate proteoglycans via its MAM domain. *Matrix Biology, 32*(3–4), 188–195. http://dx.doi.org/10.1016/j.matbio.2013.01.005.

Schulze, B., Sasaki, T., Costell, M., Mann, K., & Timpl, R. (1996). Structural and cell-adhesive properties of three recombinant fragments derived from perlecan domain III. *Matrix Biology, 15*(5), 349–357. Retrieved from http://www.ncbi.nlm.nih.gov/pubmed/8981331.

Schwartz, N. B. (1976). Chondroitin sulfate glycosyltransferases in cultured chondrocytes. Turnover, oscillatory change during growth, and suppression by 5-bromodeoxyuridine. *Journal of Biological Chemistry, 251*(11), 3346–3351. Retrieved from http://www.ncbi.nlm.nih.gov/pubmed/945267.

Schwartz, N. B. (1977). Regulation of chondroitin sulfate synthesis. Effect of beta-xylosides on synthesis of chondroitin sulfate proteoglycan, chondroitin sulfate chains, and core protein. *Journal of Biological Chemistry, 252*(18), 6316–6321. Retrieved from http://www.ncbi.nlm.nih.gov/pubmed/561070.

Shu, C., Smith, S. S., Little, C. B., & Melrose, J. (2013). Comparative immunolocalisation of perlecan, heparan sulphate, fibroblast growth factor-18, and fibroblast growth factor receptor-3 and their prospective roles in chondrogenic and osteogenic development of the human foetal spine. *European Spine Journal, 22*(8), 1774–1784. http://dx.doi.org/10.1007/s00586-013-2680-1.

Smith, C. I., Hilfer, S. R., Searls, R. L., Nathanson, M. A., & Allodoli, M. D. (1990). Effects of beta-D-xyloside on differentiation of the respiratory epithelium in the fetal mouse lung. *Developmental Biology, 138*(1), 42–52. Retrieved from http://www.ncbi.nlm.nih.gov/pubmed/2307288.

Smith, C. I., Webster, E. H., Nathanson, M. A., Searls, R. L., & Hilfer, S. R. (1990). Altered patterns of proteoglycan deposition during maturation of the fetal mouse lung. *Cell*

Differentiation and Development, 32(2), 83–96. Retrieved from http://www.ncbi.nlm.nih.gov/pubmed/2128042.
Smith, S. M., West, L. A., Govindraj, P., Zhang, X., Ornitz, D. M., & Hassell, J. R. (2007). Heparan and chondroitin sulfate on growth plate perlecan mediate binding and delivery of FGF-2 to FGF receptors. *Matrix Biology, 26*(3), 175–184. http://dx.doi.org/10.1016/j.matbio.2006.10.012.
Smith, S. M., West, L. A., & Hassell, J. R. (2007). The core protein of growth plate perlecan binds FGF-18 and alters its mitogenic effect on chondrocytes. *Archives of Biochemistry and Biophysics, 468*(2), 244–251. http://dx.doi.org/10.1016/j.abb.2007.10.006.
Smith, M. A., Yao, Y. M., Reist, N. E., Magill, C., Wallace, B. G., & McMahan, U. J. (1987). Identification of agrin in electric organ extracts and localization of agrin-like molecules in muscle and central nervous system. *Journal of Experimental Biology, 132*, 223–230. Retrieved from http://www.ncbi.nlm.nih.gov/pubmed/2828510.
Spooner, B. S., Bassett, K., & Stokes, B. (1985). Sulfated glycosaminoglycan deposition and processing at the basal epithelial surface in branching and beta-D-xyloside-inhibited embryonic salivary glands. *Developmental Biology, 109*(1), 177–183. Retrieved from http://www.ncbi.nlm.nih.gov/pubmed/3987960.
Spooner, B. S., Paulsen, A., & Sullins, M. S. (1989). beta-Xyloside effects on basal lamina structure and anionic site distribution the embryonic mouse submandibular salivary gland. *Archives of Oral Biology, 34*(7), 541–549. Retrieved from http://www.ncbi.nlm.nih.gov/pubmed/2480774.
Stearns, K., & Goetinck, P. F. (1979). Stimulation of chondroitin sulfate synthesis by beta-D-xyloside in chondrocytes of the proteoglycan deficient mutant nanomelia. *Journal of Cellular Physiology, 100*(1), 33–38. http://dx.doi.org/10.1002/jcp.1041000104.
Stow, J. L., Glasgow, E. F., Handley, C. J., & Hascall, V. C. (1983). Biosynthesis of proteoglycans by isolated rabbit glomeruli. *Archives of Biochemistry and Biophysics, 225*(2), 950–957. Retrieved from http://www.ncbi.nlm.nih.gov/pubmed/6625617.
Stow, J. L., Sawada, H., & Farquhar, M. G. (1985). Basement membrane heparan sulfate proteoglycans are concentrated in the lamina rarae and in podocytes of the rat renal glomerulus. *Proceedings of the National Academy of Sciences of the United States of America, 82*, 3296–3300.
Suleiman, H., Zhang, L., Roth, R., Heuser, J. E., Miner, J. H., Shaw, A. S., et al. (2013). Nanoscale protein architecture of the kidney glomerular basement membrane. *Elife, 2*, e01149. http://dx.doi.org/10.7554/eLife.01149.
Sugahara, K., Mikami, T., Uyama, T., Mizuguchi, S., Nomura, K., & Kitagawa, H. (2003). Recent advances in the structural biology of chondroitin sulfate and dermatan sulfate. *Current Opinion in Structural Biology, 13*(5), 612–620. Retrieved from http://www.ncbi.nlm.nih.gov/pubmed/14568617.
SundarRaj, N., Fite, D., Ledbetter, S., Chakravarti, S., & Hassell, J. R. (1995). Perlecan is a component of cartilage matrix and promotes chondrocyte attachment. *Journal of Cell Science, 108*(Pt 7), 2663–2672. Retrieved from http://www.ncbi.nlm.nih.gov/pubmed/7593307.
Talts, J. F., Andac, Z., Gohring, W., Brancaccio, A., & Timpl, R. (1999). Binding of the G domains of laminin alpha1 and alpha2 chains and perlecan to heparin, sulfatides, alpha-dystroglycan and several extracellular matrix proteins. *EMBO Journal, 18*(4), 863–870. http://dx.doi.org/10.1093/emboj/18.4.863.
Tapanadechopone, P., Hassell, J. R., Rigatti, B., & Couchman, J. R. (1999). Localization of glycosaminoglycan substitution sites on domain V of mouse perlecan. *Biochemical and Biophysical Research Communications, 265*(3), 680–690. http://dx.doi.org/10.1006/bbrc.1999.1714. pii:S0006-291X(99)91714-X.
Thompson, H. A., & Spooner, B. S. (1982). Inhibition of branching morphogenesis and alteration of glycosaminoglycan biosynthesis in salivary glands treated with

beta-D-xyloside. *Developmental Biology, 89*(2), 417–424. Retrieved from http://www.ncbi.nlm.nih.gov/pubmed/7056440.

Thompson, H. A., & Spooner, B. S. (1983). Proteoglycan and glycosaminoglycan synthesis in embryonic mouse salivary glands: effects of beta-D-xyloside, an inhibitor of branching morphogenesis. *Journal of Cell Biology, 96*(5), 1443–1450. Retrieved from http://www.ncbi.nlm.nih.gov/pubmed/6841454.

Tiedemann, K., Sasaki, T., Gustafsson, E., Gohring, W., Batge, B., Notbohm, H., et al. (2005). Microfibrils at basement membrane zones interact with perlecan via fibrillin-1. *Journal of Biological Chemistry, 280*(12), 11404–11412. http://dx.doi.org/10.1074/jbc.M409882200.

Toole, B. P., & Lowther, D. A. (1965). The isolation of a dermatan sulphate-protein complex from bovine heart valves. *Biochimica et Biophysica Acta, 101*(3), 364–366. Retrieved from http://www.ncbi.nlm.nih.gov/pubmed/4222033.

Tsen, G., Halfter, W., Kroger, S., & Cole, G. J. (1995). Agrin is a heparan sulfate proteoglycan. *Journal of Biological Chemistry, 270*(7), 3392–3399. Retrieved from http://www.ncbi.nlm.nih.gov/pubmed/7852425.

Tyree, B., Horigan, E. A., Klippenstein, D. L., & Hassell, J. R. (1984). Heterogeneity of heparan sulfate proteoglycans synthesized by PYS-2 cells. *Archives of Biochemistry and Biophysics, 231*(2), 328–335. Retrieved from http://www.ncbi.nlm.nih.gov/pubmed/6233937.

Van Den Born, J., Van Den Heuvel, L. P. W. J., Bakker, M. A. H., Veerkamp, J. H., Assmann, K. J. M., & Berden, J. H. M. (1992). A monoclonal antibody against GBM heparan sulfate induces an acute selective proteinuria in rats. *Kidney International, 411*, 115–123.

Vialli, M. (1951). [Use of Alcian blue 8 GS in the study of mucopolysaccharides]. *Bollettino Societa Italiana Biologia Sperimentale, 27*(4), 597–599. Retrieved from http://www.ncbi.nlm.nih.gov/pubmed/14869481.

Vincent, T. L., McLean, C. J., Full, L. E., Peston, D., & Saklatvala, J. (2007). FGF-2 is bound to perlecan in the pericellular matrix of articular cartilage, where it acts as a chondrocyte mechanotransducer. *Osteoarthritis Cartilage, 15*(7), 752–763. http://dx.doi.org/10.1016/j.joca.2007.01.021.

Wang, H., Julenius, K., Hryhorenko, J., & Hagen, F. K. (2007). Systematic analysis of proteoglycan modification sites in *Caenorhabditis elegans* by scanning mutagenesis. *Journal of Biological Chemistry, 282*(19), 14586–14597. http://dx.doi.org/10.1074/jbc.M609193200.

Warren, C. R., Grindel, B. J., Francis, L., Carson, D. D., & Farach-Carson, M. C. (2014). Transcriptional activation by NFkappaB increases perlecan/HSPG2 expression in the desmoplastic prostate tumor microenvironment. *Journal of Cellular Biochemistry, 115*(7), 1322–1333. http://dx.doi.org/10.1002/jcb.24788.

Wewer, U. (1982). Characterization of a rat yolk sac carcinoma cell line. *Developmental Biology, 93*, 416–421.

Wewer, U. M., Albrechtsen, R., & Hassell, J. R. (1985). Heparan sulfate proteoglycans made by different basement-membrane-producing tumors have immunological and structural similarities. *Differentiation, 30*(1), 61–67. Retrieved from http://www.ncbi.nlm.nih.gov/pubmed/2936642.

Whitelock, J. M., Melrose, J., & Iozzo, R. V. (2008). Diverse cell signaling events modulated by perlecan. *Biochemistry, 47*(43), 11174–11183. http://dx.doi.org/10.1021/bi8013938.

Wight, T. N., & Ross, R. (1975). Proteoglycans in primate arteries. I. Ultrastructural localization and distribution in the intima. *Journal of Cell Biology, 67*(3), 660–674. Retrieved from http://www.ncbi.nlm.nih.gov/pubmed/53234.

Willis, C. D., Poluzzi, C., Mongiat, M., & Iozzo, R. V. (2013). Endorepellin laminin-like globular 1/2 domains bind Ig3-5 of vascular endothelial growth factor (VEGF) receptor

2 and block pro-angiogenic signaling by VEGFA in endothelial cells. *FEBS Journal, 280*(10), 2271–2284. http://dx.doi.org/10.1111/febs.12164.
Wilusz, R. E., Defrate, L. E., & Guilak, F. (2012). A biomechanical role for perlecan in the pericellular matrix of articular cartilage. *Matrix Biology, 31*(6), 320–327. http://dx.doi.org/10.1016/j.matbio.2012.05.002.
Winzen, U., Cole, G. J., & Halfter, W. (2003). Agrin is a chimeric proteoglycan with attachment sites for heparan sulfate/chondroitin sulfate located in two multiple serine-glycine clusters. *Journal of Biological Chemistry, 278*(32), 30106–30114.
Wislocki, G. B., Fawcett, D. W., & Dempsey, E. W. (1951). Staining of stratified squamous epithelium of mucous membranes and skin of man and monkey by the periodic acid-Schiff method. *Anatomical Record, 110*(3), 359–375. Retrieved from http://www.ncbi.nlm.nih.gov/pubmed/14868888.
Woodall, B. P., Nystrom, A., Iozzo, R. A., Eble, J. A., Niland, S., Krieg, T., et al. (2008). Integrin alpha2beta1 is the required receptor for endorepellin angiostatic activity. *Journal of Biological Chemistry, 283*(4), 2335–2343. http://dx.doi.org/10.1074/jbc.M708364200.
Xu, Y. X., Ashline, D., Liu, L., Tassa, C., Shaw, S. Y., Ravid, K., et al. (2015). The glycosylation-dependent interaction of perlecan core protein with LDL: implications for atherosclerosis. *Journal of Lipid Research, 56*(2), 266–276. http://dx.doi.org/10.1194/jlr.M053017.
Xu, Z., Ichikawa, N., Kosaki, K., Yamada, Y., Sasaki, T., Sakai, L. Y., et al. (2010). Perlecan deficiency causes muscle hypertrophy, a decrease in myostatin expression, and changes in muscle fiber composition. *Matrix Biology, 29*(6), 461–470. http://dx.doi.org/10.1016/j.matbio.2010.06.001.
Yamada, K. (1963). Staining of sulphated polysaccharides by means of alcian blue. *Nature, 198*, 799–800. Retrieved from http://www.ncbi.nlm.nih.gov/pubmed/14002231.
Yamada, H., Denzer, A. J., Hori, H., Tanaka, T., Anderson, L. V., Fujita, S., et al. (1996). Dystroglycan is a dual receptor for agrin and laminin-2 in Schwann cell membrane. *Journal of Biological Chemistry, 271*(38), 23418–23423. Retrieved from http://www.ncbi.nlm.nih.gov/pubmed/8798547.
Yamada, S., Morimoto, H., Fujisawa, T., & Sugahara, K. (2007). Glycosaminoglycans in Hydra magnipapillata (Hydrozoa, Cnidaria): demonstration of chondroitin in the developing nematocyst, the sting organelle, and structural characterization of glycosaminoglycans. *Glycobiology, 17*(8), 886–894. http://dx.doi.org/10.1093/glycob/cwm051.
Yard, B. A., Kahlert, S., Engelleiter, R., Resch, S., Waldherr, R., Groffen, A. J., et al. (2001). Decreased glomerular expression of agrin in diabetic nephropathy and podocytes, cultured in high glucose medium. *Experimental Nephrology, 9*(3), 214–222. Retrieved from http://www.ncbi.nlm.nih.gov/entrez/query.fcgi?cmd=Retrieve&db=PubMed&dopt=Citation&list_uids=11340306.
Zhang, W., Coldefy, A. S., Hubbard, S. R., & Burden, S. J. (2011). Agrin binds to the N-terminal region of Lrp4 protein and stimulates association between Lrp4 and the first immunoglobulin-like domain in muscle-specific kinase (MuSK). *Journal of Biological Chemistry, 286*(47), 40624–40630. http://dx.doi.org/10.1074/jbc.M111.279307.
Zhang, L., David, G., & Esko, J. D. (1995). Repetitive Ser-Gly sequences enhance heparan sulfate assembly in proteoglycans. *Journal of Biological Chemistry, 270*(45), 27127–27135. Retrieved from http://www.ncbi.nlm.nih.gov/entrez/query.fcgi?cmd=Retrieve&db=PubMed&dopt=Citation&list_uids=7592967.
Zhang, L., & Esko, J. D. (1994). Amino acid determinants that drive heparan sulfate assembly in a proteoglycan. *Journal of Biological Chemistry, 269*(30), 19295–19299. Retrieved from http://www.ncbi.nlm.nih.gov/entrez/query.fcgi?cmd=Retrieve&db=PubMed&dopt=Citation&list_uids=8034692.

Zhang, B., Luo, S., Wang, Q., Suzuki, T., Xiong, W. C., & Mei, L. (2008). LRP4 serves as a coreceptor of agrin. *Neuron, 60*(2), 285–297. http://dx.doi.org/10.1016/j.neuron.2008.10.006.
Zhou, Z., Wang, J., Cao, R., Morita, H., Soininen, R., Chan, K. M., et al. (2004). Impaired angiogenesis, delayed wound healing and retarded tumor growth in perlecan heparan sulfate-deficient mice. *Cancer Research, 64*(14), 4699–4702. http://dx.doi.org/10.1158/0008-5472.CAN-04-0810.
Zhang, J., Wang, Y., Chu, Y., Su, L., Gong, Y., Zhang, R., et al. (2006). Agrin is involved in lymphocytes activation that is mediated by alpha-dystroglycan. *FASEB Journal, 20*(1), 50–58. http://dx.doi.org/10.1096/fj.04-3303com.

CHAPTER NINE

Building from the Ground up: Basement Membranes in *Drosophila* Development

Adam J. Isabella[1] and Sally Horne-Badovinac[1,2,*]
[1]Committee on Development, Regeneration, and Stem Cell Biology, The University of Chicago, Chicago, IL, USA
[2]Department of Molecular Genetics and Cell Biology, The University of Chicago, Chicago, IL, USA
*Corresponding author: E-mail: shorne@uchicago.edu

Contents

Abstract

Basement membranes (BMs) are sheetlike extracellular matrices found at the basal surfaces of epithelial tissues. The structural and functional diversity of these matrices within the body endows them with the ability to affect multiple aspects of cell behavior and communication; for this reason, BMs are integral to many developmental processes. The power of *Drosophila* genetics, as applied to the BM, has yielded substantial insight into how these matrices influence development. Here, we explore three facets of BM biology to which *Drosophila* research has made particularly important contributions. First, we discuss how newly synthesized BM proteins are secreted to and assembled exclusively on basal epithelial surfaces. Next, we examine how regulation of the structural properties of the BM mechanically supports and guides tissue morphogenesis.

Current Topics in Membranes, Volume 76
ISSN 1063-5823
http://dx.doi.org/10.1016/bs.ctm.2015.07.001

Finally, we explore how BMs influence development through the modulation of several major signaling pathways.

1. INTRODUCTION

Extracellular matrices (ECMs) are proteinaceous networks that accumulate nearly ubiquitously in the spaces between cells. ECMs link and coordinate cells both within and between tissues; their existence therefore likely contributed greatly to the rise and success of multicellular life, especially in the metazoan lineage (Ozbek, Balasubramanian, Chiquet-Ehrismann, Tucker, & Adams, 2010). Among the most ancient ECMs, the basement membrane (BM) is a specialized matrix that associates with the basal surfaces of epithelial tissues, as well as endothelial, fat, muscle, and Schwann cells. This chapter will focus predominantly on epithelial BMs. By electron microscopy, BMs appear as thin sheets (generally ~100 nm thick). They are composed primarily of two independent weblike networks of laminin and type IV collagen (collagen IV), which are heavily interlinked by proteins such as nidogen and the heparan sulfate proteoglycan (HSPG) perlecan (Yurchenco, 2011) (Figure 1). Adhesion of the BM to cells is achieved via interactions with transmembrane receptors, such as integrins and dystroglycan. Beyond the four core BM constituents, a large number of

Protein	*Drosophila* genes	Human homologs
Collagen IV 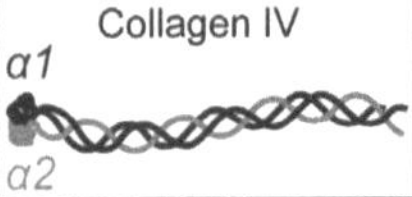	α1: *Collagen gene at 25C (Cg25C)*	*COL4A1, COL4A3, COL4A5*
	α2: *viking (vkg)*	*COL4A2, COL4A4, COL4A6*
Laminin 	α1,2: *wing blister (wb)*	*LAMA1, LAMA2*
	α3,5: *Laminin A (LanA)*	*LAMA3, LAMA4, LAMA5*
	β: *Laminin B1 (LanB1)*	*LAMB1, LAMB2, LAMB3, LAMB4*
	γ: *Laminin B2 (LanB2)*	*LAMC1, LAMC2, LAMC3*
Perlecan 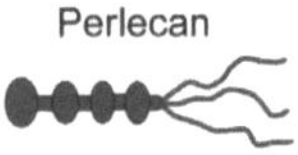	*terribly reduced optic lobes (trol)*	*HSPG2*
Nidogen 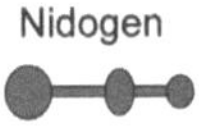	*Nidogen/entactin (Ndg)*	*NID1, NID2*

Figure 1 Overview of the core basement membrane proteins in *Drosophila*. (See color plate)

accessory proteins have been found to contribute to the network (Hynes & Naba, 2012). Differential incorporation of minor components, as well as varying isoforms and posttranslational modifications of the core proteins, lends great structural and functional diversity to the many BMs found throughout the body.

Several major roles for BMs have emerged, which will be discussed briefly here and in greater depth throughout this chapter. First, the mechanical properties of the network establish it as a physical scaffold. This property allows BMs to maintain tissue shape and integrity in the face of deformation forces and act as a substrate against which forces can be generated for cellular contraction and migration. Because of their small pore size, BMs can also provide a barrier function that helps to limit the movement of cells and large macromolecular complexes between body compartments. The ability to bind several secreted signaling molecules further allows these matrices to facilitate cell–cell communication both within and between tissues.

The functional capabilities of the BM described above make it well suited to facilitate the specification, compartmentalization, growth, and morphogenesis of distinct tissue and organ systems. Thus, it is not surprising that BMs are essential for embryonic development. The fruit fly *Drosophila melanogaster* has provided a particularly powerful system in which to dissect the specific contributions that BMs make to these processes. Nearly all developing tissue and organ systems have been well characterized and are visually and experimentally accessible. The powerful genetic techniques available, especially the ability to precisely manipulate gene expression in time and space, are also advantageous, particularly when studying a structure that plays such diverse roles in development. Moreover, the creation of functional GFP protein trap alleles of the collagen IV α2 gene *viking* and the perlecan gene *terribly reduced optic lobes* (*trol*) have transformed BM research in *Drosophila* by allowing unprecedented visual resolution of the native proteins in both fixed and living tissues (Buszczak et al., 2007; Morin, Daneman, Zavortink, & Chia, 2001).

While the core BM proteins and their receptors are well conserved between flies and humans, the fly BM can be viewed as a simplified version of its mammalian counterpart. Flies produce only two distinct laminin trimers compared to sixteen in humans, one collagen IV trimer versus three in humans, and two β and five α integrin subunits versus eight β and eighteen α subunits in humans. Although this simplicity means that flies cannot recapitulate the diversity of human BMs and ECM receptors, it increases the power to dissect protein function by limiting problems associated with redundancy.

In this chapter, we highlight important contributions that *Drosophila* research has made to our understanding of BM assembly and function during development. Because the literature on this topic is extensive, we have not attempted to provide a comprehensive summary of the data. Instead, we focus on three topic areas that exemplify the breadth and depth of BM research in this organism. First, we address the longstanding question of how BM proteins are precisely targeted to basal epithelial surfaces. We discuss how proteins produced from a variety of cellular sources achieve this goal, with a special focus on the intracellular trafficking pathway that operates within epithelial cells to transport newly synthesized BM proteins to basal regions of the plasma membrane for secretion. Second, we address the process of morphogenesis during development. We explore mechanical contributions of BMs to this process, specifically how regulated remodeling of BM structure can help to shape a tissue. In this section, we offer an in-depth discussion of the complex contributions of the BM to egg chamber elongation. Third, we address molecular signals that mediate cell–cell communication during development. We discuss contributions that BMs make to this process through the modulation of several major signaling pathways.

2. SYNTHESIS, SECRETION, AND ASSEMBLY OF BMS ON BASAL CELL SURFACES

Epithelial cells exhibit a highly polarized architecture with four distinct membrane domains—apical, junctional, lateral, and basal. To build and maintain a BM, newly synthesized components must be assembled exclusively on the basal epithelial surface. Here, we explore several ways in which BM proteins are targeted to this membrane domain, including a mechanism for the polarized secretion of BM proteins by an epithelium.

2.1 Sources of BM proteins and implications for polarized assembly

In *Drosophila*, the sources of BM proteins are complex and, in many cases, are still not clear. Some epithelia synthesize all of their own BM components, while others rely on production by other tissues. Some epithelia combine these approaches, producing a subset of their own proteins and relying on external sources for others (Figure 2). The major nonepithelial source of BM proteins also varies by developmental stage. In the embryo, BM proteins

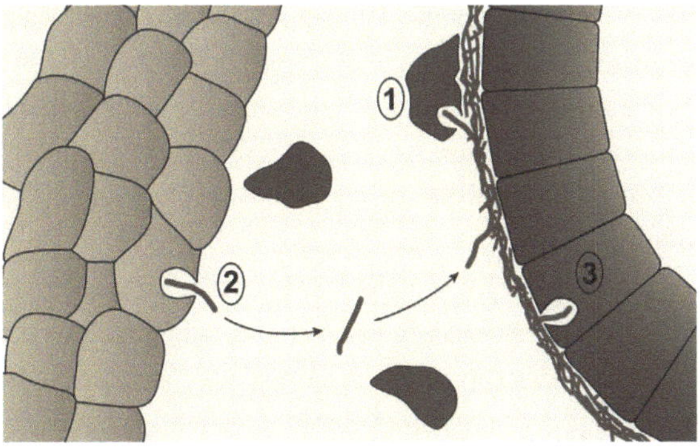

Figure 2 ***Cellular sources of basement membrane proteins in* Drosophila.** BM proteins are synthesized and secreted by three primary cell types. 1: Synthesis and local deposition by hemocytes. 2: Synthesis by and long range diffusion from the fat body. 3: Synthesis and secretion by the epithelium itself. (See color plate)

are primarily produced by hemocytes—circulating immune cells that migrate throughout the body. Hemocytes display the primary signal for transcripts encoding collagen IV and laminin by in situ hybridization (Kusche-Gullberg, Garrison, MacKrell, Fessler, & Fessler, 1992; Le Parco, Knibiehler, Cecchini, & Mirre, 1986; Mirre, Cecchini, Parco, & Knibiehler, 1988; Yasothornsrikul, Davis, Cramer, Kimbrell, & Dearolf, 1997) and, when cultured, produce large volumes of BM proteins (Fessler, Nelson, & Fessler, 1994). It should be noted, however, that some embryonic and larval epithelia appear to produce their own laminin and/or perlecan (Denef, Chen, Weeks, Barcelo, & Schüpbach, 2008; Martin et al., 1999; Sorrosal, Pérez, Herranz, & Milán, 2010).

How do proteins secreted by hemocytes assemble specifically on basal epithelial surfaces? This process must require establishment of the basal epithelial membrane as a competent surface to bind soluble BM proteins, likely by expression of cell surface BM receptors. Hemocytes also tend to cluster around BM-containing tissues, probably for the purpose of BM deposition (Kusche-Gullberg et al., 1992). This phenomenon is analogous to ECM deposition by fibroblasts in vertebrates. In this case, epithelia likely recruit hemocytes to their basal surfaces.

While hemocytes continue to produce BM proteins throughout development, during late embryogenesis and larval stages collagen IV and laminin production is also observed strongly in the fat body—a major metabolic organ in insects (Kusche-Gullberg et al., 1992; Le Parco et al., 1986; Mirre et al., 1988; Yasothornsrikul et al., 1997). The fat body appears to take over as the major production center, at least for collagen IV, in larvae. Blocking collagen IV production in the fat body results in drastic loss of this protein from BMs throughout the body, including full loss from the BM surrounding the imaginal wing disc epithelium—the pouchlike precursor to the adult wing (Pastor-Pareja & Xu, 2011).

Because the fat body is fixed in place, proteins secreted from this organ must diffuse long distances through the extracellular space to their target tissues. Yet they are still incorporated efficiently into distant BMs. As was discussed with hemocytes, this is most likely achieved by ensuring that basal epithelial surfaces have the necessary adhesive properties to capture diffusing proteins. In this case, it is equally important to prevent premature protein aggregation and promiscuous adhesion to the wrong tissues. The task of escaping the fat body is particularly onerous, as this tissue is itself surrounded by a BM through which secreted proteins bound for other tissues must pass without adhering. SPARC (secreted protein acidic and rich in cysteine) appears to promote collagen IV diffusion away from the fat body to distant epithelia, as loss of SPARC from this tissue leads to an aberrant accumulation of collagen IV between fat body cells (Pastor-Pareja & Xu, 2011; Shahab et al., 2015).

Although it is likely that the hemocytes and fat body continue to produce BM proteins throughout the life of the fly, there is one epithelium, found in adult females, that is known to synthesize and secrete all of its own major BM proteins—the follicular epithelium that surrounds the developing germ cells within the ovary. For the rest of this section, we will describe recent studies, primarily performed within this tissue, that have begun to elucidate how BM proteins synthesized within the epithelium itself are targeted exclusively to basal regions of the plasma membrane for secretion (Figure 3).

2.2 Basal localization of BM protein synthesis

In the follicular epithelium, newly synthesized BM proteins exhibit a polarized localization within the cell from the moment of translation. The mRNAs encoding both collagen IV chains (*viking and Cg25c*) and the laminin β chain (*LanB1*) show a 70% enrichment in the basal half of the cell (Lerner et al., 2013). Because the endoplasmic reticulum (ER) stretches throughout the cytoplasmic volume, this observation suggests that these transcripts are primarily translated into a specific subregion of this organelle.

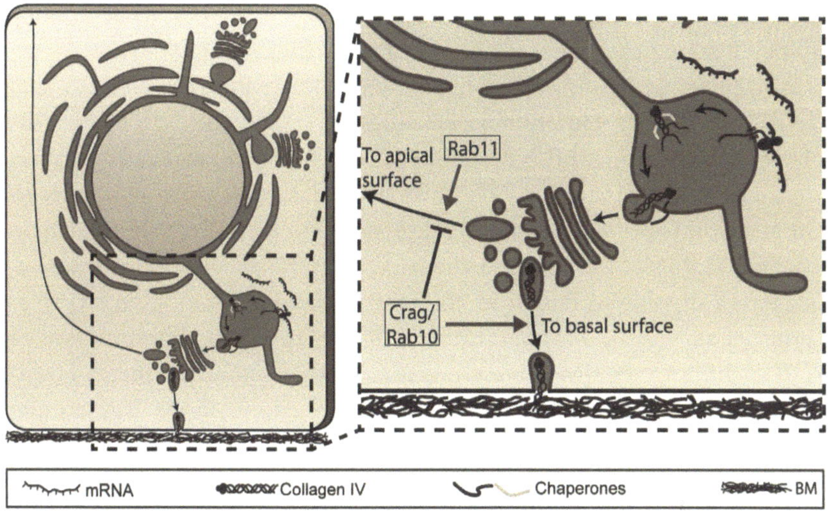

Figure 3 ***Local synthesis and polarized secretion of collagen IV in the follicular epithelium.*** Within the follicular epithelium, *collagen IV* transcripts accumulate basally and are translated and Translocated into a basal region of the endoplasmic reticulum (ER). ER-resident proteins assist in the folding and packaging of collagen IV for transport to the Golgi. After collagen IV transits through the Golgi, Crag and Rab10 promote delivery of collagen IV-containing exocytic vesicles to basal regions of the plasma membrane and prevent Rab11-dependent targeting of collagen IV-containing vesicles to the apical surface. Inset: blow-up of indicated region. (See color plate)

Thus, mRNA localization may help to establish a distinct ER compartment specialized for BM protein production.

Why might such a distinct ER compartment exist? Collagen IV places a notorious burden on the ER's protein production and transport machinery. Each collagen IV protomer is assembled from three polypeptides that wind into a triple helical structure nearly 400 nm long (Khoshnoodi, Pedchenko, & Hudson, 2008). This complex folding reaction requires a suite of ER-resident chaperones, several of which are collagen-specific. For instance, procollagen lysyl hydroxylase (Plod) and prolyl-4-hydroxylase-alpha EFB (PH4αEFB) catalyze hydroxylation of lysines and prolines, respectively, primarily within the triple helical domain; and both enzymes are required for trimer assembly and stability (Myllyharju & Kivirikko, 2004). Collagen IV's large size also prevents it from being packaged into standard CopII-coated vesicles for transport to the Golgi. The transmembrane protein Tango1 is required at ER exit sites (ERESs) to help load collagens into enlarged Golgi-bound vesicles (Pastor-Pareja & Xu, 2011; Saito et al., 2009; Venditti et al., 2012; Wilson et al., 2011). Compartmentalization of collagen IV production could, therefore, increase biosynthetic efficiency while limiting any potential negative impact on other ER processes.

In support of this idea, the mRNAs encoding Plod, PH4αEFB, and Tango1 all display enrichment in the basal cytoplasm, similar to the collagen IV-encoding mRNAs, and the Tango1 protein primarily localizes to basal ERESs (Lerner et al., 2013). Importantly, knocking down expression of any of these three proteins causes collagen IV to become trapped in a discrete region of the ER near the basal cell surface. When this happens, collagen IV does not diffuse from this location, even over long time periods (Lerner et al., 2013). This observation suggests that a mechanism exists to prohibit diffusion of BM proteins away from their site of synthesis. Localized production of ECM proteins within subregions of the ER has also been observed in vertebrates, which suggests that this may be a conserved biosynthetic strategy (Vertel, Velasco, LaFrance, Walters, & Kaczman-Daniel, 1989).

Because the ER compartment where BM proteins are synthesized is in the basal region of the cell, it is intriguing to speculate that this localization may also act as an initial step to bias secretion to the basal plasma membrane. While the Golgi typically takes the form of a singular organelle in mammalian cells, called the Golgi ribbon, *Drosophila* cells contain many dispersed Golgi stacks that each associate with a single ERES (Kondylis, Pizette, & Rabouille, 2009; Kondylis & Rabouille, 2009). This organization has led to the hypothesis that individual ERES-Golgi units could function independently of one another to facilitate polarized protein secretion. Synthesis within a basal region of the ER could, therefore, promote protein transport through basally localized ERES-Golgi units and subsequent delivery to the adjacent basal plasma membrane. However, there are several reasons to question this assertion. First, mRNA localization does not appear to be an absolute requirement for high-fidelity polarized secretion—*perlecan* (*trol*) mRNA does not display a basal bias, but the protein is still faithfully secreted to the basal surface (Lerner et al., 2013). Nor is it sufficient, as disruption of post-Golgi BM protein trafficking causes aberrant secretion to the apical surface (see next subsection). Additionally, whether such a mechanism could function in mammalian cells is unclear. Trafficking through a centralized Golgi ribbon would erase the polarity induced by mRNA localization. However, not all mammalian cells have a Golgi ribbon. Golgi outposts—analogous to the independent Golgi units in *Drosophila*—have been shown to promote polarized protein secretion, possibly in conjunction with localized mRNAs, in vertebrate neurons (Bramham & Wells, 2007; Hanus &

Ehlers, 2008; Horton et al., 2005; Lowenstein et al., 1994; Pierce, Mayer, & McCarthy, 2001; Ramírez & Couve, 2011). A distributed Golgi system has also been described in gastric parietal cells (Gunn et al., 2011).

It is important to note that the two hypotheses as to why the BM proteins are preferentially produced in a basal region of the ER are not mutually exclusive and could simultaneously promote BM formation. These observations offer interesting insight into potential mechanisms regulating BM protein production and secretion, but further study is required to understand the implications of local BM protein production. Experimentally disturbing the basal bias of BM protein production in these cells will be especially useful in discerning the veracity of these hypotheses.

2.3 Post-Golgi trafficking of BM proteins to the basal surface

Apicobasal polarity depends on the polarized trafficking of newly synthesized transmembrane proteins to either the apical or combined basal and lateral (basolateral) membrane domains. Several sorting mechanisms have been identified that direct individual proteins to each of these locations (Rodriguez-Boulan, Kreitzer, & Müsch, 2005; Stoops & Caplan, 2014). Knowledge of these pathways, however, has failed to provide insight into the polarized trafficking pathway that transports BM proteins exclusively to the basal surface. Indeed, several classic studies performed in cultured mammalian epithelial cells have indicated that distinct pathways exist for polarized trafficking of BM versus transmembrane proteins. Treatment with NH_4Cl or colchicine, which perturbs the acidification of intracellular compartments and microtubule dynamics, respectively, disrupts the polarized secretion of BM proteins without affecting basolateral transmembrane proteins. Under these conditions, BM proteins are secreted from both the apical and basal epithelial surfaces (Boll, Partin, Katz, Caplan, & Jamieson, 1991; Caplan, Stow, & Newman, 1987; De Almeida & Stow, 1991; Natori et al., 1992). Conversely, disruption of Cdc42 function disrupts polarized secretion of basolateral transmembrane proteins but has no effect on BM proteins (Cohen, Müsch, & Rodriguez-Boulan, 2001). In fact, even integrins appear to move through a different trafficking pathway than BM proteins (Boll et al., 1991). These studies revealed that a specific pathway for polarized BM secretion exists, but offered little insight into its molecular details.

Recent genetic studies in *Drosophila* have begun to identify the major molecular players that control polarized BM secretion in this system. The DENN domain-containing protein Crag (calmodulin-binding protein related to a Rab3 GDP–GTP exchange protein) was discovered in a forward genetic screen for novel regulators of apicobasal polarity in the follicular epithelium (Denef et al., 2008). Similar to the early observations in mammalian cells, BM proteins accumulate on both the apical and basal surfaces of *Crag* mutant follicle cells, whereas apical and basolateral transmembrane proteins localize normally. Apical accumulation of perlecan was also observed in the epidermis of *Crag* mutant embryos. Importantly, this paper confirmed that the apical deposition of BM proteins is not due to transcytosis of protein from the existing BM, but rather due to the aberrant secretion of newly synthesized proteins. Crag was initially observed to localize to apical and lateral cell membranes and to Rab5- and Rab11-positive endosomes, although an important population near the basal surface has since been described (see below).

DENN domain-containing proteins commonly function as guanine nucleotide exchange factors (GEFs) for Rab-family GTPases (Marat, Dokainish, & McPherson, 2011), which are molecular switches that cycle between an active GTP-bound state and an inactive GDP-bound state. GEFs transition Rabs to the GTP-bound state, whereas GTPase activating proteins (GAPs) induce transition to the GDP-bound state. The presence of specific active Rab proteins confers identity to membrane-bound compartments within the cell, such as organelles and trafficking vesicles (Barr, 2013). Rabs are also master regulators of vesicle activity, controlling their formation, sorting, targeting, fission, and fusion (Hutagalung & Novick, 2011). It was therefore speculated that Crag might activate a Rab that plays one or more roles in the polarized secretion of BM proteins.

Crag was later found to be a GEF for a known exocytic Rab, Rab10, first in mammals and then in flies (Lerner et al., 2013; Xiong et al., 2012; Yoshimura, Gerondopoulos, Linford, Rigden, & Barr, 2010). Consistent with this result, Rab10 depletion also causes BM proteins to accumulate on both the apical and basal surfaces of the follicular epithelium (Lerner et al., 2013). One key function of a GEF is to recruit its cognate Rab to the correct intracellular membranes (Blümer et al., 2013). Interestingly, Rab10 and Crag colocalize on membrane-bound compartments that are tightly associated with the basal surfaces in the follicle cells, proximal to the ER compartment where BM proteins are synthesized. This basally localized population of Rab10 is lost in *Crag* mutant cells, which suggests that BM proteins likely pass through these compartments on their way to the basal surface.

Although Rab10's exact role(s) in polarized BM secretion remains to be determined, one appealing hypothesis is that this protein functions in an endosomal recycling compartment (ERC) to help sort BM cargos into a basally directed trafficking pathway. Biosynthetic cargo sorting commonly occurs in the trans-Golgi network (Anitei & Hoflack, 2011; Santiago-Tirado & Bretscher, 2011). However, in polarized epithelial cells most exocytic traffic also passes through ERCs, where additional sorting occurs (Fölsch, Mattila, & Weisz, 2009; Gonzalez & Rodriguez-Boulan, 2009). Rab10 localizes to endocytic compartments in both *Caenorhabditis elegans* and mammalian cells (Babbey et al., 2006; Chen et al., 2006; Shi et al., 2010). Moreover, although Rab11 (another Rab that localizes to ERCs) is not normally required for BM traffic, when Rab10 or Crag are depleted, the BM proteins that travel to the apical surface do so through a Rab11-dependent mechanism (Lerner et al., 2013). This observation suggests that Crag and Rab10 promote the sorting of BM proteins away from a Rab11-dependent pathway.

Crag and Rab10 have been established as the core components of a BM-specific trafficking pathway, though how this pathway recognizes, sorts, and targets BM proteins to the basal surface remains to be discovered, as do additional components involved in this process. Two additional proteins have been identified that are required for polarized BM secretion, although the mechanisms by which they do so are less well understood. These are phosphatidylinositol synthase (Pis), an enzyme involved in the production of phosphoinositides (Devergne, Tsung, Barcelo, & Schupbach, 2014), and Scarface, a secreted serine protease-like protein that lacks catalytic activity (Sorrosal et al., 2010).

The phosphoinositides are a family of phospholipids that regulate a stunning array of cellular processes (Balla, 2013). Various phosphoinositide isoforms are created by kinase- and phosphatase-mediated interconversion between different phosphorylation states of a common phospholipid backbone, phosphatidylinositol. Pis synthesizes phosphatidylinositol and is therefore required for formation of all phosphoinositides; however, the authors focused their analyses on the role of phosphatidylinositol 4,5 bisphosphate (PIP2) in polarized BM secretion (Devergne et al., 2014). PIP2 functions as an apical determinant in the regulation of epithelial polarity (Martin-Belmonte et al., 2007). It also regulates multiple steps of polarized vesicle trafficking, including cargo sorting and membrane fusion (Balla, 2013). In the follicular epithelium, PIP2 is enriched on apical and lateral membranes. Loss of Pis or other PIP2 biosynthetic enzymes decreased PIP2 levels and caused a loss of Crag from apical and lateral membranes (Devergne et al., 2014). How the lateral and apical Crag populations would feed into basal protein secretion is difficult to say. It is also unclear whether loss of these populations upon Pis depletion is specific, or is an indirect effect of a general reduction in Crag protein levels throughout the cell.

In *scarface* mutant embryos, laminin accumulates on the apical surface of the lateral epidermis and an adjacent extraembryonic epithelium called the amnioserosa; this phenotype is also seen in *Crag* mutant embryos (Sorrosal et al., 2010). Interestingly, expression of Scarface exclusively in the lateral epidermis is sufficient to rescue the BM secretion defect in the amnioserosa of *scarface* mutants, which suggests that this protein can act at a distance. Consistent with this finding, expression of a tagged version of Scarface in a specific region of the wing disc epithelium caused the protein to accumulate on the apical surface and within the endosomal system of nonexpressing cells. Three hypotheses have been proposed for Scarface's function in polarized BM secretion (Eastburn & Mostov, 2010; Sorrosal et al., 2010). Scarface could function within an endosomal compartment to help sort BM proteins into a basally directed trafficking pathway. This possibility is appealing as it aligns well with the likely function of Rab10. Alternatively, Scarface could function at the apical plasma membrane to prevent targeting of BM-containing vesicles to this domain, a function that has also been proposed for Crag (Denef et al., 2008; Devergne et al., 2014). Finally, Scarface could remove BM proteins from the apical surface either by stimulating a proteolytic cascade or via endocytosis. Study of Scarface in the better-understood follicular epithelium may help to build a coherent model for how Crag, Rab10, and Scarface function together to control polarized BM secretion.

Finally, while the Crag/Rab10 pathway has not yet been shown to control polarized BM trafficking in vertebrates, the utilization of distinct secretory pathways for BM and basolateral transmembrane proteins in both *Drosophila* and mammalian cells is intriguing. Such a condition could arise based on special accommodations that certain BM proteins require to move through the secretory pathway (i.e. an acidic environment, enlarged vesicles, etc.), or due to differences in recognizing transmembrane and soluble proteins by the sorting machinery. Where, precisely, these proteins leave the cell could also play a role. Basolateral transmembrane proteins appear to exit the cell though an apical region of the lateral plasma membrane, just basal to the adherens junctions (Grindstaff et al., 1998). In contrast, BM proteins are more likely to exit at or very near the basal surface; indeed, Rab10 has been found on secretory vesicles bound for a basal region of the lateral plasma membrane in mammalian cells (Cao et al., 2008).

Further characterization of the proteins already known to be involved in basal targeting of BM proteins and the identification of other key players will provide a rich area for future research.

3. MECHANICAL CONTRIBUTIONS OF THE BM TO MORPHOGENESIS

Morphogenesis is the process by which cells and tissues change their shapes to create the complex form of adult tissues and organs. BM sheets are well designed to physically assist and regulate morphogenetic processes—by modifying the movement of cells via adhesive interactions, by resisting the contractile forces exerted by cells, or by restricting the expansion of growing tissues. Furthermore, a large number of studies, primarily in in vitro cell culture systems, have found that changing the physical properties of ECMs can modulate the dynamic activities of cells. For instance, changes in matrix stiffness or cell-matrix adhesion affect cellular contractile dynamics and downstream signals within the cell (Charras & Sahai, 2014). Matrix stiffness also appears to regulate cell migration, as many cells tend to migrate from softer to stiffer ECM substrates or toward ECM that is under deformation forces (Lo, Wang, Dembo, & Wang, 2000; Reinhart-King, Dembo, & Hammer, 2008; Roca-Cusachs, Sunyer, & Trepat, 2013). A softer matrix, meanwhile, can promote cellular invasion through the network (Gu et al., 2014). The organization of the matrix also influences migration dynamics (Kim, Provenzano, Smith, & Levchenko, 2012); directionally aligned matrices have been found to orient cellular migration in the direction of alignment and to increase migration speed (Diehl, Foley, Nealey, & Murphy, 2005; Provenzano et al., 2006; Provenzano, Inman, Eliceiri, Trier, & Keely, 2008; Tan & Saltzman, 2002). Confirmation of these observations in vivo remains an important task; *Drosophila* offers an enticing opportunity to explore these concepts within a developing animal.

In vivo, the BMs of developing tissues undergo heavy remodeling (Bernfield, Banerjee, Koda, & Rapraeger, 1984; Daley, Peters, & Larsen, 2008). Moreover, it is clear that different BMs, and even the same BMs over time, exhibit vastly different compositions and, therefore, physical properties. It is likely that the characteristics of these matrices are tuned to appropriately contribute to morphogenesis, although the changes in physical properties, their mechanisms, and their effects on morphogenesis are largely unclear.

3.1 Contributions of the BM to egg chamber elongation

The development of the egg chamber is perhaps the best understood example of how a coordinated progression of changes to BM structure and cellular activity interact to drive tissue morphogenesis (Horne-Badovinac, 2014). Egg chambers, which number in the hundreds within the adult ovary, are each responsible for the maturation of a single oocyte. They are composed of a central cluster of germ cells—one posteriorly localized oocyte and 15 supporting nurse cells—that are surrounded by the follicular epithelium that was discussed in section 2. This somatic tissue is made up of roughly 800–1000 cells, and the BM it produces ensheathes the entire organlike structure (Figure 4(A)). Initially small and spherical, egg chambers proceed through 14 distinct morphological stages, during which they grow to nearly 1000 times their initial volume (Cummings & King, 1969). Between stages 5 and 10, growth is channeled anisotropically to induce elongation along the anterior–posterior (A–P) axis, a process that creates the elliptical shape of the egg (Figure 4(C)).

BM structure has been examined across egg chamber development and found to shift dramatically with the onset of elongation. The BM surrounding early egg chambers does not show any obvious structure by light microscopy. However, concurrent with the onset of elongation at stage 5, dense, linear fibril-like aggregates of collagen IV, laminin, and perlecan begin to be incorporated into the existing planar matrix (Cetera et al., 2014; Gutzeit, Eberhardt, & Gratwohl, 1991; Haigo & Bilder, 2011; Schneider et al., 2006) (Figure 4(C)). These structures all align perpendicular to the A–P axis, effectively polarizing the BM. Arrays of linear actin bundles in the adjacent basal cortex of each follicle cell, which are physically coupled to the BM via integrin-based focal adhesions, align in the same direction as the BM fibrils (Bateman, Reddy, Saito, & Van Vactor, 2001; Delon & Brown, 2009; Gutzeit, 1990). Together, the BM fibrils and basal actin bundles are thought to act as a "molecular corset" that directionally constrains egg chamber growth, thereby providing the anisotropic force that drives elongation (Gutzeit et al., 1991) (Figure 4(B)). This hypothesis is supported by the observation that disruption of the tissue-level organization of these structures leads to the production of rounded eggs (Bateman et al., 2001; Cetera et al., 2014; Conder, Yu, Zahedi, & Harden, 2007; Frydman & Spradling, 2001; Gutzeit et al., 1991; Haigo & Bilder, 2011; Horne-Badovinac, Hill, Gerlach, Menegas, & Bilder, 2012; Lewellyn, Cetera, & Horne-Badovinac, 2013; Viktorinová, König, Schlichting, & Dahmann, 2009).

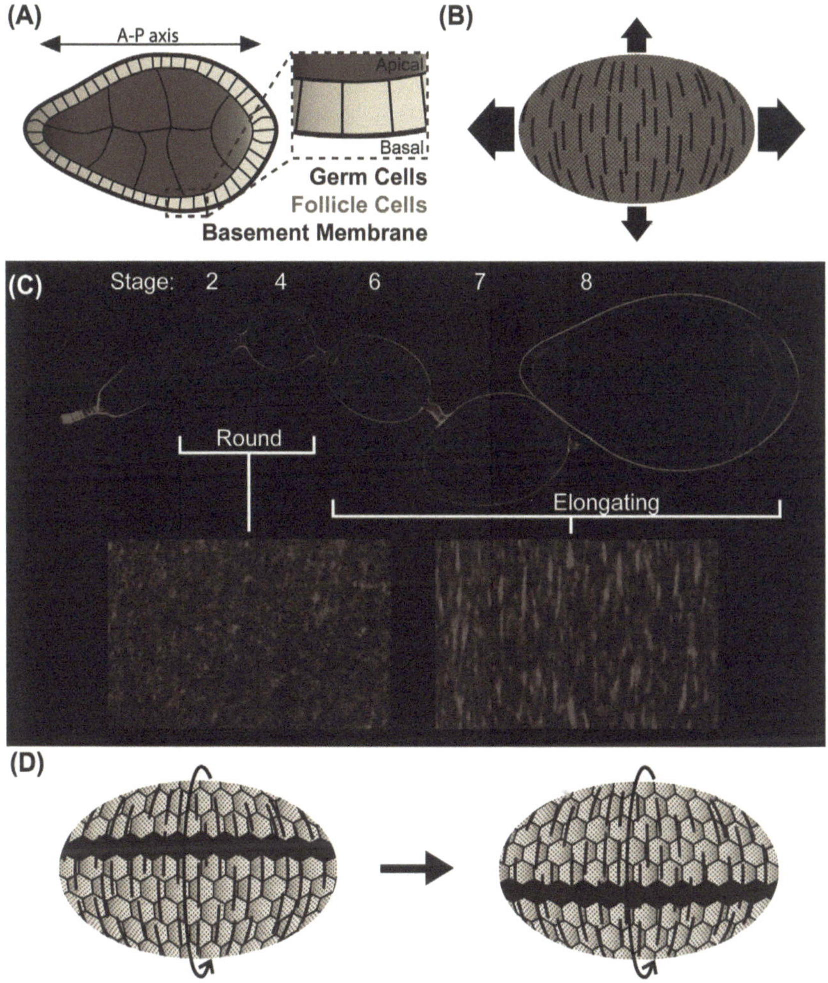

Figure 4 ***Basement membrane (BM) function in egg chamber elongation.*** (A) Illustration of a transverse section through an egg chamber. Central germ cells are surrounded by a somatic epithelium of follicle cells, which assemble a BM on their basal surfaces. (B) Model of the molecular corset. Polarized linear fibril-like structures in the BM are hypothesized to constrain egg chamber growth in the direction of polarization, biasing growth to occur along the A–P axis. Arrows indicate direction and magnitude of growth. (C) BM structural dynamics during elongation. Top: a developmental array of egg chambers showing cell outlines (actin) and the BM (Viking-GFP). Bottom: fluorescent micrographs of collagen IV (Viking-GFP) in the BM. Young, round egg chambers exhibit no obvious BM structure, while older, elongating egg chambers display polarized fibrils within the BM. (D) Overview of egg chamber rotation. In this illustration, the BM is partially transparent to reveal the cells underneath. The dark row of cells in each image represents the same cells at two different time points. The egg chamber rotates within a stationary BM, in the direction of BM fibrils. Curved arrows indicate direction of egg chamber rotation. (See color plate)

Formation of the molecular corset depends on a dramatic rotational motion of the egg chamber. Between stages 1 and 8, the basal surfaces of the follicle cells migrate along the BM, orthogonal to the A—P axis (Cetera et al., 2014; Haigo & Bilder, 2011). Because the apical surfaces of the follicle cells are physically attached to the germ cell cluster, this collective motion causes the entire egg chamber to rotate within the surrounding BM, which remains largely stationary throughout the process (Figure 4(D)). Failure of the follicle cells to migrate disrupts the polarity of the actin and BM networks and prevents egg chamber elongation (Cetera et al., 2014; Haigo & Bilder, 2011; Lerner et al., 2013; Viktorinová & Dahmann, 2013).

The presence of linear, fibril-like aggregates in the follicular BM is surprising, as a BM architecture of this type has not been described in other systems. Because flies do not produce fibril-forming collagens or fibronectin, it is intriguing to speculate as to whether these unusual BM structures might perform some of the same functions as the true fibrillar matrices found in other organisms. Future work will be required to understand the molecular organization of the BM fibrils, as well as the mechanism by which they form. Although egg chamber rotation is required for their formation (Haigo & Bilder, 2011), it is clearly not sufficient, as rotation begins at stage 1 and BM fibrils do not begin to form until stage 5 (Cetera et al., 2014). Fibril formation does, however, happen at a time when collagen IV levels are increasing in the matrix (Haigo & Bilder, 2011), which could indicate a role for new protein secretion in this process. Further, because fibrils appear to play an important functional role in this BM, it will be intriguing to explore to what extent BM superstructural elements are used in other systems to regulate tissue dynamics.

In addition to polarized fibril formation, other structural changes in the BM likely also contribute to its proposed corset function. As mentioned in the previous paragraph, collagen IV levels drastically increase in the BM between stages 5 and 8 (Haigo & Bilder, 2011). Precise control of collagen IV cross-linking also appears to influence elongation, as increasing peroxidasin-dependent collagen IV cross-linking enhances elongation while decreasing peroxidasin activity inhibits elongation without grossly affecting BM superstructure (McCall et al., 2014). In the future, it will be important to determine whether the levels or properties of other BM components are also dynamically regulated as part of the elongation program. It will also be interesting to explore whether the complex interplay that occurs between different BM components in other tissues similarly affects the physical properties of the follicular BM (Pastor-Pareja & Xu, 2011). Finally, methods will need to be developed to directly measure how each of these changes in matrix architecture alter BM strength and stiffness.

The BM also may influence the tissue-level alignment and activity of the basal actin bundles during elongation. Global actin bundle alignment depends on the rotational motion of the egg chamber at early developmental stages, but this organization becomes rotation-independent concurrent with the establishment of matrix polarity (Cetera et al., 2014). Although rotation ceases at stage 8, maintenance of basal actin bundle alignment is likely to be important for the final phase of elongation, which begins at stage 9. At this stage, periodic myosin-based contraction of the basal actin bundles contributes to elongation (He, Wang, Tang, & Montell, 2010). Proper basal actin alignment is likely required for directionality of this force. Interaction with the BM also directly influences contractile activity of the cells. Decreasing expression of the focal adhesion protein talin shortens the period of myosin contraction, while overexpressing paxillin, another focal adhesion protein, prolongs the period (He et al., 2010). Surprisingly, treatment of egg chambers with collagenase, which might be expected to mimic a decrease in cell-BM adhesion, has also been observed to prolong the period of myosin contraction (Koride et al., 2014). Myosin-based contractions, therefore, are likely regulated by complex inputs from BM structure and cell-BM adhesion.

Once growth has stopped and the elongation program is complete, the BM now plays an important role in maintaining the egg chamber's elliptical shape. When stage 13 egg chambers are treated with collagenase to disrupt the BM, they rapidly become rounder, shortening along the A—P axis, and expanding along the orthogonal axis (Haigo & Bilder, 2011). Thus the BM is required, in varying capacities, to support morphogenesis of the egg chamber for the entirety of its development.

As a final note, studies of egg chamber elongation have also revealed interesting potential effects of BM architecture and cell-BM adhesion on cell migration speed. Although the collective migration of the follicle cells that causes the egg chamber to rotate begins at stage 1, the speed of this migration increases sharply at stage 6, just after the BM fibrils begin to form and collagen IV levels begin to rise (Cetera et al., 2014). However, further investigation will be required to determine whether changes in BM architecture play a causal role in this acceleration. Altering integrin levels in the follicle cells also modifies their migration speed, such that decreasing integrins increases speed and increasing integrins decreases speed (Lewellyn et al., 2013).

3.2 Contributions of the BM to the morphogenesis of other tissues

Loss of function studies have revealed that, in addition to the egg chamber, BMs also appear to play critical roles in the morphogenesis of many other fly tissues and organs. To date, however, few of these initial observations have been followed up with mechanistic studies. Two tissues that have been examined more intensely in this regard are the wing imaginal disc epithelium and migrating glial cells in the imaginal eye disc. These studies are detailed below.

In the wing disc epithelium, local matrix degradation is required for two developmental processes. During larval development, adjacent tracheal tissue must invade through the wing disc BM and contact the underlying epithelium to form the air sac primordium. This process requires local BM degradation by matrix metalloprotease 2 (Mmp2) (Guha, Lin, & Kornberg, 2009). In the pupa, the internal wing disc protrudes through the body wall to form the external wing in a process known as disc eversion. This morphogenesis also requires Mmp2-mediated BM degradation (Srivastava, Pastor-Pareja, Igaki, Pagliarini, & Xu, 2007). A unique pattern of collagen IV cleavage is seen during disc eversion, suggesting precise, context-specific modifications of BM structure during morphogenesis (Fessler, Condic, Nelson, Fessler, & Fristrom, 1993).

Maintenance of proper cell shape in the larval wing disc epithelium also depends on interaction with an appropriately structured BM. Integrin-based adhesion to the BM is crucial for maintaining proper cell shape (Domínguez-Giménez, Brown, & Martín-Bermudo, 2007). Maintenance of cell shape also depends on balanced and opposing forces contributed by collagen IV and perlecan. Loss of collagen IV causes flattening of wing disc epithelial cells and the entire tissue, while loss of perlecan causes disc compaction and elongation of cells along their apicobasal axes (Pastor-Pareja & Xu, 2011). A similar antagonistic role for collagen IV and perlecan has been observed in the *C. elegans* neuromuscular junction, where the two molecules differentially regulate growth of presynaptic boutons (Qin, Liang, & Ding, 2014).

Finally, regulation of BM structure, specifically its stiffness, impacts the migration of glial cells within the eye disc. A stiffer matrix is known to promote cell migration in vitro (Lo et al., 2000; Roca-Cusachs et al., 2013). It was recently demonstrated that increasing the activity of the collagen cross-linking enzyme lysyl oxidase (Lox) or integrins increases BM stiffness in vivo in this system (Kim et al., 2014). This study further found that migrating glial cells upregulate Lox and integrin expression to promote their own migration. This is superficially similar to the observation, discussed above, that altering integrin levels affects cell migration in the egg chamber. However, during glial migration upregulation of integrins promotes migration, while in the egg chamber integrin upregulation slows migration or, when severe enough, inhibits it completely (Lewellyn et al., 2013). Integrin levels therefore appear to play a complex and context-specific role in migration dynamics.

4. CONTRIBUTIONS OF THE BM TO CELL–CELL SIGNALING DURING DEVELOPMENT

Cell-to-cell signaling, primarily through several major secreted molecules and their receptors, is crucial to regulate and coordinate tissue development. Although the predominant view of the BM tends to be structure-centric, it also serves as a major extracellular signaling platform. By interacting directly with secreted signaling proteins, the BM can act to limit their diffusion or modify their interactions with cell surface receptors. Genetic and biochemical evidence from flies and vertebrates indicates that the BM regulates most, if not all, major developmental signaling pathways, including TGF-β/BMP (Paralkar, Vukicevic, & Reddi, 1991; Wang, Harris, Bayston, & Ashe, 2008), FGF (Folkman et al., 1988; Klagsbrun, 1990; Lin, Buff, Perrimon, & Michelson, 1999; Park et al., 2003), Wingless/Wnt (Binari et al., 1997; Perrimon & Bernfield, 2000), and Hedgehog (Datta et al., 2006; Park et al., 2003; Rubin, Choi, & Segal, 2002; The, Bellaiche, & Perrimon, 1999). An advantage of the developmental focus of *Drosophila* research is that the early studies performed in this system offered not only evidence for these interactions, but also immediate indications of their developmental relevance. Since the connection was made between the BM and signaling, a diverse set of examples have emerged in *Drosophila* of the roles these interactions play in guiding specific developmental processes. Three such processes will be discussed here: axonal pathfinding, Malpighian tubule morphogenesis, and regulation of stem cell activity.

4.1 Modulation of Slit/Robo and Semaphorin/Plexin signaling during axonal pathfinding

As the nervous system develops, growth of axons away from neuronal cell bodies is required to appropriately innervate the body and promote connections with other neurons or tissues. As discussed above, the BM is an important permissive substrate for migration of many cell types, including neurons (Takagi et al., 1996), but it also acts to regulate the response of extending axons to multiple environmental signals that provide attractive or repulsive cues to guide their growth. Slit and Robo are a highly conserved signaling duo that were first discovered in forward genetic screens for developmental defects in *Drosophila* embryos (Nüsslein-Volhard, Wieschaus, & Kluding, 1984; Seeger, Tear, Ferres-Marco, & Goodman, 1993). Subsequent work showed that Slit is an extracellular protein that primarily functions as a repellent cue for axons that express the Robo receptor (Brose et al., 1999; Dickson & Gilestro, 2006). There is evidence that Slit may bind laminin in vertebrates (Brose et al., 1999), and laminin misexpression causes axon guidance defects in *Drosophila* (García-Alonso, Fetter, & Goodman, 1996; Kraut, Menon, & Zinn, 2001). Further, decreasing laminin or integrin expression enhances the axon pathfinding defects in a *Slit* hypomorphic allele, suggesting that the BM modulates axonal responsiveness to Slit signals (Stevens & Jacobs, 2002). Slit binding to heparin also enhances the Slit–Robo interaction (Hussain et al., 2006), and the transmembrane HSPG syndecan, which can function as a BM receptor (Carey, 1997), acts with Robo as a Slit coreceptor (Johnson et al., 2004; Smart et al., 2011; Steigemann, Molitor, Fellert, Jäckle, & Vorbrüggen, 2004). Whether this interaction occurs cooperatively with or independently of the BM is unclear, although perlecan, the BM HSPG, does not appear to exhibit similar activity (Steigemann et al., 2004).

Semaphorin-based axon guidance also relies on interactions with the BM. Semaphorin-1A is a transmembrane protein expressed at axon guidance decision points that signals to the axonal receptor Plexin A, which mediates repulsion at sites of Semaphorin contact (He, Wang, Koprivica, Ming, & Song, 2002). Similar to Slit, vertebrate Semaphorin activity is enhanced by heparin (De Wit, De Winter, Klooster, & Verhaagen, 2005). In this case, however, the HSPG utilized appears to be perlecan, which is heavily deposited at axon branch points and is required to augment a *Semaphorin-1A* gain-of-function mutation in *Drosophila* (Cho, Chak, Andreone, Wooley, & Kolodkin, 2012). *Syndecan* showed no genetic interaction with *Semaphorin-1A* in this study. It is intriguing that two signaling pathways, Slit/Robo and Semaphorin-1A/Plexin A, which regulate similar axonal guidance events by different molecular means, exhibit nonoverlapping reliance on two HSPG proteins. The coincidence of HSPG utilization among these pathways is mysterious, although utilization of different HSPG cofactors may enhance signaling diversity while maintaining signal distinction and resolution in the crowded neuronal milieu.

4.2 Modulation of BMP signaling during Malpighian tubule morphogenesis

Epithelial tubes are organized with their apical surfaces surrounding an internal lumen and their basal surfaces covered by a BM. Tubular outgrowth and branching are important in the development of several organs, including lungs, kidney, and salivary and mammary glands (Andrew & Ewald, 2010). Work in several vertebrate systems has identified mechanical roles for the BM and other ECMs in branching morphogenesis (Varner & Nelson, 2014). In *Drosophila*, the BM has been implicated in signaling during tubule morphogenesis as well. The *Drosophila* kidney ortholog, the Malpighian tubules, exhibits stereotyped outgrowth guided by a leading group of "kink cells" at the anterior of the tissue (Denholm, 2013). Local expression of the BMP homolog decapentaplegic (Dpp) in tissues along the tubule outgrowth route is necessary and sufficient to guide this morphogenesis (Bunt et al., 2010). Further, local deposition of collagen IV by hemocytes on the outgrowing tubules is required to transduce the Dpp signal in kink cells (Figure 5(A)). Supporting this observation, Dpp directly binds the C-terminus of collagen IV, which enhances interaction of Dpp with its receptors (Wang et al., 2008). This interaction appears to be conserved in mammals (Paralkar et al., 1991).

In the case of the Malpighian tubules, collagen IV helps to localize and concentrate the Dpp signal for its reception by target cells. Two other examples where Dpp and collagen IV interact in a similar manner will be discussed in the next section. Interestingly, Dpp is also known to regulate several developmental processes in *Drosophila* by formation of precise morphogen gradients. Because collagen IV appears to restrict diffusion of Dpp, it has been proposed that the BM may play an important role in these instances of Dpp signaling as well (Sawala, Sutcliffe, & Ashe, 2012; Umulis, Shimmi, O'Connor, & Othmer, 2010; Wang et al., 2008). It will therefore be intriguing to examine the role of the BM during Dpp signaling in other contexts.

4.3 Regulation of stem cell maintenance, differentiation, and division

Stem cells are multipotent progenitors that provide a key source of new cells during development. They also play important roles in tissue maintenance

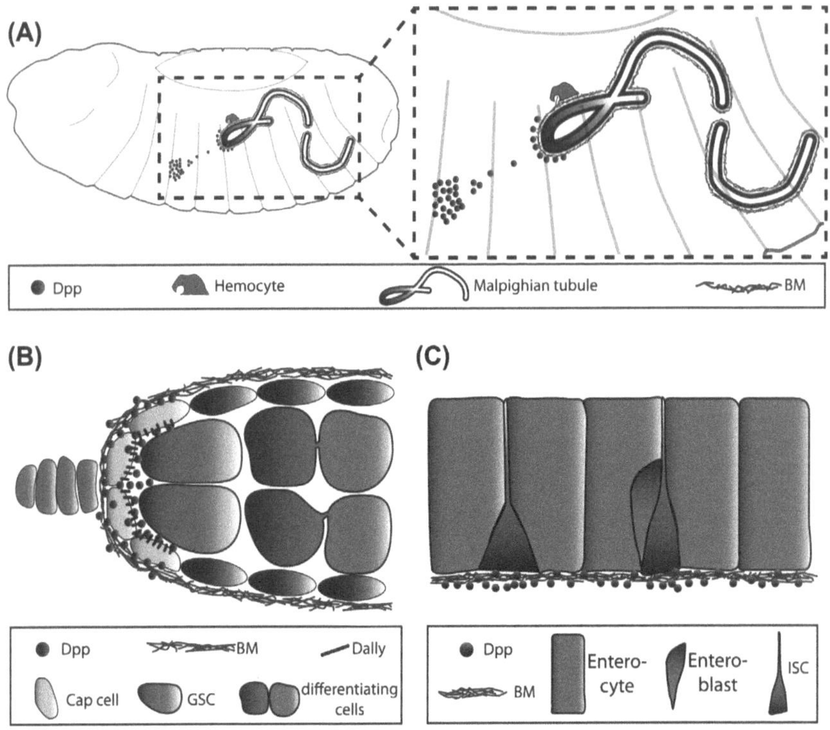

Figure 5 ***Basement membrane (BM) regulation of Dpp signaling in three developmental contexts.*** (A) Embryonic Malpighian tubule morphogenesis. Diffusible Dpp protein attracts the growing Malpighian tubule. Hemocyte-deposited collagen IV around the Malpighian tubule promotes reception of the Dpp signal by tubule cells. (B) Ovarian germ line stem cell (GSC) maintenance. GSCs are maintained within the stem cell niche via interaction with cap cells. Collagen IV in the BM and the HSPG Dally concentrate cap cell-derived Dpp to promote signal reception by GSCs but not differentiating daughter cells. (C) Intestinal stem cell (ISC) maintenance. ISCs exhibit basally positioned cell bodies that adhere to the BM. Dpp is concentrated by collagen IV within the BM to promote a high level of signal reception by ISCs but not the more apically localized, differentiating enteroblasts. (See color plate)

and repair in adults. The defining feature of stem cells is that they divide asymmetrically to produce two distinct cells: one stem cell to replace the mother and one cell that will differentiate. In *Drosophila*, the capacity to identify and observe stem cells in vivo under different genetic conditions has revealed distinct roles for the BM in several tissues, three of which are discussed here.

In *Drosophila*, the female germ line stem cells (GSCs) sit at the anterior end of the ovary in a specialized signaling environment deemed the niche (Losick, Morris, Fox, & Spradling, 2011; Xie & Spradling, 1998). The primary cells that make up the niche, the cap cells, bind GSCs through cadherin-based adhesions (Song, Zhu, Doan, & Xie, 2002). When GSCs divide, one daughter remains attached to the cap cells, while the other is expelled from this environment and, lacking the niche signals, differentiates. One key niche signal, secreted by the cap cells, is Dpp (López-Onieva, Fernández-Miñán, & González-Reyes, 2008; Wang et al., 2008). Collagen IV in the BM surrounding the niche binds Dpp and restricts its diffusion (Figure 5(B)); in the absence of collagen IV, the Dpp signaling field expands to reach GSC daughter cells outside of the niche, which prevents them from differentiating and leads to an overabundance of stem cells (Wang et al., 2008). The HSPG Dally is also enriched around the niche and appears to concentrate Dpp and promote its reception by GSCs (Guo & Wang, 2009; Hayashi, Kobayashi, & Nakato, 2009).

A mechanism similar to that seen with the GSCs also regulates adult intestinal stem cells (ISCs). While the ISC niche is not clearly understood, these cells are scattered throughout the intestinal epithelium and adhere directly to the BM (Micchelli & Perrimon, 2006; Ohlstein & Spradling, 2006). The BM appears to play a role in defining the niche, as it was recently discovered that Dpp maintains stem cell identity and that this protein is confined to the basal surface by collagen IV. This localization allows higher signal reception by the ISCs than by the differentiating ISC daughter cells, the enteroblasts, whose cell bodies are located more apically within the epithelium (Tian & Jiang, 2014) (Figure 5(C)). Dpp has also been proposed to regulate ISC proliferation, although the role of the BM in this case has not been elucidated (Guo, Driver, & Ohlstein, 2013; Li, Zhang, Han, Shi, & Lin, 2013). The BM may therefore play a common role in regulating signals within stem cell niches.

Asymmetric division of *Drosophila* neuroblasts, the stem cells that give rise to the nervous system, occurs via an intrinsic mechanism that can, at least in part, occur independently of the cells' external environment (Broadus & Doe, 1997). This is achieved by uneven segregation of fate-determining factors into one of the two daughter cells (Knoblich, 2008). While the importance of the environment in neuroblast fate is not well understood, the BM does play a role in determining when and to what extent neuroblasts divide. *Drosophila* perlecan was first identified as a factor that promoted activation of quiescent neuroblasts in larvae by counteracting the antiproliferative activity of the secreted glycoprotein Anachronism—hence its name, *terribly reduced optic lobes* (Datta, 1995; Voigt, Pflanz, Schäfer, & Jäckle, 2002). Perlecan was also found to promote neuroblast proliferation by binding to and enhancing the signaling of Hedgehog and the FGF homolog Branchless (Park et al., 2003). The BM, therefore, appears to modulate several signaling pathways to ensure proper stem cell function in many contexts.

5. CONCLUSION

Developing tissues require precise control of their size, shape, activity, and signaling environment to robustly create the adult organism. The BM has been found to integrally regulate all of these processes, thereby contributing an important external input to guide coordinated cellular activity. While studies of the BM to this point have been informative, they have also revealed how much we still have to learn. Regarding the polarized deposition and assembly of BM proteins, it will be crucial to continue to identify new factors that regulate this process. Proteins that are already known to function with Rab10 in *C. elegans* and mammalian cells are excellent candidates in this regard, and forward genetic screening strategies in *Drosophila* are likely to identify even more. In terms of understanding the functional properties of BMs once they are built, it will be important to better map their structural diversity. Although evidence from *Drosophila* and vertebrates suggests that BM architecture can vary from one tissue to the next, the precise structures of individual matrices are still largely unexplored. It will therefore be interesting to examine the composition (both of core and accessory proteins) and architecture of a diverse set of BMs, establish the levels of heterogeneity between them, and connect the physical characteristics of individual BMs to their functional properties. Expanding the library of fluorescently tagged BM proteins using modern genome editing techniques will greatly assist this endeavor. Additionally, recent evidence suggests that a BM may be more than the sum of its parts—that individual elements modify and collaborate with other proteins in the BM, on the cell surface, and within the local environment to create a complex interactive network. Understanding the nature of such interactions, in synergy with an enhanced understanding of BM structure, will further reveal the dynamic inputs of these matrices to cellular activity.

ACKNOWLEDGMENTS

We are grateful to Joel Collier and to members of the Horne-Badovinac Lab for helpful discussions and critical comments on the manuscript. This work was supported by NIH T32 HD055164 and a National Science Foundation Graduate Research Fellowship to A.J.I., and grants from the National Institutes of Health (R01-GM094276 and the American Cancer Society to S.H-B.

REFERENCES

Andrew, D. J., & Ewald, A. J. (2010). Morphogenesis of epithelial tubes: insights into tube formation, elongation, and elaboration. *Developmental Biology, 341*(1), 34–55.

Anitei, M., & Hoflack, B. (2011). Exit from the trans-Golgi network: from molecules to mechanisms. *Current Opinion in Cell Biology, 23*(4), 443–451.

Babbey, C., Ahktar, N., Wang, E., Chen, C., Grant, B., & Dunn, K. (2006). Rab10 regulates membrane transport through early endosomes of polarized Madin-Darby canine. *Molecular Biology of the Cell, 17*, 3156–3175.

Balla, T. (2013). Phosphoinositides: tiny lipids with giant impact on cell regulation. *Physiological Reviews, 93*(3), 1019–1137.

Barr, F. A. (2013). Rab GTPases and membrane identity: causal or inconsequential? *The Journal of Cell Biology, 202*(2), 191–199.

Bateman, J., Reddy, R. S., Saito, H., & Van Vactor, D. (2001). The receptor tyrosine phosphatase Dlar and integrins organize actin filaments in the *Drosophila* follicular epithelium. *Current Biology, 11*(17), 1317–1327.

Bernfield, M., Banerjee, S. D., Koda, J. E., & Rapraeger, A. C. (1984). Remodelling of the basement membrane: morphogenesis and maturation. In *Basement membranes and cell movement* (pp. 179–196).

Binari, R., Staveley, B., Johnson, W., Godavarti, R., Sasisekharan, R., & Manoukian, A. (1997). Genetic evidence that heparin-like glycosaminoglycans are involved in wingless signaling. *Development, 124*, 2623–2632.

Blümer, J., Rey, J., Dehmelt, L., Mazel, T., Wu, Y.-W., Bastiaens, P., et al. (2013). RabGEFs are a major determinant for specific Rab membrane targeting. *The Journal of Cell Biology, 200*(3), 287–300.

Boll, W., Partin, J. S., Katz, A. I., Caplan, M. J., & Jamieson, J. D. (1991). Distinct pathways for basolateral targeting of membrane and secretory proteins in polarized epithelial cells. *Proceedings of the National Academy of Sciences of the United States of America, 88*(19), 8592–8596.

Bramham, C. R., & Wells, D. G. (2007). Dendritic mRNA: transport, translation and function. *Nature Reviews Neuroscience, 8*(10), 776–789.

Broadus, J., & Doe, C. Q. (1997). Extrinsic cues, intrinsic cues and microfilaments regulate asymmetric protein localization in *Drosophila* neuroblasts. *Current Biology, 7*(11), 827–835.

Brose, K., Bland, K., Wang, K., Arnott, D., Henzel, W., Goodman, C., et al. (1999). Slit proteins bind Robo receptors and have an evolutionarily conserved role in repulsive axon guidance. *Cell, 96*, 795–806.

Bunt, S., Hooley, C., Hu, N., Scahill, C., Weavers, H., & Skaer, H. (2010). Hemocyte-secreted type IV collagen enhances BMP signaling to guide renal tubule morphogenesis in *Drosophila*. *Developmental Cell, 19*(2), 296–306.

Buszczak, M., Paterno, S., Lighthouse, D., Bachman, J., Planck, J., Owen, S., et al. (2007). The Carnegie protein trap library: a versatile tool for *Drosophila* developmental studies. *Genetics, 175*, 1505–1531.

Cao, Z., Li, C., Higginbotham, J. N., Franklin, J. L., Tabb, D. L., Graves-Deal, R., et al. (2008). Use of fluorescence-activated vesicle sorting for isolation of Naked2-associated, basolaterally targeted exocytic vesicles for proteomics analysis. *Molecular & Cellular Proteomics, 7*(9), 1651–1667.

Caplan, M., Stow, J., & Newman, A. (1987). Dependence on pH of polarized sorting of secreted proteins. *Nature, 329*, 632–635.

Carey, D. (1997). Syndecans: multifunctional cell-surface co-receptors. *Biochemical Journal, 327*, 1–16.

Cetera, M., Ramirez-San Juan, G. R., Oakes, P. W., Lewellyn, L., Fairchild, M. J., Tanentzapf, G., et al. (2014). Epithelial rotation promotes the global alignment of

contractile actin bundles during *Drosophila* egg chamber elongation. *Nature Communications, 5*, 5511.
Charras, G., & Sahai, E. (2014). Physical influences of the extracellular environment on cell migration. *Nature Reviews Molecular Cell Biology, 15*(12), 813–824.
Chen, C., Schweinsberg, P., Vashist, S., Mareiniss, D., Lambie, E., & Grant, B. (2006). RAB-10 is required for endocytic recycling in the *Caenorhabditis elegans* intestine. *Molecular Biology of the Cell, 17*, 1286–1297.
Cho, J. Y., Chak, K., Andreone, B. J., Wooley, J. R., & Kolodkin, A. L. (2012). The extracellular matrix proteoglycan perlecan facilitates transmembrane semaphorin-mediated repulsive guidance. *Genes & Development, 26*(19), 2222–2235.
Cohen, D., Müsch, A., & Rodriguez-Boulan, E. (2001). Selective control of basolateral membrane protein polarity by cdc42. *Traffic, 2*(8), 556–564.
Conder, R., Yu, H., Zahedi, B., & Harden, N. (2007). The serine/threonine kinase dPak is required for polarized assembly of F-actin bundles and apical-basal polarity in the *Drosophila* follicular epithelium. *Developmental Biology, 305*(2), 470–482.
Cummings, M., & King, R. (1969). The cytology of the vitellogenic stages of oogenesis in *Drosophila melanogaster* I. General staging characteristics. *Journal of Morphology, 128*, 427–442.
Daley, W. P., Peters, S. B., & Larsen, M. (2008). Extracellular matrix dynamics in development and regenerative medicine. *Journal of Cell Science, 121*(Pt 3), 255–264.
Datta, M. W., Hernandez, A. M., Schlicht, M. J., Kahler, A. J., DeGueme, A. M., Dhir, R., et al. (2006). Perlecan, a candidate gene for the CAPB locus, regulates prostate cancer cell growth via the Sonic Hedgehog pathway. *Molecular Cancer, 5*(9).
Datta, S. (1995). Control of proliferation activation in quiescent neuroblasts of the *Drosophila* central nervous system. *Development, 121*, 1173–1182.
De Almeida, J. B., & Stow, J. L. (1991). Disruption of microtubules alters polarity of basement membrane proteoglycan secretion in epithelial cells. *The American Journal of Physiology, 261*, C691–C700.
De Wit, J., De Winter, F., Klooster, J., & Verhaagen, J. (2005). Semaphorin 3A displays a punctate distribution on the surface of neuronal cells and interacts with proteoglycans in the extracellular matrix. *Molecular and Cellular Neurosciences, 29*(1), 40–55.
Delon, I., & Brown, N. H. (2009). The integrin adhesion complex changes its composition and function during morphogenesis of an epithelium. *Journal of Cell Science, 122*, 4363–4374.
Denef, N., Chen, Y., Weeks, S. D., Barcelo, G., & Schüpbach, T. (2008). Crag regulates epithelial architecture and polarized deposition of basement membrane proteins in *Drosophila*. *Developmental Cell, 14*(3), 354–364.
Denholm, B. (2013). Shaping up for action: the path to physiological maturation in the renal tubules of *Drosophila*. *Organogenesis, 9*(1), 40–54.
Devergne, O., Tsung, K., Barcelo, G., & Schupbach, T. (2014). Polarized deposition of basement membrane proteins depends on Phosphatidylinositol synthase and the levels of Phosphatidylinositol 4,5-bisphosphate. *Proceedings of the National Academy of Sciences of the United States of America, 111*(21), 7689–7694.
Dickson, B. J., & Gilestro, G. F. (2006). Regulation of commissural axon pathfinding by slit and its Robo receptors. *Annual Review of Cell and Developmental Biology, 22*, 651–675.
Diehl, K. A., Foley, J. D., Nealey, P. F., & Murphy, C. J. (2005). Nanoscale topography modulates corneal epithelial cell migration. *Journal of Biomedical Materials Research. Part A, 75*(3), 603–611.
Domínguez-Giménez, P., Brown, N. H., & Martín-Bermudo, M. D. (2007). Integrin-ECM interactions regulate the changes in cell shape driving the morphogenesis of the *Drosophila* wing epithelium. *Journal of Cell Science, 120*, 1061–1071.

Eastburn, D. J., & Mostov, K. E. (2010). Laying the foundation for epithelia: insights into polarized basement membrane deposition. *EMBO Reports, 11*(5), 329–330.

Fessler, L., Nelson, R., & Fessler, J. (1994). *Drosophila* extracellular matrix. *Methods in Enzymology, 245*, 271–294.

Fessler, L. I., Condic, M. L., Nelson, R. E., Fessler, J. H., & Fristrom, J. W. (1993). Site-specific cleavage of basement membrane collagen IV during *Drosophila* metamorphosis. *Development, 117*(3), 1061–1069.

Folkman, J., Klagsbrun, M., Sasse, J., Wadzinski, M., Ingber, D., & Vlodavsky, I. (1988). A heparin-binding angiogenic protein – basic fibroblast growth factor – is stored within basement membrane. *The American Journal of Pathology, 130*(2), 393–400.

Fölsch, H., Mattila, P. E., & Weisz, O. A. (2009). Taking the scenic route: biosynthetic traffic to the plasma membrane in polarized epithelial cells. *Traffic, 10*(8), 972–981.

Frydman, H. M., & Spradling, A. C. (2001). The receptor-like tyrosine phosphatase lar is required for epithelial planar polarity and for axis determination within *Drosophila* ovarian follicles. *Development, 128*(16), 3209–3220.

García-Alonso, L., Fetter, R., & Goodman, C. (1996). Genetic analysis of Laminin A in *Drosophila*: extracellular matrix containing laminin A is required for ocellar axon pathfinding. *Development, 122*, 2611–2621.

Gonzalez, A., & Rodriguez-Boulan, E. (2009). Clathrin and AP1B: key roles in basolateral trafficking through trans-endosomal routes. *FEBS Letters, 583*(23), 3784–3795.

Grindstaff, K. K., Yeaman, C., Anandasabapathy, N., Hsu, S.-C., Rodriguez-Boulan, E., Scheller, R. H., et al. (1998). Sec6/8 complex is recruited to cell–cell contacts and specifies transport vesicle delivery to the basal-lateral membrane in epithelial cells. *Cell, 93*(5), 731–740.

Gu, Z., Liu, F., Tonkova, E. A., Lee, S. Y., Tschumperlin, D. J., & Brenner, M. B. (2014). Soft matrix is a natural stimulator for cellular invasiveness. *Molecular Biology of the Cell, 25*(4), 457–469.

Guha, A., Lin, L., & Kornberg, T. B. (2009). Regulation of *Drosophila* matrix metalloprotease Mmp2 is essential for wing imaginal disc:trachea association and air sac tubulogenesis. *Developmental Biology, 335*(2), 317–326.

Gunn, P. A., Gliddon, B. L., Londrigan, S. L., Lew, A. M., van Driel, I. R., & Gleeson, P. A. (2011). The Golgi apparatus in the endomembrane-rich gastric parietal cells exist as functional stable mini-stacks dispersed throughout the cytoplasm. *Biology of the Cell, 103*(12), 559–572.

Guo, Z., Driver, I., & Ohlstein, B. (2013). Injury-induced BMP signaling negatively regulates *Drosophila* midgut homeostasis. *The Journal of Cell Biology, 201*(6), 945–961.

Guo, Z., & Wang, Z. (2009). The glypican Dally is required in the niche for the maintenance of germline stem cells and short-range BMP signaling in the *Drosophila* ovary. *Development, 136*(21), 3627–3635.

Gutzeit, H. (1990). The microfilament pattern in the somatic follicle cells of mid-vitellogenic ovarian follicles of *Drosophila*. *European Journal of Cell Biology, 53*(2), 349–356.

Gutzeit, H., Eberhardt, W., & Gratwohl, E. (1991). Laminin and basement membrane-associated microfilaments in wild-type and mutant *Drosophila* ovarian follicles. *Journal of Cell Science, 100*, 781–788.

Haigo, S. L., & Bilder, D. (2011). Global tissue revolutions in a morphogenetic movement controlling elongation. *Science, 331*(6020), 1071–1074.

Hanus, C., & Ehlers, M. D. (2008). Secretory outposts for the local processing of membrane cargo in neuronal dendrites. *Traffic, 9*, 1437–1445.

Hayashi, Y., Kobayashi, S., & Nakato, H. (2009). *Drosophila* glypicans regulate the germline stem cell niche. *The Journal of Cell Biology, 187*(4), 473–480.

He, L., Wang, X., Tang, H. L., & Montell, D. J. (2010). Tissue elongation requires oscillating contractions of a basal actomyosin network. *Nature Cell Biology, 12*(12), 1133–1142.

He, Z., Wang, K. C., Koprivica, V., Ming, G., & Song, H.-J. (2002). Knowing how to navigate: mechanisms of semaphorin signaling in the nervous system. *Science STKE*, re1.

Horne-Badovinac, S. (2014). The *Drosophila* egg chamber-a new spin on how tissues elongate. *Integrative and Comparative Biology, 54*(4), 667–676.

Horne-Badovinac, S., Hill, J., Gerlach, G., Menegas, W., & Bilder, D. (2012). A screen for round egg mutants in *Drosophila* identifies tricornered, furry, and misshapen as regulators of egg chamber elongation. *G3, 2*(3), 371–378.

Horton, A. C., Rácz, B., Monson, E. E., Lin, A. L., Weinberg, R. J., & Ehlers, M. D. (2005). Polarized secretory trafficking directs cargo for asymmetric dendrite growth and morphogenesis. *Neuron, 48*(5), 757–771.

Hussain, S.-A., Piper, M., Fukuhara, N., Strochlic, L., Cho, G., Howitt, J. A., et al. (2006). A molecular mechanism for the heparan sulfate dependence of slit-robo signaling. *The Journal of Biological Chemistry, 281*(51), 39693–39698.

Hutagalung, A. H., & Novick, P. J. (2011). Role of Rab GTPases in membrane traffic and cell physiology. *Physiological Reviews, 91*, 119–149.

Hynes, R. O., & Naba, A. (2012). Overview of the matrisome—an inventory of extracellular matrix constituents and functions. *Cold Spring Harbor Perspectives in Biology, 4*, a004903.

Johnson, K., Ghose, A., Epstein, E., Lincecum, J., O'Connor, M., & Van Vactor, D. (2004). Axonal heparan sulfate proteoglycans regulate the distribution and efficiency of the repellent slit during midline axon guidance. *Current Biology, 14*, 499–504.

Khoshnoodi, J., Pedchenko, V., & Hudson, B. G. (2008). Mammalian collagen IV. *Microscopy Research and Technique, 71*(5), 357–370.

Kim, D.-H., Provenzano, P. P., Smith, C. L., & Levchenko, A. (2012). Matrix nanotopography as a regulator of cell function. *The Journal of Cell Biology, 197*(3), 351–360.

Kim, S. N., Jeibmann, A., Halama, K., Witte, H. T., Wälte, M., Matzat, T., et al. (2014). ECM stiffness regulates glial migration in *Drosophila* and mammalian glioma models. *Development, 141*(16), 3233–3242.

Klagsbrun, M. (1990). The affinity of fibroblast growth factors (FGFs) for heparin; FGF-heparan sulfate interactions in cells and extracellular matrix. *Current Opinion in Cell Biology, 2*, 857–863.

Knoblich, J. A. (2008). Mechanisms of asymmetric stem cell division. *Cell, 132*(4), 583–597.

Kondylis, V., Pizette, S., & Rabouille, C. (2009). The early secretory pathway in development: a tale of proteins and mRNAs. *Seminars in Cell & Developmental Biology, 20*(7), 817–827.

Kondylis, V., & Rabouille, C. (2009). The Golgi apparatus: lessons from *Drosophila*. *FEBS Letters, 583*(23), 3827–3838.

Koride, S., He, L., Xiong, L., Lan, G., Montell, D., & Sun, S. (2014). Mechanochemical regulation of oscillatory follicle cell dynamics in the developing *Drosophila* egg chamber. *Molecular Biology of the Cell, 25*(22), 3709–3716.

Kraut, R., Menon, K., & Zinn, K. (2001). A gain-of-function screen for genes controlling motor axon guidance and synaptogenesis in *Drosophila*. *Current Biology, 11*(6), 417–430.

Kusche-Gullberg, M., Garrison, K., MacKrell, A., Fessler, L., & Fessler, J. (1992). Laminin A chain: expression during *Drosophila* development and genomic sequence. *The EMBO Journal, 11*(12), 4519–4527.

Le Parco, Y., Knibiehler, B., Cecchini, J., & Mirre, C. (1986). Stage and tissue-specific expression of a collagen gene during *Drosophila melanogaster* development. *Experimental Cell Research, 163*, 405–412.

Lerner, D. W., McCoy, D., Isabella, A. J., Mahowald, A. P., Gerlach, G. F., Chaudhry, T. A., et al. (2013). A Rab10-dependent mechanism for polarized basement membrane secretion during organ morphogenesis. *Developmental Cell, 24*(2), 159–168.

Lewellyn, L., Cetera, M., & Horne-Badovinac, S. (2013). Misshapen decreases integrin levels to promote epithelial motility and planar polarity in *Drosophila*. *The Journal of Cell Biology, 200*(6), 721–729.
Li, Z., Zhang, Y., Han, L., Shi, L., & Lin, X. (2013). Trachea-derived dpp controls adult midgut homeostasis in *Drosophila*. *Developmental Cell, 24*(2), 133–143.
Lin, X., Buff, E., Perrimon, N., & Michelson, A. (1999). Heparan sulfate proteoglycans are essential for FGF receptor signaling during *Drosophila* embryonic development. *Development, 126*, 3715–3723.
Lo, C.-M., Wang, H.-B., Dembo, M., & Wang, Y.-L. (2000). Cell movement is guided by the rigidity of the substrate. *Biophysical Journal, 79*(1), 144–152.
López-Onieva, L., Fernández-Miñán, A., & González-Reyes, A. (2008). Jak/Stat signalling in niche support cells regulates dpp transcription to control germline stem cell maintenance in the *Drosophila* ovary. *Development, 135*(3), 533–540.
Losick, V. P., Morris, L. X., Fox, D. T., & Spradling, A. (2011). *Drosophila* stem cell niches: a decade of discovery suggests a unified view of stem cell regulation. *Developmental Cell, 21*(1), 159–171.
Lowenstein, P., Morrison, E., Bain, D., Shering, A., Banding, G., Douglas, P., et al. (1994). Polarized distribution of the trans-Golgi network marker TGN38 during the in vitro development of neocortical neurons: effects of nocodazole and brefeldin A. *European Journal of Neuroscience, 6*, 1453–1465.
Marat, A. L., Dokainish, H., & McPherson, P. S. (2011). DENN domain proteins: regulators of Rab GTPases. *The Journal of Biological Chemistry, 286*(16), 13791–13800.
Martin, D., Zusman, S., Li, X., Williams, E., Khare, N., DaRocha, S., et al. (1999). Wing blister, a new *Drosophila* laminin alpha chain required for cell adhesion and migration during embryonic and imaginal development. *The Journal of Cell Biology, 145*(1), 191–201.
Martin-Belmonte, F., Gassama, A., Datta, A., Yu, W., Rescher, U., Gerke, V., et al. (2007). PTEN-mediated apical segregation of phosphoinositides controls epithelial morphogenesis through Cdc42. *Cell, 128*(2), 383–397.
McCall, A. S., Cummings, C. F., Bhave, G., Vanacore, R., Page-McCaw, A., & Hudson, B. G. (2014). Bromine is an essential trace element for assembly of collagen IV scaffolds in tissue development and architecture. *Cell, 157*(6), 1380–1392.
Micchelli, C. A., & Perrimon, N. (2006). Evidence that stem cells reside in the adult *Drosophila* midgut epithelium. *Nature, 439*, 475–479.
Mirre, C., Cecchini, J., Le Parco, Y., & Knibiehler, B. (1988). De novo expression of a type IV collagen gene in *Drosophila* embryos is restricted to mesodermal derivatives and occurs at germ band shortening. *Development, 102*, 369–376.
Morin, X., Daneman, R., Zavortink, M., & Chia, W. (2001). A protein trap strategy to detect GFP-tagged proteins expressed from their endogenous loci in *Drosophila*. *Proceedings of the National Academy of Sciences of the United States of America, 98*(26), 15050–15055.
Myllyharju, J., & Kivirikko, K. I. (2004). Collagens, modifying enzymes and their mutations in humans, flies and worms. *Trends in Genetics, 20*(1), 33–43.
Natori, Y., O'Meara, Y., Manning, E., Minto, A., Levine, J., Weise, W., et al. (1992). Production and polarized secretion of basement membrane components by glomerular epithelial cells. *The American Journal of Physiology, 262*, 131–137.
Nüsslein-Volhard, C., Wieschaus, E., & Kluding, H. (1984). Mutations affecting the pattern of the larval cuticle in *Drosophila melanogaster* I. Zygotic loci on the second chromosome. *Roux's Archives of Developmental Biology, 193*, 267–282.
Ohlstein, B., & Spradling, A. (2006). The adult *Drosophila* posterior midgut is maintained by pluripotent stem cells. *Nature, 439*, 470–474.
Ozbek, S., Balasubramanian, P. G., Chiquet-Ehrismann, R., Tucker, R. P., & Adams, J. C. (2010). The evolution of extracellular matrix. *Molecular Biology of the Cell, 21*(24), 4300–4305.

Paralkar, V. M., Vukicevic, S., & Reddi, A. H. (1991). Transforming growth factor β type 1 binds to collagen IV of basement membrane matrix: implications for development. *Developmental Biology, 143*(2), 303–308.
Park, Y., Rangel, C., Reynolds, M. M., Caldwell, M. C., Johns, M., Nayak, M., et al. (2003). *Drosophila* perlecan modulates FGF and hedgehog signals to activate neural stem cell division. *Developmental Biology, 253*(2), 247–257.
Pastor-Pareja, J. C., & Xu, T. (2011). Shaping cells and organs in *Drosophila* by opposing roles of fat body-secreted collagen IV and perlecan. *Developmental Cell, 21*, 245–256.
Perrimon, N., & Bernfield, M. (2000). Specificities of heparan sulphate proteoglycans in developmental processes. *Nature, 404*, 725–728.
Pierce, J. P., Mayer, T., & McCarthy, J. B. (2001). Evidence for a satellite secretory pathway in neuronal dendritic spines. *Current Biology, 11*(5), 351–355.
Provenzano, P. P., Eliceiri, K. W., Campbell, J. M., Inman, D. R., White, J. G., & Keely, P. J. (2006). Collagen reorganization at the tumor-stromal interface facilitates local invasion. *BMC Medicine, 4*(1), 38.
Provenzano, P. P., Inman, D. R., Eliceiri, K. W., Trier, S. M., & Keely, P. J. (2008). Contact guidance mediated three-dimensional cell migration is regulated by Rho/ROCK-dependent matrix reorganization. *Biophysical Journal, 95*(11), 5374–5384.
Qin, J., Liang, J., & Ding, M. (2014). Perlecan antagonizes collagen IV and ADAMTS9/GON-1 in restricting the growth of presynaptic boutons. *The Journal of Neuroscience, 34*(31), 10311–10324.
Ramírez, O. A., & Couve, A. (2011). The endoplasmic reticulum and protein trafficking in dendrites and axons. *Trends in Cell Biology, 21*(4), 219–227.
Reinhart-King, C. A., Dembo, M., & Hammer, D. A. (2008). Cell-cell mechanical communication through compliant substrates. *Biophysical Journal, 95*(12), 6044–6051.
Roca-Cusachs, P., Sunyer, R., & Trepat, X. (2013). Mechanical guidance of cell migration: lessons from chemotaxis. *Current Opinion in Cell Biology, 25*(5), 543–549.
Rodriguez-Boulan, E., Kreitzer, G., & Müsch, A. (2005). Organization of vesicular trafficking in epithelia. *Nature Reviews Molecular Cell Biology, 6*, 233–247.
Rubin, J., Choi, Y., & Segal, R. (2002). Cerebellar proteoglycans regulate sonic hedgehog responses during development. *Development, 129*, 2223–2232.
Saito, K., Chen, M., Bard, F., Chen, S., Zhou, H., Woodley, D., et al. (2009). TANGO1 facilitates cargo loading at endoplasmic reticulum exit sites. *Cell, 136*(5), 891–902.
Santiago-Tirado, F. H., & Bretscher, A. (2011). Membrane-trafficking sorting hubs: cooperation between PI4P and small GTPases at the trans-Golgi network. *Trends in Cell Biology, 21*(9), 515–525.
Sawala, A., Sutcliffe, C., & Ashe, H. L. (2012). Multistep molecular mechanism for bone morphogenetic protein extracellular transport in the *Drosophila* embryo. *Proceedings of the National Academy of Sciences of the United States of America, 109*(28), 11222–11227.
Schneider, M., Khalil, A. A., Poulton, J., Castillejo-Lopez, C., Egger-Adam, D., Wodarz, A., et al. (2006). Perlecan and Dystroglycan act at the basal side of the *Drosophila* follicular epithelium to maintain epithelial organization. *Development, 133*, 3805–3815.
Seeger, M., Tear, G., Ferres-Marco, D., & Goodman, C. (1993). Mutations affecting growth cone guidance in *Drosophila*: genes necessary for guidance toward or away from the midline. *Neuron, 10*, 409–426.
Shahab, J., Baratta, C., Scuric, B., Godt, D., Venken, K. J. T., & Ringuette, M. R. (2015). Loss of SPARC dysregulates basal lamina assembly to disrupt larval fat body homeostasis in Drosophila melanogaster. *Developmental Dynamics, 244*(4), 540–552.
Shi, A., Chen, C. C., Banerjee, R., Glodowski, D., Audhya, A., Rongo, C., et al. (2010). EHBP-1 functions with RAB-10 during endocytic recycling in *Caenorhabditis elegans*. *Molecular Biology of the Cell, 21*, 2930–2943.

Smart, A. D., Course, M. M., Rawson, J., Selleck, S., Van Vactor, D., & Johnson, K. G. (2011). Heparan sulfate proteoglycan specificity during axon pathway formation in the *Drosophila* embryo. *Developmental Neurobiology, 71*(7), 608–618.

Song, X., Zhu, C., Doan, C., & Xie, T. (2002). Germline stem cells anchored by adherens junctions in the *Drosophila* ovary niches. *Science, 296*, 1855–1857.

Sorrosal, G., Pérez, L., Herranz, H., & Milán, M. (2010). Scarface, a secreted serine protease-like protein, regulates polarized localization of laminin A at the basement membrane of the *Drosophila* embryo. *EMBO Reports, 11*(5), 3–9.

Srivastava, A., Pastor-Pareja, J. C., Igaki, T., Pagliarini, R., & Xu, T. (2007). Basement membrane remodeling is essential for *Drosophila* disc eversion and tumor invasion. *Proceedings of the National Academy of Sciences of the United States of America, 104*(8), 2721–2726.

Steigemann, P., Molitor, A., Fellert, S., Jäckle, H., & Vorbrüggen, G. (2004). Heparan sulfate proteoglycan syndecan promotes axonal and myotube guidance by slit/robo signaling. *Current Biology, 14*(3), 225–230.

Stevens, A., & Jacobs, J. (2002). Integrins regulate responsiveness to slit repellent signals. *The Journal of Neuroscience, 22*(11), 4448–4455.

Stoops, E. H., & Caplan, M. J. (2014). Trafficking to the apical and basolateral membranes in polarized epithelial cells. *Journal of the American Society of Nephrology, 25*(7), 1375–1386.

Takagi, Y., Nomizu, M., Gullberg, D., MacKrell, A. J., Keene, D. R., Yamada, Y., et al. (1996). Conserved neuron promoting activity in *Drosophila* and vertebrate laminin alpha1. *Journal of Biological Chemistry, 271*(30), 18074–18081.

Tan, J., & Saltzman, W. M. (2002). Topographical control of human neutrophil motility on micropatterned materials with various surface chemistry. *Biomaterials, 23*(15), 3215–3225.

The, I., Bellaiche, Y., & Perrimon, N. (1999). Hedgehog movement is regulated through tout velu – dependent synthesis of a heparan sulfate proteoglycan. *Molecular Cell, 4*, 633–639.

Tian, A., & Jiang, J. (2014). Intestinal epithelium-derived BMP controls stem cell self-renewal in *Drosophila* adult midgut. *eLife, 3*, e01857.

Umulis, D. M., Shimmi, O., O'Connor, M. B., & Othmer, H. G. (2010). Organism-scale modeling of early *Drosophila* patterning via bone morphogenetic proteins. *Developmental Cell, 18*(2), 260–274.

Varner, V. D., & Nelson, C. M. (2014). Cellular and physical mechanisms of branching morphogenesis. *Development, 141*, 2750–2759.

Venditti, R., Scanu, T., Santoro, M., Di Tullio, G., Spaar, A., Gaibisso, R., et al. (2012). Sedlin controls the ER export of procollagen by regulating the Sar1 cycle. *Science, 337*, 1668–1672.

Vertel, B., Velasco, A., LaFrance, S., Walters, L., & Kaczman-Daniel, K. (1989). Precursors of chondroitin sulfate proteoglycan are segregated within a subcompartment of the chondrocyte endoplasmic reticulum. *The Journal of Cell Biology, 109*, 1827–1836.

Viktorinová, I., & Dahmann, C. (2013). Microtubule polarity predicts direction of egg chamber rotation in *Drosophila*. *Current Biology, 23*(15), 1472–1477.

Viktorinová, I., König, T., Schlichting, K., & Dahmann, C. (2009). The cadherin Fat2 is required for planar cell polarity in the *Drosophila* ovary. *Development, 136*(24), 4123–4132.

Voigt, A., Pflanz, R., Schäfer, U., & Jäckle, H. (2002). Perlecan participates in proliferation activation of quiescent *Drosophila* neuroblasts. *Developmental Dynamics, 224*, 403–412.

Wang, X., Harris, R. E., Bayston, L. J., & Ashe, H. L. (2008). Type IV collagens regulate BMP signalling in *Drosophila*. *Nature, 455*(7209), 72–77.

Wilson, D. G., Phamluong, K., Li, L., Sun, M., Cao, T. C., Liu, P. S., et al. (2011). Global defects in collagen secretion in a Mia3/TANGO1 knockout mouse. *The Journal of Cell Biology, 193*(5), 935–951.

Xie, T., & Spradling, A. C. (1998). decapentaplegic is essential for the maintenance and division of germline stem cells in the *Drosophila* ovary. *Cell, 94*(2), 251–260.

Xiong, B., Bayat, V., Jaiswal, M., Zhang, K., Sandoval, H., Charng, W.-L., et al. (2012). Crag is a GEF for Rab11 required for rhodopsin trafficking and maintenance of adult photoreceptor cells. *PLoS Biology, 10*(12), e1001438.

Yasothornsrikul, S., Davis, W. J., Cramer, G., Kimbrell, D. A., & Dearolf, C. R. (1997). viking: identification and characterization of a second type IV collagen in *Drosophila. Gene, 198*, 17–25.

Yoshimura, S., Gerondopoulos, A., Linford, A., Rigden, D. J., & Barr, F. A. (2010). Family-wide characterization of the DENN domain Rab GDP-GTP exchange factors. *The Journal of Cell Biology, 191*(2), 367–381.

Yurchenco, P. D. (2011). Basement membranes: cell scaffoldings and signaling platforms. *Cold Spring Harbor Perspectives in Biology, 3*(2), a004911.

CHAPTER TEN

Basement Membranes in the Worm: A Dynamic Scaffolding that Instructs Cellular Behaviors and Shapes Tissues

Matthew R. Clay and David R. Sherwood*
Department of Biology, Duke University, Durham, NC, USA
*Corresponding author: E-mail: david.sherwood@duke.edu

Contents

Current Topics in Membranes, Volume 76
ISSN 1063-5823
http://dx.doi.org/10.1016/bs.ctm.2015.08.001

Abstract

The nematode worm *Caenorhabditis elegans* has all the major basement membrane proteins found in vertebrates, usually with a smaller gene family encoding each component. With its powerful forward genetics, optical clarity, simple tissue organization, and the capability to functionally tag most basement membrane components with fluorescent proteins, *C. elegans* has facilitated novel insights into the assembly and function of basement membranes. Although basement membranes are generally thought of as static structures, studies in *C. elegans* have revealed their active properties and essential functions in tissue formation and maintenance. Here, we review discoveries from *C. elegans* development that highlight dynamic aspects of basement membrane assembly, function, and regulation during organ growth, tissue polarity, cell migration, cell invasion, and tissue attachment. These studies have helped transform our view of basement membranes from static support structures to dynamic scaffoldings that play broad roles in regulating tissue organization and cellular behavior that are essential for development and have important implications in human diseases.

1. INTRODUCTION

Basement membranes are specialized, cell-associated networks of extracellular matrix that underlie epithelia, endothelia, and enwrap muscles, fat, and Schwann cells (Yurchenco, 2011). The emergence of basement membranes coincided with the origin of metazoan life, suggesting that basement membranes were essential for the construction of complex, multicellular animals (Hynes, 2012; Ozbek, Balasubramanian, Chiquet-Ehrismann, Tucker, & Adams, 2010; Whittaker et al., 2006). Consistent with this notion, loss of key basement membrane components results in early embryonic lethality, and disruption of basement membranes is associated with human diseases involving most organ systems (reviews by Van Agtmael & Bruckner-Tuderman, 2010; Yurchenco & Patton, 2009). Traditionally, basement membranes have been thought of as static assemblies of proteins, glycoproteins, and proteoglycans that function mainly to support tissues. However, recent in vivo studies in developmental contexts have revealed that basement membrane components are dynamically secreted, assembled, and rearranged to affect cell shape, tissue polarization, and morphogenesis (reviews by Daley & Yamada, 2013; Morrissey & Sherwood, 2015). These findings have begun to transform our view of the basement membrane into an active participant in many important biological processes.

Studies in cell lines, mice, flies, and worms have supported the idea that a cell-associated network of laminin is the building block of all basement membranes (Urbano et al., 2009; Yurchenco, 2011). Laminin is a secreted heterotrimeric protein composed of an α, β, and γ subunit that binds to receptors and lipids at the cell surface and self-assembles into a sheetlike polymeric network. Following laminin assembly, an independent network of cross-linked type IV collagen, assembled from heterotrimers containing two α1-like and one α2-like chains, is overlaid onto basement membranes (reviews by Hohenester & Yurchenco, 2013; Yurchenco, 2011). Type IV collagen is often the most abundant protein in basement membranes, and its cross-linking provides basement membranes the ability to withstand mechanical load (Fidler et al., 2014; Hohenester & Yurchenco, 2013; LeBleu, MacDonald, & Kalluri, 2007; Vanacore et al., 2009). The supramolecular grid of laminin and collagen is thought to provide basement membranes with their thin, dense structure, and several proteins, including perlecan and nidogen, have been implicated in connecting the independent laminin and collagen networks (Yurchenco, 2011). Elaborating their shared, core architecture, basement membranes can contain different forms of laminin that associate with collagen, nidogen, and perlecan at different densities (Breitkreutz, Koxholt, Thiemann, & Nischt, 2013; Rohrbach & Murrah, 1993). Various combinations of other proteins, including fibulins, collagen XVIII, agrin, hemicentin, and SPARC can be present in basement membranes, creating diverse structures with unique biochemical and biophysical properties (Candiello, Cole, & Halfter, 2010; LeBleu et al., 2007; Tzu & Marinkovich, 2008). The mechanisms that direct incorporation of specific proteins into basement membranes at particular sites, however, have remained largely unclear.

Several factors have limited the study of how vertebrate basement membranes are assembled, and how they ultimately function, in specific tissues. First, laminin and type IV collagen are essential for embryonic survival, making mechanistic loss of function studies difficult (Miner, Cunningham, & Sanes, 1998; Rozario & DeSimone, 2010; Smyth et al., 1999). Second, the expanded families of basement membrane components found in vertebrates further complicate loss of function studies. For example, vertebrates have at least 16 laminin and 3 type IV collagen trimers (Aumailley et al., 2005; Boutaud et al., 2000; Kobayashi & Uchiyama, 2003). Mutation of an individual laminin or collagen subunit frequently affects more than one trimer, thus disrupting basement membranes of multiple tissues (reviewed by Rozario & DeSimone, 2010). Third, the visual inaccessibility of most vertebrate tissues to live-cell microscopy has made it difficult to image basement membranes and basement membrane dynamics during development. The *Caenorhabditis elegans* model system avoids many of these complications. The *C. elegans* genome encodes orthologs of the structural basement membrane components laminin, type IV collagen, type XVIII collagen, nidogen, perlecan, and agrin; the basement membrane-associated proteins fibulin-1, hemicentin, SPARC, F-spondin, and papilin (Hrus et al., 2007; Kramer, 2005; Woo et al., 2008); and the major basement membrane receptors integrin and dystroglycan (Hutter et al., 2000). Importantly, these gene families have not undergone large expansions in the *C. elegans* lineage, thus simplifying experimental analysis. Further, most basement membrane proteins have been fluorescently tagged to create functional fusion proteins, allowing them to be directly visualized in the optically transparent, rapidly developing, simple body plan of the worm (Figure 1). These features combined with forward genetics (including viable mutant alleles of most basement membrane components) and reverse genetics (including postembryonic RNAi to bypass embryonic lethal effects) make *C. elegans* uniquely suited for studying basement

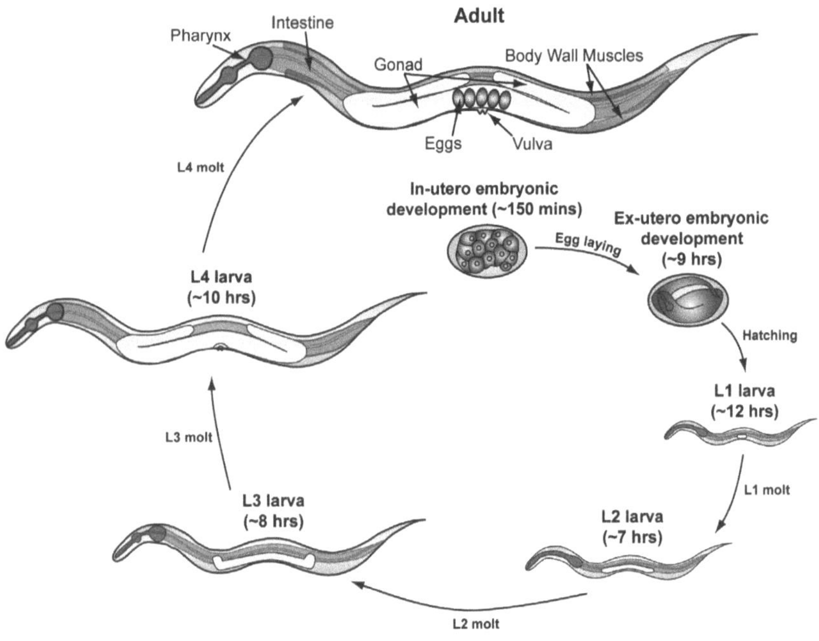

Figure 1 ***Overview of*** **Caenorhabditis elegans** ***development.*** A schematic diagram depicting *C. elegans* development, life cycle, and several adult tissues. The times indicate the duration spent in each developmental stage when grown at 22 °C. Embryogenesis involves two stages: (1) a proliferative stage that includes gastrulation and (2) organogenesis and morphogenesis. Larval development consists of four stages (L1–L4), each separated by a molt of the stage-specific cuticle. During larval development, several tissues continue morphogenesis and growth. Two notable examples are the expansion and shaping of the gonad (yellow), and establishment of the egg-laying apparatus, which includes morphogenesis of the vulva and formation of the uterine–vulval attachment. (See color plate)

membranes in vivo. This chapter highlights recent work from *C. elegans* development that has advanced our understanding of the assembly, dynamic nature, and diverse functions of basement membranes.

2. BASEMENT MEMBRANE ASSEMBLY IN THE WORM: DIVERSE COMPOSITIONS DETERMINED BY LOCAL INTERACTIONS AND RECRUITMENT OF MATRIX COMPONENTS

The two primary structural components of basement membranes, laminin and type IV collagen, are assembled into almost all basement membranes. *Caenorhabditis elegans* has two laminin α subunits (*lam-3* and *epi-1*), one β subunit (*lam-1*), and one γ subunit (*lam-2*) that form two distinct laminin heterotrimers with unique α subunits (Kramer, 2005). Gene expression and immunolocalization analyses revealed that laminin trimers containing the different α subunits are secreted between the primary tissue layers at the onset of gastrulation and become differentially enriched along organs as they develop (Huang et al., 2003). Many—but not all—*C. elegans* tissues express the laminin α subunit that is found in the tissue's basement membrane. For example, during larval development, the *C. elegans* body wall muscles express the α subunit *epi-1* whereas the epidermis expresses the α subunit *lam-3*, and these basement membranes contain only the laminin-α subunit expressed locally (Huang et al., 2003). This is similar to the mouse embryonic kidney, where the laminin α5 subunit shows a restricted localization pattern that is distinct from α1, and each correlates strongly with their respective mRNA expression patterns (Ekblom et al., 1990; Sorokin, Pausch, Durbeej, & Ekblom, 1997). Together, these data suggest that tissue-specific expression of laminins may be an important factor in determining basement membrane composition.

In addition to local production, laminin secreted from distant sites can be incorporated into basement membranes from the extracellular fluid. For example, mutants of the laminin β subunit (*lam-1*) have pharyngeal defects that can be rescued by expressing *lam-1* only in the intestine (Rasmussen, Reddy, & Priess, 2012). Furthermore, many tissues do not express either laminin α subunit, but are nevertheless surrounded by laminin-containing basement membranes. For example, the sublateral nerves do not express laminin, but they are encased by a basement membrane specifically containing the α subunit LAM-3 (Huang et al., 2003). Similar situations have been described in vertebrate tissues, such as the mouse neural tube basement membrane, which contains the laminin α5 subunit that is not expressed by any cells within the neural tube (Copp et al., 2011). Together, these results show that laminin secreted into the extracellular space can be incorporated into basement membranes in a selective manner.

Similar to laminin, type IV collagen is also found in basement membranes of *C. elegans* tissues that do not express it. The *C. elegans* genome encodes one α1-like collagen chain (*emb-9*/αA), and one α2-like chain (*let-2*/αB), which are thought to make a single heterotrimeric type IV collagen molecule, composed of two EMB-9 chains and one LET-2 chain. Nearly all *C. elegans* basement membranes contain type IV collagen, but it is mainly expressed in the body wall muscle and distal tip cells (DTCs) (specialized cells in the somatic gonad, see below) (Graham et al., 1997). This suggests that collagen secreted into the extracellular space by one tissue can be incorporated into the basement membranes of other tissues. Supporting this notion, expression of *emb-9* in only the body wall muscle of *emb-9* mutants leads to accumulation of type IV collagen in most tissues and rescues the embryonic lethality associated with loss of *emb-9* (Graham et al., 1997). During *Drosophila* larval development, the collagen in most basement membranes is also secreted from a distant source, implying this might be a conserved aspect of type IV collagen addition to and incorporation into basement membranes (Pastor-Pareja & Xu, 2011).

The incorporation oflaminin and collagen produced from distant sites into basement membranes suggests that tissues are exposed to multiple isoforms of these proteins circulating in the extracellular space. Thus, cells must have mechanisms that target the deposition of specific basement membrane components. Indeed, data from many experimental models suggest that basement membrane assembly is initiated by cell-anchored components such as receptors. Specifically, it is thought that cell surface binding of laminin induces self-assembly, which builds an initial scaffolding for basement membrane formation (reviews by Hohenester & Yurchenco, 2013; Yurchenco, 2011). Integrins are transmembrane heterodimer complexes composed of an α and a β subunit and are thought to induce basement membrane assembly by binding laminins (Yurchenco, 2011). In *C. elegans,* loss of integrin disrupts basement membrane integrity, similar to weak mutant alleles of the laminin α subunit *epi-1* (Baum & Garriga, 1997; Huang et al., 2003). Vertebrate integrins display some differential laminin-binding characteristics in vitro, but most nonetheless bind multiple laminin trimers; therefore, differential binding of integrins to laminin is unlikely to fully account for how laminin is targeted to specific tissues (Hohenester & Yurchenco, 2013; Nishiuchi et al., 2006). Different combinations of integrins and other cell surface receptors may mediate tissue-specific basement membrane assembly. Dystroglycan is another cell surface receptor that binds laminin (Yurchenco, 2011). Supporting a role for dystroglycan as a laminin receptor, in *C. elegans* loss of dystroglycan disrupts the gonadal basement membrane, similar to loss of laminin (Johnson, 2006). However, dystroglycan is not necessary for the initial localization of laminin to the gonadal basement membrane, suggesting it might be primarily required to maintain the basement membrane as the gonad expands during development (Johnson, 2006). A role for the transmembrane glycoprotein teneurin-1 (encoded by *C. elegans ten-1*; ortholog of *Drosophila ten-m* family and vertebrate ten-m-1–m-4) in regulating basement membrane integrity was also discovered in *C. elegans* (Trzebiatowska, Topf, Sauder, Drabikowski, & Chiquet-Ehrismann, 2008). Loss of teneurin-1 disrupts the basement membranes of tissues where it is expressed such as the pharynx, gonad, and epidermis (Drabikowski, Trzebiatowska, & Chiquet-Ehrismann, 2005; Topf & Chiquet-Ehrismann, 2011; Trzebiatowska et al., 2008). Teneurin-1 interacts genetically with both integrin and dystroglycan, and loss of teneurin-1 enhances the embryonic lethality of hypomorphic type IV collagen mutants, indicating teneurin-1 could act as a receptor for type IV collagen (Topf & Chiquet-Ehrismann, 2011; Trzebiatowska et al., 2008). However, it has not yet been shown whether teneurin-1 binds or directly recruits collagen or any other basement membrane components. Integrin, dystroglycan, and teneurin-1 clearly have functions in mediating basement membrane integrity in *C. elegans*; however, their roles in the recruitment of specific laminin isoforms as well as type IV collagen to basement membrane remain unclear.

In *C. elegans*, the basement membrane components nidogen, fibulin-1, hemicentin, SPARC, and F-spondin have cell-specific expression patterns, but are also secreted into the extracellular fluid and undergo localized assembly on distant tissues (Fitzgerald & Schwarzbauer, 1998; Kawano et al., 2009; Kim & Wadsworth, 2000; Muriel, Dong, Hutter, & Vogel, 2005; Vogel & Hedgecock, 2001; Woo et al., 2008). Although the mechanisms that regulate the tissue-specific distributions of these proteins are largely unknown, several studies have revealed that local interactions between basement membrane components are important for their tissue-specific distribution. For example, at sites where mechanosensory neurons and uterine cells attach to basement membranes, fibulin-1 localization depends on hemicentin and a unique splice variant of perlecan (Muriel et al., 2005; Muriel, Xu, Kramer, & Vogel, 2006). Fibulin-1 in turn regulates hemicentin by refining and concentrating hemicentin localization in these regions (Muriel et al., 2005). In another case, the extracellular protease MIG-17 is secreted from body wall muscles and localizes to the gonadal basement membrane where it recruits fibulin-1, which in turn recruits nidogen (see below) (Kubota, Kuroki, & Nishiwaki, 2004; Kubota, Ohkura, Tamai, Nagata, & Nishiwaki, 2008). Interestingly, fibulin-1 is also required to retain type IV collagen within the gonadal basement membrane during late larval stages (Kubota, Nagata, Sugimoto, & Nishiwaki, 2012). Together, these observations suggest that dynamic and complex interactions between basement membrane proteins and extracellular proteases might be a common mechanisms to help generate the unique composition of basement membranes surrounding tissues.

3. BASEMENT MEMBRANES INSTRUCT DIVERSE CELLULAR BEHAVIORS AND SHAPE TISSUES

The role of basement membranes in establishing and maintaining tissue architecture has long been recognized. Recent studies have also begun to reveal active roles for basement membranes in controlling morphogenesis in many organisms (reviews by Daley & Yamada, 2013; Morrissey & Sherwood, 2015). In particular, pioneering studies in *C. elegans* have shown that basement membranes play dynamic roles in regulating numerous cellular behaviors as well as the shaping of tissues.

3.1 Nidogen regulates axon guidance and synapse formation

Nidogen is a ubiquitous basement membrane glycoprotein that binds both laminin and type IV collagen in vitro (Aumailley, Wiedemann, Mann, & Timpl, 1989; Fox et al., 1991; Paulsson et al., 1987). These characteristics made it an intriguing candidate to mediate the initial connection between laminin and collagen networks. In *C. elegans*, however, nidogen (*nid-1*) mutants undergo embryogenesis normally, are viable, and form basement membranes containing both laminin and type IV collagen (Kang & Kramer, 2000; Kim & Wadsworth, 2000). Although not required for basement membrane formation or collagen recruitment, nidogen is enriched within the sublateral nerves and near the nerve cords at the lateral edges of the body wall muscle where it helps guide longitudinal axons (Hutter, 2003; Kang & Kramer, 2000; Kim & Wadsworth, 2000; Unsoeld, Park, & Hutter, 2013). Genetic studies showed that nidogen in the SDQR sublateral neuron influences signaling through the netrin receptor UNC-40 (Deleted in Colon Cancer, DCC ortholog) to regulate axon guidance (Kim & Wadsworth,

2000; Kim, Ren, Fox, & Wadsworth, 1999). Additionally, nidogen functions in the same axon guidance pathway as the discoidin domain receptors, *ddr-1* and *ddr-2,* which are expressed in the PVPR neurons of the ventral nerve cord (Unsoeld et al., 2013). Thus, nidogen can act in multiple pathways to direct axon guidance of specific longitudinal nerves. Discoidin domain receptors are established collagen receptors in mammals (Leitinger & Hohenester, 2007), so it is possible that nidogen regulates DDR-1 and DDR-2 indirectly through alterations in type IV collagen in the basement membrane. Nidogen is also enriched near neuromuscular junction synapses where it acts with the intracellular adapter protein α liprin (encoded by *C. elegans syd-2*) to maintain proper synapse size and function (Ackley et al., 2005, 2003). Vertebrates have two nidogens, and mice lacking nidogen-2 have defects in neuromuscular junctions (Fox, Ho, Smyth, & Sanes, 2008). Thus, much like in *C. elegans,* vertebrate nidogens may have instructive roles in regulating cellular behaviors.

3.2 Perlecan instructs muscle cell attachment

Perlecan (encoded by *C. elegans unc-52*), a basement membrane heparan sulfate proteoglycan, regulates a number of cellular behaviors including axon guidance (Tang & Wadsworth, 2014; Yang, Lee, Tang, & Wadsworth, 2014), sensory dendrite branching (Liang, Dong, Moerman, Shen, & Wang, 2015), and DTC migration (see below) (Merz, Alves, Kawano, Zheng, & Culotti, 2003). Perlecan is also essential for embryonic viability through its role in organizing muscle cell attachments to the epidermis. Movement in *C. elegans* is generated through contraction of the body wall muscles. To allow efficient motility, the force of muscle contraction must be mechanically transmitted to the epidermis. This force transmission is accomplished through specialized attachments, known as dense bodies and M-lines, two related integrin-mediated adhesions that anchor the muscle sarcomeres to the basement membrane between the muscle and epidermis that are enriched for and require perlecan (Francis & Waterston, 1991; Moerman & Williams, 2006; Mullen, Rogalski, Bush, Gorji, & Moerman, 1999; Rogalski, Williams, Mullen, & Moerman, 1993). In animals lacking perlecan, there is a complete failure to form muscle attachments, resulting in a paralyzed, lethal embryonic arrest (Hresko, Williams, & Waterston, 1994; Rogalski, Gilchrist, Mullen, & Moerman, 1995; Williams & Waterston, 1994). The muscle cells express perlecan, but how it becomes enriched at specific sites in the basement membrane to organize attachment between the muscles and epidermis is unknown.

3.3 Laminin mediates coordinated cell polarity

Work from cell culture and organotypic models has shown that the basement membrane component laminin induces cell polarity (Ekblom, Vestweber, & Kemler, 1986; Klein, Langegger, Timpl, & Ekblom, 1988; O'Brien et al., 2001), but due to embryonic lethality of mutants, it has remained unclear if and how laminin regulates tissue polarization in vivo. Recent studies in *C. elegans*, have revealed a specific and highly regulated role for laminin in coordinating tissue polarity. The *C. elegans* digestive tract is composed of three connected, specialized epithelial tubes: the pharynx, which is responsible for capturing and grinding food; the intestine, which digests the food; and specialized valve cells, which link the pharynx and intestine (Mango, 2007). The polarization of the pharyngeal primordium, which also contains the valve cell precursors, depends on signals from laminin. The pharyngeal primordium aligns in two bilaterally symmetric rows of cells then undergoes epithelialization in response to laminin to form a polarized cyst (Rasmussen et al., 2012). The intestinal primordium polarizes independently of laminin at a different time, thus exhibiting a distinct axis of polarity (Leung, Hermann, & Priess, 1999; Rasmussen et al., 2012; Rasmussen, Feldman, Reddy, & Priess, 2013). As the tissue segments connect, the valve cells reorganize to align with the polarity of the intestinal cells (Rasmussen et al., 2013). Throughout attachment, the majority of the pharyngeal and intestinal primordia are covered with laminin; however, laminin is specifically absent at the interface where the intestinal cells contact the posterior valve cells (Rasmussen et al., 2013). In animals where intestinal cells are ablated, or in mutants where intestinal cells do not send out the protrusions that normally contact the valve cells, laminin is ectopically deposited on the posterior valve cells, and they do not develop the normal, radial axis of polarity (Rasmussen et al., 2013). These results suggest that the intestine prevents the accumulation of laminin on the posterior valve cells and that in contrast to the pharyngeal primordium, the absence of laminin is necessary to properly polarize these cells. Therefore, laminin deposition is precisely controlled to regulate specific polarity events within tissues.

3.4 Laminin functions to regulate germ stem cell niche establishment

Many complex tissues arise from and are maintained by stem cells. Adult stem cells are typically found in a niche that supports stem cell survival and pluripotency. Extracellular matrices, including basement membranes, provide an adhesive substrate that retains the stem cell in the niche and regulates signaling that promotes self-renewal (reviews by Brizzi, Tarone, & Defilippi, 2012; Marthiens, Kazanis, Moss, Long, & Ffrench-Constant, 2010). Work in *C. elegans* has revealed a role for laminin in the establishment of the germ stem cell niche. The germ stem cell niche is formed when the two somatic gonad precursors, which arise in anterior regions of the embryo, migrate posteriorly to enwrap the two primordial germ cells (Figure 2(A) and (B)) (Harrell & Goldstein, 2011; Kimble & White, 1981; Nance & Priess, 2002; Rohrschneider & Nance, 2013; Sulston, Schierenberg, White, & Thomson, 1983). Somatic gonad precursors extend protrusions as they migrate along the surface of the endoderm, which is surrounded by laminin. Interestingly, the endodermal cells just posterior to the primordial germ cells have additional laminin around their cell surfaces, and when the migrating somatic gonad precursors contact these endodermal cells, they stop migrating and enwrap the germ cells (Rohrschneider & Nance, 2013). RNAi-mediated reduction of laminin does not affect initial somatic gonad precursor migration. However, the somatic gonad precursor cells migrate just beyond the primordial germ cells in the absence of laminin, suggesting that laminin provides an important cue for somatic gonad precursors to stop migrating (Figure 2(C)) (Rohrschneider & Nance, 2013). Since somatic gonad precursors eventually retract and enwrap the germ cells despite laminin knockdown, additional signals likely promote enwrapment and may regulate where the somatic gonad precursors stop. Laminin is also increased in areas where germ cells stop migrating in mice, and *Drosophila* laminin mutants have primordial germ cells and somatic gonad precursors in ectopic locations (García-Castro, Anderson, Heasman, & Wylie, 1997; Weyers, Milutinovich, Takeda, Jemc, & Van Doren, 2011), suggesting that laminin may be a conserved instructive cue in positioning these cells. Laminin may carry out this function by localizing another cue that halts cell migration, or laminin itself could provide a direct signal to stop migrating cells by binding a cell surface receptor.

3.5 Integrins, extracellular signals, and basement membrane components regulate DTC migration

The establishment and shaping of tissues in many organisms often relies on cell migration. For example, in *C. elegans* the migration of the DTCs helps to shape the gonad during larval development (see Figure 1 for staging) (reviewed by Wong & Schwarzbauer, 2012). The entire gonad, including the DTCs and germ cells, is surrounded by a basement membrane

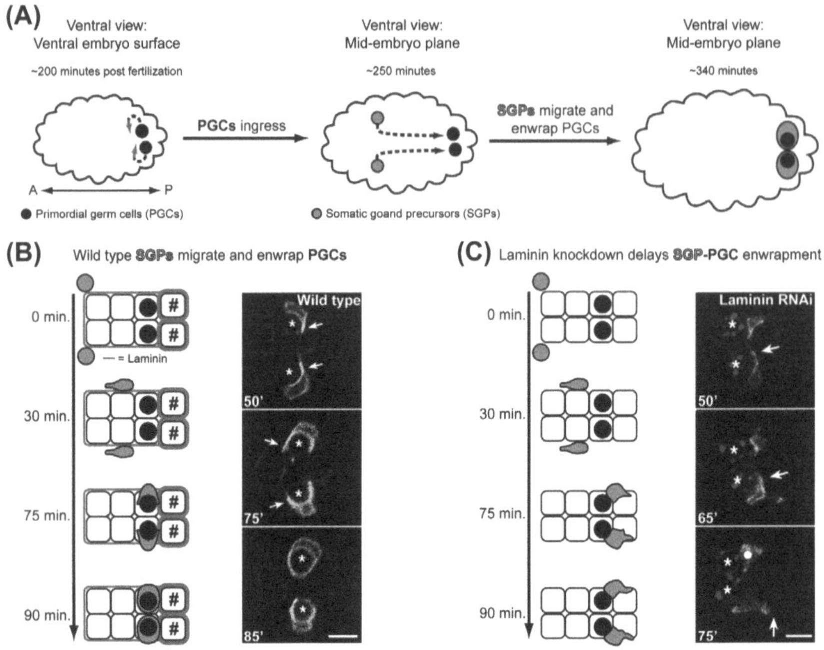

Figure 2 ***A laminin-dependent signal halts somatic gonad precursor cell migration.*** (A) A schematic diagram of germ stem cell niche establishment. During embryogenesis the primordial germ cells (PGCs; black circles) are specified in the posterior region of the ventral embryo surface before ingressing during gastrulation. Somatic gonad precursors (SGPs; gray circles) are born in anterior regions of the embryo and migrate along the endoderm to meet the primordial germ cells. Upon reaching the germ cells, somatic gonad precursors halt their migration and establish the niche by enwrapping the primordial germ cells. (B) Cell behaviors during somatic gonad precursor migration and primordial germ cell enwrapment. The schematic diagrams (left; normal distribution of laminin is shown in green) and confocal time-lapse images (right) depict somatic gonad precursors (gray circles (left), green cells (right)) migrating along the endoderm (white boxes (left), red cells (right)). Somatic gonad precursors stop migrating when they contact endoderm cells that have increased laminin around their cell surfaces (#) that lie just posterior to the primordial germ cells (black circles (left), marked with asterisk (right)). After stopping, somatic gonad precursors extend cellular processes (arrows) that enwrap primordial germ cells to establish the niche. (C) When laminin is decreased or absent, somatic gonad precursors migrate further and extend projections past the primordial germ cells, resulting in delayed enwrapment. PGCs, primordial germ cells; SGPs, somatic gonad precursors. *Confocal images in (B) and (C) originally appeared in Rohrschneider and Nance (2013) and were adapted with permission.* (See color plate)

(Hall et al., 1999); thus when the DTCs migrate, they are encased by the gonadal basement membrane. During migration, the DTCs secrete many basement membrane components, including laminin, collagen, nidogen, agrin, papilin, and hemicentin (Graham et al., 1997; Hrus et al., 2007; Huang et al., 2003; Kawano et al., 2009; Kim & Wadsworth, 2000; Vogel & Hedgecock, 2001). The DTC has the complex task of laying down new basement membrane, remodeling deposited basement membrane to allow gonad expansion, and using the basement membrane as a substrate to promote or orient migration. In addition to the gonadal basement membrane, DTC migration is influenced by the neighboring muscle and epidermal basement membranes during different stages of migration (Figure 3). Migration is initiated in the L2 larval stage when the two DTCs move in anterior and posterior directions adjacent to the ventral body wall muscle basement membrane (Figure 3(B)). In the mid-to-late L3 stage, DTCs turn toward the dorsal side of the animal and migrate lateral to the epidermal basement membrane (Figure 3(C)). When DTCs reach the dorsal body wall muscle basement membrane, they turn and move back toward the midbody (Figure 3(D)), where they continue to migrate until early adulthood. A complex interplay of integrin receptor activity, extracellular signaling, and basement membrane components directs various aspects of DTC migration.

The *C. elegans* genome encodes two α integrin subunits (*ina-1* and *pat-2*) and one β integrin subunit (*pat-3*) and form two heterodimeric integrin receptors (INA-1/PAT-3 and PAT-2/PAT-3), which are thought to bind basement membrane components (Kramer, 2005). Both integrin complexes are expressed in the DTCs and regulate distinct aspects of DTC migration. The *ina-1* gene (ortholog of laminin-binding integrin α subunits) promotes DTC migration, is expressed throughout migration, and is downregulated as migration ceases (Figure 3(B)–(D)) (Baum & Garriga, 1997; Kikuchi et al., 2015; Meighan & Schwarzbauer, 2007). In mutants of the *vab-3* transcription factor (vertebrate paired box Pax6 ortholog), *ina-1* expression is maintained in DTCs and they continue migrating into adulthood, resulting in long, looping gonad arms (Meighan & Schwarzbauer, 2007). INA-1 also regulates DTC pathfinding along the dorsal side of the animal. In animals with reduced *ina-1* function, DTCs undergo multiple, ectopic turns when they reach the dorsal muscle basement membrane. The *pat-2* gene (ortholog of RGD binding α subunits) is expressed beginning at the L3 stage when DTCs turn to migrate dorsally. In contrast to

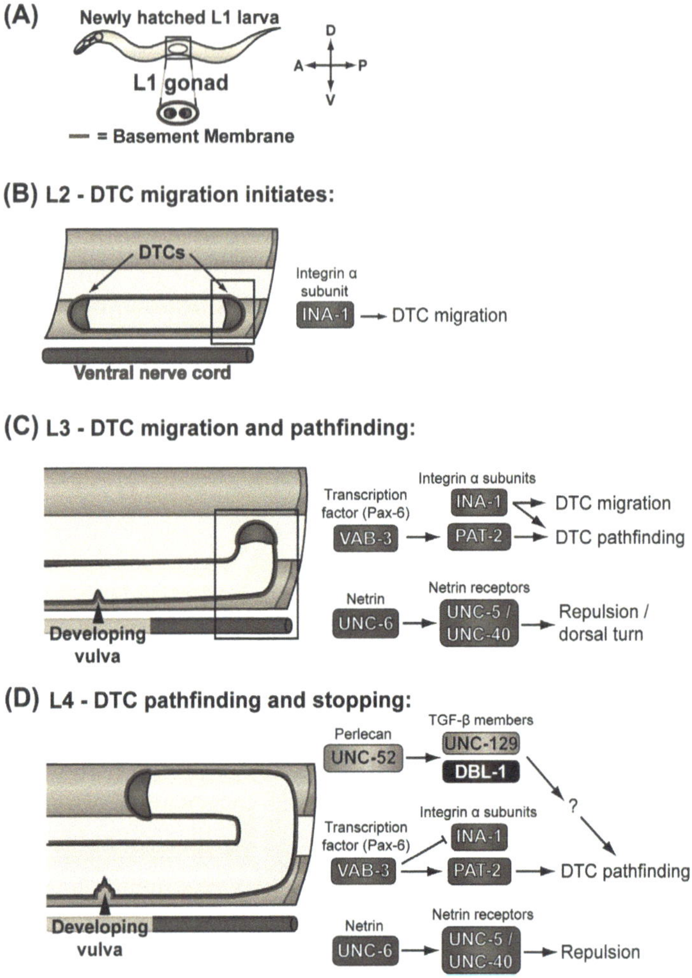

Figure 3 ***Larval morphogenesis of the gonad and molecular pathways controlling distal tip cell (DTC) migration.*** (A) A schematic diagram of a newly hatched L1 larva and gonad. The gonadal basement membrane is shown in red in all illustrations. (B) DTC migration initiates at the L2 stage, and DTCs migrate adjacent to the ventral body wall muscle basement membrane (orange). The integrin α subunit *ina-1* is expressed in DTCs beginning at L2 and promotes DTC migration. (C) In L3, the DTCs turn and migrate toward the dorsal body wall muscle along the epidermal basement membrane (light blue). During L3, the transcription factor VAB-3 (Pax6 ortholog) induces expression of the integrin α subunit *pat-2*. Both INA-1 and PAT-2 regulate DTC turning and proper pathfinding. In addition, the action of netrin (UNC-6) secreted from the ventral nerve cord binds the netrin receptors UNC-5 and UNC-40 (DCC) in the DTCs to mediate repulsion from the ventral body wall. (D) Once the DTCs reach the dorsal body wall muscle in the L4 stage, the DTCs turn back toward the animal midbody. This turn is regulated by the integrins INA-1 and PAT-2 in the DTCs, the basement membrane protein perlecan (UNC-52), and the TGF-β family members UNC-129 and DBL-1. Perlecan and *unc-129* are expressed in the body wall muscle, and perlecan might sequester TGF-β members within the dorsal body wall muscle basement membrane. During L4, VAB-3 downregulates *ina-1* integrin expression, which is necessary to stop DTC migration in early adulthood. (See color plate)

ina-1, which is turned off by *vab-3, pat-2* expression is initiated by *vab-3* (Figure 3(C) and (D)) (Meighan & Schwarzbauer, 2007). Knocking down *pat-2* results in DTCs that execute multiple turns, or DTCs that do not turn toward the dorsal muscle and instead migrate back toward the midbody along the ventral side of the animal (Meighan & Schwarzbauer, 2007). Thus, integrins are crucial players in DTC migration, with INA-1 promoting DTC migration and both INA-1 and PAT-2 cooperating to regulate DTC pathfinding.

In addition to the coordinated activity of integrins, DTC migration depends on extracellular signaling that can be influenced by the basement membrane component perlecan. Perlecan is expressed in the body wall muscle and appears to influence DTC migration by sequestering or localizing the TGF-β family members UNC-129 and DBL-1 (Figure 3(D)) (Merz et al., 2003). This interaction between perlecan and TGF-β might affect netrin signaling, since the loss of perlecan, as well as *dbl-1* or *unc-129,* enhances the DTC migration defects seen in weak mutants for the netrin receptor *unc-5* (Merz et al., 2003). The netrin receptors UNC-5 and UNC-40 (DCC) act in the DTC to initiate the turn toward the dorsal muscle by mediating repulsion from the ventrally secreted cue netrin (UNC-6) (Figure 3(C) and (D)).

The functions of integrins, extracellular signals, and basement membrane components are spatiotemporally regulated to control DTC migration, and these pathways may exhibit extensive cross talk. In addition to the signaling interactions between netrin and TGF-β mentioned above, interactions between integrins and netrin or TGF-β signaling have been characterized in other contexts. For example, during anchor cell (AC) invasion in *C. elegans,* integrin regulates localization of the netrin receptor UNC-40 (DCC) (see below) (Hagedorn et al., 2009; Ziel, Hagedorn, Audhya, & Sherwood, 2009), and in vertebrates integrins play important roles in controlling the expression and activation of TGF-β (reviewed by Ivaska & Heino, 2011; Margadant & Sonnenberg, 2010). In the DTC, the loss of the netrin receptor *unc-5* results in DTCs that fail to migrate to the dorsal muscle (Hedgecock, Culotti, & Hall, 1990; Su et al., 2000), which is a phenotype similar to loss of the integrin α subunit *pat-2* (Meighan & Schwarzbauer, 2007) and it will be interesting to determine if netrin and integrin interact during DTC migration. Further, it will be important to determine if other basement membrane components influence DTCs similar to perlecan, and to explore how other molecules, such as secreted proteases (see below), regulate these pathways.

3.6 Extracellular metalloproteases within the gonadal basement membrane regulate DTC migration and gonad shape

Another family of proteins associated with the basement membrane that regulate DTC migration and gonad shape are the ADAMTS (A Disintegrin And Metalloprotease with ThromboSpondin repeats) proteins—conserved, secreted proteases that cleave the extracellular matrix and have roles in development, physiological function, and disease (reviewed by Dubail & Apte, 2015; Tang, 2001). GON-1 is a *C. elegans* ADAMTS protein that controls gonad elongation and shape (Blelloch & Kimble, 1999; Hesselson, Newman, Kim, & Kimble, 2004; Kubota et al., 2012). Strong expression of *gon-1* is observed in the DTCs throughout their migration, and *gon-1* is also expressed by the body wall muscles (Figure 4(A)) (Blelloch & Kimble, 1999). Site of action experiments indicated that GON-1 secreted from the DTCs allows gonad arm elongation via DTC migration, whereas GON-1 secreted from the body wall muscles allows the gonad to expand radially during DTC outgrowth (Figure 4(B) and (C)) (Blelloch & Kimble, 1999). Genetic and localization studies showed that GON-1 opposes the function of the basement membrane-associated protein fibulin-1 and type IV collagen in the gonad, and that fibulin-1 maintains collagen within the basement membrane at late larval stages (Hesselson et al., 2004; Kubota et al., 2012). Whereas gonads in *gon-1* mutants fail to elongate and expand, mutants for fibulin-1 or type IV collagen have an opposite phenotype of widened gonad arms (Hesselson et al., 2004; Kubota et al., 2004, 2012; Muriel et al., 2005). The catalytic activity of GON-1 is required for its function, suggesting that it may cleave fibulin-1 or type IV collagen (Blelloch & Kimble, 1999), however, no such cleavage has been detected (Hesselson et al., 2004). Given the function of collagen in providing structural stability to basement membranes, it is possible that GON-1 might lower type IV collagen levels or reduce collagen function to allow for a more pliable basement membrane during DTC migration and gonad expansion.

MIG-17 is an ADAMTS that regulates the directionality of DTC migration. Loss of MIG-17 leads to a "wandering" DTC phenotype where DTCs make multiple, incorrect turns throughout their migration (Nishiwaki, Hisamoto, & Matsumoto, 2000). MIG-17, along with MIG-18, a novel cofactor required for MIG-17 activation, is secreted from the body wall muscle (Figure 4(A)) (Kim et al., 2014; Nishiwaki et al., 2000). MIG-17 is secreted in a pro-form that is incorporated into the gonadal basement

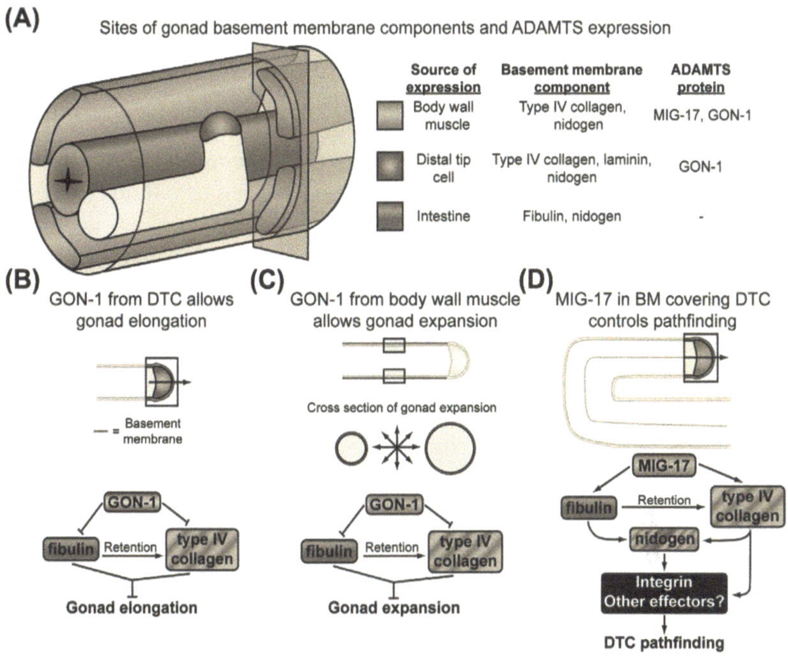

Figure 4 ***A Disintegrin And Metalloprotease with ThromboSpondin (ADAMTS) proteins act in the gonadal basement membrane to control gonad elongation, expansion, and distal tip cell (DTC) migration.*** (A) A schematic diagram depicting a portion of a *Caenorhabditis elegans* larva showing the tissues that express gonadal basement membrane components and ADAMTS proteins. The body wall muscles (orange) express and secrete MIG-17, GON-1, and type IV collagen; the DTCs (gray) secrete GON-1, type IV collagen, laminin, and nidogen; and the intestine (green) secretes fibulin-1 and nidogen. (B–D) Mechanisms utilized by ADAMTS proteins to control gonad elongation (B), gonad expansion (C), and DTC pathfinding (C). Colors represent the sites of protein secretion and correspond to the key in (A). (B) The action of GON-1 secreted from the DTC antagonizes fibulin and type IV collagen and allows gonad arm elongation. (C) GON-1 secreted from the body wall muscle acts within the gonadal basement membrane to allow gonad expansion. (D) MIG-17 secreted from the body wall muscle acts in the basement membrane covering the DTC to regulate a functional network of fibulin-1, type IV collagen, and nidogen that regulates integrin or other effectors to control DTC pathfinding. (See color plate)

membrane where the pro-domain is removed to activate MIG-17 at the L3 stage (Ihara & Nishiwaki, 2007, 2008). MIG-17 functions in the basement membrane covering the DTC where it recruits fibulin-1 (Kubota et al., 2004; Nishiwaki et al., 2000). Genetic and overexpression studies suggest that MIG-17 "activates" fibulin-1 in the basement membrane—possibly through direct proteolytic cleavage or through indirect interactions with another substrate—such that fibulin-1 recruits nidogen, which has a subtle role in DTC pathfinding (Figure 4(D)) (Kubota et al., 2008). In addition to regulating fibulin-1, genetic studies suggest that MIG-17 also activates type IV collagen within the gonadal basement membrane to promote DTC pathfinding (Kubota et al., 2012). Notably, type IV collagen determines the level of the INA-1/PAT-3 integrin complex in the DTC (Kubota et al., 2012). As integrins are also known regulators of DTC pathfinding (see above), they may be key effectors of the MIG-17 pathway (Figure 4(D)) (Baum & Garriga, 1997; Kubota et al., 2012; Lee, Cram, Shen, & Schwarzbauer, 2001; Meighan & Schwarzbauer, 2007).

Together these studies have indicated that ADAMTS proteins within the basement membrane control cell migration and organ shape in gonad formation. These proteins appear to be important in shaping other organs as well. For example, MIG-17 and several basement membrane components are important for pharyngeal morphogenesis (Jafari et al., 2010), and GON-1 and type IV collagen play a role in the formation and growth of motor neuron presynaptic boutons (Qin, Liang, & Ding, 2014). The functions of ADAMTS proteins may be context dependent, however, as GON-1 restrains growth of presynaptic boutons, whereas it promotes growth in the gonad.

4. CREATING GAPS IN—OR LINKS BETWEEN—BASEMENT MEMBRANES HELPS SHAPE TISSUES

Basement membranes act as formidable barriers that contain and separate tissues. During morphogenesis, distinct tissues often join together, and dramatic structural changes in basement membranes must occur to facilitate the migration or exchange of cells from one tissue to another. Recent studies in *C. elegans* have identified important molecular mechanisms that control the creation of large openings in basement membranes that allow the direct connection of cells between tissues. In addition, work in the worm has identified a new adhesion system that links neighboring tissues through adjoining basement membranes.

4.1 Creating de novo gaps in the basement membrane

Formation of de novo gaps in basement membranes occurs frequently during development and also underlies the progression of diseases such as cancer (reviewed by Kelley, Lohmer, Hagedorn, & Sherwood, 2014). In spite of this widespread importance, how openings in the basement membrane are created is not well understood. During *C. elegans* larval development, the uterine and vulval tissues are initially separated by their respective basement membranes (Sherwood & Sternberg, 2003). To initiate uterine–vulval attachment, the AC, a specialized uterine cell, invades across the uterine and vulval basement membranes. AC invasion creates an initial breach in the basement membrane that is widened as the vulval cells invaginate to connect the tissues (Figure 5) (reviewed by Hagedorn & Sherwood, 2011; see Hagedorn et al., 2013; Sherwood & Sternberg, 2003; Ziel et al., 2009). Studies of uterine–vulval attachment have begun to reveal the cellular and molecular mechanisms that create gaps in basement membranes.

AC invasion is highly stereotyped and occurs in synchrony with the divisions of the underlying vulval precursor cell (P6.p; Figure 5(A)) (Sherwood & Sternberg, 2003). The initial breach in the basement membrane is generated by invadopodia—small, membrane-associated, actin-rich structures (Figure 5(A) and (B)) that are also present in metastatic cancer cells (Hagedorn et al., 2014, 2013; Murphy & Courtneidge, 2011). Similar to invadopodia in cancer cells, the AC's invadopodia are dependent on integrin–basement membrane interactions (Figure 5(A)) (Hagedorn et al., 2014, 2009, 2013; Lohmer, Kelley, Hagedorn, & Sherwood, 2014). Therefore, the cellular machinery responsible for breaching the basement membrane is organized, in part, by the basement membrane itself. After breach, the basement membrane is removed underneath the entire AC footprint. A variety of mechanisms for opening holes in basement membrane exist (reviewed by Kelley et al., 2014), but in many cases the activity of matrix metalloproteinases (MMPs) are thought to dissolve basement membrane (Overall & Kleifeld, 2006; Shay, Lynch, & Fingleton, 2015; Valastyan & Weinberg, 2011). Consistent with this notion, AC invasion depends on the action of *fos-1*—the single *C. elegans* ortholog of the fos family of bZIP transcription factors—which induces expression of the MMP *zmp-1* (Sherwood, Butler, Kramer, & Sternberg, 2005). However, animals lacking *zmp-1* still have efficient AC invasion (Sherwood et al., 2005), suggesting that removal of basement membrane may depend on redundant MMPs or might involve proteolysis-independent mechanisms (see below). Prior to AC invasion the netrin receptor UNC-40 (DCC) is polarized to the ACs invasive surface through the action of integrin and the UNC-40 ligand UNC-6 (netrin) (Hagedorn et al., 2009; Wang et al., 2014; Ziel et al., 2009). UNC-40 (DCC) becomes enriched at the site of initial breach where it directs formation of an invasive protrusion that crosses the basement membrane (Figure 5(A)) (Hagedorn et al., 2013). Photoconversion of laminin under the AC revealed that the

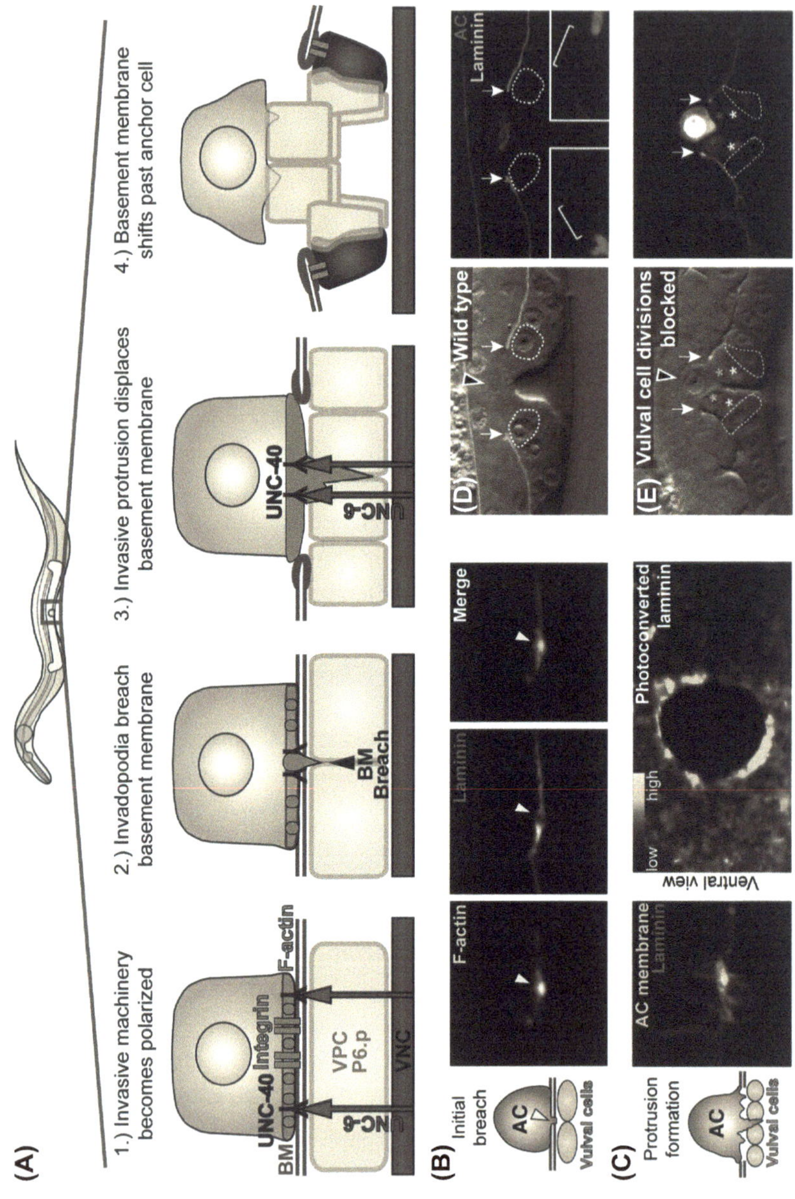
(A)
1.) Invasive machinery becomes polarized
2.) Invadopodia breach basement membrane
3.) Invasive protrusion displaces basement membrane
4.) Basement membrane shifts past anchor cell
BM
UNC-40
Integrin
F-actin
UNC-6
VPC
P6.p
VNC
BM Breach
UNC-40
UNC-6
(B)
Initial breach
AC
Vulval cells
F-actin
Laminin
Merge
(C)
Protrusion formation
AC
Vulval cells
AC membrane
Laminin
Ventral view
low
high
Photoconverted laminin
(D)
Wild type
AC
Laminin
(E)
Vulval cell divisions blocked

Figure 5 ***Creating a de novo gap in basement membrane during uterine–vulval attachment.*** (A) A schematic diagram of basement membrane breach and hole widening during uterine–vulval attachment. 1.) The anchor cell (AC) invasive cellular machinery is polarized by integrin–basement membrane interactions and netrin (UNC-6) localized within the basement membrane that binds to the netrin receptor UNC-40 (DCC) in the AC. Netrin is secreted from the ventral nerve cord (VNC). 2.) UNC-40 (DCC) traffics to the site where invadopodia (marked by F-actin) breach the basement membrane. 3.) UNC-6 (netrin) binding UNC-40 (DCC) stimulates the formation of an invasive protrusion that displaces basement membrane. 4.) Vulval cell invagination shifts the basement membrane sheet until integrin adhesion in the vulval vulD cell (purple), which is triggered by increased laminin at the edges of the hole, stops basement membrane sliding. (B) Confocal images showing invadopodia (F-actin) breaching the basement membrane (yellow arrowheads), which has been labeled with fluorescent laminin. (C) Confocal images showing the invasive protrusion. Ventral view shows accumulation of laminin that was photoconverted under the AC before being displaced by the invasive protrusion. (D) The basement membrane slides past the AC (black arrowhead) and stops (arrows mark position) over the vulval vulD cell (circled). Brackets mark space between the AC and the basement membrane. (E) Basement membrane does not slide when cell divisions of vulval cells (marked by asterisks) are blocked. AC, anchor cell; BM, basement membrane; VPC, vulval precursor cell; VNC, ventral nerve cord. *All confocal images were adapted with permission. Images in (B) and (C) originally appeared in Hagedorn et al. (2013); images in (D) appeared in Ihara et al. (2011); images in (E) appeared in Matus et al. (2014).* (See color plate)

invasive protrusion displaces the basement membrane to expand the existing hole (Figure 5(C)) (Hagedorn et al., 2013). Approximately 30% of the laminin is displaced, with the remainder removed presumably by proteolysis. Thus, the combined action of proteases and the physical actions of the AC create and widen a gap in the basement membrane.

Once AC invasion is complete, the gap in the basement membrane continues to widen past the footprint of the AC to allow direct contact between the uterine and vulval cells that form the mature uterine–vulval attachment. Live imaging of animals with ablated ACs or vulval cells showed that this further expansion requires only the vulval cells (Ihara et al., 2011). Optical highlighting of laminin and type IV collagen revealed that as the gap widens the basement membrane slides as a sheet over the dividing and invaginating vulval cells. These observations suggest that forces created by vulval cell invagination physically shift the basement membrane. Strikingly, the basement membrane always stops over vulD—the only vulval cell that does not divide during invagination (Figure 5(A) and (D)) (Ihara et al., 2011). Manipulations of vulval cell divisions revealed that dividing vulval cells lose attachment with the basement membrane and allow it to slide (Figure 5(E)) (Matus et al., 2014). At the edge of the gap, laminin is enriched from the original displacement of basement membrane by the AC (Hagedorn et al., 2013; Matus et al., 2014). This higher level of laminin plays a key role in stabilizing the position of the sliding basement membrane by increasing the INA-1/PAT-3 integrin complex within the cell membrane of vulD, which binds laminin and halts basement membrane sliding (Ihara et al., 2011; Matus et al., 2014).

These studies in *C. elegans* revealed that forming and widening the gap in basement membrane during uterine–vulval attachment requires the coordinated actions of a number of cellular behaviors, including cell invasion, cell division, and tissue invagination. Creating the basement membrane gap also depends on instructive cues from the basement membrane itself that promote basement membrane breaching by invadopodia and that stabilize the position of the widening gap by signaling for increased adhesion.

4.2 Adhesion between basement membranes

There are several biological processes where basement membranes of adjacent tissues come into contact and appear connected. For example, the basement membranes of kidney glomerular podocytes and lung alveolar epithelial cells appear to firmly connect or possibly fuse with the basement membrane of the neighboring vasculature; basement membranes of the invaginating optic cup meet and link before being removed to form a continuous epithelium; and the basement membranes of tissues at the blood–brain barrier appear to stably attach (Abrahamson, 1985; Morrissey & Sherwood, 2015; Obermeier, Daneman, & Ransohoff, 2013; Tsuji, Kita, Ozaki, Narama, & Matsuura, 2012; Vaccaro & Brody, 1981). It has remained unclear if specific adhesions between basement membranes occur to regulate these basement membrane–basement membrane interactions. Recent work in *C. elegans* has identified the molecular components of a new adhesion system between basement membranes of adjacent tissues and provided insight into several roles for this adhesion.

The uterine and vulval basement membranes under the AC are in close apposition to one another before invasion (Figure 6(A)) (Morrissey et al., 2014; Sherwood & Sternberg, 2003). Studies combining photobleaching of the basement membrane and forced tissue shifting revealed that the uterine and vulval basement membranes move independently of one another several hours before invasion (Morrissey et al., 2014). However, just prior to AC invasion, the basement membranes under the AC no longer move independently, suggesting they become linked. Consistent with this notion, electron microscopy revealed punctate structures between the uterine and vulval basement membranes, indicative of adhesions between the basement membranes (Figure 6(A)) (Morrissey et al., 2014). Hemicentin (encoded by *C. elegans him-4*) is a conserved, large, extracellular protein with 48 tandem immunoglobulin repeats that is associated with basement membranes (Vogel & Hedgecock, 2001; Xu, Dong, & Vogel, 2007). Hemicentin is secreted from the AC and assembles into stable puncta in basement membranes underneath the AC (Morrissey et al., 2014; Sherwood et al., 2005; Sherwood & Sternberg, 2003), suggesting it may mediate linkage between uterine and vulval basement membranes. Indeed, in hemicentin mutants the basement membranes are not linked, and the AC breaches the uterine and vulval basement membranes sequentially rather than simultaneously, resulting in a delay in invasion (Figure 6(B)) (Morrissey et al., 2014). Additionally, the cytolinker plakin (*vab-10a*) and integrin (*ina-1/pat-3*) are expressed in the AC and are essential for hemicentin puncta organization and basement membrane connection. These newly identified, cell-associated linkages between juxtaposed basement membranes are termed B-LINKs for Basement membrane LINKage (Figure 6(C)) (Morrissey et al., 2014). In the AC, B-LINKs are transient adhesions, but stable B-LINKs also form between the basement membranes of the uterus and epidermis to maintain uterine structure during egg laying

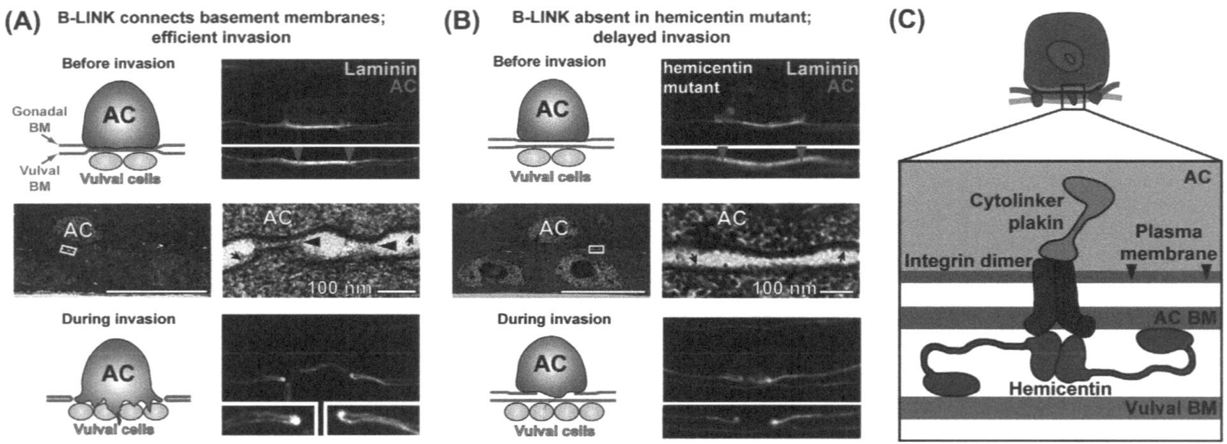

Figure 6 ***Basement membrane LINKages (B-LINKs) connect adjacent basement membranes.*** (A) During anchor cell (AC) invasion, the gonadal and vulval basement membranes (BMs; labeled with fluorescent laminin) under the anchor cell are connected together (red arrowheads). Tight adhesion between the basement membranes is mediated by the B-LINK, seen as punctate structures (black arrowheads) between the gonadal and vulval basement membranes (black arrows) in transmission electron micrographs. The B-LINK allows efficient invasion. (B) In hemicentin mutants the B-LINK is absent and the gonadal and vulval basement membranes are no longer linked, evidenced by space between the basement membranes seen by fluorescent microscopy (red arrowheads) and lack of punctate structures between the basement membranes (black arrows) in electron micrographs. Disrupting the B-LINK leads to stepwise breaching of the gonadal and vulval basement membranes and delayed invasion. (C) A schematic diagram showing molecular components of the B-LINK. Hemicentin is secreted from the AC and forms punctate structures in the BM. The cytolinker plakin (VAB-10A; blue) and the transmembrane integrin complex (INA-1/PAT-3; purple) are necessary for B-LINK formation. AC, anchor cell; BM, basement membrane. *Confocal images, electron microscopy images, and illustration of molecular mechanism originally appeared in Morrissey et al. (2014) and were adapted with permission.* (See color plate)

(Morrissey et al., 2014; Newman, White, & Sternberg, 1996; Vogel & Hedgecock, 2001). Further, B-LINKs may be conserved across taxa, as zebra fish hemicentins regulate the interaction between neighboring epithelia in the fin epidermis, as well as the connection between somites and the epidermis (Carney et al., 2010; Feitosa et al., 2012). While work from *C. elegans* definitively identified B-LINKs, these studies suggest that B-LINKs likely have broad roles in connecting various tissues during development and in forming stable structures that mediate long-term tissue connections in adults.

5. FUTURE DIRECTIONS AND PERSPECTIVES

Throughout this chapter, we have highlighted studies from *C. elegans* that have advanced our understanding of basement membranes. In *C. elegans,* as in vertebrates, early loss of laminin, the key building block of all basement membranes, results in gross disruptions in all tissues and early embryonic lethality (Kao, Huang, Hedgecock, Hall, & Wadsworth, 2006). These observations clearly highlight the widespread and essential importance of basement membranes to organogenesis and development. By using the worm system's strengths including RNAi, null and hypomorphic genetic mutations, and live imaging of basement membrane components, work in *C. elegans* has extended our understanding of basement membranes beyond their essential role in tissue support. In particular, studies in *C. elegans* have provided new insights into how basement membranes are built in vivo, how individual components within the basement membrane actively regulate numerous cellular behaviors, and how basement membranes are remodeled to facilitate tissue connections. Many aspects of basement membrane assembly, function, and mechanics, however, remain unclear.

5.1 Identifying key determinants of basement membrane composition and selective deposition

One important area of future study is to further understand how extracellular matrix components are localized to specific basement membranes in vivo, as the specific localization or concentration of basement membrane proteins is crucial in regulating many cellular processes. It is clear that localized secretion of basement membrane components is often important. For example, the AC secretes hemicentin to link basement membranes (Morrissey et al., 2014). However, in many cases matrix components are recruited from the extracellular fluid to specific basement membranes or concentrated within specific cellular regions, and in most of these instances we do not have a clear understanding of how this occurs. One glaring example is type IV collagen. Studies in cell culture have suggested that basement membrane proteins nidogen, perlecan, and agrin act redundantly to recruit collagen to basement membranes (reviews by Hohenester & Yurchenco, 2013; Yurchenco, 2011). Due to the difficulties associated with removing all of these basement membrane components in vertebrates, this has not yet been verified in vivo. With single genes encoding nidogen, perlecan, and agrin in *C. elegans*, it should be possible to definitively determine if these proteins act redundantly, or in various combinations, to incorporate collagen into basement membranes. Furthermore, it is conceivable that known extracellular matrix receptors play a role in recruiting collagen to basement membranes in vivo. With small gene families encoding basement membrane receptors in *C. elegans* (Kramer, 2005) it is feasible to knock down known receptors individually, in different combinations, and in a tissue-specific manner to determine whether they affect collagen distribution. It is also possible that unidentified or uncharacterized collagen-specific receptors, interactions between collagen and other basement membrane components, or the action of enzymes such as proteases are important for directing incorporation of basement membrane collagen. Unbiased screens in *C. elegans* could identify such genes, and the genetic, cellular, and molecular tools of *C. elegans* would allow in-depth analysis of how these molecules function in a cell-specific manner. Similar approaches could be taken to identify genes involved in generating the specific localization patterns of other basement membrane proteins we have discussed: for example, the presence of laminin along the sublateral nerves, the enrichment nidogen at muscle edges, the concentrated perlecan at muscle cell basement membrane attachments, the increased laminin around the endoderm cells posterior to the primordial germ cells, and hemicentin specifically at the attachments of mechanosensory and uterine cells (Huang et al., 2003; Kang & Kramer, 2000; Kim & Wadsworth, 2000; Muriel et al., 2005; Rogalski et al., 1993; Rohrschneider & Nance, 2013).

5.2 Physical forces on basement membranes

We have also highlighted diverse aspects of organ growth and morphogenesis for which the mechanical properties of basement membranes are likely important in maintaining tissue shape and integrity. The biomechanical properties of the basement membrane change over time, and basement membranes are thought to sense and respond to external physical forces

(Candiello et al., 2007; Halfter et al., 2013; Ingber, 2003; Moore et al., 2005). How basement membranes detect and mediate responses to force is largely unknown. During morphogenesis, basement membranes likely experience changes in force as a result of expanding or contracting tissues, changes in basement membrane composition, or the actions of enzymes such as proteases. For example, the *C. elegans* gonad undergoes a dramatic expansion during larval development (Hirsh, Oppenheim, & Klass, 1976; Kimble & White, 1981; Keeley and Sherwood, unpublished data), and the gonadal basement membrane likely experiences increased tension during this process. This tension could be relieved by addition of new basement membrane from sources such as the DTCs. Alternatively, the protease GON-1, which is important for gonad expansion (see above), may relieve basement membrane tension through reducing type IV collagen levels or function, thus allowing flexibility in the basement membrane for the gonad to expand and take on complex shapes. The forces borne by the basement membrane have not been precisely measured, and how basement membranes bear load is poorly understood. FRET-based sensors have been used to visualize tension across various molecules (Borghi et al., 2012; Grashoff et al., 2010; Meng, Suchyna, Lazakovitch, Gronostajski, & Sachs, 2011). These sensors could be adapted to *C. elegans* basement membrane components to determine the load that basement membranes bear, and if these loads are altered during morphogenetic processes, by proteases, or by alterations in basement membrane composition. These tools could also be important for determining whether the physical properties of the basement membrane are altered before it is breached and displaced by the AC during invasion, or as it is shifted by invaginating cells during uterine–vulval attachment.

6. CONCLUSION

Relative to the millions of years of evolution that have shaped basement membrane structure and function, our study and understanding of basement membrane biology is in its infancy. Our knowledge of basement membranes has transformed greatly since they were first described in 1840 (Bowman, 1840), and there is an emerging appreciation for basement membranes as dynamic structures that interact with cells and influence various aspects of their biology. In recent decades, the study of basement membranes in *C. elegans* has led to significant advancements in our understanding of these important biological structures. We expect that further studies harnessing the strengths of the *C. elegans* system will continue to provide a more complete view of the diverse, conserved, and crucial functions of basement membranes during development, which will also provide insight into the many human diseases where basement membranes are misregulated.

ACKNOWLEDGMENTS

We would like to thank Kacy Gordon, Dan Keeley, Lauren Lilley Lohmer, and Meghan Morrissey for their insightful comments on the manuscript. The authors acknowledge support from the Pew Scholars Program in the Biomedical Sciences and the National Institutes of Health (GM079320, GM100083 to D.R.S); M.R.C. is an HHMI Fellow of the Damon Runyon Cancer Research Foundation (DRG-2225-15).

REFERENCES

Abrahamson, D. R. (1985). Origin of the glomerular basement membrane visualized after in vivo labeling of laminin in newborn rat kidneys. *Journal of Cell Biology, 100*(6), 1988–2000.

Ackley, B. D., Harrington, R. J., Hudson, M. L., Williams, L., Kenyon, C. J., Chisholm, A. D., et al. (2005). The two isoforms of the *Caenorhabditis elegans* leukocyte-common antigen related receptor tyrosine phosphatase PTP-3 function independently in axon guidance and synapse formation. *Journal of Neuroscience, 25*(33), 7517–7528.

Ackley, B. D., Kang, S. H., Crew, J. R., Suh, C., Jin, Y., & Kramer, J. M. (2003). The basement membrane components nidogen and type XVIII collagen regulate organization of neuromuscular junctions in *Caenorhabditis elegans*. *Journal of Neuroscience, 23*(9), 3577–3587.

Aumailley, M., Bruckner-Tuderman, L., Carter, W. G., Deutzmann, R., Edgar, D., Ekblom, P., et al. (2005). A simplified laminin nomenclature. *Matrix Biology: Journal of the International Society for Matrix Biology, 24*(5), 326–332.

Aumailley, M., Wiedemann, H., Mann, K., & Timpl, R. (1989). Binding of nidogen and the laminin-nidogen complex to basement membrane collagen type IV. *European Journal of Biochemistry/FEBS, 184*(1), 241–248.

Baum, P. D., & Garriga, G. (1997). Neuronal migrations and axon fasciculation are disrupted in ina-1 integrin mutants. *Neuron, 19*(1), 51–62.

Blelloch, R., & Kimble, J. (1999). Control of organ shape by a secreted metalloprotease in the nematode *Caenorhabditis elegans*. *Nature, 399*(6736), 586–590.

Borghi, N., Sorokina, M., Shcherbakova, O. G., Weis, W. I., Pruitt, B. L., Nelson, W. J., et al. (2012). E-cadherin is under constitutive actomyosin-generated tension that is increased at cell-cell contacts upon externally applied stretch. *Proceedings of the National Academy of Sciences of the United States of America, 109*(31), 12568–12573.

Boutaud, A., Borza, D. B., Bondar, O., Gunwar, S., Netzer, K. O., Singh, N., et al. (2000). Type IV collagen of the glomerular basement membrane. Evidence that the chain specificity of network assembly is encoded by the noncollagenous NC1 domains. *Journal of Biological Chemistry, 275*(39), 30716–30724.

Bowman, W. (1840). On the minute structure and movements of voluntary muscles. *Philosophical Transactions of the Royal Society of London, 130*, 457–501.

Breitkreutz, D., Koxholt, I., Thiemann, K., & Nischt, R. (2013). Skin basement membrane: the foundation of epidermal integrity—BM functions and diverse roles of bridging molecules nidogen and perlecan. *BioMed Research International, 2013*, 179784.

Brizzi, M. F., Tarone, G., & Defilippi, P. (2012). Extracellular matrix, integrins, and growth factors as tailors of the stem cell niche. *Current Opinion in Cell Biology, 24*(5), 645–651.

Candiello, J., Balasubramani, M., Schreiber, E. M., Cole, G. J., Mayer, U., Halfter, W., et al. (2007). Biomechanical properties of native basement membranes. *The FEBS Journal, 274*(11), 2897–2908.
Candiello, J., Cole, G. J., & Halfter, W. (2010). Age-dependent changes in the structure, composition and biophysical properties of a human basement membrane. *Matrix Biology: Journal of the International Society for Matrix Biology, 29*(5), 402–410.
Carney, T. J., Feitosa, N. M., Sonntag, C., Slanchev, K., Kluger, J., Kiyozumi, D., et al. (2010). Genetic analysis of fin development in zebrafish identifies furin and hemicentin1 as potential novel fraser syndrome disease genes. *PLoS Genetics, 6*(4), e1000907.
Copp, A. J., Carvalho, R., Wallace, A., Sorokin, L., Sasaki, T., Greene, N. D. E., et al. (2011). Regional differences in the expression of laminin isoforms during mouse neural tube development. *Matrix Biology: Journal of the International Society for Matrix Biology, 30*(4), 301–309.
Daley, W. P., & Yamada, K. M. (2013). ECM-modulated cellular dynamics as a driving force for tissue morphogenesis. *Current Opinion in Genetics and Development, 23*(4), 408–414.
Drabikowski, K., Trzebiatowska, A., & Chiquet-Ehrismann, R. (2005). ten-1, an essential gene for germ cell development, epidermal morphogenesis, gonad migration, and neuronal pathfinding in *Caenorhabditis elegans. Developmental Biology, 282*(1), 27–38.
Dubail, J., & Apte, S. S. (2015). Insights on ADAMTS proteases and ADAMTS-like proteins from mammalian genetics. *Matrix Biology: Journal of the International Society for Matrix Biology, 44–46*, 24–37.
Ekblom, M., Klein, G., Mugrauer, G., Fecker, L., Deutzmann, R., Timpl, R., et al. (1990). Transient and locally restricted expression of laminin A chain mRNA by developing epithelial cells during kidney organogenesis. *Cell, 60*(2), 337–346.
Ekblom, P., Vestweber, D., & Kemler, R. (1986). Cell-matrix interactions and cell adhesion during development. *Annual Review of Cell Biology, 2*, 27–47.
Feitosa, N. M., Zhang, J., Carney, T. J., Metzger, M., Korzh, V., Bloch, W., et al. (2012). Hemicentin 2 and Fibulin 1 are required for epidermal-dermal junction formation and fin mesenchymal cell migration during zebrafish development. *Developmental Biology, 369*(2), 235–248.
Fidler, A. L., Vanacore, R. M., Chetyrkin, S. V., Pedchenko, V. K., Bhave, G., Yin, V. P., et al. (2014). A unique covalent bond in basement membrane is a primordial innovation for tissue evolution. *Proceedings of the National Academy of Sciences of the United States of America, 111*(1), 331–336.
Fitzgerald, M. C., & Schwarzbauer, J. E. (1998). Importance of the basement membrane protein SPARC for viability and fertility in *Caenorhabditis elegans. Current Biology, 8*(23), 1285–1288.
Fox, M. A., Ho, M. S. P., Smyth, N., & Sanes, J. R. (2008). A synaptic nidogen: developmental regulation and role of nidogen-2 at the neuromuscular junction. *Neural Development, 3*, 24.
Fox, J. W., Mayer, U., Nischt, R., Aumailley, M., Reinhardt, D., Wiedemann, H., et al. (1991). Recombinant nidogen consists of three globular domains and mediates binding of laminin to collagen type IV. *EMBO Journal, 10*(11), 3137–3146.
Francis, R., & Waterston, R. H. (1991). Muscle cell attachment in *Caenorhabditis elegans. Journal of Cell Biology, 114*(3), 465–479.
García-Castro, M. I., Anderson, R., Heasman, J., & Wylie, C. (1997). Interactions between germ cells and extracellular matrix glycoproteins during migration and gonad assembly in the mouse embryo. *Journal of Cell Biology, 138*(2), 471–480.
Graham, P. L., Johnson, J. J., Wang, S., Sibley, M. H., Gupta, M. C., & Kramer, J. M. (1997). Type IV collagen is detectable in most, but not all, basement membranes of *Caenorhabditis elegans* and assembles on tissues that do not express it. *Journal of Cell Biology, 137*(5), 1171–1183.

Grashoff, C., Hoffman, B. D., Brenner, M. D., Zhou, R., Parsons, M., Yang, M. T., et al. (2010). Measuring mechanical tension across vinculin reveals regulation of focal adhesion dynamics. *Nature, 466*(7303), 263–266.

Hagedorn, E. J., Kelley, L. C., Naegeli, K. M., Wang, Z., Chi, Q., & Sherwood, D. R. (2014). ADF/cofilin promotes invadopodial membrane recycling during cell invasion in vivo. *Journal of Cell Biology, 204*(7), 1209–1218.

Hagedorn, E. J., & Sherwood, D. R. (2011). Cell invasion through basement membrane: the anchor cell breaches the barrier. *Current Opinion in Cell Biology, 23*(5), 589–596.

Hagedorn, E. J., Yashiro, H., Ziel, J. W., Ihara, S., Wang, Z., & Sherwood, D. R. (2009). Integrin acts upstream of netrin signaling to regulate formation of the anchor cell's invasive membrane in *C. elegans*. *Developmental Cell, 17*(2), 187–198.

Hagedorn, E. J., Ziel, J. W., Morrissey, M. A., Linden, L. M., Wang, Z., Chi, Q., et al. (2013). The netrin receptor DCC focuses invadopodia-driven basement membrane transmigration in vivo. *Journal of Cell Biology, 201*(6), 903–913.

Halfter, W., Candiello, J., Hu, H., Zhang, P., Schreiber, E., & Balasubramani, M. (2013). Protein composition and biomechanical properties of in vivo-derived basement membranes. *Cell Adhesion & Migration, 7*(1), 64–71.

Hall, D. H., Winfrey, V. P., Blaeuer, G., Hoffman, L. H., Furuta, T., Rose, K. L., et al. (1999). Ultrastructural features of the adult hermaphrodite gonad of *Caenorhabditis elegans*: relations between the germ line and soma. *Developmental Biology, 212*(1), 101–123.

Harrell, J. R., & Goldstein, B. (2011). Internalization of multiple cells during *C. elegans* gastrulation depends on common cytoskeletal mechanisms but different cell polarity and cell fate regulators. *Developmental Biology, 350*(1), 1–12.

Hedgecock, E. M., Culotti, J. G., & Hall, D. H. (1990). The unc-5, unc-6, and unc-40 genes guide circumferential migrations of pioneer axons and mesodermal cells on the epidermis in *C. elegans*. *Neuron, 4*(1), 61–85.

Hesselson, D., Newman, C., Kim, K. W., & Kimble, J. (2004). GON-1 and fibulin have antagonistic roles in control of organ shape. *Current Biology, 14*(22), 2005–2010.

Hirsh, D., Oppenheim, D., & Klass, M. (1976). Development of the reproductive system of *Caenorhabditis elegans*. *Developmental Biology, 49*(1), 200–219.

Hohenester, E., & Yurchenco, P. D. (2013). Laminins in basement membrane assembly. *Cell Adhesion & Migration, 7*(1), 56–63.

Hresko, M. C., Williams, B. D., & Waterston, R. H. (1994). Assembly of body wall muscle and muscle cell attachment structures in *Caenorhabditis elegans*. *Journal of Cell Biology, 124*(4), 491–506.

Hrus, A., Lau, G., Hutter, H., Schenk, S., Ferralli, J., Brown-Luedi, M., et al. (2007). *C. elegans* agrin is expressed in pharynx, IL1 neurons and distal tip cells and does not genetically interact with genes involved in synaptogenesis or muscle function. *PLoS One, 2*(8), e731.

Huang, C.-C., Hall, D. H., Hedgecock, E. M., Kao, G., Karantza, V., Vogel, B. E., et al. (2003). Laminin alpha subunits and their role in *C. elegans* development. *Development, 130*(14), 3343–3358.

Hutter, H. (2003). Extracellular cues and pioneers act together to guide axons in the ventral cord of *C. elegans*. *Development, 130*(22), 5307–5318.

Hutter, H., Vogel, B. E., Plenefisch, J. D., Norris, C. R., Proenca, R. B., Spieth, J., et al. (2000). Conservation and novelty in the evolution of cell adhesion and extracellular matrix genes. *Science, 287*(5455), 989–994.

Hynes, R. O. (2012). The evolution of metazoan extracellular matrix. *Journal of Cell Biology, 196*(6), 671–679.

Ihara, S., Hagedorn, E. J., Morrissey, M. A., Chi, Q., Motegi, F., Kramer, J. M., et al. (2011). Basement membrane sliding and targeted adhesion remodels tissue boundaries during uterine-vulval attachment in *Caenorhabditis elegans*. *Nature Cell Biology, 13*(6), 641–651.

Ihara, S., & Nishiwaki, K. (2007). Prodomain-dependent tissue targeting of an ADAMTS protease controls cell migration in *Caenorhabditis elegans*. *EMBO Journal, 26*(11), 2607–2620.
Ihara, S., & Nishiwaki, K. (2008). Stage-specific activation of MIG-17/ADAMTS controls cell migration in *Caenorhabditis elegans*. *The FEBS Journal, 275*(17), 4296–4305.
Ingber, D. E. (2003). Tensegrity II. How structural networks influence cellular information processing networks. *Journal of Cell Science, 116*(Pt 8), 1397–1408.
Ivaska, J., & Heino, J. (2011). Cooperation between integrins and growth factor receptors in signaling and endocytosis. *Annual Review of Cell and Developmental Biology, 27*, 291–320.
Jafari, G., Burghoorn, J., Kawano, T., Mathew, M., Mörck, C., Axäng, C., et al. (2010). Genetics of extracellular matrix remodeling during organ growth using the *Caenorhabditis elegans* pharynx model. *Genetics, 186*(3), 969–982.
Johnson, R. P. (2006). *C. elegans* dystroglycan DGN-1 functions in epithelia and neurons, but not muscle, and independently of dystrophin. *Development, 133*(10), 1911–1921.
Kang, S. H., & Kramer, J. M. (2000). Nidogen is nonessential and not required for normal type IV collagen localization in *Caenorhabditis elegans*. *Molecular Biology of the Cell, 11*(11), 3911–3923.
Kao, G., Huang, C.-C., Hedgecock, E. M., Hall, D. H., & Wadsworth, W. G. (2006). The role of the laminin β subunit in laminin heterotrimer assembly and basement membrane function and development in *C. elegans*. *Developmental Biology, 290*(1), 211–219.
Kawano, T., Zheng, H., Merz, D. C., Kohara, Y., Tamai, K. K., Nishiwaki, K., et al. (2009). *C. elegans* mig-6 encodes papilin isoforms that affect distinct aspects of DTC migration, and interacts genetically with mig-17 and collagen IV. *Development, 136*(9), 1433–1442.
Kelley, L. C., Lohmer, L. L., Hagedorn, E. J., & Sherwood, D. R. (2014). Traversing the basement membrane in vivo: a diversity of strategies. *Journal of Cell Biology, 204*(3), 291–302.
Kikuchi, T., Shibata, Y., Kim, H.-S., Kubota, Y., Yoshina, S., Mitani, S., et al. (2015). The BED finger domain protein MIG-39 halts migration of distal tip cells in *Caenorhabditis elegans*. *Developmental Biology, 397*(2), 151–161.
Kim, H.-S., Kitano, Y., Mori, M., Takano, T., Harbaugh, T. E., Mizutani, K., et al. (2014). The novel secreted factor MIG-18 acts with MIG-17/ADAMTS to control cell migration in *Caenorhabditis elegans*. *Genetics, 196*(2), 471–479.
Kim, S., Ren, X. C., Fox, E., & Wadsworth, W. G. (1999). SDQR migrations in *Caenorhabditis elegans* are controlled by multiple guidance cues and changing responses to netrin UNC-6. *Development, 126*(17), 3881–3890.
Kim, S., & Wadsworth, W. G. (2000). Positioning of longitudinal nerves in *C. elegans* by nidogen. *Science, 288*(5463), 150–154.
Kimble, J. E., & White, J. G. (1981). On the control of germ cell development in *Caenorhabditis elegans*. *Developmental Biology, 81*(2), 208–219.
Klein, G., Langegger, M., Timpl, R., & Ekblom, P. (1988). Role of laminin A chain in the development of epithelial cell polarity. *Cell, 55*(2), 331–341.
Kobayashi, T., & Uchiyama, M. (2003). Characterization of assembly of recombinant type IV collagen alpha3, alpha4, and alpha5 chains in transfected cell strains. *Kidney International, 64*(6), 1986–1996.
Kramer, J. M. (2005). Basement membranes. *WormBook*, ed. The *C. elegans* Research Community, http://www.wormbook.org.
Kubota, Y., Kuroki, R., & Nishiwaki, K. (2004). A fibulin-1 homolog interacts with an ADAM protease that controls cell migration in *C. elegans*. *Current Biology, 14*(22), 2011–2018.
Kubota, Y., Nagata, K., Sugimoto, A., & Nishiwaki, K. (2012). Tissue architecture in the *Caenorhabditis elegans* gonad depends on interactions among fibulin-1, type IV collagen and the ADAMTS extracellular protease. *Genetics, 190*(4), 1379–1388.
Kubota, Y., Ohkura, K., Tamai, K. K., Nagata, K., & Nishiwaki, K. (2008). MIG-17/ADAMTS controls cell migration by recruiting nidogen to the basement membrane

in *C. elegans*. *Proceedings of the National Academy of Sciences of the United States of America, 105*(52), 20804–20809.
LeBleu, V. S., MacDonald, B., & Kalluri, R. (2007). Structure and function of basement membranes. *Experimental Biology and Medicine, 232*(9), 1121–1129.
Lee, M., Cram, E. J., Shen, B., & Schwarzbauer, J. E. (2001). Roles for beta(pat-3) integrins in development and function of *Caenorhabditis elegans* muscles and gonads. *Journal of Biological Chemistry, 276*(39), 36404–36410.
Leitinger, B., & Hohenester, E. (2007). Mammalian collagen receptors. *Matrix Biology: Journal of the International Society for Matrix Biology, 26*(3), 146–155.
Leung, B., Hermann, G. J., & Priess, J. R. (1999). Organogenesis of the *Caenorhabditis elegans* intestine. *Developmental Biology, 216*(1), 114–134.
Liang, X., Dong, X., Moerman, D. G., Shen, K., & Wang, X. (2015). Sarcomeres pattern proprioceptive sensory dendritic endings through UNC-52/Perlecan in *C. elegans*. *Developmental Cell, 33*(4), 388–400.
Lohmer, L. L., Kelley, L. C., Hagedorn, E. J., & Sherwood, D. R. (2014). Invadopodia and basement membrane invasion in vivo. *Cell Adhesion & Migration, 8*(3).
Mango, S. (2007). The *C. elegans* pharynx: a model for organogenesis. *WormBook*, ed. The *C. elegans* Research Community, http://www.wormbook.org.
Margadant, C., & Sonnenberg, A. (2010). Integrin-TGF-beta crosstalk in fibrosis, cancer and wound healing. *EMBO Reports, 11*(2), 97–105.
Marthiens, V., Kazanis, I., Moss, L., Long, K., & Ffrench-Constant, C. (2010). Adhesion molecules in the stem cell niche—more than just staying in shape? *Journal of Cell Science, 123*(Pt 10), 1613–1622.
Matus, D. Q., Chang, E., Makohon-Moore, S. C., Hagedorn, M. A., Chi, Q., & Sherwood, D. R. (2014). Cell division and targeted cell cycle arrest opens and stabilizes basement membrane gaps. *Nature Communications, 5*, 4184.
Meighan, C. M., & Schwarzbauer, J. E. (2007). Control of *C. elegans* hermaphrodite gonad size and shape by vab-3/Pax6-mediated regulation of integrin receptors. *Genes & Development, 21*(13), 1615–1620.
Meng, F., Suchyna, T. M., Lazakovitch, E., Gronostajski, R. M., & Sachs, F. (2011). Real time FRET based detection of mechanical stress in cytoskeletal and extracellular matrix proteins. *Cellular and Molecular Bioengineering, 4*(2), 148–159.
Merz, D. C., Alves, G., Kawano, T., Zheng, H., & Culotti, J. G. (2003). UNC-52/perlecan affects gonadal leader cell migrations in *C. elegans* hermaphrodites through alterations in growth factor signaling. *Developmental Biology, 256*(1), 173–186.
Miner, J. H., Cunningham, J., & Sanes, J. R. (1998). Roles for laminin in embryogenesis: exencephaly, syndactyly, and placentopathy in mice lacking the laminin α5 chain. *The Journal of Cell Biology, 143*(6), 1713–1723.
Moerman, D. G., & Williams, B. D. (2006). Sarcomere assembly in *C. elegans* muscle. *WormBook*, ed. The *C. elegans* Research Community, http://www.wormbook.org.
Moore, K. A., Polte, T., Huang, S., Shi, B., Alsberg, E., Sunday, M. E., et al. (2005). Control of basement membrane remodeling and epithelial branching morphogenesis in embryonic lung by Rho and cytoskeletal tension. *Developmental Dynamics, 232*(2), 268–281.
Morrissey, M. A., Keeley, D. P., Hagedorn, E. J., McClatchey, S. T. H., Chi, Q., Hall, D. H., et al. (2014). B-LINK: a hemicentin, plakin, and integrin-dependent adhesion system that links tissues by connecting adjacent basement membranes. *Developmental Cell, 31*(3), 319–331.
Morrissey, M. A., & Sherwood, D. R. (2015). An active role for basement membrane assembly and modification in tissue sculpting. *Journal of Cell Science, 128*(9), 1661–1668.
Mullen, G. P., Rogalski, T. M., Bush, J. A., Gorji, P. R., & Moerman, D. G. (1999). Complex patterns of alternative splicing mediate the spatial and temporal distribution of perlecan/UNC-52 in *Caenorhabditis elegans*. *Molecular Biology of the Cell, 10*(10), 3205–3221.

Muriel, J. M., Dong, C., Hutter, H., & Vogel, B. E. (2005). Fibulin-1C and Fibulin-1D splice variants have distinct functions and assemble in a hemicentin-dependent manner. *Development, 132*(19), 4223–4234.

Muriel, J. M., Xu, X., Kramer, J. M., & Vogel, B. E. (2006). Selective assembly of fibulin-1 splice variants reveals distinct extracellular matrix networks and novel functions for perlecan/UNC-52 splice variants. *Developmental Dynamics, 235*(10), 2632–2640.

Murphy, D. A., & Courtneidge, S. A. (2011). The 'ins' and 'outs' of podosomes and invadopodia: characteristics, formation and function. *Nature Reviews Molecular Cell Biology, 12*(7), 413–426.

Nance, J., & Priess, J. R. (2002). Cell polarity and gastrulation in *C. elegans*. *Development, 129*(2), 387–397.

Newman, A. P., White, J. G., & Sternberg, P. W. (1996). Morphogenesis of the *C. elegans* hermaphrodite uterus. *Development, 122*(11), 3617–3626.

Nishiuchi, R., Takagi, J., Hayashi, M., Ido, H., Yagi, Y., Sanzen, N., et al. (2006). Ligand-binding specificities of laminin-binding integrins: a comprehensive survey of laminin-integrin interactions using recombinant alpha3beta1, alpha6beta1, alpha7beta1 and alpha6beta4 integrins. *Matrix Biology: Journal of the International Society for Matrix Biology, 25*(3), 189–197.

Nishiwaki, K., Hisamoto, N., & Matsumoto, K. (2000). A metalloprotease disintegrin that controls cell migration in *Caenorhabditis elegans*. *Science, 288*(5474), 2205–2208.

Obermeier, B., Daneman, R., & Ransohoff, R. M. (2013). Development, maintenance and disruption of the blood–brain barrier. *Nature Medicine, 19*(12), 1584–1596.

O'Brien, L. E., Jou, T. S., Pollack, A. L., Zhang, Q., Hansen, S. H., Yurchenco, P., et al. (2001). Rac1 orientates epithelial apical polarity through effects on basolateral laminin assembly. *Nature Cell Biology, 3*(9), 831–838.

Overall, C. M., & Kleifeld, O. (2006). Tumour microenvironment – opinion: validating matrix metalloproteinases as drug targets and anti-targets for cancer therapy. *Nature Reviews. Cancer, 6*(3), 227–239.

Ozbek, S., Balasubramanian, P. G., Chiquet-Ehrismann, R., Tucker, R. P., & Adams, J. C. (2010). The evolution of extracellular matrix. *Molecular Biology of the Cell, 21*(24), 4300–4305.

Pastor-Pareja, J. C., & Xu, T. (2011). Shaping cells and organs in Drosophila by opposing roles of fat body-secreted collagen IV and perlecan. *Developmental Cell, 21*(2), 245–256.

Paulsson, M., Aumailley, M., Deutzmann, R., Timpl, R., Beck, K., & Engel, J. (1987). Laminin-nidogen complex. Extraction with chelating agents and structural characterization. *European Journal of Biochemistry/FEBS, 166*(1), 11–19.

Qin, J., Liang, J., & Ding, M. (2014). Perlecan antagonizes collagen IV and ADAMTS9/GON-1 in restricting the growth of presynaptic boutons. *Journal of Neuroscience, 34*(31), 10311–10324.

Rasmussen, J. P., Feldman, J. L., Reddy, S. S., & Priess, J. R. (2013). Cell interactions and patterned intercalations shape and link epithelial tubes in *C. elegans*. *PLoS Genetics, 9*(9), e1003772.

Rasmussen, J. P., Reddy, S. S., & Priess, J. R. (2012). Laminin is required to orient epithelial polarity in the *C. elegans* pharynx. *Development, 139*(11), 2050–2060.

Rogalski, T. M., Gilchrist, E. J., Mullen, G. P., & Moerman, D. G. (1995). Mutations in the unc-52 gene responsible for body wall muscle defects in adult *Caenorhabditis elegans* are located in alternatively spliced exons. *Genetics, 139*(1), 159–169.

Rogalski, T. M., Williams, B. D., Mullen, G. P., & Moerman, D. G. (1993). Products of the unc-52 gene in *Caenorhabditis elegans* are homologous to the core protein of the mammalian basement membrane heparan sulfate proteoglycan. *Genes & Development, 7*(8), 1471–1484.

Rohrbach, D. H., & Murrah, V. A. (1993). Molecular aspects of basement membrane pathology. In D. H. R. Timpl (Ed.), *Molecular and cellular aspects of basement membrane* (pp. 385–419). Academic Press.

Rohrschneider, M. R., & Nance, J. (2013). The union of somatic gonad precursors and primordial germ cells during *Caenorhabditis elegans* embryogenesis. *Developmental Biology, 379*(2), 139–151.

Rozario, T., & DeSimone, D. W. (2010). The extracellular matrix in development and morphogenesis: a dynamic view. *Developmental Biology, 341*(1), 126–140.

Shay, G., Lynch, C. C., & Fingleton, B. (2015). Moving targets: emerging roles for MMPs in cancer progression and metastasis. *Matrix Biology: Journal of the International Society for Matrix Biology, 44–46*, 200–206.

Sherwood, D. R., Butler, J. A., Kramer, J. M., & Sternberg, P. W. (2005). FOS-1 promotes basement-membrane removal during anchor-cell invasion in *C. elegans*. *Cell, 121*(6), 951–962.

Sherwood, D. R., & Sternberg, P. W. (2003). Anchor cell invasion into the vulval epithelium in *C. elegans*. *Developmental Cell, 5*(1), 21–31.

Smyth, N., Vatansever, H. S., Murray, P., Meyer, M., Frie, C., Paulsson, M., et al. (1999). Absence of basement membranes after targeting the LAMC1 gene results in embryonic lethality due to failure of endoderm differentiation. *Journal of Cell Biology, 144*(1), 151–160.

Sorokin, L. M., Pausch, F., Durbeej, M., & Ekblom, P. (1997). Differential expression of five laminin α (1-5) chains in developing and adult mouse kidney. *Developmental Dynamics, 210*(4), 446–462.

Su, M., Merz, D. C., Killeen, M. T., Zhou, Y., Zheng, H., Kramer, J. M., et al. (2000). Regulation of the UNC-5 netrin receptor initiates the first reorientation of migrating distal tip cells in *Caenorhabditis elegans*. *Development, 127*(3), 585–594.

Sulston, J. E., Schierenberg, E., White, J. G., & Thomson, J. N. (1983). The embryonic cell lineage of the nematode *Caenorhabditis elegans*. *Developmental Biology, 100*(1), 64–119.

Tang, B. L. (2001). ADAMTS: a novel family of extracellular matrix proteases. *International Journal of Biochemistry and Cell Biology, 33*(1), 33–44.

Tang, X., & Wadsworth, W. G. (2014). SAX-3 (Robo) and UNC-40 (DCC) regulate a directional bias for axon guidance in response to multiple extracellular cues. *PLoS One, 9*(10), e110031.

Topf, U., & Chiquet-Ehrismann, R. (2011). Genetic interaction between *Caenorhabditis elegans* teneurin ten-1 and prolyl 4-hydroxylase phy-1 and their function in collagen IV-mediated basement membrane integrity during late elongation of the embryo. *Molecular Biology of the Cell, 22*(18), 3331–3343.

Trzebiatowska, A., Topf, U., Sauder, U., Drabikowski, K., & Chiquet-Ehrismann, R. (2008). *Caenorhabditis elegans* teneurin, ten-1, is required for gonadal and pharyngeal basement membrane integrity and acts redundantly with integrin ina-1 and dystroglycan dgn-1. *Molecular Biology of the Cell, 19*(9), 3898–3908.

Tsuji, N., Kita, K., Ozaki, K., Narama, I., & Matsuura, T. (2012). Organogenesis of mild ocular coloboma in FLS mice: failure of basement membrane disintegration at optic fissure margins. *Experimental Eye Research, 94*(1), 174–178.

Tzu, J., & Marinkovich, M. P. (2008). Bridging structure with function: structural, regulatory, and developmental role of laminins. *International Journal of Biochemistry and Cell Biology, 40*(2), 199–214.

Unsoeld, T., Park, J.-O., & Hutter, H. (2013). Discoidin domain receptors guide axons along longitudinal tracts in *C. elegans*. *Developmental Biology, 374*(1), 142–152.

Urbano, J. M., Torgler, C. N., Molnar, C., Tepass, U., López-Varea, A., Brown, N. H., et al. (2009). Drosophila laminins act as key regulators of basement membrane assembly and morphogenesis. *Development, 136*(24), 4165–4176.

Vaccaro, C. A., & Brody, J. S. (1981). Structural features of alveolar wall basement membrane in the adult rat lung. *Journal of Cell Biology, 91*(2 Pt 1), 427–437.

Valastyan, S., & Weinberg, R. A. (2011). Tumor metastasis: molecular insights and evolving paradigms. *Cell, 147*(2), 275–292.

Van Agtmael, T., & Bruckner-Tuderman, L. (2010). Basement membranes and human disease. *Cell and Tissue Research, 339*(1), 167–188.

Vanacore, R., Ham, A.-J. L., Voehler, M., Sanders, C. R., Conrads, T. P., Veenstra, T. D., et al. (2009). A sulfilimine bond identified in collagen IV. *Science, 325*(5945), 1230–1234.

Vogel, B. E., & Hedgecock, E. M. (2001). Hemicentin, a conserved extracellular member of the immunoglobulin superfamily, organizes epithelial and other cell attachments into oriented line-shaped junctions. *Development, 128*(6), 883–894.

Wang, Z., Linden, L. M., Naegeli, K. M., Ziel, J. W., Chi, Q., Hagedorn, E. J., et al. (2014). UNC-6 (netrin) stabilizes oscillatory clustering of the UNC-40 (DCC) receptor to orient polarity. *Journal of Cell Biology, 206*(5), 619–633.

Weyers, J. J., Milutinovich, A. B., Takeda, Y., Jemc, J. C., & Van Doren, M. (2011). A genetic screen for mutations affecting gonad formation in Drosophila reveals a role for the slit/robo pathway. *Developmental Biology, 353*(2), 217–228.

Whittaker, C. A., Bergeron, K.-F., Whittle, J., Brandhorst, B. P., Burke, R. D., & Hynes, R. O. (2006). The echinoderm adhesome. *Developmental Biology, 300*(1), 252–266.

Williams, B. D., & Waterston, R. H. (1994). Genes critical for muscle development and function in *Caenorhabditis elegans* identified through lethal mutations. *Journal of Cell Biology, 124*(4), 475–490.

Wong, M.-C., & Schwarzbauer, J. E. (2012). Gonad morphogenesis and distal tip cell migration in the *Caenorhabditis elegans* hermaphrodite. *Wiley Interdisciplinary Reviews. Developmental Biology, 1*(4), 519–531.

Woo, W.-M., Berry, E. C., Hudson, M. L., Swale, R. E., Goncharov, A., & Chisholm, A. D. (2008). The *C. elegans* F-spondin family protein SPON-1 maintains cell adhesion in neural and non-neural tissues. *Development, 135*(16), 2747–2756.

Xu, X., Dong, C., & Vogel, B. E. (2007). Hemicentins assemble on diverse epithelia in the mouse. *The Journal of Histochemistry and Cytochemistry: Official Journal of the Histochemistry Society, 55*(2), 119–126.

Yang, Y., Lee, W. S., Tang, X., & Wadsworth, W. G. (2014). Extracellular matrix regulates UNC-6 (netrin) axon guidance by controlling the direction of intracellular UNC-40 (DCC) outgrowth activity. *PLoS One, 9*(5), e97258.

Yurchenco, P. D. (2011). Basement membranes: cell scaffoldings and signaling platforms. *Cold Spring Harbor Perspectives in Biology, 3*(2).

Yurchenco, P. D., & Patton, B. L. (2009). Developmental and pathogenic mechanisms of basement membrane assembly. *Current Pharmaceutical Design, 15*(12), 1277–1294.

Ziel, J. W., Hagedorn, E. J., Audhya, A., & Sherwood, D. R. (2009). UNC-6 (netrin) orients the invasive membrane of the anchor cell in *C. elegans*. *Nature Cell Biology, 11*(2), 183–189.

Index

Note: Page numbers followed by "f" and "t" indicate figures and tables, respectively.

D

E

F

H

I

M

N

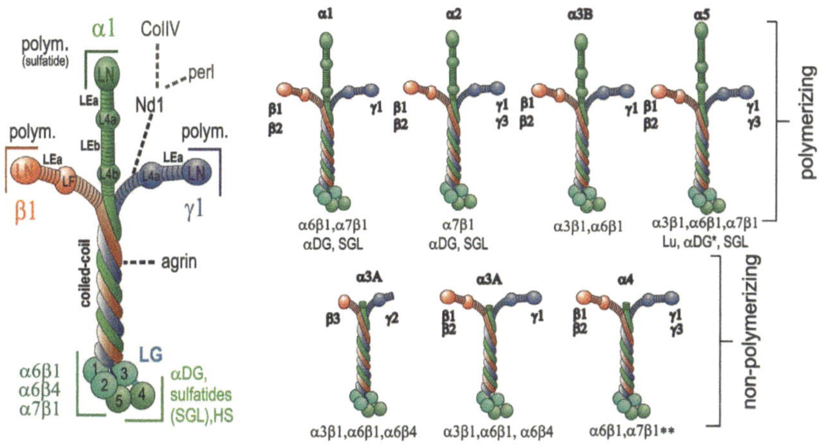

Peter D. Yurchenco, Figure 1 ***Laminin Family.*** Laminin subunits shown are the α1, α2, α3, α4, and α5, β1, β2, and β3 and the γ1, γ2, and γ3 chains (green, red, and blue). Functional loci are shown adjacent to domain designations for laminin-111 on the left side (polym, polymerization domain; Nd1, nidogen-1-binding site; αDG, α-dystroglycan; αxβy integrins; SGL, sulfated glycolipids); Lu, Lutheran glycoprotein receptor; HS, heparan sulfates). Laminin isoforms (right side) are grouped by their domain similarities. Polymerizing laminins each possess three different LN domains, while non-polymerizing laminins lack one or more LN domains. Laminin-411 (**) is unusual among laminins in that it binds weakly to integrins and poorly to dystroglycan.

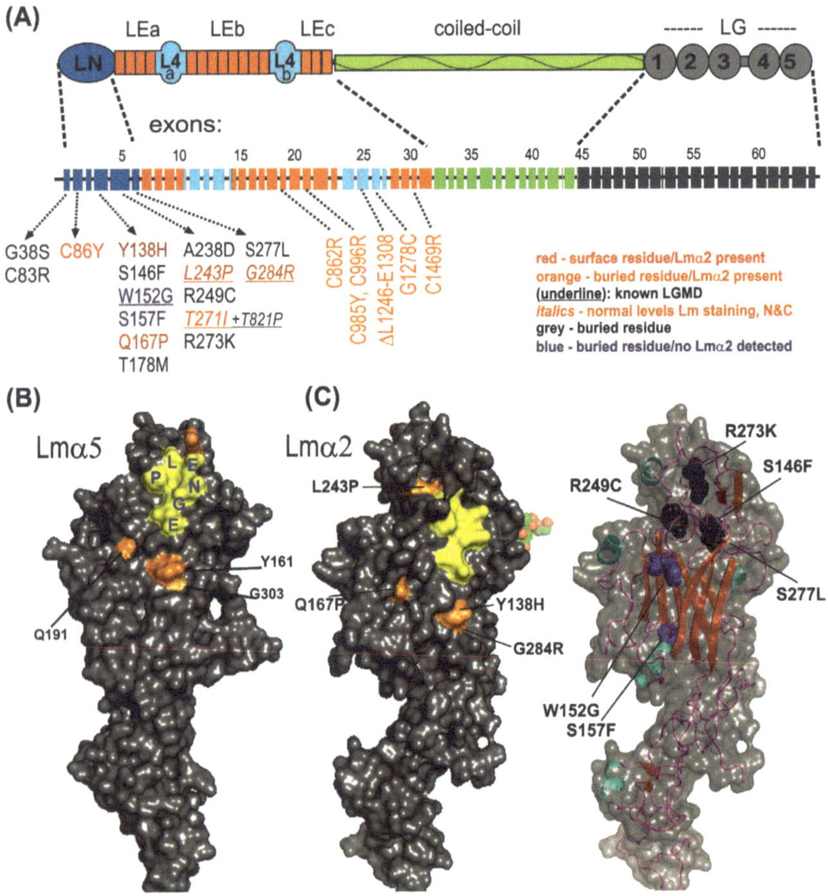

(A)
LEa
LEb
LEc
coiled-coil
------ LG ------
LN
L4 a
L4 b
1 2 3 4 5
exons:
5 10 15 20 25 30 35 40 45 50 55 60
G38S C83R
C86Y
Y138H S146F W152G S157F Q167P T178M
A238D L243P R249C T271I +T821P R273K
S277L G284R
C862R
C985Y, C996R
ΔL1246-E1308
G1278C
C1469R
red - surface residue/Lmα2 present
orange - buried residue/Lmα2 present
(underline): known LGMD
italics - normal levels Lm staining, N&C
grey - buried residue
blue - buried residue/no Lmα2 detected
(B)
Lmα5
P L E N G E
Y161
G303
Q191
(C)
Lmα2
L243P
Q167P
Y138H
G284R
R273K
S146F
R249C
S277L
W152G
S157F

Peter D. Yurchenco, Figure 4 ***Laminin-α2-deficient muscular dystrophy mutations and Lmα-subunit short-arm structures.*** Panel A: *Map of missense single amino acid substitutions and in-frame deletions causing a muscular dystrophy with absent, reduced or normal expression of laminin-α2.* Mutations causing absence (blue text), or either reduced or normal (red, orange italics) laminin-α2 expression based on biopsy immunostaining are distinguished for the indicated mutations (Leiden Open Variation Database, www.dmd.nl/LAMA2_seqvar.html; Chan et al., 2014; Di Blasi et al., 2005; Gavassini et al., 2011; Geranmayeh et al., 2010; Marques et al., 2014; Oliveira et al., 2008; Rajakulendran, Parton, Holton, & Hanna, 2011). Panel B: *Lmα5–LN–LEa domains.* The mouse laminin-α5 LN and adjacent LEa domain structure has been determined by X-ray crystallography by Hohenester and colleagues (Hussain et al., 2011). The surface structure was rendered in the PyMOL Molecular Graphics System (v.1.5.0.4 Schrödinger, LLC) from the published coordinates (Protein Data Bank, 2Y38). The P-L-E-N-G-E polymerization patch is indicated in yellow. Residues that correspond to ones associated with laminin-α2 deficiency are indicated in orange. Panel C: *Lmα2 LN–LEa domains, homology fit, with dystrophic mutations.* A homology fit using the mouse laminin-α2 sequence was generated in Modeller 9.11 *(kindly provided by David Briggs and Erhard Hohenester, Imperial College, London)* and rendered in PyMOL. The left side shows a surface distribution with solvent-accessible LAMA2 mutations (P-L-E-N-G-E polymerization patch in yellow). The right side shows a cartoon of the core structure with buried residues (blue and gray spheres) within a surface shell. C86Y, S146F, W152G, C199S, L243P, and R249C are mostly buried and seem likely to disrupt the domain fold, resulting either in secretion of a globally modified domain or in reduced/absent secretion. Other indicated mutations might selectively affect polymerization.

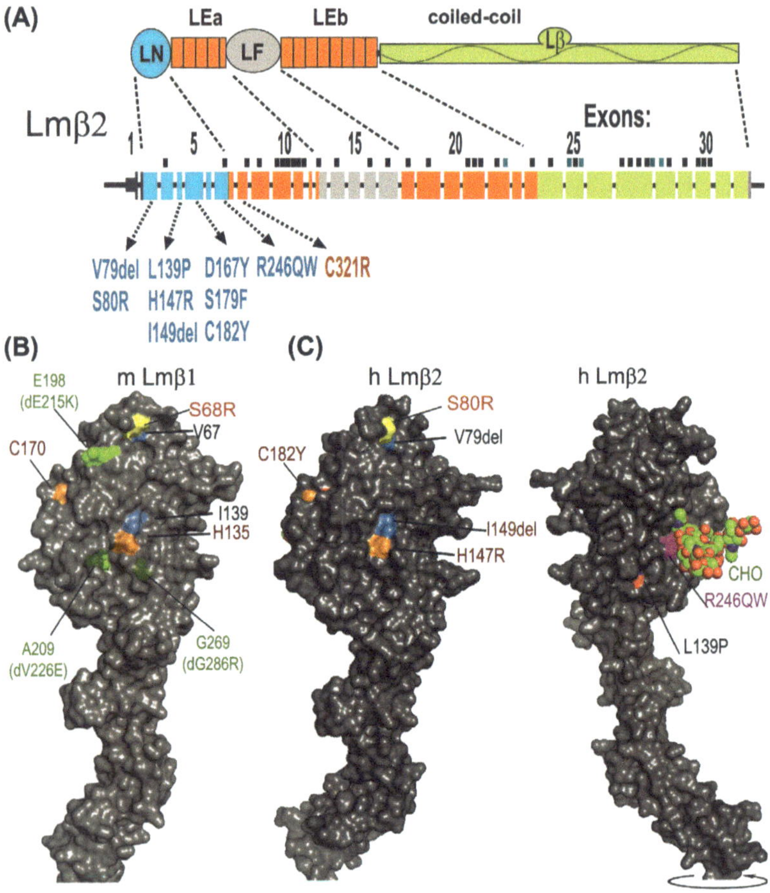

Peter D. Yurchenco, Figure 5 ***Pierson syndrome mutations and laminin-β-short-arm domain structures.*** Panel A: *Map of Pierson syndrome mutations*. Lmβ2 domains and exons are shown. Missense and short deletions, largely confined to LN domain, indicated with dotted arrows. Nonsense and other truncating mutations indicated with square dots over the exons. *(Map based on Matejas et al. (2010).)* Panel B: *Laminin-β1 LN–LEa domains*. The domains were rendered in PyMOL using the coordinates (PDB-4AQS) for the crystal structure of the mouse laminin-β1 tip (Purvis & Hohenester, 2012). Mutations corresponding to Pierson mutations labeled in red (likely ablating polymerization) and black (unknown effect). Surface and deeper (spheres) mutations causing failure of basement membrane in *Drosophila* labeled in green. Panel C: *Lmβ2 LN–LEa domains, homology fit, surface, with Pierson mutations indicated*. The human laminin-β1 sequence was homology fitted to Protein Data Bank structures using the Bioinformatics Toolkit (Max Planck Institute for Developmental Biology, HHpred). A rotated structure containing N-linked carbohydrate (CHO) is shown on the right. The R246Q and R246W mutations that cause a failure of secretion map adjacent to the asparagine near the carbohydrate. S80R, on the other hand, is located on the converse side on the same face as several Pierson and *Drosophila* mutations.

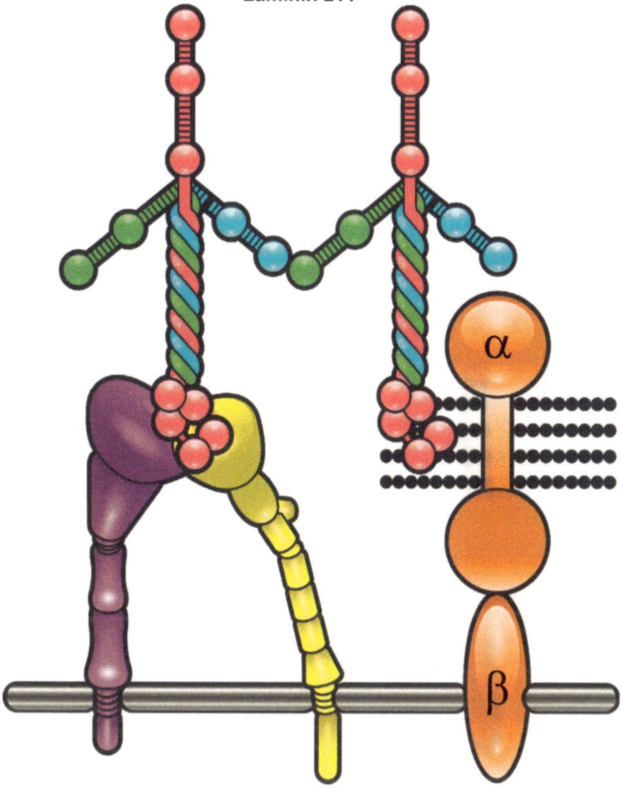

Madeleine Durbeej, Figure 1 Laminin-211 binding to cell surface receptors dystroglycan and integrin α7β1 in skeletal muscle. Laminin-211 is a heterotrimeric protein composed of α2 (pink), β1 (green), and γ1 (blue) chains. Laminin-211 binds to dystroglycan and integrin α7β1 via the C-terminal laminin G-like domains of laminin-α2 chain. Dystroglycan is composed of two subunits; α- and β-dystroglycan. α-Dystroglycan is highly glycosylated and binds to laminin-211 via its O-linked glycan chain moieties (chains with black circles) (Yoshida-Moriguchi et al., 2010). α-Dystroglycan is also attached to transmembrane β-dystroglycan. Integrins are heterodimeric and consist of one α (purple) and one β (yellow) subunit. The predominant integrin in skeletal muscle is α7β1. Note that the figure is not drawn to scale.

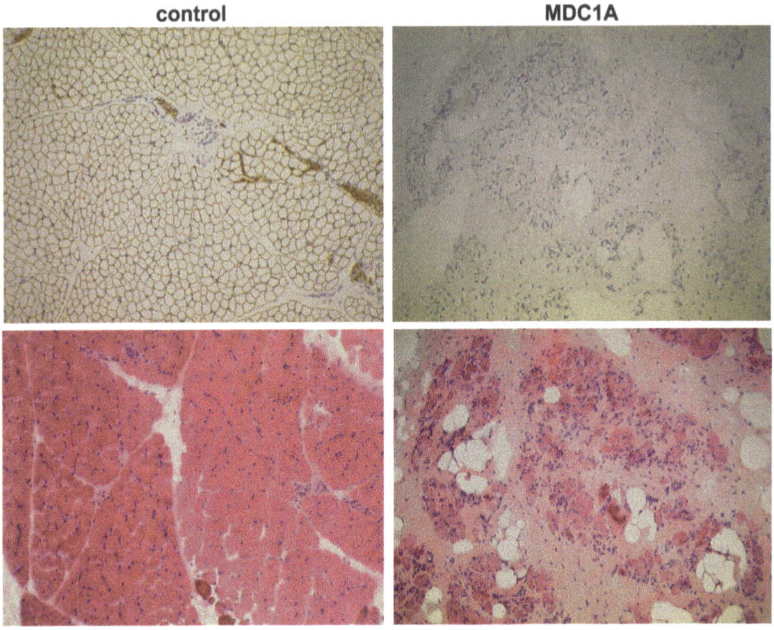

Madeleine Durbeej, Figure 2 Immunohistochemical and histological assessment of MDC1A biopsy. Upper panel: sections were stained with monoclonal antibody 22B2 detecting the 300 kDa N-terminal fragment of laminin-α2 chain. Note complete absence of laminin-α2 chain in MDC1A muscle. Lower panel: hematoxylin and eosin staining reveals massive muscle degeneration, regenerating fibers with centrally located nuclei, fibrosis, and fat cell infiltration.

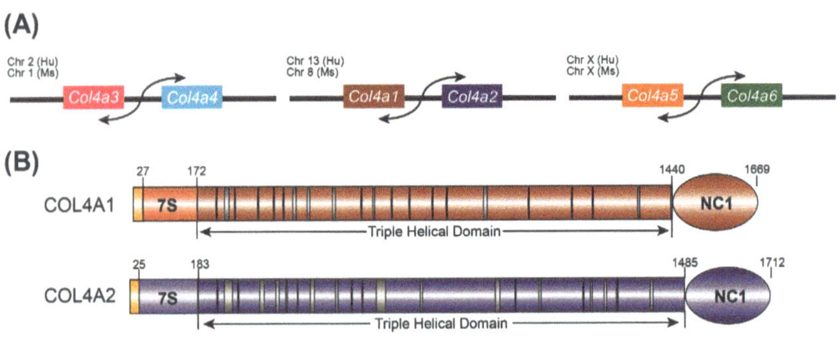

Mao Mao *et al.*, Figure 1 ***Chromosomal arrangements for type IV collagens and domain structures for COL4A1 and COL4A2.*** (A) Human (Hu) and mouse (Ms) type IV collagens are located on three distinct chromosomes as three pairs of genes transcribed from shared bidirectional promoters. (B) Type IV collagens have three functional domains. Following the signal peptide (yellow box), type IV collagens contain a 7S domain at the N-terminus, a triple helical domain and an NC1 domain at the C-terminus. Numbers above the schematics indicate amino acids in human COL4A1 or COL4A2. Gray boxes indicate repeat interruptions in the triple helical domain. Chr, chromosome. The program DOG v2.0 was used to draw the protein structure (Ren et al., 2009).

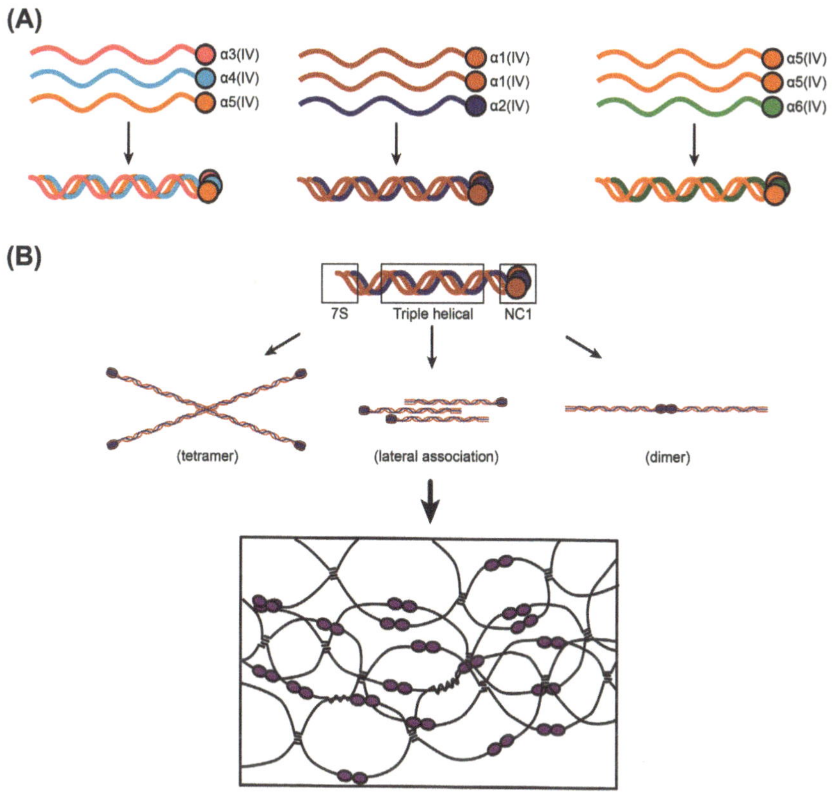

Mao Mao *et al.*, Figure 2 ***Type IV collagen heterotrimer and network formation.*** (A) Among 56 possible combinations, only α1α1α2, α3α4α5, and α5α5α6 heterotrimers are formed. (B) In the basement membrane, type IV collagens form an irregular and complex polygonal network. The 7S domains of collagen molecules interact to form tetramers and the NC1 domains interact to form dimers. Lateral associations along the triple helical domain allow branching and further strengthening of the type IV collagen network.

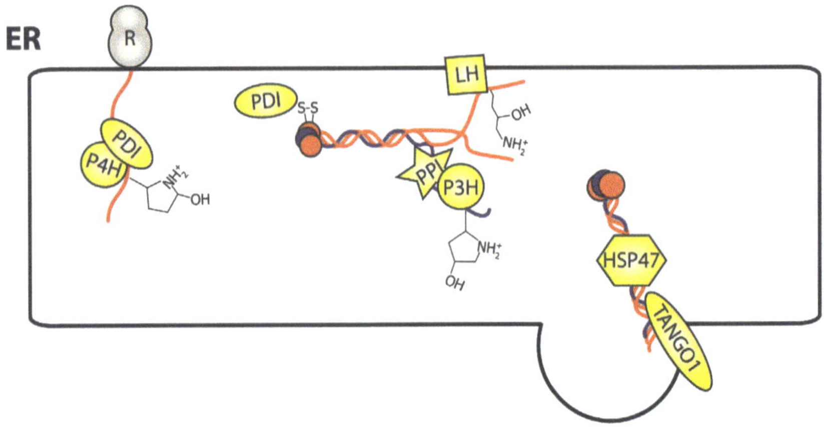

Mao Mao *et al.*, Figure 3 ***Type IV collagen biosynthetic pathway.*** Various enzymes posttranslationally modify nascent type IV collagens, and chemical chaperones prevent their aggregation in the ER. After heterotrimer formation, type IV collagen protomers are packed into specialized cargo vesicles to be transported via the Golgi to the extracellular matrix. HSP47, heat shock protein 47; LH, lysyl hydroxylase; PDI, protein disulfide isomerase; PPI, peptidylprolyl isomerase; P3H, prolyl 3-hydroxylase; P4H, prolyl 4-hydroxylase; R, ribosome; TANGO1, transport and Golgi organization 1.

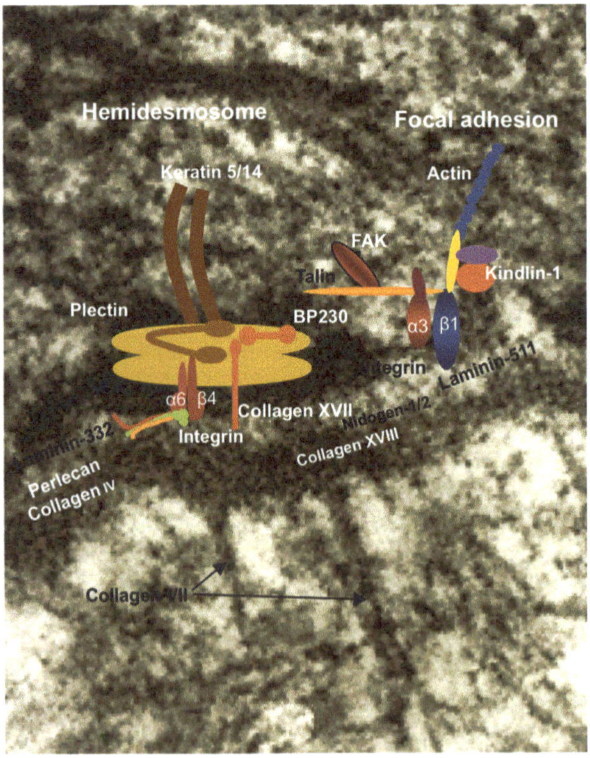

Cristina Has and Alexander Nyström, Figure 1 ***Schematic representation of the major molecules of the epidermal basement membrane and associated structures.*** An electron microscopic picture representing the dermal–epidermal junction serves as background.

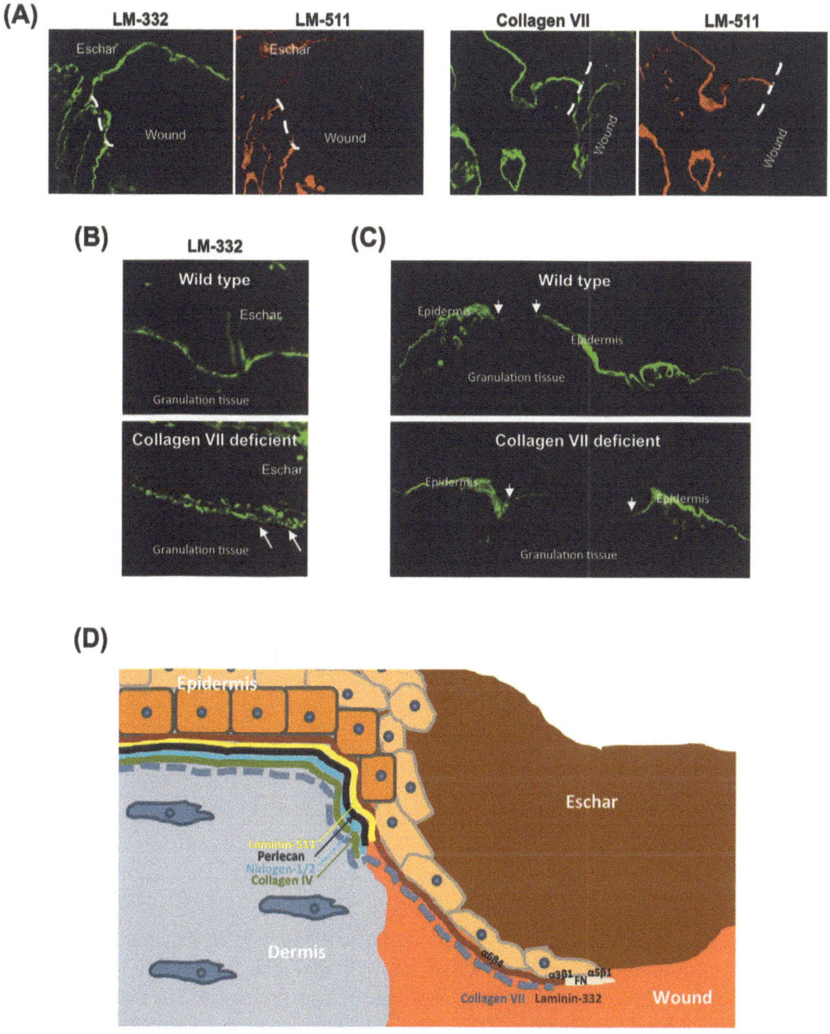

Cristina Has and Alexander Nyström, Figure 2 ***Epidermal basement membrane components in skin wound closure.*** (A) Three-day-old full-thickness mouse wounds stained for laminin-332, laminin-511, and collagen VII as indicated. (B) Laminin-332 deposition in wild-type- and collagen VII-deficient seven-day-old mouse wounds. Note the altered irregular laminin-332 deposition in absence of collagen VII. (C) Wounds as in (B) stained for keratin 1 to visualize the delayed reepithelialization and subsequent epithelial maturation of collagen VII-deficient wounds. (D) Schematic representation of expression of key epidermal basement membrane components during wound healing.

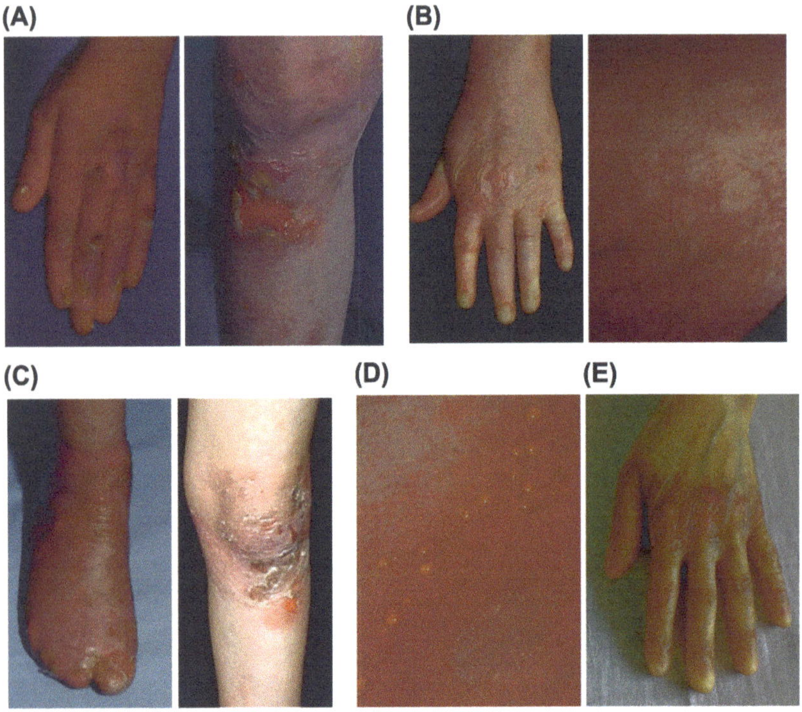

Cristina Has and Alexander Nyström, Figure 4 ***Clinical spectrum of genetic and acquired disorders of the epidermal basement membrane and associated structures.*** (A) Five-year-old boy with junctional epidermolysis bullosa (EB) due to a homozygous *COL17A1* splice site mutation. Note blisters, erosions, crusts, atrophic scarring dyspigmentation, and nail dystrophy. (B) 44-year-old patient with Kindler syndrome. Note skin atrophy, fibrotic changes of the fingers and poikiloderma on the limbs. (C) Five-year-old boy with dystrophic EB due to a homozygous *COL7A1* frameshift mutation. Note wounds, crusts, scarring, and mutilating deformity of the foot. (D) Erythema and tens blisters in a patient with bullous pemphigoid. (E) Scarring and nail loss in a patient with EB acquisita (Otten et al., 2014).

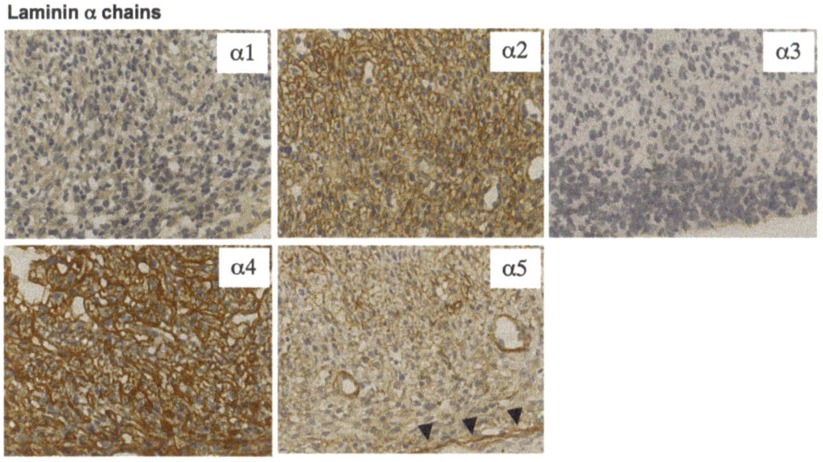

Masashi Yamada and Kiyotoshi Sekiguchi, Figure 2 Distinct localization of laminin α chains in the heart. (*Data are taken from the Mouse Basement Membrane Database (http://www.matrixome.com/bm/).*) Frozen sections of heart ventricles of E16.5 mouse embryos immunostained with antibodies against laminin α1, α2, α3, α4, and α5 chains. Sections were counterstained with hematoxylin. Positive signals for α2 were observed surrounding myocardial cells, those for α4 around both myocardial cells and blood vessels, and those for α5 around blood vessels and within the epicardium (arrowheads).

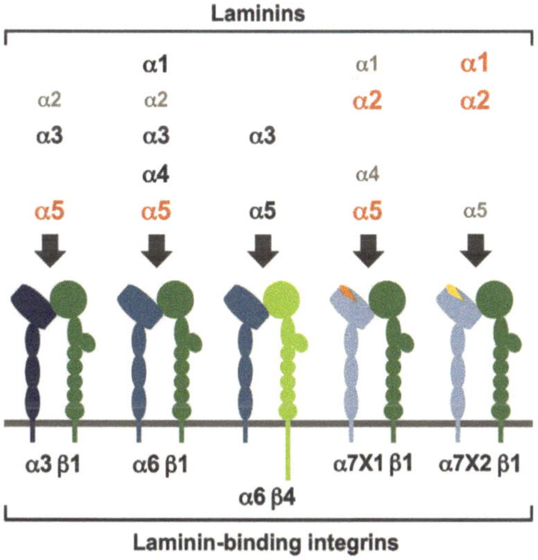

Masashi Yamada and Kiyotoshi Sekiguchi, Figure 3 Binding specificities of laminin-binding integrins. Affinities of laminins containing distinct α chains to laminin-binding integrins are indicated by different colors: Red, high; black, medium; gray, low. Font size is proportional to the affinity strength.

Protein	*Drosophila* genes	Human homologs
Collagen IV 	α1: *Collagen gene at 25C (Cg25C)*	*COL4A1, COL4A3, COL4A5*
	α2: *viking (vkg)*	*COL4A2, COL4A4, COL4A6*
Laminin 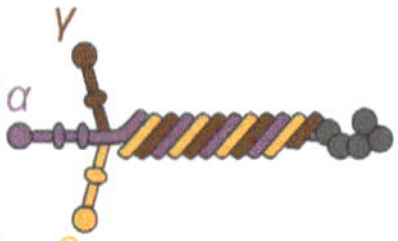	α1,2: *wing blister (wb)*	*LAMA1, LAMA2*
	α3,5: *Laminin A (LanA)*	*LAMA3, LAMA4, LAMA5*
	β: *Laminin B1 (LanB1)*	*LAMB1, LAMB2, LAMB3, LAMB4*
	γ: *Laminin B2 (LanB2)*	*LAMC1, LAMC2, LAMC3*
Perlecan 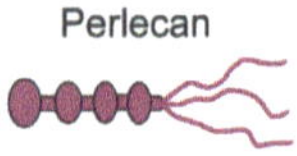	*terribly reduced optic lobes (trol)*	*HSPG2*
Nidogen 	*Nidogen/entactin (Ndg)*	*NID1, NID2*

Adam J. Isabella and Sally Horne-Badovinac, Figure 1 Overview of the core basement membrane proteins in *Drosophila*.

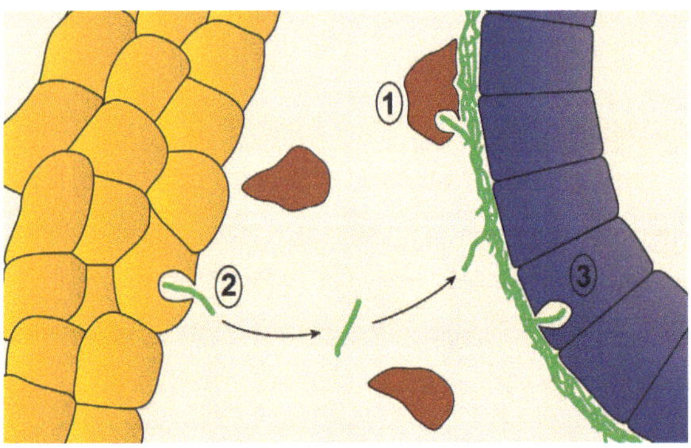

Adam J. Isabella and Sally Horne-Badovinac, Figure 2 ***Cellular sources of basement membrane proteins in* Drosophila.** BM proteins are synthesized and secreted by three primary cell types. 1: Synthesis and local deposition by hemocytes. 2: Synthesis by and long range diffusion from the fat body. 3: Synthesis and secretion by the epithelium itself.

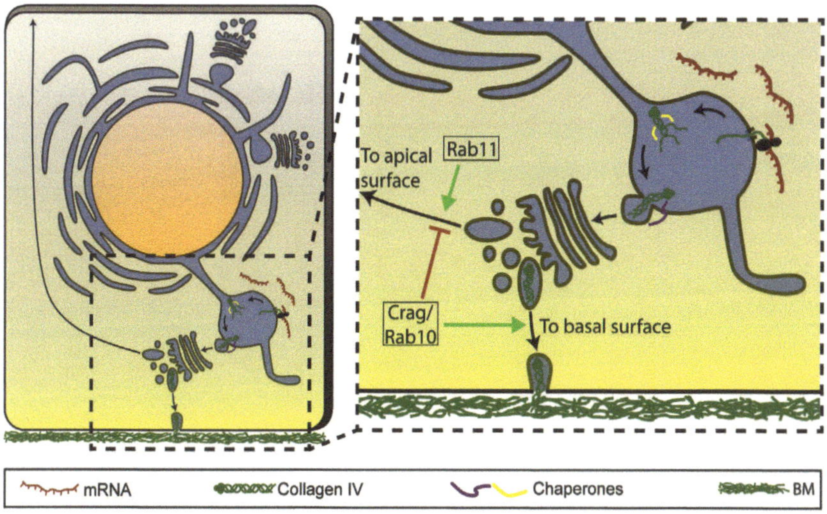

Adam J. Isabella and Sally Horne-Badovinac, Figure 3 ***Local synthesis and polarized secretion of collagen IV in the follicular epithelium.*** Within the follicular epithelium, *collagen IV* transcripts accumulate basally and are translated and Translocated into a basal region of the endoplasmic reticulum (ER). ER-resident proteins assist in the folding and packaging of collagen IV for transport to the Golgi. After collagen IV transits through the Golgi, Crag and Rab10 promote delivery of collagen IV-containing exocytic vesicles to basal regions of the plasma membrane and prevent Rab11-dependent targeting of collagen IV-containing vesicles to the apical surface. Inset: blow-up of indicated region.

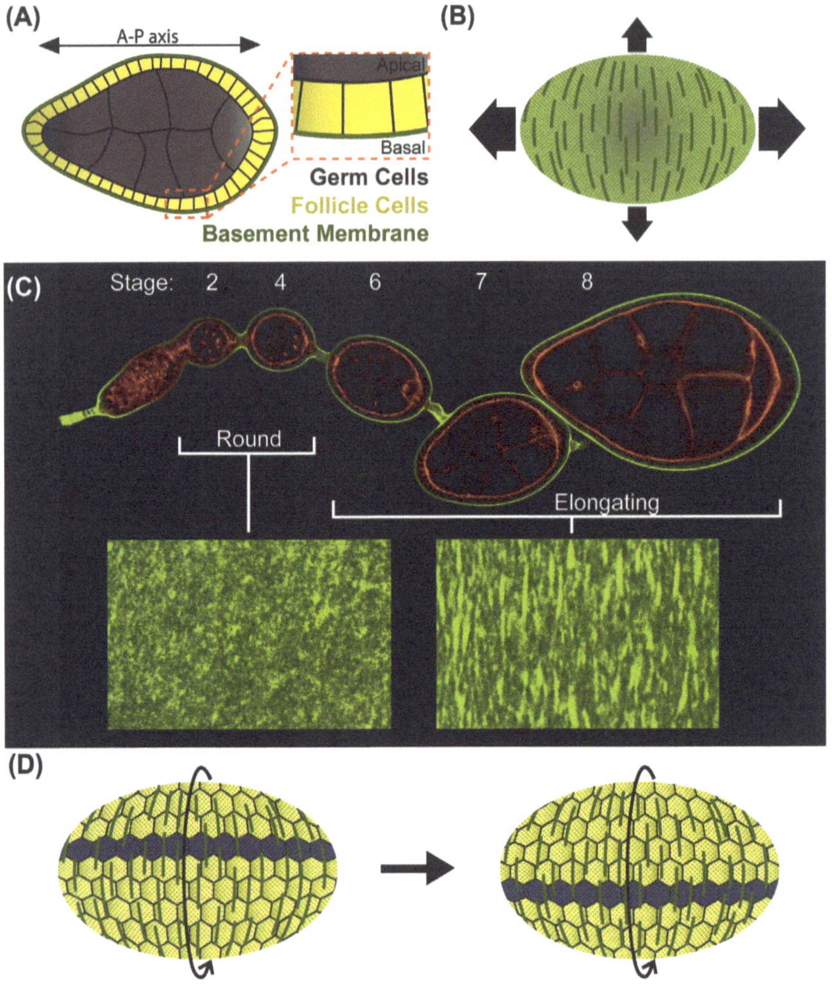

Adam J. Isabella and Sally Horne-Badovinac, Figure 4 ***Basement membrane (BM) function in egg chamber elongation.*** (A) Illustration of a transverse section through an egg chamber. Central germ cells are surrounded by a somatic epithelium of follicle cells, which assemble a BM on their basal surfaces. (B) Model of the molecular corset. Polarized linear fibril-like structures in the BM are hypothesized to constrain egg chamber growth in the direction of polarization, biasing growth to occur along the A–P axis. Arrows indicate direction and magnitude of growth. (C) BM structural dynamics during elongation. Top: a developmental array of egg chambers showing cell outlines (actin) and the BM (Viking-GFP). Bottom: fluorescent micrographs of collagen IV (Viking-GFP) in the BM. Young, round egg chambers exhibit no obvious BM structure, while older, elongating egg chambers display polarized fibrils within the BM. (D) Overview of egg chamber rotation. In this illustration, the BM is partially transparent to reveal the cells underneath. The dark row of cells in each image represents the same cells at two different time points. The egg chamber rotates within a stationary BM, in the direction of BM fibrils. Curved arrows indicate direction of egg chamber rotation.

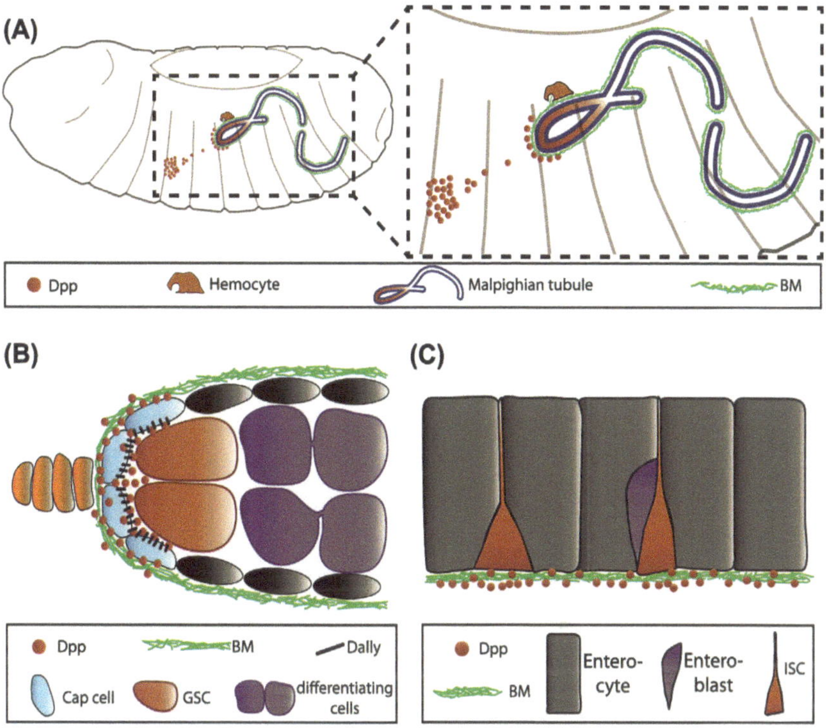

Adam J. Isabella and Sally Horne-Badovinac, Figure 5 ***Basement membrane (BM) regulation of Dpp signaling in three developmental contexts.*** (A) Embryonic Malpighian tubule morphogenesis. Diffusible Dpp protein attracts the growing Malpighian tubule. Hemocyte-deposited collagen IV around the Malpighian tubule promotes reception of the Dpp signal by tubule cells. (B) Ovarian germ line stem cell (GSC) maintenance. GSCs are maintained within the stem cell niche via interaction with cap cells. Collagen IV in the BM and the HSPG Dally concentrate cap cell-derived Dpp to promote signal reception by GSCs but not differentiating daughter cells. (C) Intestinal stem cell (ISC) maintenance. ISCs exhibit basally positioned cell bodies that adhere to the BM. Dpp is concentrated by collagen IV within the BM to promote a high level of signal reception by ISCs but not the more apically localized, differentiating enteroblasts.

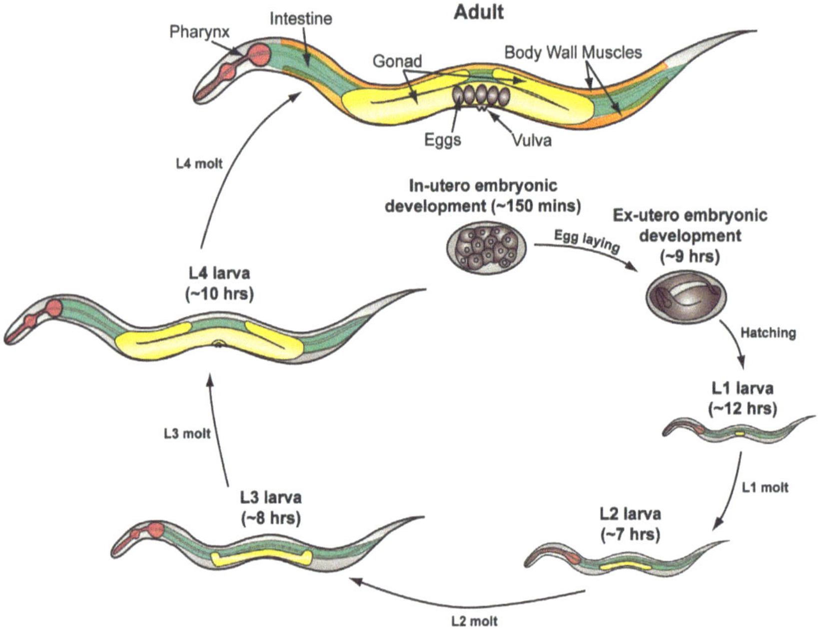

Matthew R. Clay and David R. Sherwood, Figure 1 ***Overview of* Caenorhabditis elegans *development.*** A schematic diagram depicting *C. elegans* development, life cycle, and several adult tissues. The times indicate the duration spent in each developmental stage when grown at 22 °C. Embryogenesis involves two stages: (1) a proliferative stage that includes gastrulation and (2) organogenesis and morphogenesis. Larval development consists of four stages (L1–L4), each separated by a molt of the stage-specific cuticle. During larval development, several tissues continue morphogenesis and growth. Two notable examples are the expansion and shaping of the gonad (yellow), and establishment of the egg-laying apparatus, which includes morphogenesis of the vulva and formation of the uterine–vulval attachment.

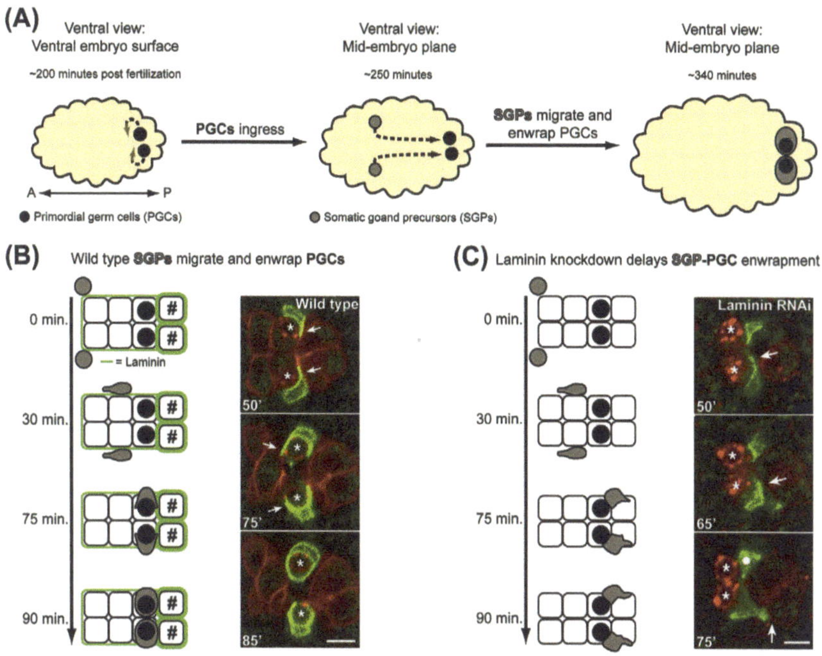

Matthew R. Clay and David R. Sherwood, Figure 2 ***A laminin-dependent signal halts somatic gonad precursor cell migration.*** (A) A schematic diagram of germ stem cell niche establishment. During embryogenesis the primordial germ cells (PGCs; black circles) are specified in the posterior region of the ventral embryo surface before ingressing during gastrulation. Somatic gonad precursors (SGPs; gray circles) are born in anterior regions of the embryo and migrate along the endoderm to meet the primordial germ cells. Upon reaching the germ cells, somatic gonad precursors halt their migration and establish the niche by enwrapping the primordial germ cells. (B) Cell behaviors during somatic gonad precursor migration and primordial germ cell enwrapment. The schematic diagrams (left; normal distribution of laminin is shown in green) and confocal time-lapse images (right) depict somatic gonad precursors (gray circles (left), green cells (right)) migrating along the endoderm (white boxes (left), red cells (right)). Somatic gonad precursors stop migrating when they contact endoderm cells that have increased laminin around their cell surfaces (#) that lie just posterior to the primordial germ cells (black circles (left), marked with asterisk (right)). After stopping, somatic gonad precursors extend cellular processes (arrows) that enwrap primordial germ cells to establish the niche. (C) When laminin is decreased or absent, somatic gonad precursors migrate further and extend projections past the primordial germ cells, resulting in delayed enwrapment. PGCs, primordial germ cells; SGPs, somatic gonad precursors. *Confocal images in (B) and (C) originally appeared in Rohrschneider and Nance (2013) and were adapted with permission.*

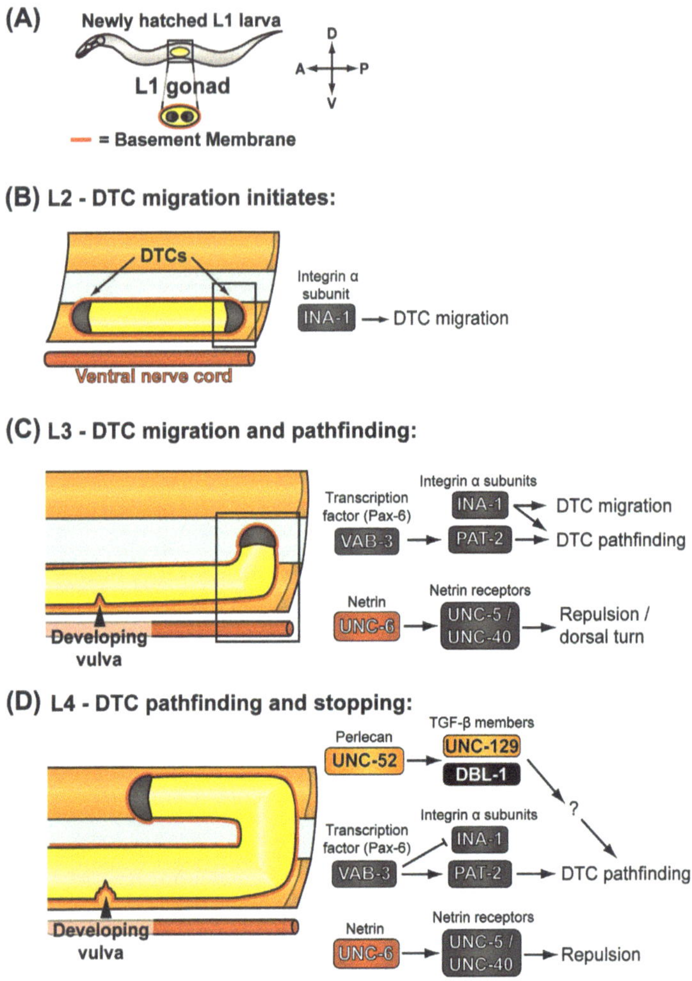

Matthew R. Clay and David R. Sherwood, Figure 3 ***Larval morphogenesis of the gonad and molecular pathways controlling distal tip cell (DTC) migration.*** (A) A schematic diagram of a newly hatched L1 larva and gonad. The gonadal basement membrane is shown in red in all illustrations. (B) DTC migration initiates at the L2 stage, and DTCs migrate adjacent to the ventral body wall muscle basement membrane (orange). The integrin α subunit *ina-1* is expressed in DTCs beginning at L2 and promotes DTC migration. (C) In L3, the DTCs turn and migrate toward the dorsal body wall muscle along the epidermal basement membrane (light blue). During L3, the transcription factor VAB-3 (Pax6 ortholog) induces expression of the integrin α subunit *pat-2*. Both INA-1 and PAT-2 regulate DTC turning and proper pathfinding. In addition, the action of netrin (UNC-6) secreted from the ventral nerve cord binds the netrin receptors UNC-5 and UNC-40 (DCC) in the DTCs to mediate repulsion from the ventral body wall. (D) Once the DTCs reach the dorsal body wall muscle in the L4 stage, the DTCs turn back toward the animal midbody. This turn is regulated by the integrins INA-1 and PAT-2 in the DTCs, the basement membrane protein perlecan (UNC-52), and the TGF-β family members UNC-129 and DBL-1. Perlecan and *unc-129* are expressed in the body wall muscle, and perlecan might sequester TGF-β members within the dorsal body wall muscle basement membrane. During L4, VAB-3 downregulates *ina-1* integrin expression, which is necessary to stop DTC migration in early adulthood.

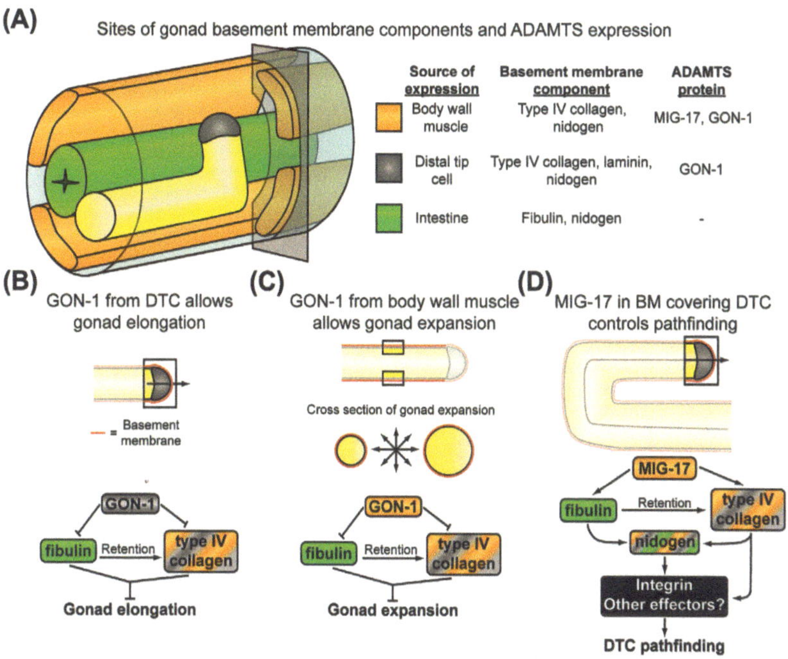

Matthew R. Clay and David R. Sherwood, Figure 4 ***A Disintegrin And Metalloprotease with ThromboSpondin (ADAMTS) proteins act in the gonadal basement membrane to control gonad elongation, expansion, and distal tip cell (DTC) migration.*** (A) A schematic diagram depicting a portion of a *Caenorhabditis elegans* larva showing the tissues that express gonadal basement membrane components and ADAMTS proteins. The body wall muscles (orange) express and secrete MIG-17, GON-1, and type IV collagen; the DTCs (gray) secrete GON-1, type IV collagen, laminin, and nidogen; and the intestine (green) secretes fibulin-1 and nidogen. (B–D) Mechanisms utilized by ADAMTS proteins to control gonad elongation (B), gonad expansion (C), and DTC pathfinding (C). Colors represent the sites of protein secretion and correspond to the key in (A). (B) The action of GON-1 secreted from the DTC antagonizes fibulin and type IV collagen and allows gonad arm elongation. (C) GON-1 secreted from the body wall muscle acts within the gonadal basement membrane to allow gonad expansion. (D) MIG-17 secreted from the body wall muscle acts in the basement membrane covering the DTC to regulate a functional network of fibulin-1, type IV collagen, and nidogen that regulates integrin or other effectors to control DTC pathfinding.

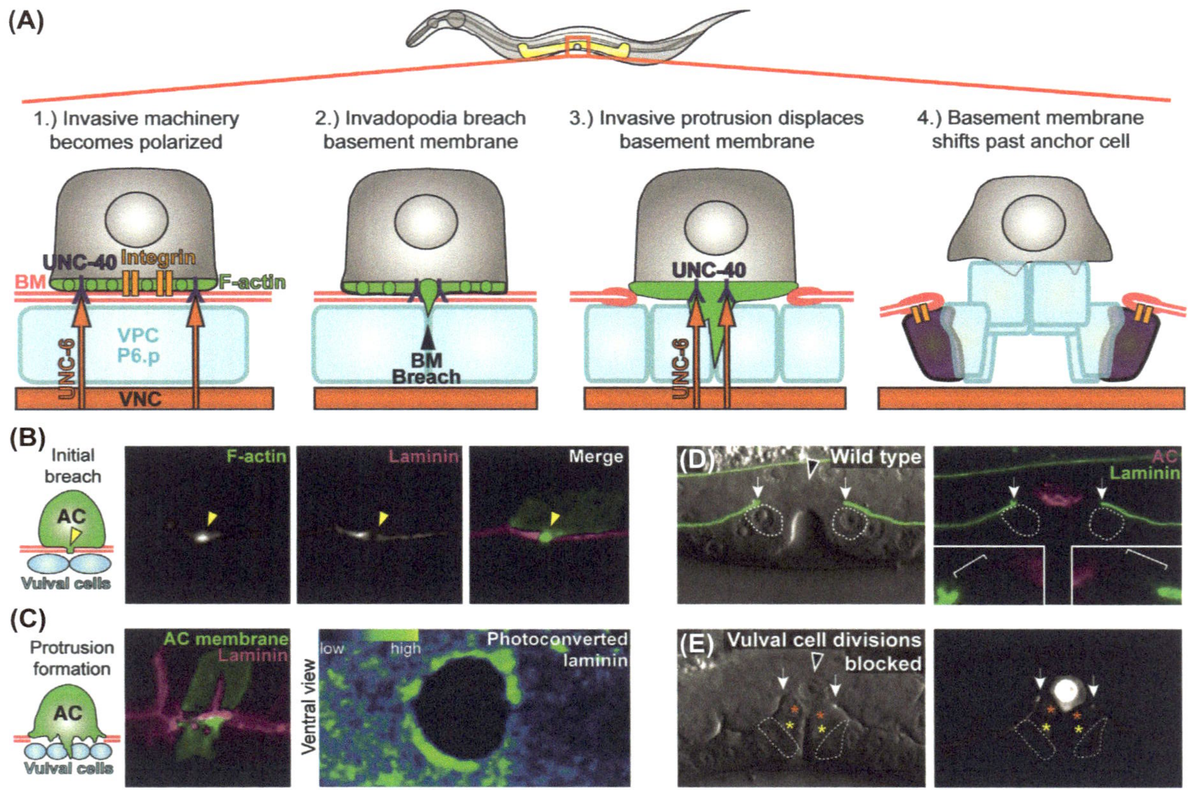

(A)
1.) Invasive machinery becomes polarized
2.) Invadopodia breach basement membrane
3.) Invasive protrusion displaces basement membrane
4.) Basement membrane shifts past anchor cell
UNC-40
Integrin
BM
F-actin
VPC
P6.p
UNC-6
VNC
BM
Breach
UNC-40
UNC-6
(B)
Initial breach
AC
Vulval cells
F-actin
Laminin
Merge
(C)
Protrusion formation
AC
Vulval cells
AC membrane
Laminin
Ventral view
low
high
Photoconverted laminin
(D)
Wild type
AC
Laminin
(E)
Vulval cell divisions blocked

Matthew R. Clay and David R. Sherwood, Figure 5 ***Creating a de novo gap in basement membrane during uterine–vulval attachment.*** (A) A schematic diagram of basement membrane breach and hole widening during uterine–vulval attachment. 1.) The anchor cell (AC) invasive cellular machinery is polarized by integrin–basement membrane interactions and netrin (UNC-6) localized within the basement membrane that binds to the netrin receptor UNC-40 (DCC) in the AC. Netrin is secreted from the ventral nerve cord (VNC). 2.) UNC-40 (DCC) traffics to the site where invadopodia (marked by F-actin) breach the basement membrane. 3.) UNC-6 (netrin) binding UNC-40 (DCC) stimulates the formation of an invasive protrusion that displaces basement membrane. 4.) Vulval cell invagination shifts the basement membrane sheet until integrin adhesion in the vulval vulD cell (purple), which is triggered by increased laminin at the edges of the hole, stops basement membrane sliding. (B) Confocal images showing invadopodia (F-actin) breaching the basement membrane (yellow arrowheads), which has been labeled with fluorescent laminin. (C) Confocal images showing the invasive protrusion. Ventral view shows accumulation of laminin that was photoconverted under the AC before being displaced by the invasive protrusion. (D) The basement membrane slides past the AC (black arrowhead) and stops (arrows mark position) over the vulval vulD cell (circled). Brackets mark space between the AC and the basement membrane. (E) Basement membrane does not slide when cell divisions of vulval cells (marked by asterisks) are blocked. AC, anchor cell; BM, basement membrane; VPC, vulval precursor cell; VNC, ventral nerve cord. *All confocal images were adapted with permission. Images in (B) and (C) originally appeared in Hagedorn et al. (2013); images in (D) appeared in Ihara et al. (2011); images in (E) appeared in Matus et al. (2014).*

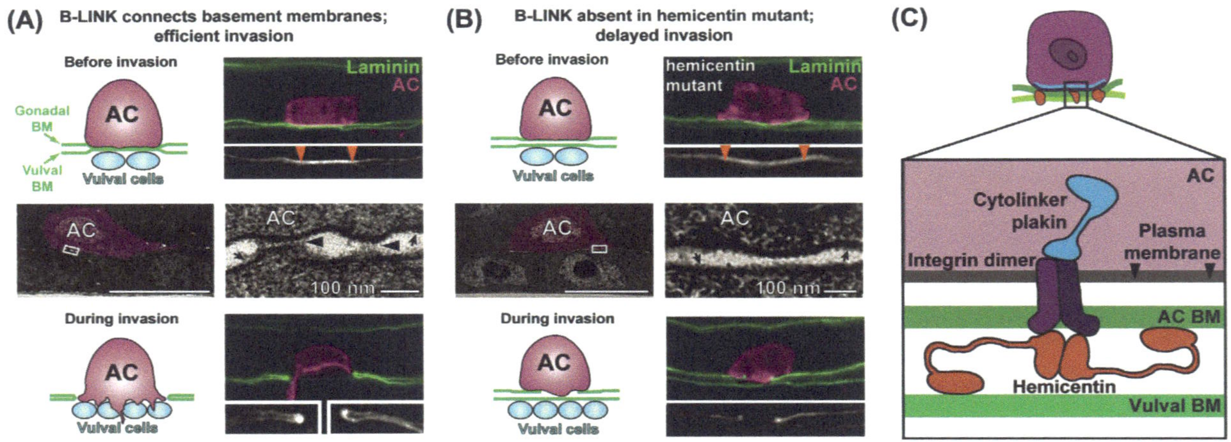

Matthew R. Clay and David R. Sherwood, Figure 6 ***Basement membrane LINKages (B-LINKs) connect adjacent basement membranes.*** (A) During anchor cell (AC) invasion, the gonadal and vulval basement membranes (BMs; labeled with fluorescent laminin) under the anchor cell are connected together (red arrowheads). Tight adhesion between the basement membranes is mediated by the B-LINK, seen as punctate structures (black arrowheads) between the gonadal and vulval basement membranes (black arrows) in transmission electron micrographs. The B-LINK allows efficient invasion. (B) In hemicentin mutants the B-LINK is absent and the gonadal and vulval basement membranes are no longer linked, evidenced by space between the basement membranes seen by fluorescent microscopy (red arrowheads) and lack of punctate structures between the basement membranes (black arrows) in electron micrographs. Disrupting the B-LINK leads to stepwise breaching of the gonadal and vulval basement membranes and delayed invasion. (C) A schematic diagram showing molecular components of the B-LINK. Hemicentin is secreted from the AC and forms punctate structures in the BM. The cytolinker plakin (VAB-10A; blue) and the transmembrane integrin complex (INA-1/PAT-3; purple) are necessary for B-LINK formation. AC, anchor cell; BM, basement membrane. *Confocal images, electron microscopy images, and illustration of molecular mechanism originally appeared in Morrissey et al. (2014) and were adapted with permission.*

Kolibri 360 GmbH
Ferdinand-Jühlke-Straße 7
99095 Erfurt, Deutschland
produktsicherheit@kolibri360.de